Graphs and Applications

Springer
London
Berlin
Heidelberg
New York
Barcelona
Hong Kong
Milan
Paris
Singapore
Tokyo

Joan M. Aldous and Robin J. Wilson

Graphs and Applications

An Introductory Approach

With 644 illustrations by Steve Best

Joan Aldous, BSc, PhD
Robin Wilson, MA, PhD

Faculty of Mathematics and Computing, The Open University,
Walton Hall, Milton Keynes MK7 6AA, UK

ISBN 1-85233-259-X Springer-Verlag London Berlin Heidelberg

British Library Cataloguing in Publication Data
Aldous, Joan
Graphs and applications : an introductory approach
1.Graph theory
I.Title II.Wilson, Robin J. (Robin James), 1943-
511.5
ISBN 185233259X

Library of Congress Cataloging-in-Publication Data
Aldous, Joan M., 1938-
Graphs and applications : an introductory approach /
Joan M. Aldous and Robin J. Wilson
p. cm.
Includes bibliographical references and index.
ISBN 1-85233-259-X (alk. paper)
1. Graph theory. I. Wilson, Robin J. II. Title.
QA166 .A425 2000
511'.5--dc21 99-056960

Printed in Great Britain

This book is based upon The Open University course MT365: *Graphs, Networks and Design*, first published in 1995. Material is reproduced by permission of The Open University.

To find out more about The Open University, visit the following website: http://www.open.ac.uk

Typeset by Ian Kingston Editorial Services, Nottingham
Printed and bound at the Athenæum Press Ltd, Gateshead, Tyne & Wear
12/3830-543210 Printed on acid-free paper SPIN 10752269

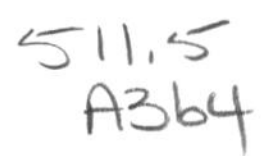

Preface

In recent years, there has been a significant movement away from traditional calculus courses toward courses on discrete mathematics. The impetus for this has undoubtedly been due in part to the increasing importance of the computer, and the consequence has been a proliferation of courses and books entitled *Discrete Mathematics*, *Finite Mathematics*, *Mathematics for Computer Science*, and other similar titles.

It is an unfortunate feature of some of these courses that many different topics are covered at a superficial level, leaving the student frustrated and confused and having little understanding of the underlying reasons for introducing so many seemingly unrelated areas. Our experience is that students benefit more from an introductory course based in just one area, chosen to link with other subjects whenever the instructor considers it appropriate to do so. Graph theory is an ideal topic for such an introductory course. It is fun, students enjoy it, they can 'get their hands dirty' drawing pictures, and it is an excellent stepping stone to a wide range of courses in mathematics and computer science.

This book arose out of a British Open University course, *Graphs, Networks and Design*, which first appeared in 1995 and regularly attracts about 500 students per year. It supersedes an earlier book *Graphs: an Introductory Approach* (John Wiley & Sons, 1990) with similar content, produced by Robin J. Wilson and John J. Watkins and based on a previous Open University course. As with other Open University courses, *Graphs, Networks and Design* consists mainly of correspondence material, supported by audio-cassette tapes, computer software, and BBC videos. Having produced this material, the course team felt that parts of it would be ideally suited to the classroom situation, and could successfully be converted into book form appropriate for an international audience.

A related volume, *Networks and Algorithms: an Introductory Approach* (John Wiley & Sons, 1994), was prepared by A. K. Dolan and J. M. Aldous. Each book is self-contained, and is suitable for a semester course on discrete mathematics in the first, second or third year of a college or university. The approach, terminology and notation are the same for both books, so an instructor wishing to teach both graphs and networks may use the two books concurrently.

Chapters 1–6 contain the basic definitions relating to graphs and digraphs, together with a number of examples. Chapters 7–13 contain a number of topics from which an instructor can select, depending upon the length of the course. Case Studies are given at the ends of selected chapters.

Included with this text is a CD-ROM for use with Windows on a PC. This contains a database of 1252 graphs (all the simple unlabelled graphs with up to seven vertices). It also contains software that enables the user to construct graphs and digraphs, and to perform simple operations on vertices and edges. Brief notes on how to use the software and some suggested activities are described in the Computing Notes at the end of the book.

The book contains a large number of problems and exercises. An instructor can also use the graph database as a resource for setting further exercises.

The adaptation and general editing of this book is by Joan Aldous. Much of the material is based on the work of Robin Wilson. The Computing Notes are based on the work of Richard Scott, Roger Lowry and Keith Cavanagh. The original software was designed by Adam Gawronski and Jon Rosewell, and adapted by Adam Gawronski. The original artwork, also used for the book, was drawn by Steve Best and Howard Taylor, and adapted by Steve Best. Others who contributed to the original course are Alan Dolan, Jennifer Harding (editor), Jeff Johnson, Fred Holroyd, Rob Lyon (designer), Roy Nelson and Joe Rooney.

Joan M. Aldous
Robin J. Wilson

Study Guide

An important part of learning graph theory is problem solving, and for this reason a large number of problems are included within and at the end of each main chapter. Most of these are routine exercises, designed to test understanding of the material in the text. The problems in the text are designed to be tackled as they are encountered: full solutions are given at the back of the book. The exercises at the ends of chapters are intended for revision or assessment purposes; solutions to these are not given.

An appendix on the methods of proof used in the book is included after the final chapter. The discussion is given in the context of graph theory. The appendix may be studied at any time after the study of Chapter 2.

The following diagram indicates possible study paths.

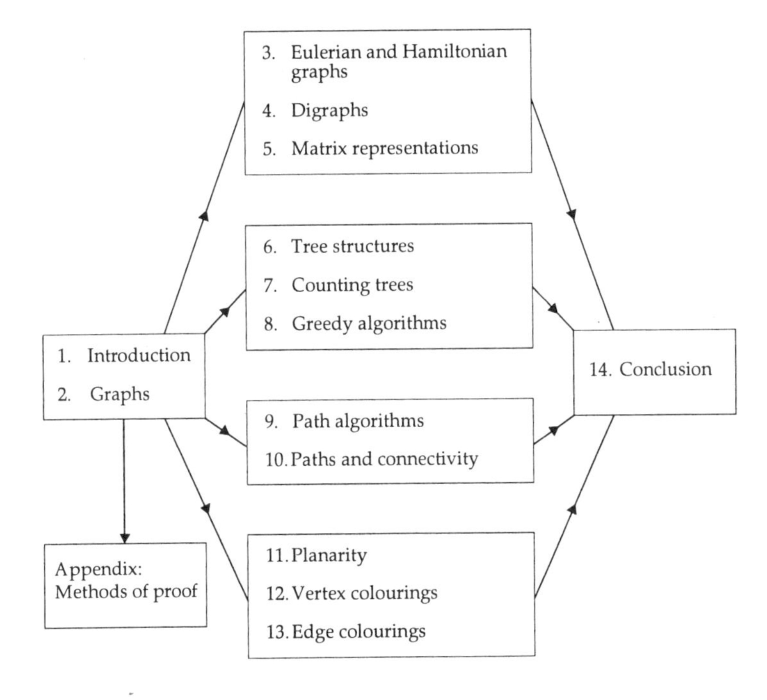

Contents

Chapter 1

Introduction

After studying this chapter, you should be able to:

- explain what are meant by the terms *graph, weighted graph, digraph, weighted digraph, vertex, edge, arc, weight* and *network;*
- explain what are meant by *the utilities problem, optimal path problems, Königsberg bridges problem, braced rectangular frameworks* and *the travelling salesman problem.*

In this introductory chapter, we aim to give you some idea of what this book is about. We describe a number of problems and invite you to try to solve simple instances of them. Several of these are straightforward, and are included in order to give you an idea of some of the topics covered in the book. Others are more difficult, and illustrate the need for a more systematic approach, to be given in later chapters.

The primary aim of this chapter is to show that an important step in the process of solving a problem may be to represent the situation by a diagram, such as a *graph*. We also introduce some terminology associated with graphs that is used throughout the book.

1.1 Graphs, Digraphs and Networks

Graphs

In order to introduce the idea of a *graph*, we study some problems and their diagrammatic representations. We begin by considering some problems for which appropriate representations are supplied or are natural.

Finding Routes

Consider the problem of finding a route between two stations of the London Underground. The following diagram represents the central part of the London Underground.

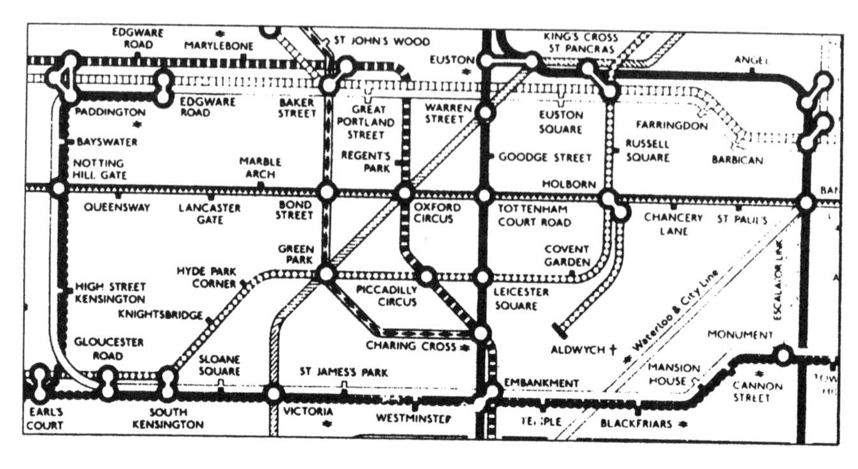

Like all maps, it does not represent every feature of the area, but only those of relevance to the people that use it. The Underground map does not represent the exact geographical locations of the stations, or the precise distances between the stations, but does depict the way in which the stations are interconnected, so that passengers can plan their routes from one station to another.

Problem 1.1

On the London Underground map, find the 'best' route between Marble Arch and Westminster. What interpretations can be given to the word 'best'?

However, when we use a road map, not only are the interconnections between towns important, but so are the distances or travel times between them. For example, the following map shows some of the major routes between a number of cities in the USA, where the numbers indicate the driving times (in hours) between pairs of cities.

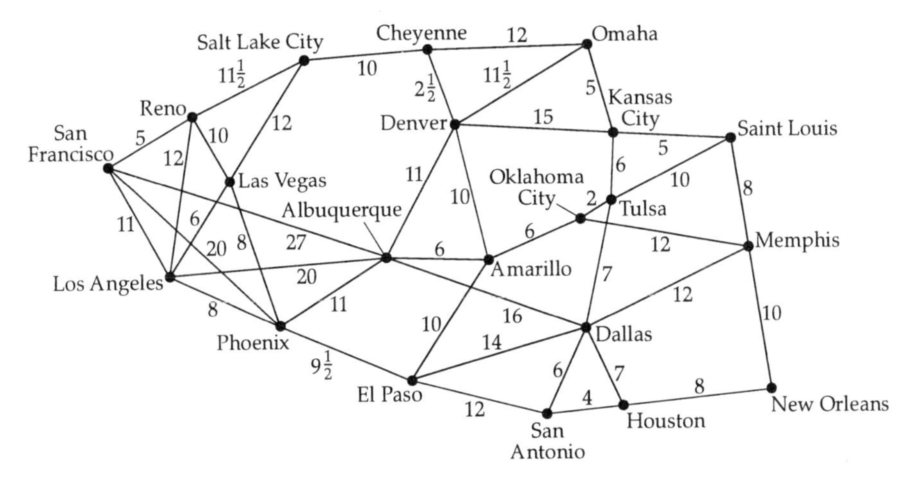

Problem 1.2

On the above map, find the shortest time taken to drive from Los Angeles to Amarillo, and from San Francisco to Denver.

Chemistry

A molecule consists of a number of atoms linked by chemical bonds. For example, a methane molecule (CH_4) consists of a carbon atom (C) bonded to four hydrogen atoms (H), and may be represented by diagram (a) below. Similarly, an ethane molecule (C_2H_6) consists of two carbon atoms bonded to six hydrogen atoms, and may be represented by diagram (b) below.

```
    H              H   H
    |              |   |
H — C — H      H — C — C — H
    |              |   |
    H              H   H
   (a)              (b)
```

More generally, an alkane (or paraffin) is a molecule with formula C_nH_{2n+2}; for every alkane, each carbon atom is bonded to exactly four atoms, and each hydrogen atom is bonded to exactly one atom (a carbon atom). Such a molecule can be represented diagrammatically, with atoms indicated by their chemical symbols, and chemical bonds shown by lines linking these symbols.

Diagrams of this sort do not tell us how the atoms are aligned in space; for example, the hydrogen atoms of methane do not all lie in a plane, but are situated at the vertices of a regular tetrahedron, with the carbon atom at the centre.

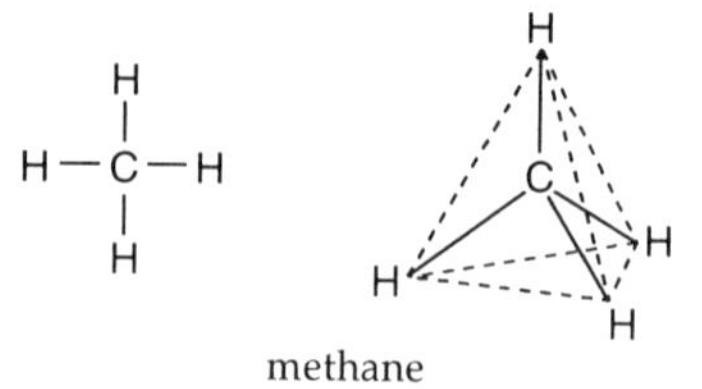

methane

Nevertheless, such diagrams are useful for illustrating how the various atoms are connected, and we can obtain much information about the likely chemical behaviour of a molecule by studying its diagram. For example, the alkanes have the formula C_nH_{2n+2} and their graphs have a branching tree-like structure.

If there are four or more carbon atoms, then there may exist different molecules (known as *isomers*) with the same formula, as illustrated by the following diagram for C_5H_{12}.

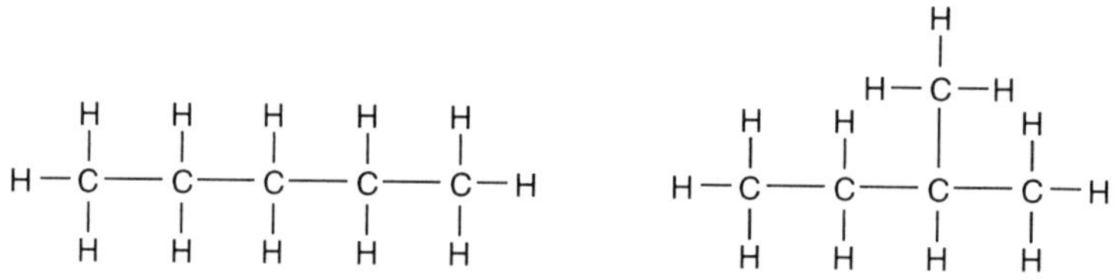

Problem 1.3

Show that there is just one alkane with formula C_3H_8, but that there are two alkanes with formula C_4H_{10}.

Utilities Problem

In this problem three quarrelsome neighbours wish to connect their houses to the three utilities *gas, water* and *electricity* in such a way that the nine connections do not cross each other in the plane. In the following diagram, eight connections appear, but house *B* is not connected to *water*. Can you insert the missing connection as required?

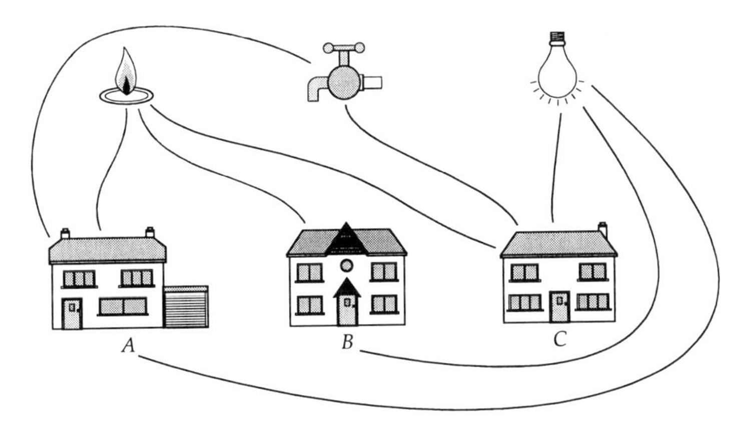

The utilities problem is related to a number of practical problems arising in the study of printed circuits. In these problems, electronic components are constructed by means of conducting strips printed directly on to a flat board of insulating material. Such printed connectors may not cross, since this would lead to undesirable electrical contact at crossing points.

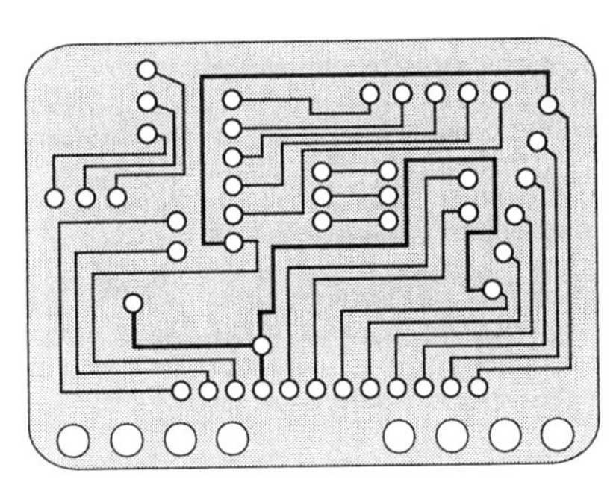

Problem 1.4

For which of the following circuits can you redraw the wires in such a way that no crossing points occur?

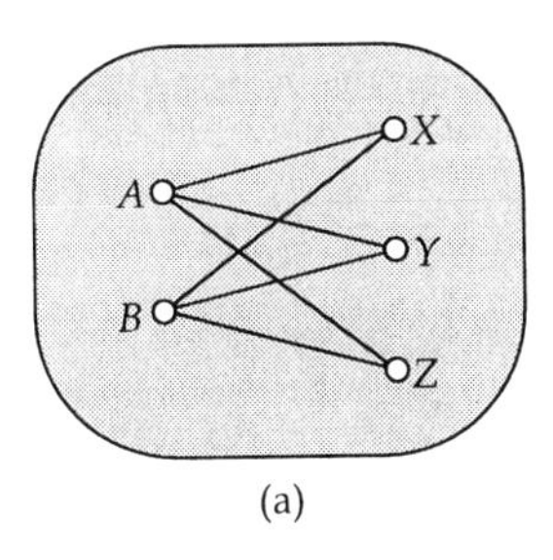

(a)

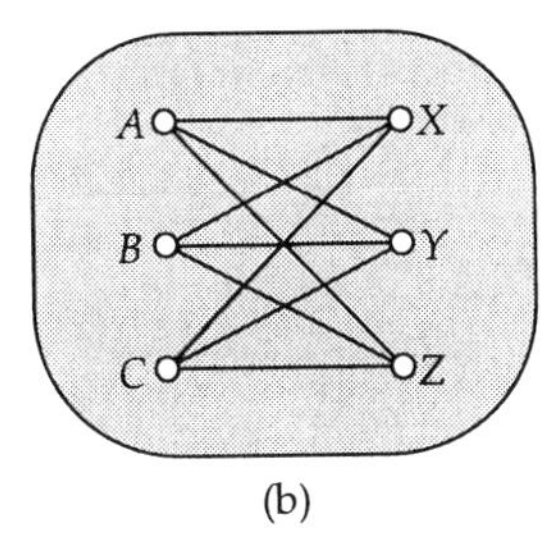

(b)

In each of these examples, we have a system of objects that are interrelated in some way – stations interconnected by rails, atoms connected by bonds, and houses connected to utilities. In each case, we can draw a diagram in which we represent the objects by points, and the interconnections between pairs of objects by lines (not necessarily straight) between the corresponding points. Such a diagram is called a *graph*; the points representing the objects are called *vertices*, and the lines representing the interconnections are called *edges*.

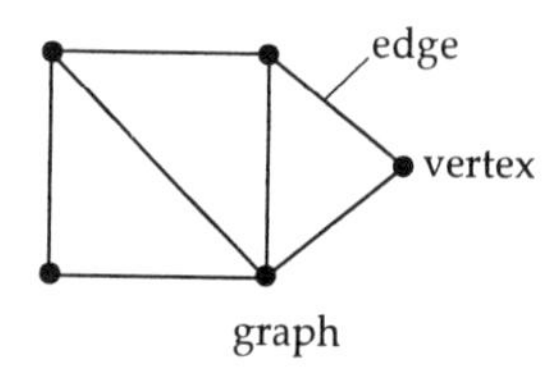

We express these ideas as follows.

A **graph** is a diagram consisting of points, called **vertices**, joined by lines, called **edges**; each edge **joins** exactly two vertices.

Remark This terminology is not completely standard; some authors use *node* or *point* for what we call a vertex, and *arc* or *line* for what we call an edge.

The utilities problem can be represented by a graph with six vertices, corresponding to the three houses and the three utilities, and nine edges, corresponding to the nine possible connections. (Note that the three houses are not joined to each other, and nor are the three utilities.) The following diagrams illustrate two possible drawings of this graph. The utilities problem is that of finding yet another drawing that involves no crossing of edges.

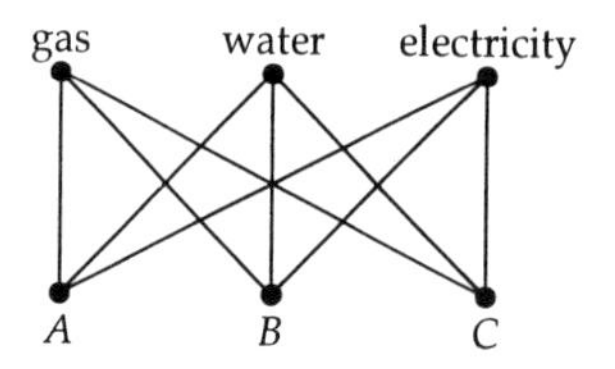

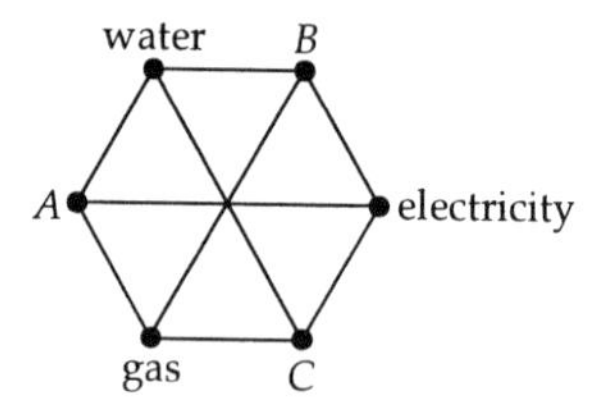

Such a graph, in which the vertices split naturally into two sets (corresponding here to houses and utilities), with edges joining vertices in different sets, is called a *bipartite graph*. Bipartite graphs feature throughout this book, particularly in applications.

The concept of a graph is simple, and graphs can be used whenever we wish to depict interconnections or relationships between objects. For example, any of the following graphs with four vertices and five edges can be used to depict the five football games played in a certain period among four teams – Arsenal has played once against Chelsea and Everton, but not against Liverpool; Chelsea and Everton have played each other twice; and Liverpool has played Everton once.

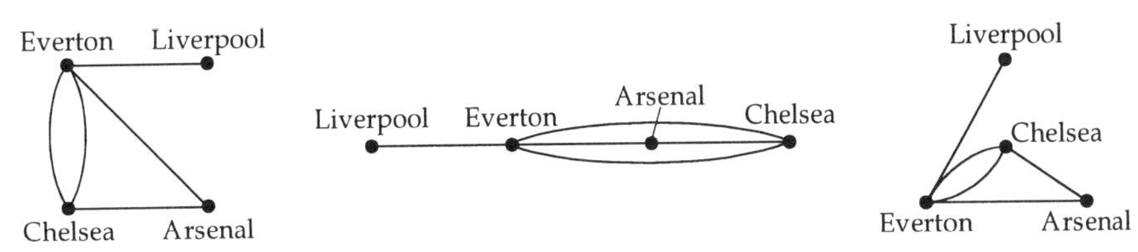

We can draw a graph in many ways, as long as it represents the same interconnections; each of the above drawings represents the same information about which teams have played which, and we regard them as the same graph.

Problem 1.5

Draw a graph that represents the following friendships among four people:

John is friends with Joan and Jill, but not Jack;
Jack is friends with Jill, but not Joan;
Joan is friends with Jill.

Problem 1.6

Suppose that there are six people at a party. Prove that it is always possible to find either three people who all know each other, or three people none of whom knows either of the other two.

Hint Represent the six people by the vertices of a graph, and consider the five possible edges emerging from one of the vertices.

Problem 1.7

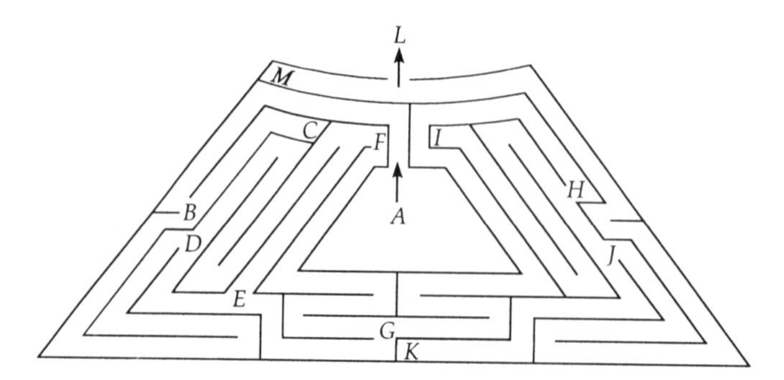

The Hampton Court maze may be represented by a graph that shows the available choices at each junction; for example, at the junction *B* there are two choices – to go to *C* or to *D*. We thus obtain the following graph:

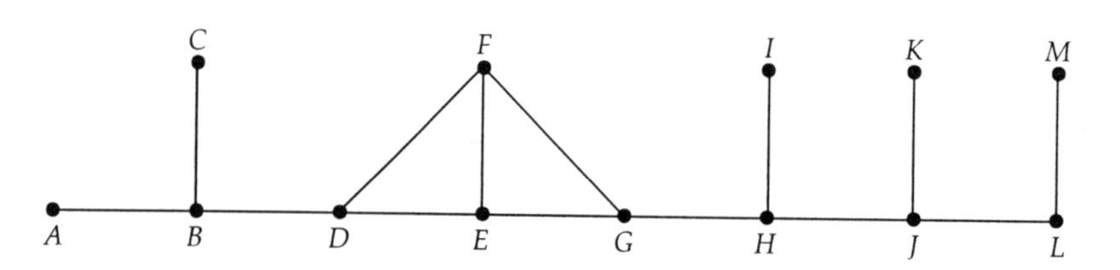

Use this graph to list all the routes from the centre (*A*) to the exit (*L*) that do not involve retracing steps.

Königsberg Bridges Problem

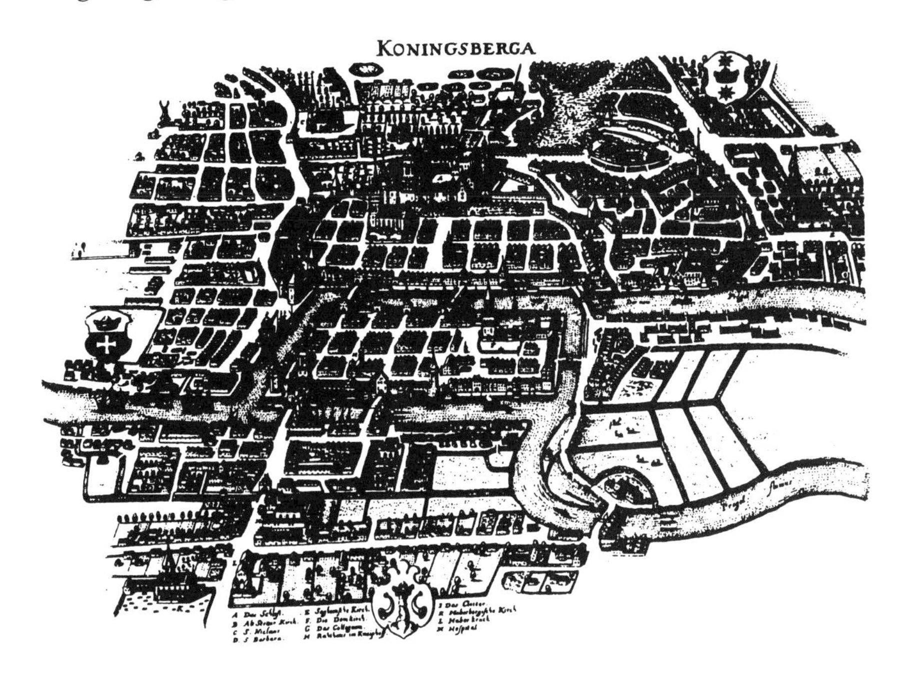

In the early eighteenth century the medieval city of Königsberg in Eastern Prussia contained a central island called Kneiphof, around which the river Pregel flowed before dividing into two. The four parts of the city (A, B, C, D) were interconnected by seven bridges (a, b, c, d, e, f, g), as shown in the following diagram:

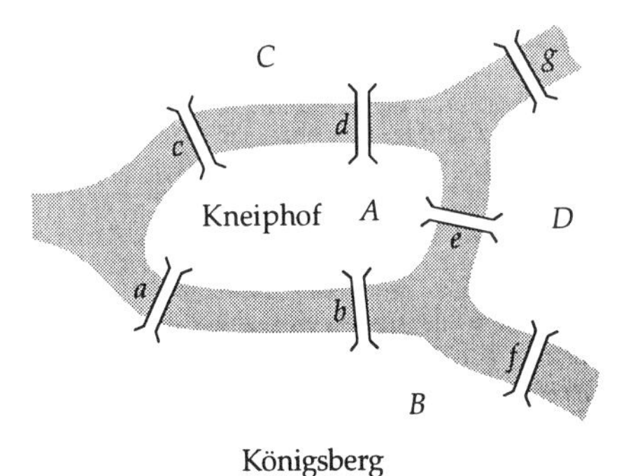

Königsberg

It is said that the citizens of Königsberg entertained themselves by trying to find a route that crosses each bridge exactly once, and returns to the starting point. Try as they might, they could find no such route, and they began to believe the task impossible.

Problem 1.8

In the above Königsberg bridges diagram, try to find a route that crosses each bridge exactly once and returns to the starting point. Do you think that such a route exists?

In this problem, there are four land areas interconnected by seven bridges. We can represent these interconnections by drawing a graph with four vertices, corresponding to the four land areas, and seven edges, corresponding to the seven bridges.

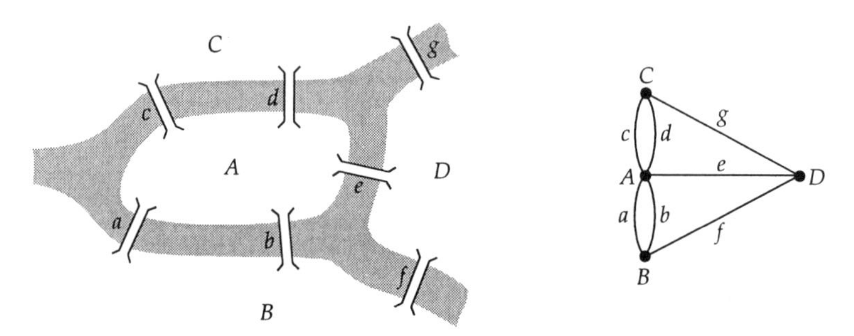

The problem of crossing each of the seven bridges exactly once has now been transformed into a graph problem:

can you draw the above graph and return to your starting point, without lifting your pen from the paper and without tracing any edge twice?

Braced Rectangular Frameworks

Many buildings are supported by rectangular steel frameworks, and it is important that such frameworks should remain rigid under heavy loads. One way to achieve this is to add *braces*, to prevent distortion in the plane.

For example, the following diagram shows how a simple unbraced rectangular framework can be distorted.

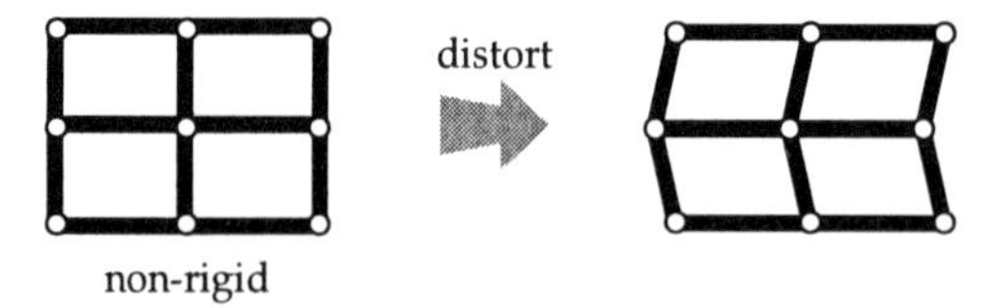

The addition of only two braces, in the form of rectangular plates (indicated by shading), cannot make this framework rigid, as the following diagrams illustrate:

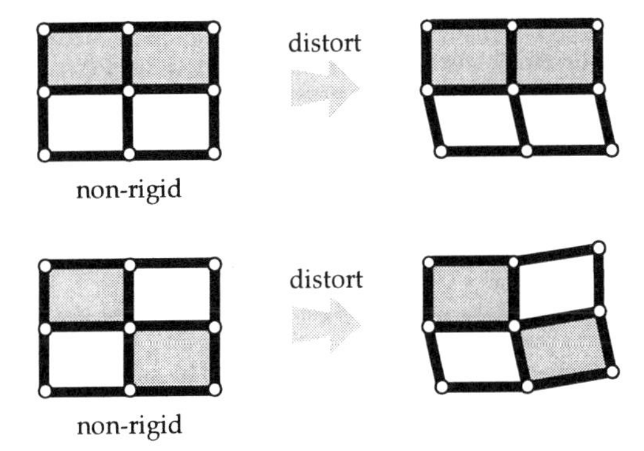

The minimum number of braces that must be added to make this framework rigid is 3.

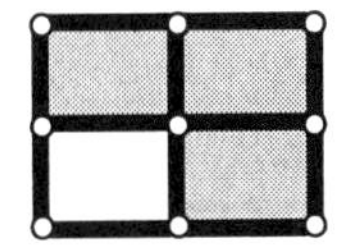

Now consider the following three frameworks:

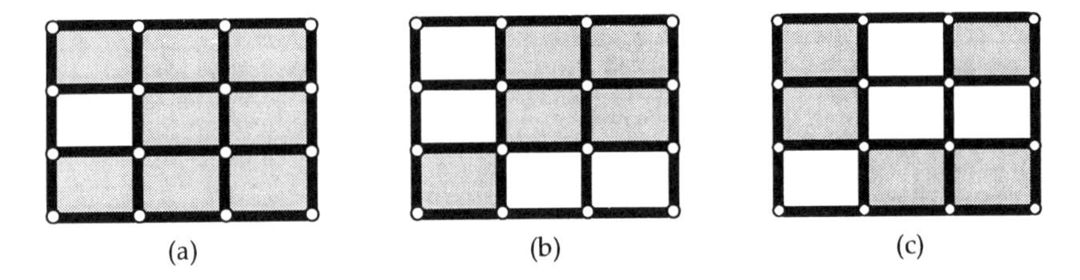
(a) (b) (c)

Framework (a) is rigid, but is over-braced, since some braces can be removed without affecting the rigidity of the framework.

Framework (b) is not rigid, since it can be distorted as shown below:

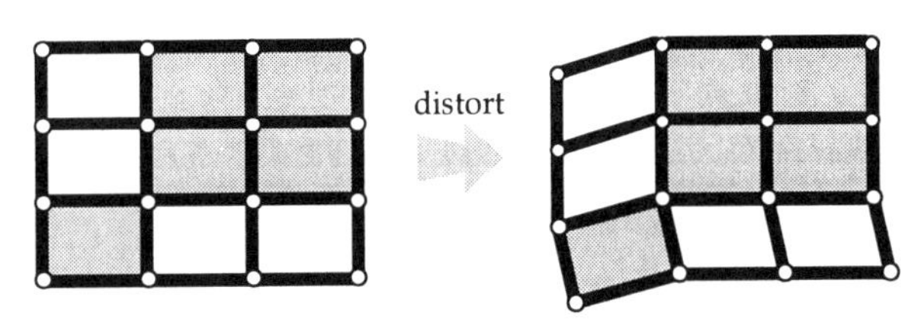

But how about framework (c)? is it rigid?

If so, can any braces be removed without affecting the rigidity?

If not, how can it be made rigid?

Problem 1.9

(a) Which braces can be removed from framework (a) without losing the rigidity?
(b) Which braces could be added to framework (b) to achieve rigidity?
(c) Try to answer the questions about framework (c) in the above text.

We can represent a braced framework by a graph as follows. Each brace occurs in one row and one column of the framework, so we can record the positions of the various braces by drawing a graph whose vertices correspond to the rows and columns of the framework and whose edges correspond to those rows and columns where a brace appears. For example, in framework (c) there is a brace in row 1, column 1, but no brace in row 1, column 2; in the corresponding graph, there is an edge joining the row 1 vertex to the column 1 vertex, but no edge joining the row 1 vertex to the column 2 vertex.

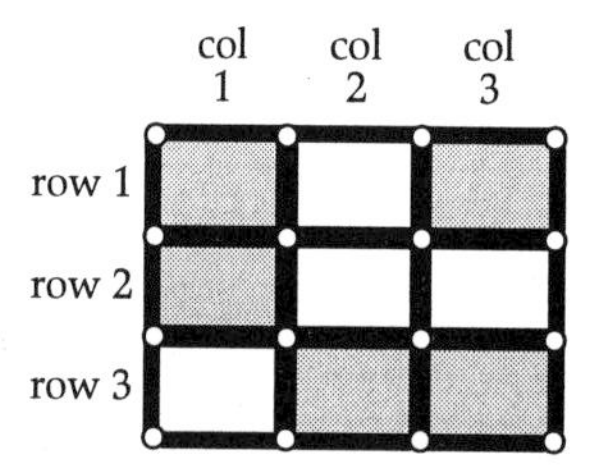

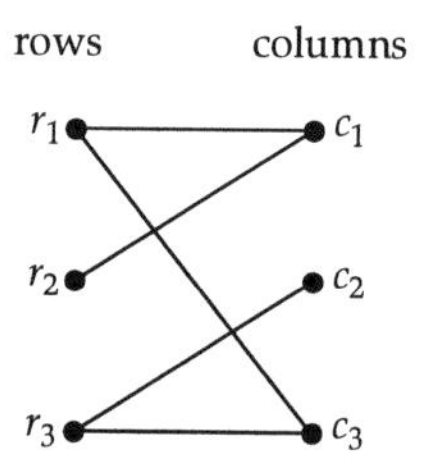

Notice that we obtain a bipartite graph in which the two sets correspond to the rows and columns.

Problem 1.10

(a) Draw the graph for framework (b) above.
(b) By considering frameworks (b) and (c) and their graphs, try to determine a property of a graph that corresponds to rigidity of the corresponding framework.

In Chapter 6 you will see how the problems of determining whether a given braced rectangular framework is rigid, and whether any braces can be removed without affecting the rigidity, can be answered directly by studying the corresponding graph.

Next, we introduce two important variations of the concept of a graph – *weighted graphs* and *digraphs*.

Weighted Graphs

Earlier, we gave the following road map of cities in the USA. This map has the form of a graph, with vertices representing cities and edges representing roads joining them in pairs. Each edge has a number associated with it, representing the *driving time* (in hours) between neighbouring pairs of cities. These numbers are called **weights**, and a graph with a weight associated with each edge is a **weighted graph**.

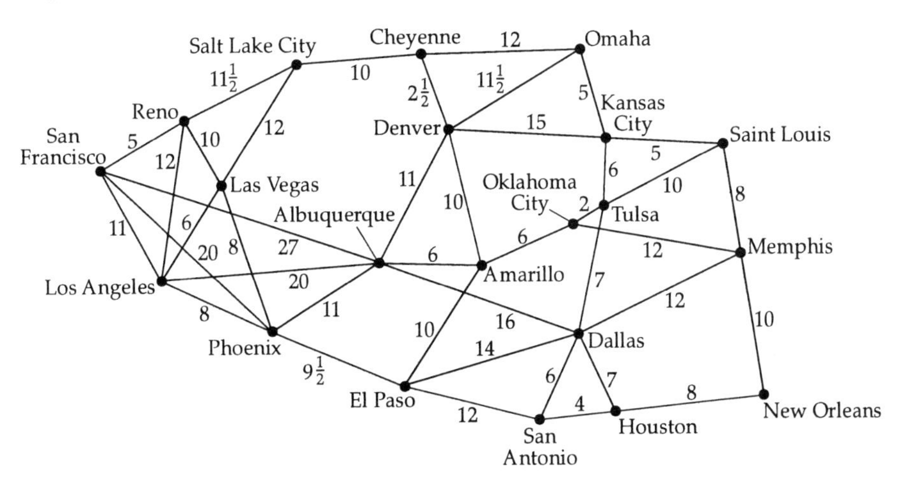

The weight on an edge can refer to many things. For example, on a road map, it may represent the *distance*, *time* or *cost* involved in travelling along the corresponding road.

Another example of a weighted graph appears in the following problem.

Travelling Salesman Problem

A travelling salesman wishes to visit a number of cities and return to his starting point, selling his wares as he goes. He wants to select the route with the least total length. Which route should he choose? And what is its length?

As a particular instance of this problem, consider the four cities illustrated below.

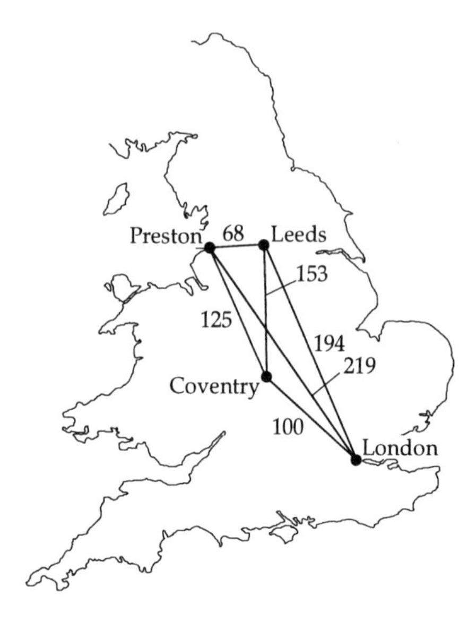

This weighted graph has four vertices, corresponding to London, Coventry, Preston and Leeds, and six edges joining them. Here, the weight on each edge represents the *distance* (in miles) between the corresponding pair of cities. The problem is to select a minimum-weight route that visits all the vertices and returns to the starting point, London.

It is easy to show by trial-and-error methods that a solution of this instance of the travelling salesman problem is the route

London – Coventry – Preston – Leeds – London

(in either direction), with total distance $100 + 125 + 68 + 194 = 487$ miles. Any other route involves a longer total distance.

Unfortunately, as soon as we increase the number of cities significantly, we run into difficulties. There is no convenient method known that provides a solution for *any* travelling salesman problem, although there are several *ad hoc* procedures that yield approximate solutions. Indeed, the only known procedure that is *guaranteed* to solve *any* given travelling salesman problem is the *exhaustion* method that involves looking at all possible routes and choosing the shortest. This is feasible if there are ten cities, since the number of possible routes is then $\frac{1}{2}(9!) = \frac{1}{2}(9 \times 8 \times 7 \times 6 \times 5 \times 4 \times 3 \times 2 \times 1) = 181440$, and a computer sorting through these at the rate of one thousand per second would find the best route in about 3 minutes. On the other hand, if there are twenty cities, then the number of possible routes is about 12.16×10^{17}, and a computer sorting through them at the same rate would take roughly 38 million years! This rapid increase in time as the number of cities grows is known as the *combinatorial explosion*.

Problem 1.11

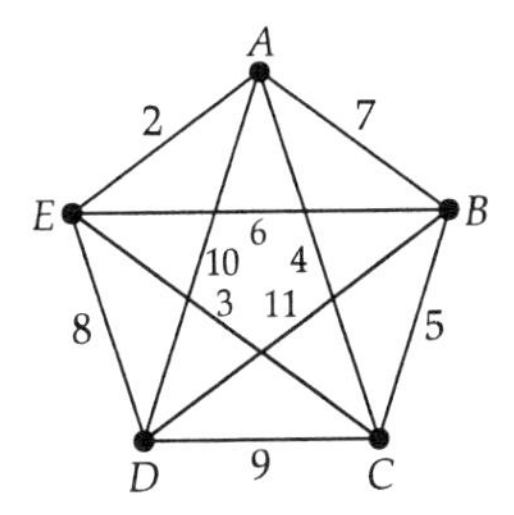

A zoo-keeper visiting the antelopes wishes to visit the bears, camels, dingos and elephants. The locations are denoted by *A*, *B*, *C*, *D*, *E*, and the distances are as shown on the diagram. Using trial and error, find a route that involves the least possible total distance and returns to the starting point *A*.

Although there is no known efficient procedure that is guaranteed to solve *any* given instance of the travelling salesman problem, several efficient procedures are known that give a good *approximation* to the minimum length of a route that visits all the cities and returns to the starting point. Such a procedure is presented in Chapter 8.

Digraphs

Consider the following one-way street system in a town.

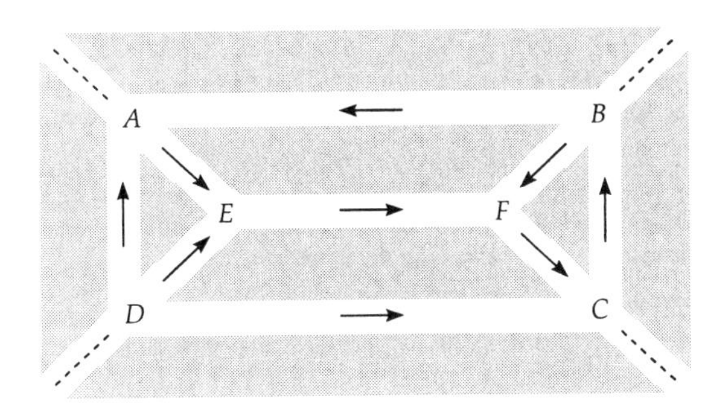

Because the streets are all one-way, we cannot represent this system by a graph, but we can represent it by a similar diagram in which we put arrows on the edges to indicate the directions of the one-way streets. Such a diagram is called a *digraph* – an abbreviation of *di*rected *graph*.

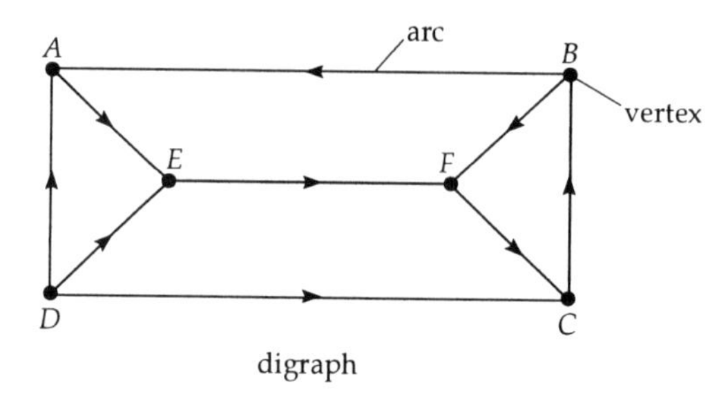

digraph

> A **digraph** is a diagram consisting of points, called **vertices**, joined by directed lines, called **arcs**; each arc **joins** exactly two vertices.

Another example of a digraph is the structural diagram of the book, given in the Study Guide; here each vertex corresponds to a group of chapters, and the arcs can be used to trace possible study paths through the book.

Problem 1.12

Represent the following one-way street system by a digraph.

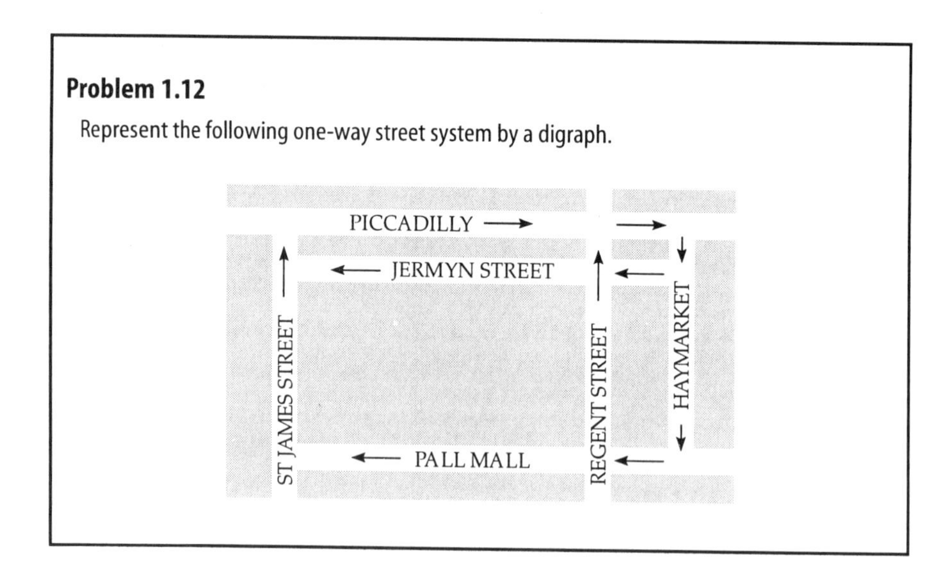

Networks

Just as we can assign weights to the edges of a graph to form a weighted graph, so we can assign weights to the arcs of a digraph to form a weighted digraph. This leads to the idea of a *network*.

The word *network* commonly arises in everyday life. We speak of a *rail network* as a system of stations interlinked by railway lines, or a *road network* as a system of towns interlinked by roads. We speak of a programme produced by a television company being *networked* around the country, and an *electrical network* involving terminals interconnected by wires or a *telecommunication network* involving interlinked telephone exchanges and subscribers.

We also use the word *network* in a more specialized sense to mean a graph or digraph that carries some numerical information. This information depends on the particular application under consideration, but may consist of weights associated with the edges or arcs, and possibly with some of the vertices.

Note that a graph or digraph represents only the interconnections of a system, and gives no further information about its elements. A network conveys additional quantitative information.

Road Networks

In a road network, the weight of each arc (road) may correspond to any of the following: its length in miles or kilometres; the estimated or actual time taken to travel along it; the cost involved in travelling along it (fuel, tolls, etc.). Other possible weights may relate to the attractiveness of the scenery, the danger of travelling along the road, or the quality of the road surface.

Now suppose that we wish to find the shortest route between two particular towns. In this problem we are concerned not with the amount of traffic that each road can take, but only with the length of each section of the road, so we assign the length as the weight on the corresponding edge of our graph. The problem of finding the shortest route between two towns on our map is thus reduced to that of finding the shortest route between the two corresponding vertices on our graph – that is, the route with the lowest total weight.

A simple shortest path problem like the following can be solved by inspection, but for larger scale problems we need a systematic step-by-step method. We describe such a method in Chapter 9.

Problem 1.13

By inspection, find the shortest route from S to T in the following network:

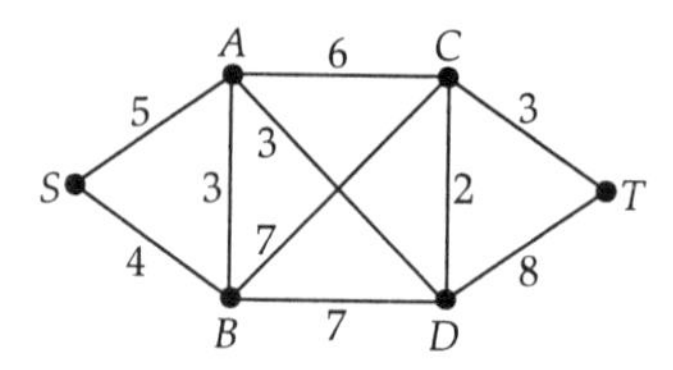

Hint First find the shortest route from A to T, then from B to T, then from S to T.

Pipeline Networks

The following diagram represents a network of pipelines along which a fluid (for example, gas, oil or water) flows from a source S to a terminal (or *sink*) T. Each of the intermediate points A–I represents a pipe junction at which the total flow into the junction must equal the total flow out (so no fluid is 'lost' along the way). Each line between two junctions represents a pipeline, and the number next to it is the *capacity* of that pipeline (in some appropriate units); the flow along a pipeline must not exceed the capacity, and must be in the direction indicated.

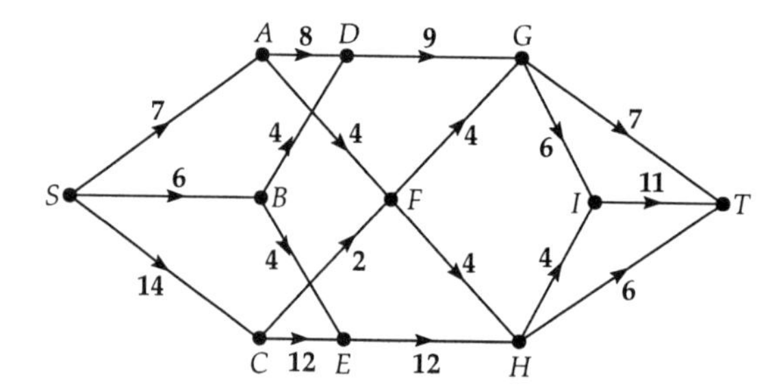

Inspection of the above diagram shows that a *flow* of at most 7 units of fluid can be sent along the route $S \rightarrow A \rightarrow D \rightarrow G \rightarrow T$ without exceeding the capacity of any of the pipelines SA, AD, DG, GT. This is illustrated in the following diagram, where the first number on each line represents the flow along that pipeline and the second number – in bold – is the capacity.

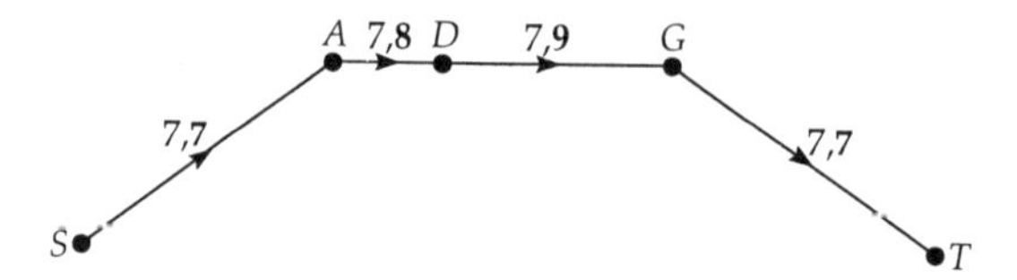

Properties of Graph Models

We have indicated how we can use graphs to organize our knowledge of structural features in a variety of situations. Such a graph representation of a problem is called a **model**, because it models the essential features of the situation for the problem in hand.

The models that we develop in this book are based upon the formal structures of combinatorial mathematics, and on graph theory in particular. In the situations we investigate, we select those features that can be represented by graphs.

What kinds of features can be represented by graphs? Broadly speaking, we represent *structural* features. This is done in one of the following two ways.

In many of our examples,

the elements of the structure are represented by vertices

and

the connections between them are represented by edges.

It is these relations that essentially constitute the structural features.

Another way in which graphs can represent structural features is, in a sense, the *dual* of the first:

the elements of the structure are represented by edges

and

the connections between them are represented by vertices.

Network problems are often modelled in this way. For example, the edges may represent roads or pipes, and the vertices represent the junctions or connections between the elements.

For many applications, such a pictorial representation may be all that is needed. By representing a a situation in such a simple diagrammatic form, we may be able to derive all the information we require. For other applications, we may need to analyse the problems carefully and study the properties of the graphs that arise as diagrammatic representations. In particular, we investigate the properties of graphs of the following types:

bipartite graphs (Chapter 2);
graphs with a branching tree-like structure (Chapter 6);
graphs that are in one piece (Chapter 10);
graphs in which no edges cross (Chapter 11).

1.2 Classifying Problems

It is also useful to consider a classification of problems based on four types of question that commonly arise. Most of the problems you will meet in this book can be described under one or more of the following interrelated headings:

Existence problems	Does there exist ...? Is it possible to ...?
Construction problems	If ... exists, how can we construct it?
Enumeration problems	How many ... are there? Can we list them all?
Optimization problems	If there are several ..., which one is the best?

Before discussing each of these types of problem in detail, we interpret all four types of problem in the context of braced rectangular frameworks.

In investigating a given rectangular framework, we may wish to consider any of the following questions:

Existence problem	Is it possible to brace the framework so as to make it rigid?
Construction problem	If a rigid bracing exists, how can we construct one?
Enumeration problem	How many rigid bracings are there? Can we list them all?
Optimization problem	Which rigid bracings involve fewest braces?

We now look at each type of problem in turn.

Existence Problems

Faced with any problem, it is natural to ask *does a solution exist?* This question is not always easy to answer.

For example, in the Königsberg bridges problem, the question faced by the citizens of Königsberg was:

> is it possible to go for a walk crossing each of the seven bridges once and only once, and ending at the starting point?

This is clearly an example of an *existence* problem, since it is concerned with whether or not a suitable walk exists.

Note that, in order to demonstrate that such a walk exists, it is sufficient to produce a specific walk. However, to show that such a walk does not exist, we must actually *prove* that the problem has no solution.

Another celebrated example of an existence problem is the utilities problem: does there exist a way of putting in all nine connections?

It is easy to convince oneself by trial and error that the answer to this existence problem is no, but actually *proving* that no solution exists is more difficult.

We can sometimes prove that something must exist, even though we may not know how to find a specific instance of it. For example, it is easy to see that

> in any group of eight people, there must exist at least two people who were born on the same day of the week,

since there are only seven days in a week, so the eight people cannot all have been born on different days. However, although we know that there *exist* two people that were born on the same day of the week, the above explanation does not help us to *find* them.

Construction Problems

Once we know that a solution to a problem exists, we may wish to find a way of constructing it. For example, it is not particularly helpful to know that it is possible to get out of a maze if you are stuck in the middle of it – what you really need is a method for getting out! As we saw above, we can often show that an existence problem can be solved by *constructing* a solution, but this is not always the case, as you saw at the end of the previous subsection. In such problems we know that solutions exist because there are theoretical reasons for this, but these reasons may not give us any clue as to how a specific solution may be constructed.

Some of the problems in this book can be solved by trial and error or by examining all possible cases, but many problems are too complex to be solved in this way, because the time involved may make such methods impractical. For example, it is easy to analyse a telecommunication system that interconnects only five or six subscribers, since most problems that arise can be solved by trial and error. On the other hand, a modern telephone exchange may involve the interconnections of tens of thousands of subscribers, and any *ad hoc* approach is out of the question.

For constructing solutions to complex problems we need a systematic step-by-step procedure that can be applied to large systems. Such a procedure is called an *algorithm*. You may find it helpful to think of an algorithm as similar to a cookery recipe. A recipe consists of a list of ingredients, corresponding to the input data, and a list of instructions to be carried out in a particular order. To make a particular dish, you take the ingredients and follow the recipe.

Definition

An **algorithm** is a systematic step-by-step procedure consisting of:
- a description of appropriate input data;
- a finite, ordered list of instructions, to be carried out one at a time;
- a STOP instruction, to indicate when the procedure is complete;
- a description of appropriate output data.

In order to solve a given problem, we input the data and carry out the instructions one at a time in the given order, until we obtain either a solution to the problem in hand or an indication that the problem cannot be solved by the algorithm for the given data.

The form in which an algorithm is presented varies from problem to problem, and may consist of instructions written in everyday language or a computer language. Often, if the input and/or output data are obvious, we do not state them explicitly in our description of the algorithm.

In many cases we can apply algorithms without the aid of a computer. On the other hand, many practical problems are far too large or complex to be dealt with in this way. In such circumstances, it is necessary to express the algorithm in a form that can be implemented by a computer. This means that the instructions must be precisely and unambiguously stated, and that the algorithm must terminate after a finite number of steps.

Several of the problems described in this book can be solved by means of an *efficient* algorithm – one that can solve any instance of the problem in a 'reasonable' amount of time. However, for some problems, no efficient algorithm is currently known.

Enumeration Problems

Once we know that a particular problem has a solution, and we can construct such solutions, the next questions are *how many solutions are there?* and *what are they?*

We distinguish between these two types of problem. A *counting* problem is one in which we wish to know *how many* objects of a certain kind there are; a *listing* problem is one in which we wish to produce a *list* of all these objects.

In general, the problems of counting and listing may be closely related. For example, the easiest way of counting something is often to construct a list of all possibilities and then to count how many there are. Indeed, for some counting problems this may be the *only* known method of solution – for example, when counting the number of times the digit 3 appears in the first hundred significant figures of π.

In many instances the listing problem is much harder to solve than the corresponding counting problem. In fact, there are many problems for which the answer to the counting problem is known, but no one has been able or willing to list all the possibilities. For example, in chemistry, the problem of *counting* the number of different alkanes C_nH_{2n+2} for any given value of n has been solved, although there is no simple formula for the answer. However, the problem of *listing* such molecules for a given value of n is intractable except when n is small. For example, the number of different molecules with the formula $C_{25}H_{52}$ is known – it is over a million – but no one has ever made a complete list of them.

Optimization Problems

For many combinatorial problems it is not enough to know that a given problem has a solution. It may not even be enough to be able to construct a solution using an efficient algorithm, or to list all the possible solutions. In many cases we need to find the 'best' solution, and part of the problem may be in deciding what is meant by the word 'best'.

For example, in Problem 1.1, you were asked to find the 'best' route between two stations of the London Underground. In this case, your 'best' route might be a route that involves going through the smallest number of stations on the way, or a route that involves the smallest number of changes, or some mixture of the two. In any case, when solving such a problem, you may need to seek some further information, such as the time required to change lines, before you can determine the best solution.

In Problem 1.2 you were asked to find the shortest route (in hours) between two American cities. Here, the meaning of 'best' is clear, as any route that takes the shortest amount of time is a 'best' solution. However, even for this problem, you may want to take other factors into account, such as the amount of expressway driving, the attractiveness of the scenery, and so on.

Another optimization problem is the *travelling salesman problem*; here, we wish to find a route of minimum total length joining a number of cities. Again, the meaning of 'best' is clear, although there may be external considerations that indicate one particular solution as the most appropriate.

Optimization problems of this kind occur frequently throughout the book. In such problems, we usually want to maximize or minimize some given parameter (distance, time, etc.), and we can sometimes do this by carrying out an appropriate algorithm.

1.3 Seeking Solutions

Once we have represented a situation by a graph, digraph or network, and identified the nature of the problem, what then? It is useful to pose some further questions such as those outlined below.

What hope have we of finding a solution?

Can we show that a solution exists?
Can we show that there is no solution?
Is there only one solution?
Are there many solutions? If so, is there a 'best' solution?
Do we need to find a solution exactly, or is it sufficient to find an approximate solution? or to find upper and lower bounds between which the correct solution must lie?

How can we find a solution?

Are we going to *construct* the solution?
Can we develop a systematic method – an algorithm – that gives the correct answer to *any* instance of the given problem in a reasonable time, regardless of the size of the problem?
If not, can we develop a method that gives a reasonably accurate solution with a known error bound?

The questions above indicate two types of approach.

One approach is *theoretical*. For graph representations, we develop the theory of graphs. We define different types of graph, study their properties, make conjectures, and prove theorems. Many graph theorems tell us whether a graph has a certain property. However, many properties discovered theoretically do not tell us anything directly useful; a theoretical method may solve an *existence* problem, but not help with the corresponding *construction* problem.

Another approach is concerned directly with showing how the property can arise in the graph structure. Such methods construct objects with the required property, and may be called *practical* methods. In this book we encounter such methods in the form of graph algorithms. For many optimization problems, it is not difficult to find suitable algorithms that can be applied quickly and efficiently; one example is the shortest route problem. Unfortunately, there are also a large number of problems for which no efficient algorithms are known, although there may be 'heuristic' methods based on experience or intuition that work fairly well in practice; these are methods that are quick to apply, but do not usually lead to a correct solution. An important example of such a problem is the travelling salesman problem, where we can easily find approximate solutions that are close to the correct value.

Finally, a combination of the two approaches leads to a classification of algorithms in terms of the times required for their execution, which in turn enables us to classify many of the problems themselves as 'easy' or 'hard'. For example, under this classification, the shortest route problem is 'easy': the travelling salesman problem is 'hard'.

Chapter 2
Graphs

After studying this chapter, you should be able to:

- explain the terms *graph, labelled graph, unlabelled graph, vertex, edge, adjacent, incident, multiple edges, loop, simple graph* and *subgraph*;
- determine whether two given graphs are *isomorphic*;
- explain the terms *degree, degree sequence* and *regular graph*;
- state and use the *handshaking lemma*;
- explain the terms *walk, trail, path, closed walk, closed trail, cycle, connected graph, disconnected graph* and *component*;
- explain what are meant by *complete graphs, null graphs, cycle graphs*, the *Platonic graphs, cubes* and the *Petersen graph*;
- explain what are meant by *bipartite graphs, complete bipartite graphs, trees, path graphs* and *cubes*;
- describe the use of graphs in the solution to the four-cubes problem, and in the social sciences.

The intuitive idea of a graph is already familiar to you from the previous chapter. In this chapter we treat the subject more formally, introducing the basic definitions and examples that will be needed throughout the book.

2.1 Graphs and Subgraphs

We start by recalling the definition of a *graph*.

Definitions

A **graph** consists of a set of elements called **vertices** and a set of elements called **edges**. Each edge **joins** two vertices.

For example, the graph shown below has four vertices $\{u, v, w, x\}$ and six edges $\{1, 2, 3, 4, 5, 6\}$. Edge 1 joins the vertices u and x, edge 2 joins the vertices u and w, edges 3 and 4 join the vertices v and w, edge 5 joins the vertices w and x, and edge 6 joins the vertex x to itself.

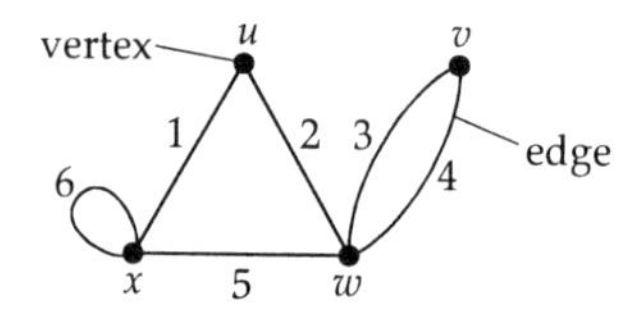

We often denote an edge by specifying its two vertices; for example, edge 1 is denoted by ux or xu, edges 3 and 4 are denoted by vw or wv, and edge 6 is denoted by xx.

The above graph contains more than one edge joining v and w, and an edge joining the vertex x to itself. The following terminology is useful when discussing such graphs.

Definitions

In a graph, two or more edges joining the same pair of vertices are **multiple edges**. An edge joining a vertex to itself is a **loop**.

A graph with no multiple edges or loops is a **simple graph**.

For example, graph (a) below has multiple edges and graph (b) has a loop, so neither is a simple graph. Graph (c) has no multiple edges or loops, and is therefore a simple graph.

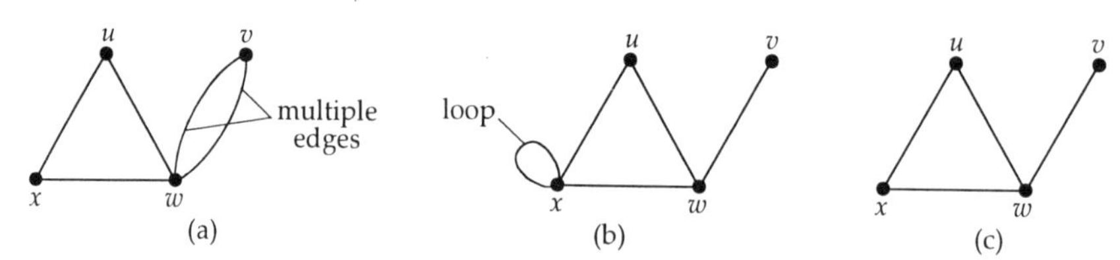

Problem 2.1

Write down the vertices and edges of each of the following graphs. Are these graphs simple graphs?

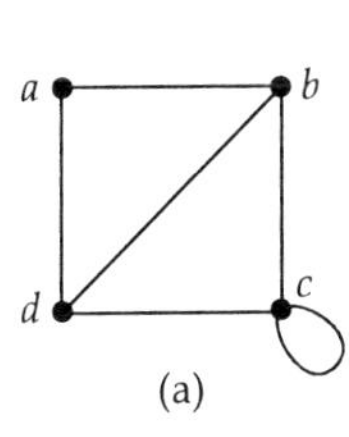

(a)

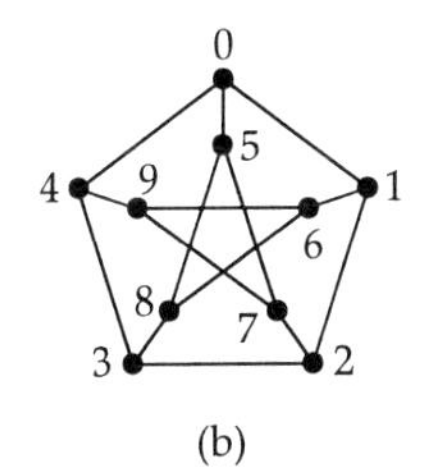

(b)

Problem 2.2

Draw the graphs whose vertices and edges are as follows. Are these graphs simple graphs?

(a) vertices: $\{u, v, w, x\}$ edges: $\{uv, vw, vx, wx\}$
(b) vertices: $\{1, 2, 3, 4, 5, 6, 7, 8\}$ edges: $\{12, 22, 23, 34, 35, 67, 68, 78\}$

Adjacency and Incidence

Since graph theory is primarily concerned with relationships between objects, it is convenient to introduce some terminology that indicates when certain vertices and edges are 'next to each other' in a graph.

Definitions

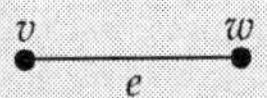

The vertices v and w of a graph are **adjacent** vertices if they are joined by an edge e. The vertices v and w are **incident** with the edge e, and the edge e is **incident** with the vertices v and w.

For example, in the graph below, the vertices u and x are adjacent, vertex w is incident with edges 2, 3, 4 and 5, and edge 6 is incident with the vertex x.

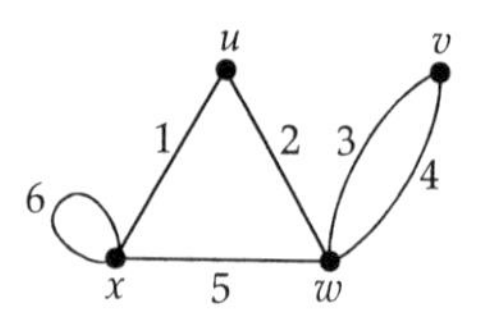

Problem 2.3

Which of the following statements hold for the graph on the right?

(a) vertices v and w are adjacent;
(b) vertices v and x are adjacent;
(c) vertex u is incident with edge 2;
(d) edge 5 is incident with vertex x.

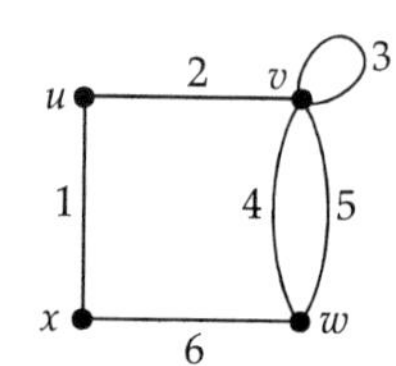

Isomorphism

It follows from the definition that a graph is completely determined when we know its vertices and edges, and that two graphs are *the same* if they have the same vertices and edges. Once we know the vertices and edges, we can draw the graph and, in principle, any picture we draw is as good as any other; the actual way in which the vertices and edges are drawn is irrelevant – although some pictures are easier to use than others!

For example, recall the *utilities graph*, in which three houses A, B and C are joined to the three utilities gas (g), water (w) and electricity (e). This graph is specified completely by the following sets:

vertices: $\{A, B, C, g, w, e\}$
edges: $\{Ag, Aw, Ae, Bg, Bw, Be, Cg, Cw, Ce\}$,

and can be drawn in many ways, such as the following:

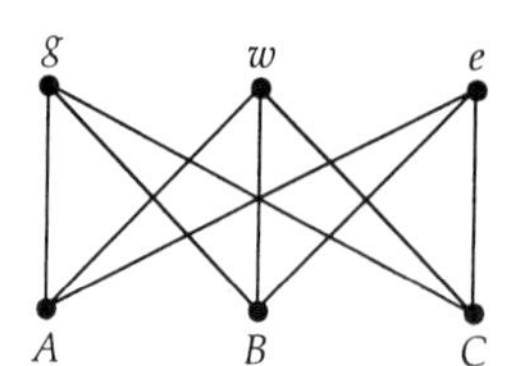

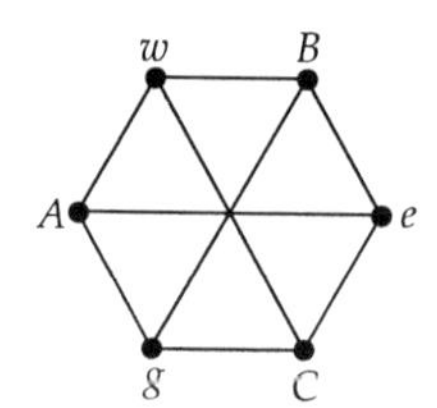

Each of these diagrams has six vertices and nine edges, and conveys the same information – each house is joined to each utility, but no two houses are joined, and no two utilities are joined. It follows that these two dissimilar diagrams represent the same graph.

On the other hand, two diagrams may look similar, but represent different graphs. For example, the diagrams below look similar, but they are not the same graph: for example, AB is an edge of the second graph, but not the first.

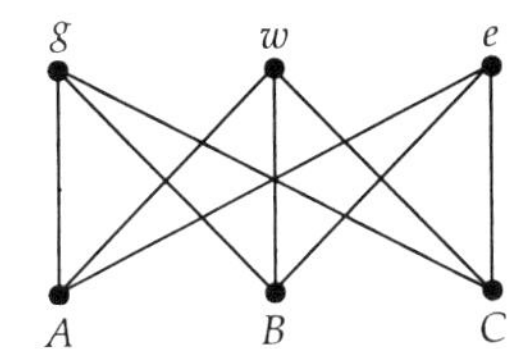

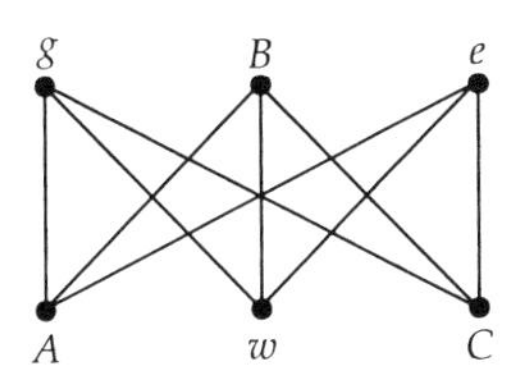

We express this similarity by saying that the graphs represented by these two diagrams are *isomorphic*. This means that *the two graphs have essentially the same structure*: we can relabel the vertices in the first graph to get the second graph – in this case, we simply interchange the labels w and B.

This leads to the following definition.

> **Definition**
>
> Two graphs G and H are **isomorphic** if H can be obtained by relabelling the vertices of G – that is, if there is a one-one correspondence between the vertices of G and those of H, such that the number of edges joining each pair of vertices in G is equal to the number of edges joining the corresponding pair of vertices in H. Such a one–one correspondence is an **isomorphism**.

For example, the graphs G and H represented by the diagrams

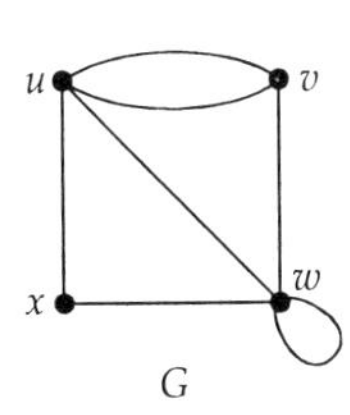

and

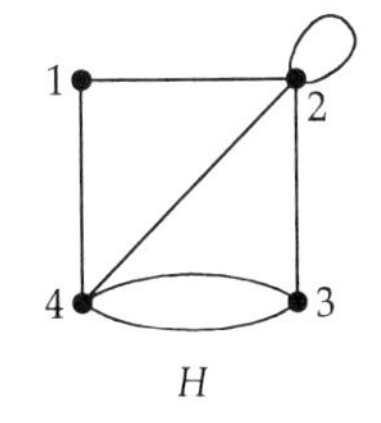

are not the same, but they are isomorphic, since we can relabel the vertices in the graph G to get the graph H, using the following one–one correspondence:

G	$\leftrightarrow$	H
u	$\leftrightarrow$	4
v	$\leftrightarrow$	3
w	$\leftrightarrow$	2
x	$\leftrightarrow$	1

Note that edges in G correspond to edges in H – for example:

the two edges joining u and v in G correspond to the two edges joining 4 and 3 in H;
the edge uw in G corresponds to the edge 42 in H;
the loop ww in G corresponds to the loop 22 in H.

To check whether two graphs are *the same*, we must check whether all the vertex labels correspond. However, to check whether two graphs are *isomorphic*, we must investigate whether we can relabel the vertices of one graph to give those of the other. In order to do this, we first check that the graphs have the same numbers of vertices and edges, and then look for special features in the two graphs, such as a loop, multiple edges, or the number of edges meeting at a vertex. For example, the following two graphs both have five vertices and six edges, but are not isomorphic, as the first has two vertices where just two edges meet, whereas the second has only one.

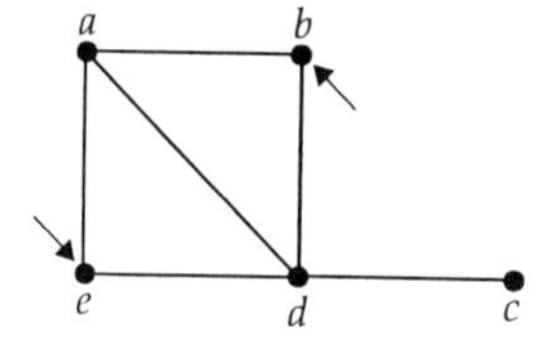

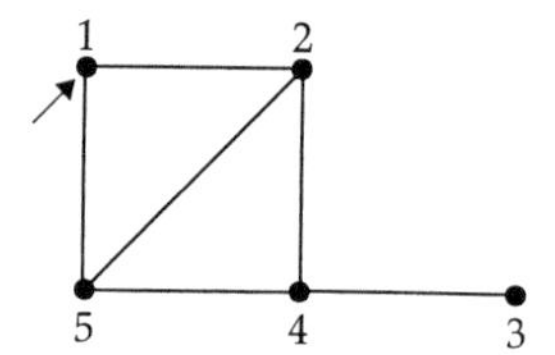

Problem 2.4

By suitably relabelling the vertices, show that the following pairs of graphs are isomorphic:

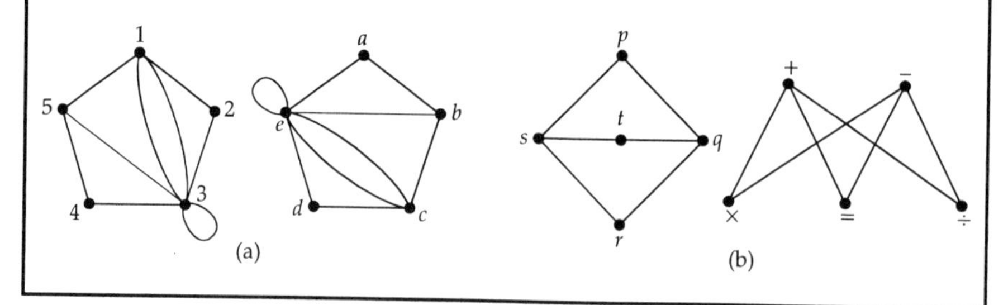

Problem 2.5

Are the following two graphs isomorphic? If so, find a suitable one-one correspondence between the vertices of the first and those of the second; if not, explain why no such one-one correspondence exists.

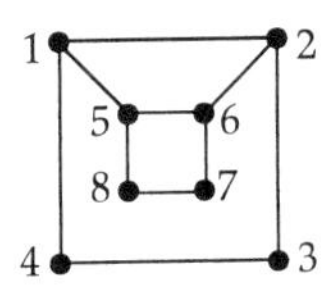

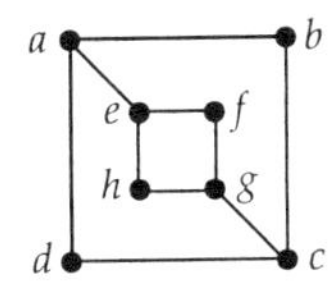

Sometimes it is unnecessary to have labels on the graphs. In such cases, we omit the labels and refer to the resulting object as an *unlabelled graph*. For example, the unlabelled graph

corresponds to either of the following isomorphic graphs:

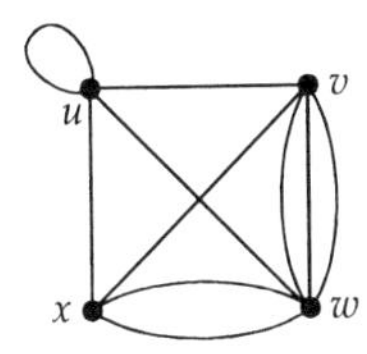

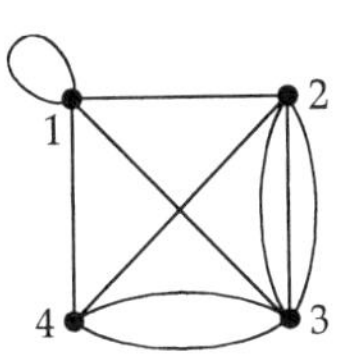

Indeed, it also corresponds to either of the following graphs, which are isomorphic to the above two:

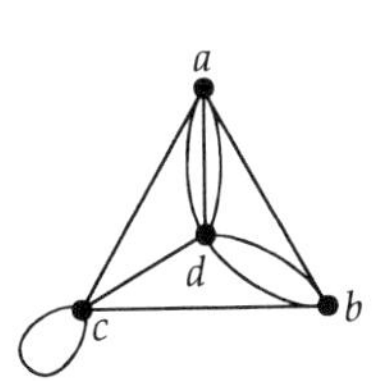

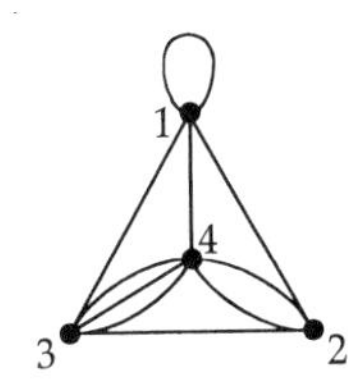

We say that two unlabelled graphs such as

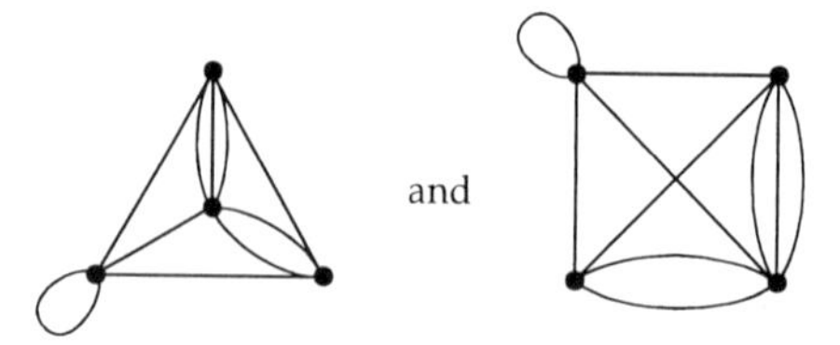

are *isomorphic* if labels can be attached to their vertices so that they become the same graph.

Problem 2.6

By suitably labelling the vertices, show that the following unlabelled graphs are isomorphic:

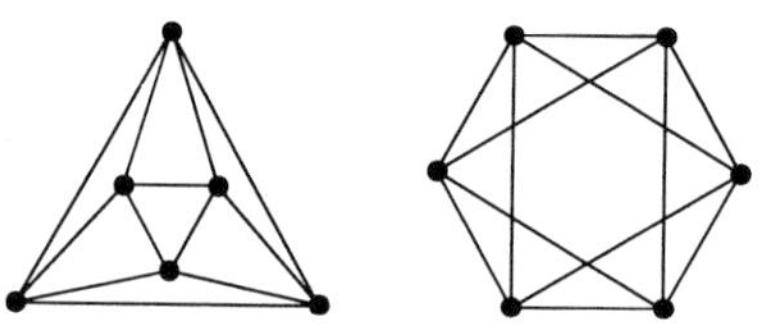

From now on, we use the term *graph* to indicate either a graph with labels on the vertices and/or edges, or an unlabelled graph. The meaning is usually clear from the context, but if there is any possibility of confusion, we insert the word *labelled* or *unlabelled*, as appropriate.

Counting Graphs

What are the relative numbers of labelled and unlabelled graphs with the same number of vertices?

When counting *labelled* graphs, we distinguish between any two that are not *the same*. For example, there are eight different labelled simple graphs with three vertices:

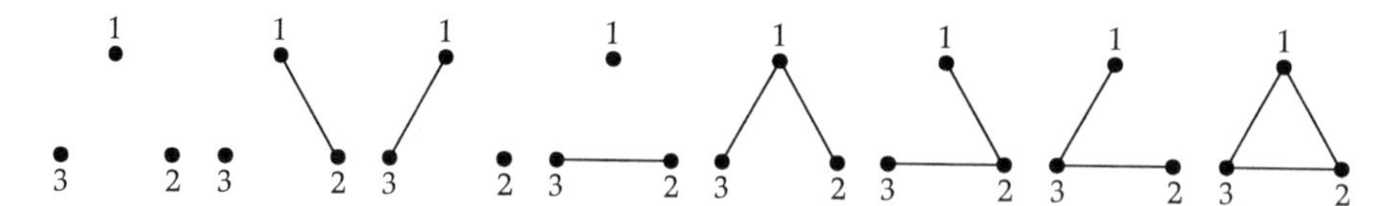

When counting *unlabelled* graphs, we distinguish between any two that are not *isomorphic*. For example, there are just four different unlabelled simple graphs with three vertices:

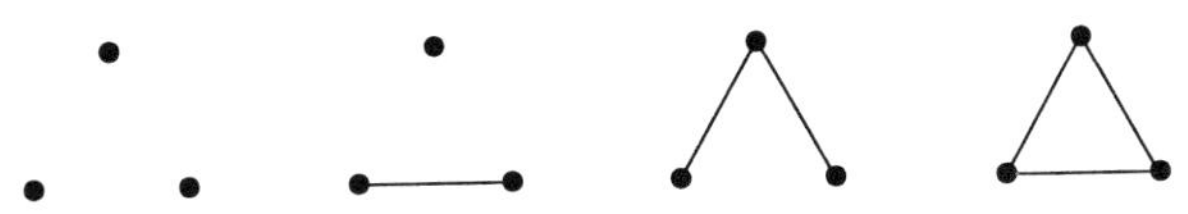

The following table lists the numbers of labelled and unlabelled simple graphs with up to eight vertices:

n	1	2	3	4	5	6	7	8
labelled graphs	1	2	8	64	1024	32768	2097152	268435456
unlabelled graphs	1	2	4	11	34	156	1044	12346

Remark Notice how fast these numbers grow. This is another example of the *combinatorial explosion*.

In general, counting problems for labelled graphs are much easier to solve than their counterparts for unlabelled graphs. In fact, there are certain types of graph for which the former problem has been solved while the latter remains unsolved.

Historical Note

In 1935, the Hungarian mathematician Georg Pólya obtained a general formula from which one can calculate the number of unlabelled graphs with any numbers of vertices and edges. Pólya's methods have since been applied to several other graph-counting problems.

Subgraphs

In mathematics we often study complicated objects by looking at simpler objects of the same type contained in them – subsets of sets, subgroups of groups, and so on. In graph theory we make the following definition.

Definition

A **subgraph** of a graph G is a graph all of whose vertices are vertices of G and all of whose edges are edges of G.

Remark Note that G is a subgraph of itself.

For example, the following graphs are all subgraphs of the graph G on the left, with vertices $\{u, v, w, x\}$ and edges $\{1, 2, 3, 4, 5\}$.

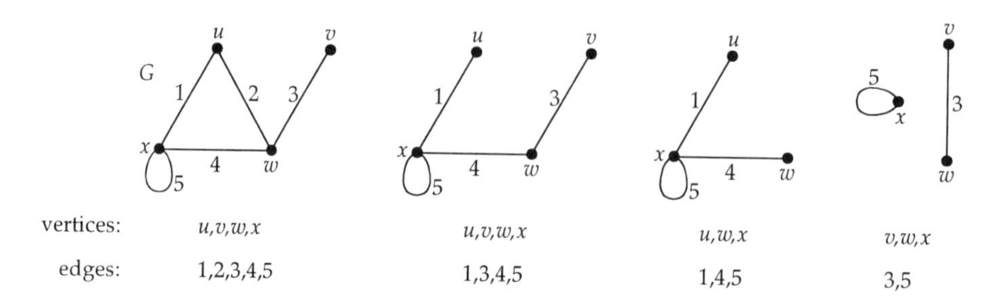

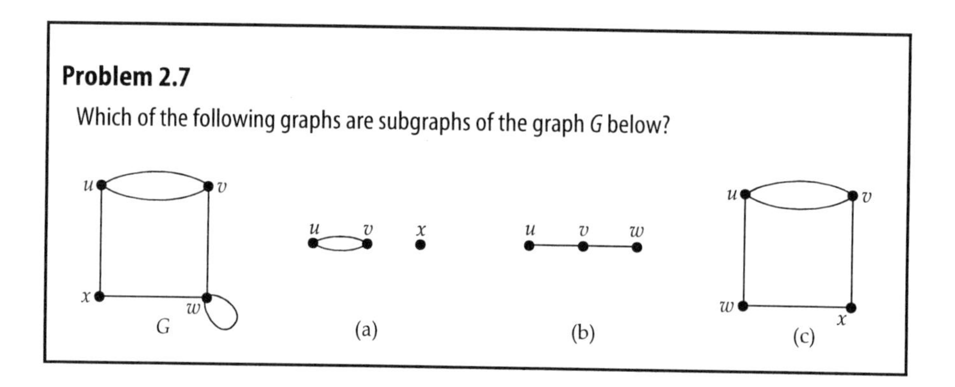

Problem 2.7

Which of the following graphs are subgraphs of the graph G below?

The idea of a subgraph can be extended to unlabelled graphs. For example, the following graphs are all subgraphs of the unlabelled graph H on the left; the configuration in graph (c) occurs at each corner of H.

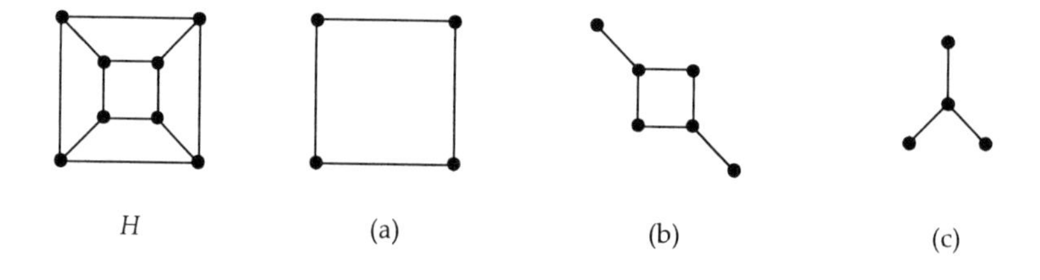

Problem 2.8

Which of the following graphs are subgraphs of the graph H below?

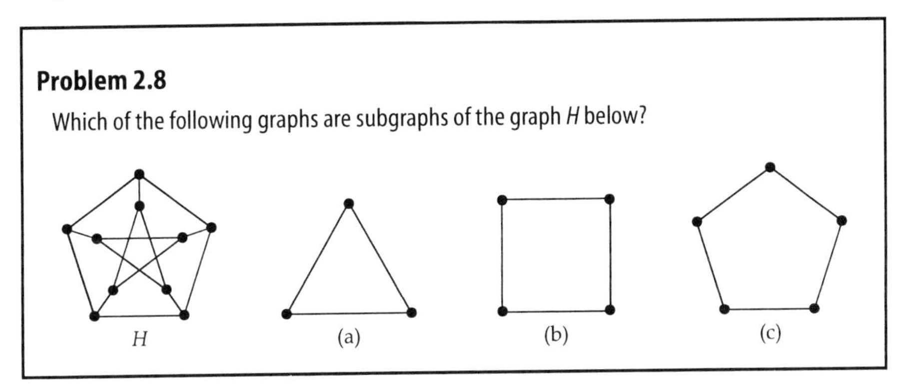

2.2 Vertex Degrees

In many applications of graph theory we need a term for the number of edges meeting at a vertex. For example, we may wish to specify the number of roads meeting at a particular intersection, the number of wires meeting at a given terminal of an electrical network, or the number of chemical bonds joining a given atom to its neighbours. These situations are illustrated below:

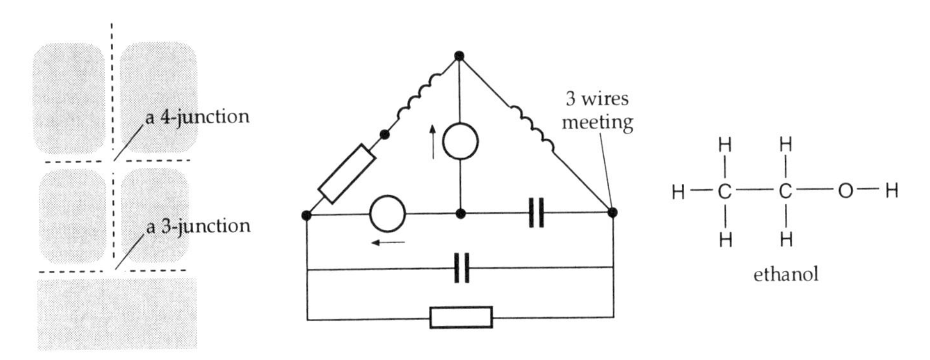

In chemistry, the term *valency* is used to indicate the number of bonds connecting an atom to its neighbours. For example, a carbon atom C has valency 4, an oxygen atom O has valency 2, and a hydrogen atom H has valency 1, as illustrated in the above diagram representing the molecule *ethanol*. For graphs, we usually use the word *degree*.

Definition

In a graph, the **degree** of a vertex v is the number of edges incident with v, with each loop counted twice, and is denoted by **deg** $\boldsymbol{v}$.

Remark Each loop contributes 2 to the degree of the corresponding vertex because it has two ends joined to that vertex.

For example, consider the following graphs:

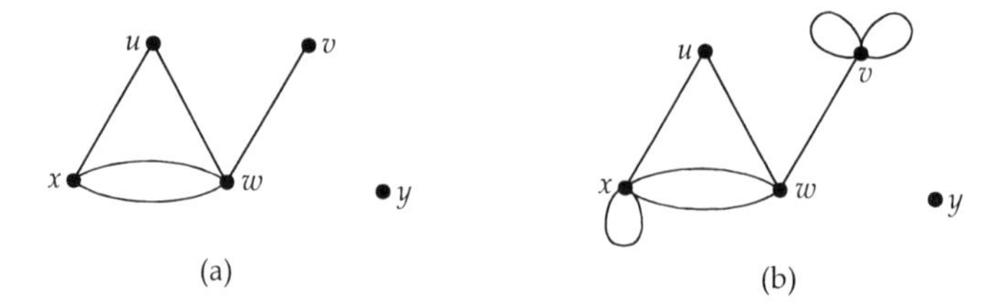

(a) (b)

Graph (a) has vertex degrees

$$\deg u = 2, \deg v = 1, \deg w = 4, \deg x = 3, \deg y = 0,$$

and graph (b) has vertex degrees

$$\deg u = 2, \deg v = 5, \deg w = 4, \deg x = 5, \deg y = 0.$$

We sometimes need to list the degrees of all the vertices in a graph, and this is usually done by writing them down in increasing order, with repeats where necessary. Accordingly, we make the following definition.

Definition

The **degree sequence** of a graph G is the sequence obtained by listing the vertex degrees of G in increasing order, with repeats as necessary.

For example,

graph (a) above has degree sequence (0, 1, 2, 3, 4);
graph (b) above has degree sequence (0, 2, 4, 5, 5).

Problem 2.9

Write down the degree sequence of each of the following graphs:

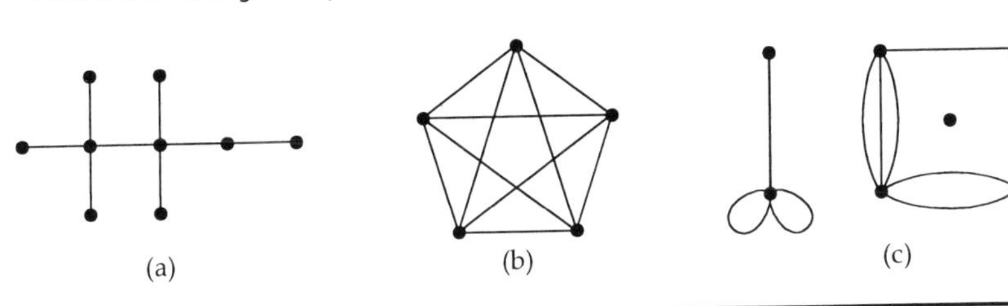

Problem 2.10

For each of the graphs in Problem 2.9, write down:

the number of edges;
the sum of the degrees of all the vertices.

What is the connection between your answers? Can you explain why this connection arises?

Handshaking Lemma

In the solution to Problem 2.10, you should have noticed that the sum of the vertex degrees of each graph is exactly twice the number of edges. A corresponding result holds for *any* graph, and is often called the *handshaking lemma*.

Theorem 2.1: Handshaking Lemma

In any graph, the sum of all the vertex degrees is equal to twice the number of edges.

Proof In any graph, each edge has two ends, so it contributes exactly 2 to the sum of the vertex degrees. The result follows immediately. ■

The name *handshaking lemma* arises from the fact that a graph can be used to represent a group of people shaking hands.

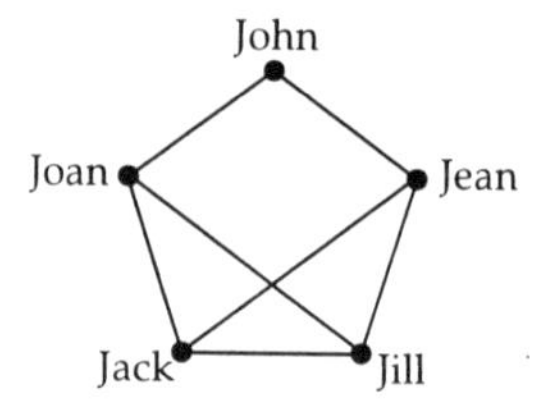

In such a graph, the vertices represent the people and an edge appears whenever the corresponding people have shaken hands. With this interpretation, the number of edges represents the total number of handshakes, the degree of a vertex is the number of hands shaken by the corresponding person, and the sum of the degrees is the total number of hands shaken. The handshaking lemma states that the total number of hands shaken is twice the number of handshakes – the reason being that exactly two hands are involved in each handshake.

Historical note

The handshaking lemma first appeared in 1736 in Leonhard Euler's paper on the Königsberg bridges problem. This important paper, although not in the language of graph theory, is widely regarded as 'the earliest paper in graph theory'.

Problem 2.11

(a) Use the handshaking lemma to prove that, in any graph, the number of vertices of odd degree is even.

(b) Verify that the result of part (a) holds for each of the graphs in Problem 2.9.

2.3 Paths and Cycles

Many applications of graphs involve 'getting from one vertex to another'. For example, you may wish to find the shortest route between one town and another. Other examples include the routeing of a telephone call between one subscriber and another, the flow of current between two terminals of an electrical network, and the tracing of a maze. We now make this idea precise by defining a *walk* in a graph.

> **Definition**
>
> A **walk of length *k*** in a graph is a succession of *k* edges of the form
>
> *uv*, *vw*, *wx*, ..., *yz*.
>
> This walk is denoted by *uvwx...yz*, and is referred to as a **walk between *u* and *z*.**

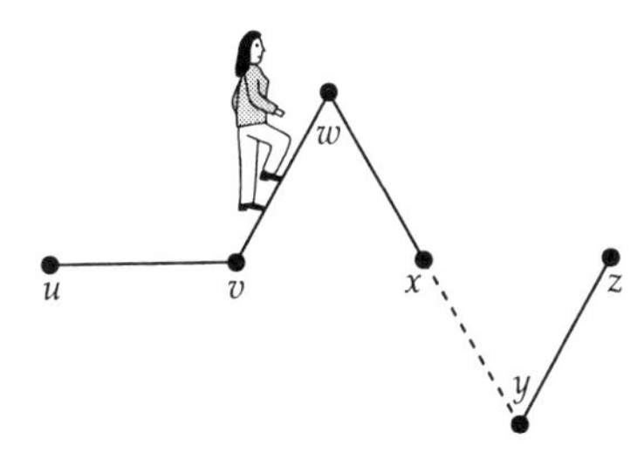

We can think of such a walk as going from u to v, then from v to w, then from w to x, and so on, until we arrive eventually at the vertex z. Since the edges are undirected, we can also think of it as a walk from z to y and on, eventually, to x, w, v and u. So we can equally well denote this walk by $zy...xwvu$, and refer to it as a walk between z and u.

Note that we do not require all the edges or vertices in a walk to be different. For example, in the following graph, $uvwxywvzzy$ is a walk of length 9 between the vertices u and y, which includes the edge vw twice and the vertices v, w, y and z twice.

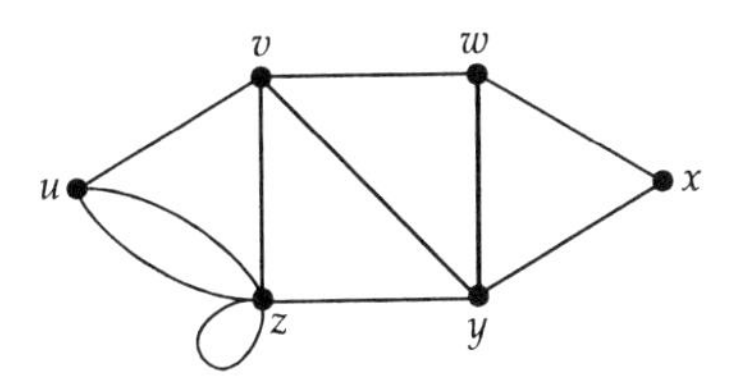

Paths, Trails and Connected Graphs

It is sometimes useful to be able to refer to a walk under more restrictive conditions in which we require all the edges, or all the vertices, to be different.

Definitions

A **trail** is a walk in which all the edges, but not necessarily all the vertices, are different.
A **path** is a walk in which all the edges and all the vertices are different.

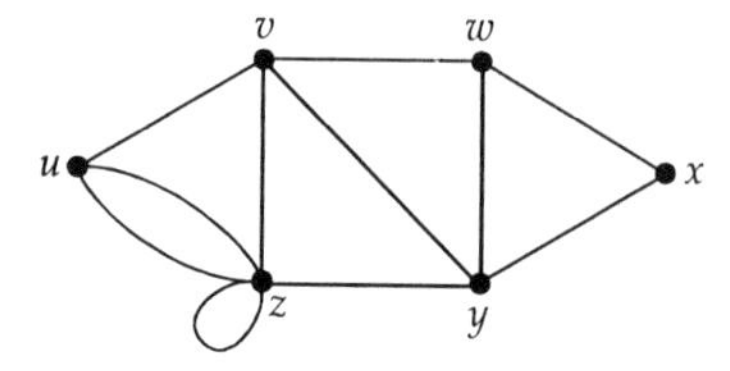

For example, in the graph above, the walk *vzzywxy* is a trail which is not a path, since the vertices *y* and *z* both occur twice, whereas the walk *vwxyz* has no repeated vertices, and is therefore a path.

Problem 2.12

Complete the following statements concerning the above graph:

(a) *xyzzvy* is a .... of length ...... between ..... and ..... ;

(b) *uvyz* is a .... of length ...... between ..... and ..... .

Problem 2.13

Write down all the paths between *s* and *y* in the following graph:

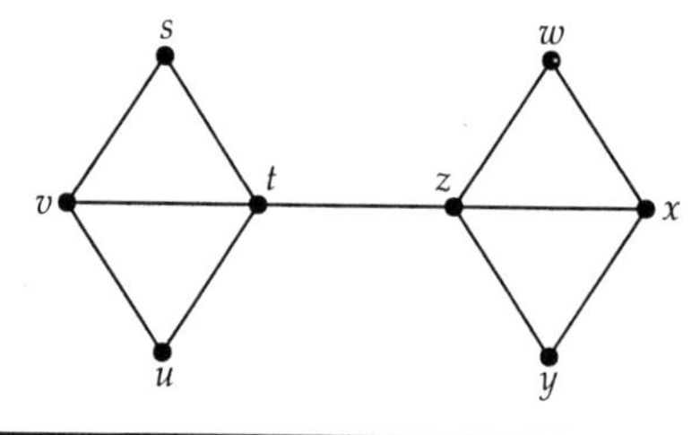

We can use the concept of a path to define a *connected graph*. Intuitively, a graph is *connected* if it is 'in one piece'; for example, the following graph is not connected, but can be split into four connected subgraphs.

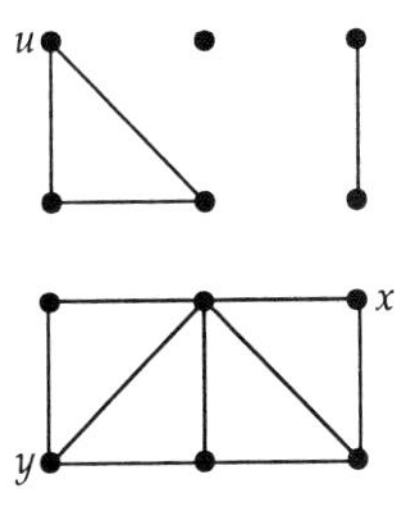

The observation that there is a path between x and y (which lie in the same subgraph), but not between u and y (which lie in different subgraphs), leads to the following definitions.

Definitions

A graph is **connected** if there is a path between each pair of vertices, and is **disconnected** otherwise.
An edge in a connected graph is a **bridge** if its removal leaves a disconnected graph.
Every disconnected graph can be split up into a number of connected subgraphs, called **components**.

For example, in the graph in Problem 2.13, the edge tz is a bridge; and the following disconnected graph has three components:

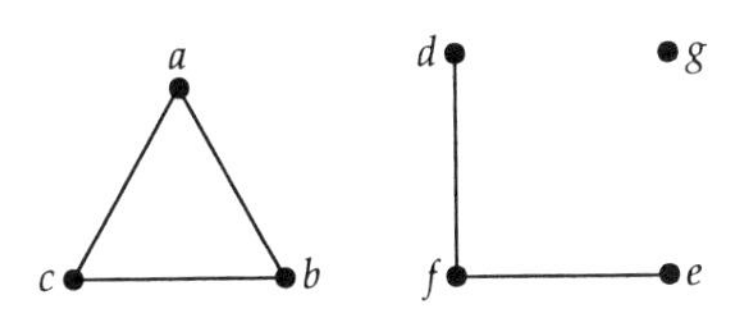

Problem 2.14

Draw:

(a) a connected graph with eight vertices;
(b) a disconnected graph with eight vertices and two components;
(c) a disconnected graph with eight vertices and three components.

Closed Trails and Cycles

It is also useful to have a special term for those walks or trails that start and finish at the same vertex. We say that they are *closed*.

Definitions

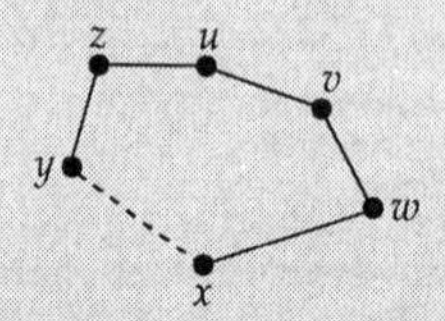

A **closed walk** in a graph is a succession of edges of the form

uv, vw, wx, ..., yz, zu,

that starts and ends at the same vertex.

A **closed trail** is a closed walk in which all the edges are different.

A **cycle** is a closed walk in which all the edges are different and all the intermediate vertices are different.

A walk or trail is **open** if it starts and finishes at different vertices.

For example, consider the following graph:

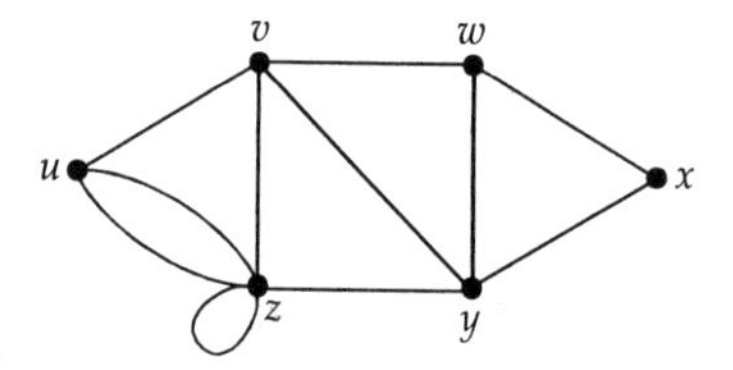

The closed walk *vywxyzv* is a closed trail which is not a cycle, whereas the closed trails *zz*, *vwxyv* and *vwxyzv* are all cycles. A cycle of length 3, such as *vwyv* or *wxyw*, is called a **triangle**. In describing closed walks, we can allow any vertex to be the starting vertex. For example, the triangle *vwyv* can equally well be written as *wyvw* or *yvwy* or (since the direction is immaterial) by *vywv*, *wvyw* or *ywvy*.

Problem 2.15

For the graph on the right, write down:

(a) a closed walk that is not a closed trail;
(b) a closed trail that is not a cycle;
(c) all the cycles of lengths 1, 2, 3 and 4.

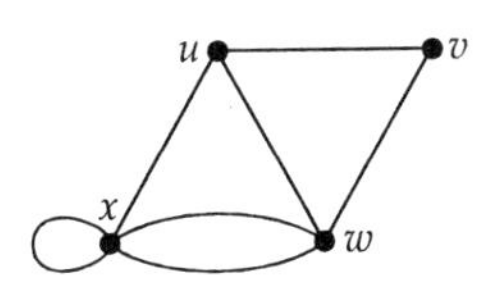

2.4 Regular and Bipartite Graphs

In this section we give some examples of two important classes of graphs: regular and bipartite graphs.

Regular Graphs

A graph in which all the vertex degrees are the same is given a special name.

Definitions

A graph is **regular** if its vertices all have the same degree.
A regular graph is ***r*-regular**, or **regular of degree *r***, if the degree of each vertex is *r*.

In the following diagrams we illustrate some r-regular graphs, for various values of r:

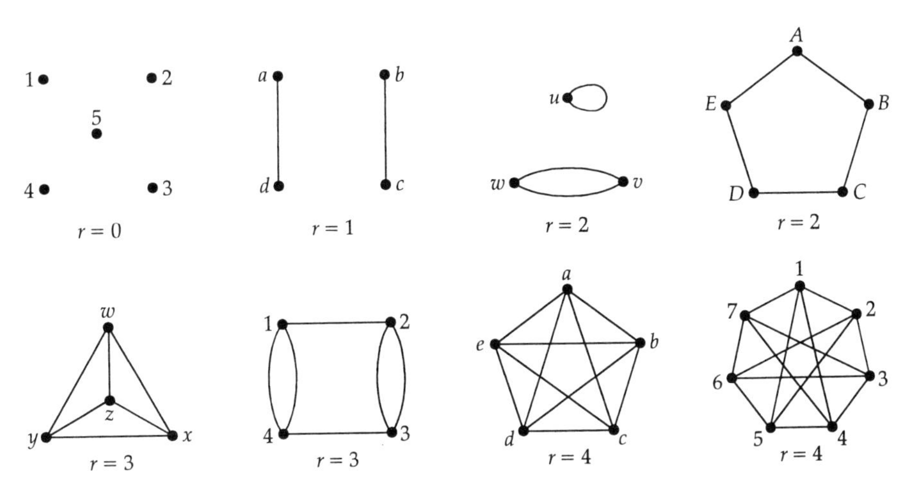

Problem 2.16

Draw an r-regular graph with eight vertices when:

(a) $r = 3$; (b) $r = 4$; (c) $r = 5$.

A useful consequence of the handshaking lemma is the following result.

Theorem 2.2

Let G be an r-regular graph with n vertices. Then G has $nr/2$ edges.

Proof Let G be a graph with n vertices, each of degree r; then the sum of the degrees of all the vertices is nr. By the handshaking lemma, the number of edges is one-half of this sum, which is $nr/2$. ■

Problem 2.17

Verify that Theorem 2.2 holds for each of the following regular graphs:

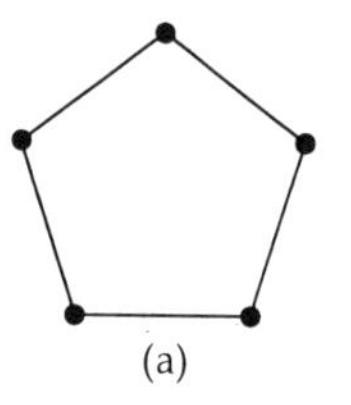
(a)

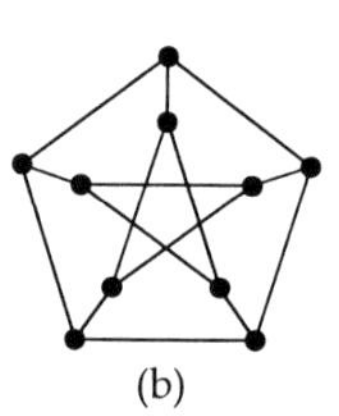
(b)

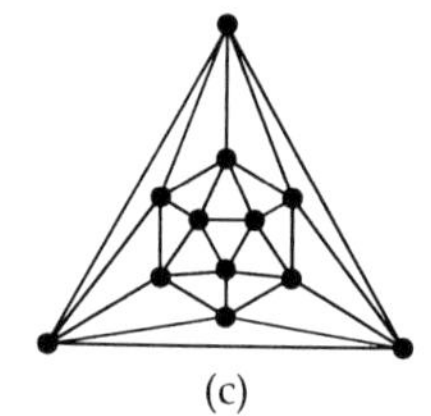
(c)

Problem 2.18

(a) Prove that there are no 3-regular graphs with seven vertices.
(b) Prove that, if n and r are both odd, then there are no r-regular graphs with n vertices.

Examples of Regular Graphs

We now consider some important classes of regular graphs.

Complete Graphs

A **complete graph** is a graph in which each vertex is joined to each of the others by exactly one edge.
The complete graph with n vertices is denoted by K_n.

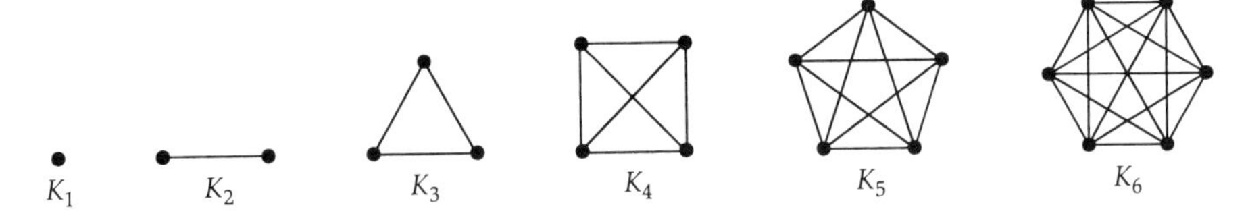

The graph K_n is regular of degree $n - 1$, and therefore has $n(n - 1)/2$ edges, by Theorem 2.2.

Null Graphs

A **null graph** is a graph with no edges.
The null graph with n vertices is denoted by N_n.

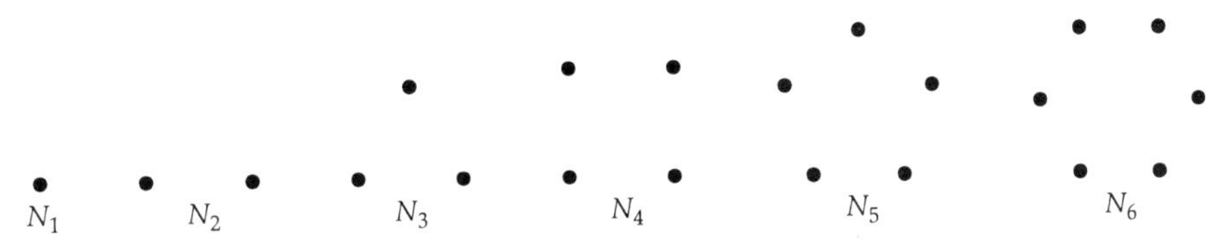

The graph N_n is regular of degree 0.

Cycle Graphs

A **cycle graph** is a graph consisting of a single cycle of vertices and edges.
The cycle graph with n vertices is denoted by C_n.

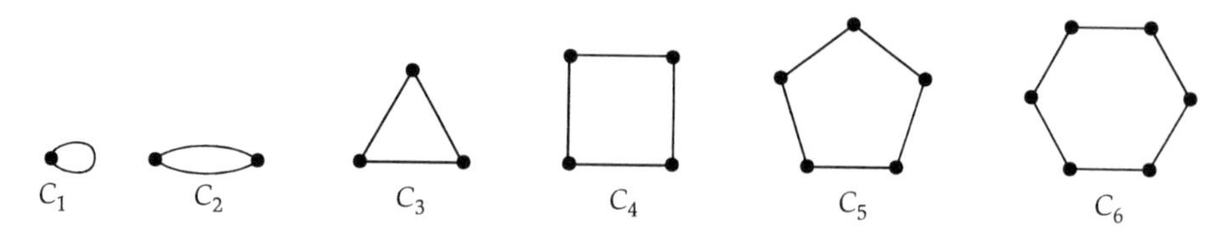

The graph C_n is regular of degree 2, and has n edges. For $n \geq 3$, C_n can be drawn in the form of a regular polygon.

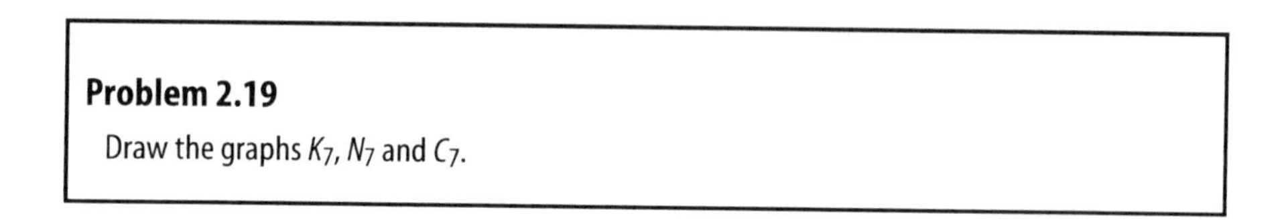

Problem 2.19

Draw the graphs K_7, N_7 and C_7.

The Platonic Graphs

The following five regular solids are known as the *Platonic solids*.

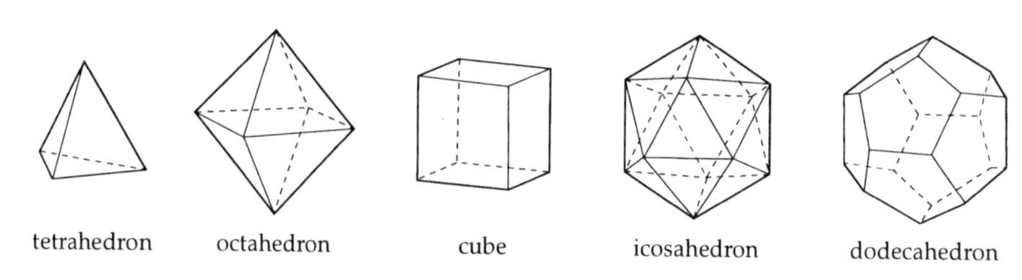

We can regard the vertices and edges of each solid as the vertices and edges of a regular graph. The resulting five graphs are known as the **Platonic graphs**, and are often drawn as follows:

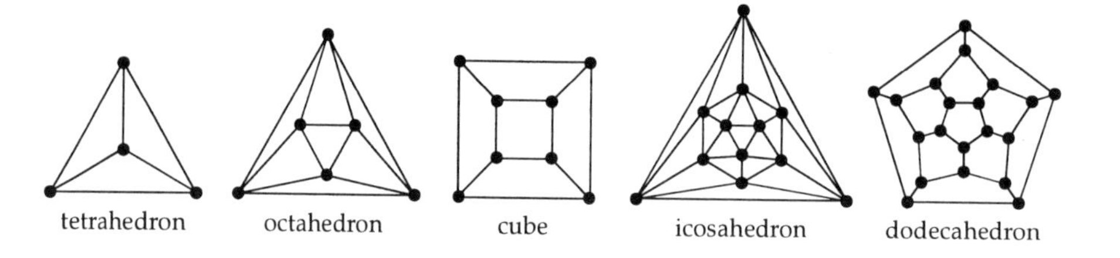

The tetrahedron, cube and dodecahedron are 3-regular, the octahedron graph is 4-regular and the icosahedron graph is 5-regular.

Historical note

The name *Platonic* arises from the fact that these solids were mentioned in Plato's *Timaeus*. These solids are also discussed in Book XIII of Euclid's *Elements*.

The Petersen Graph

The **Petersen graph** is named after the Danish mathematician Julius Petersen; he discussed this graph in a paper of 1898. The Petersen graph is a 3-regular graph with 10 vertices and 15 edges; it may be drawn in various ways, two of which are

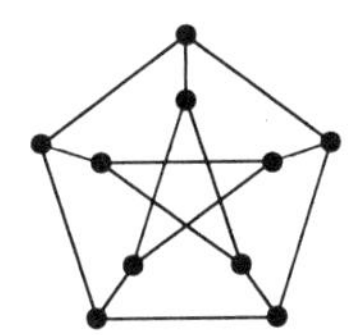

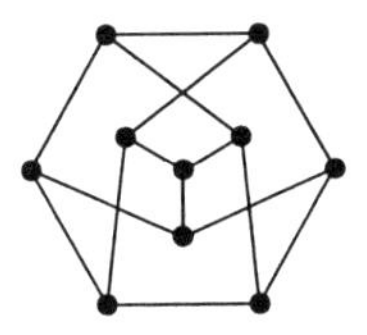

This graph has many interesting properties that we shall meet in later chapters.

Bipartite Graphs

Of particular importance in applications are the *bipartite* graphs.

Definition

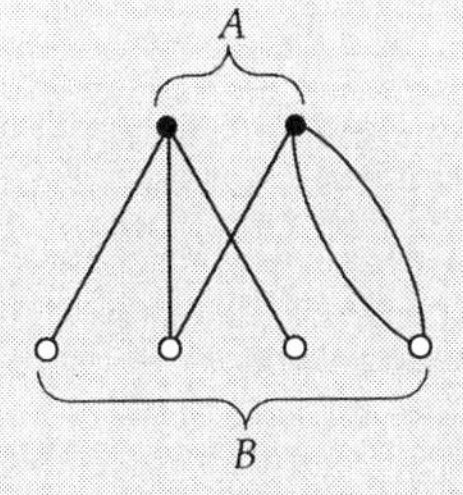

A **bipartite graph** is a graph whose set of vertices can be split into two subsets A and B in such a way that each edge of the graph joins a vertex in A and a vertex in B.

We can distinguish the vertices in A from those in B by drawing one set in black and one set in white; then each edge is incident with a black vertex and a white vertex. Two examples of bipartite graphs are:

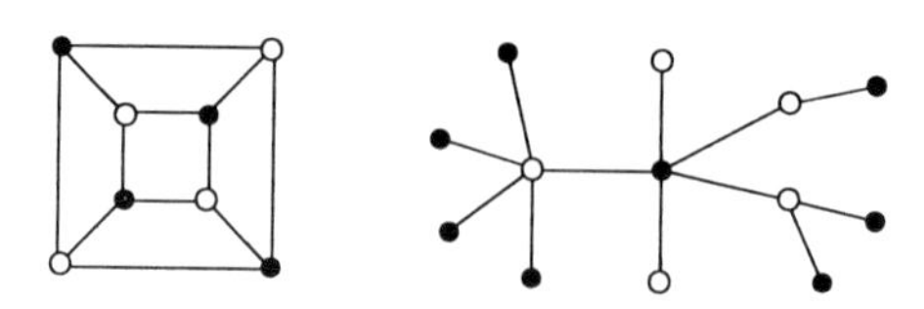

Problem 2.20

Prove that, in a bipartite graph, every cycle has an even number of edges.

Examples of Bipartite Graphs

As with regular graphs, there are several important classes of bipartite graphs.

Complete Bipartite Graphs

A **complete bipartite graph** is a bipartite graph in which each vertex in A is joined to each vertex in B by just one edge.

The complete bipartite graph with r vertices in A and s vertices in B is denoted by $K_{r,s}$. Some examples of complete bipartite graphs are:

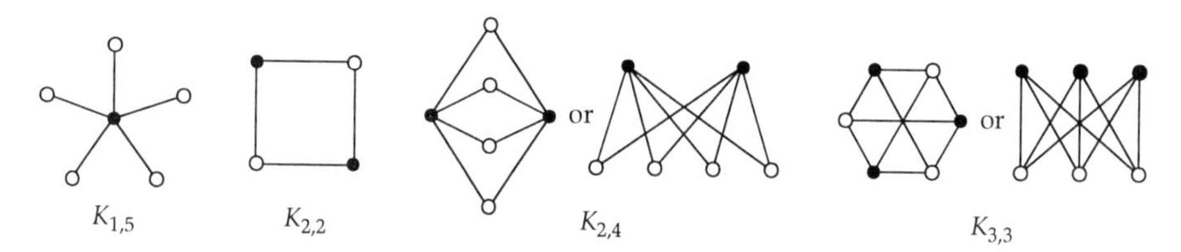

The graph $K_{r,s}$ is the same as $K_{s,r}$ (exchange the roles of A and B); it has $r+s$ vertices (r vertices of degree s and s vertices of degree r) and rs edges.

Problem 2.21

(a) Draw the graphs $K_{2,3}$, $K_{1,7}$ and $K_{4,4}$.
How many vertices and edges does each have?

(b) Under what condition on r and s is $K_{r,s}$ a regular graph?

Trees

One of the most important classes of bipartite graphs is the class of *trees*. A **tree** is a connected graph with no cycles. Some examples of trees are:

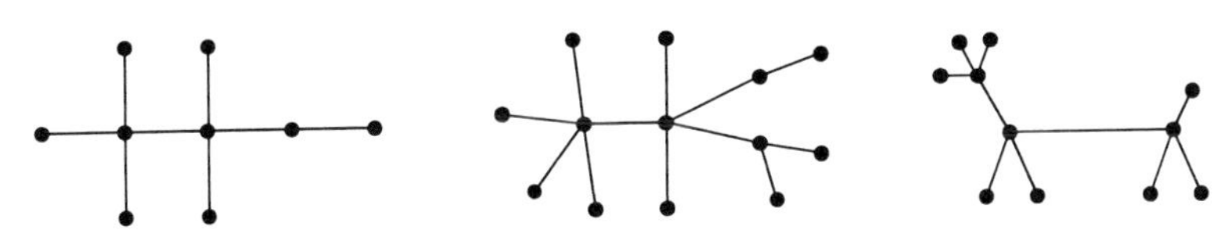

Since a tree is connected, there is at least one path between each pair of vertices. Suppose that there are two vertices of the tree joined by *two* paths. Then these paths create a cycle that may include all the edges in both paths, or only some of them:

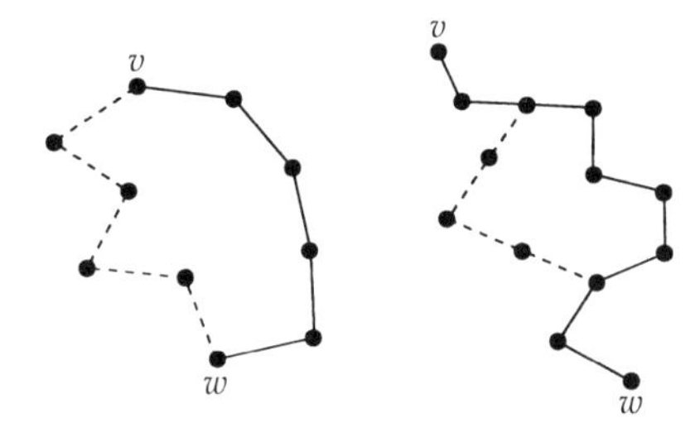

But this contradicts our definition of a tree, so no such pair of vertices exists. It follows that

in a tree, there is just *one* path between each pair of vertices.

Problem 2.22

There are eight unlabelled trees with five or fewer vertices. Draw them.

Problem 2.23

(a) Explain why every tree is a bipartite graph, by colouring alternate vertices black and white.

(b) Explain why a tree with n vertices has $n - 1$ edges.

Path Graphs

A **path graph** is a tree consisting of a single path through all its vertices. The path graph with n vertices is denoted by P_n.

The graph P_n has $n - 1$ edges, and is obtained from the cycle graph C_n by removing any edge.

Cubes

Of particular interest among the bipartite graphs are the *cubes*. They have important applications in coding theory, and may be constructed by taking as vertices all binary words (sequences of 0s and 1s) of a given length and joining two of these vertices whenever the corresponding sequences differ in just one place. The graph thus obtained from the binary words of length k is called the ***k*-cube** or ***k*-dimensional cube**, and is denoted by Q_k.

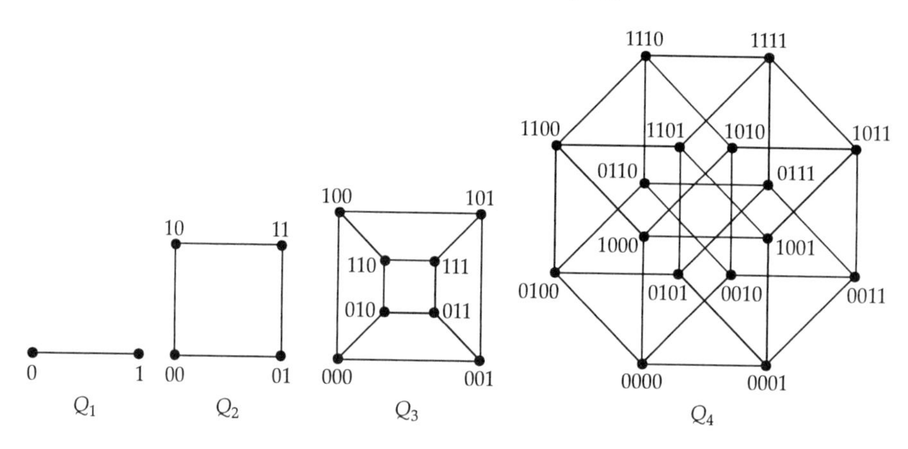

The graph Q_k has 2^k vertices, and is regular of degree k. It follows from Theorem 2.2 that Q_k has $k \times 2^{k-1}$ edges.

2.5 Case Studies

We conclude this chapter with two case studies – the four cubes problem and social networks.

Four Cubes Problem

An intriguing recreational puzzle, which has been marketed under the name of *Instant Insanity*, concerns four cubes whose faces are coloured red, blue, green and yellow. These cubes are depicted in flattened-out form below.

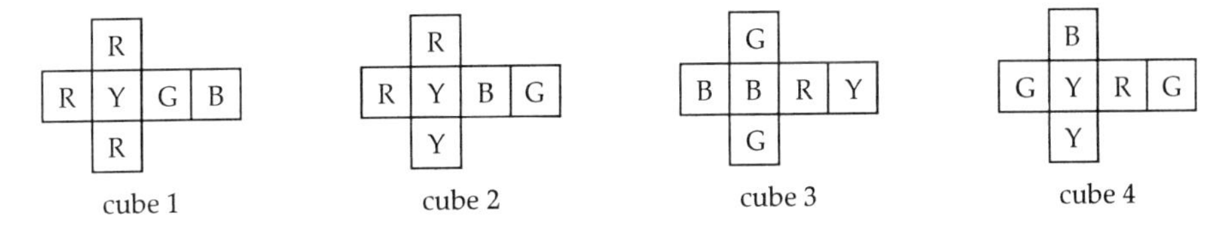

The problem is

> to pile the cubes on top of each other so that all four colours appear on each side of the resulting 'stack'.

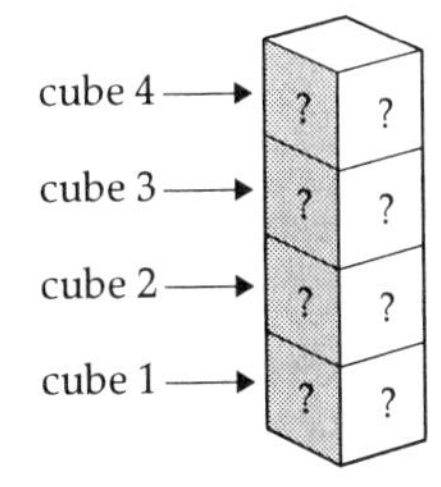

As we shall see, there is essentially only one way in which this can be done for this particular set of cubes.

A trial-and-error approach to this problem is inadvisable, since there are many thousands of different ways of stacking the cubes.

Now, if one face of a cube appears on one side of the stack, then the opposite face of the cube must appear on the opposite side of the stack. It follows that our primary concern is with opposite pairs of faces, and that we must decide, for each cube, which two of the three opposite pairs should appear on the sides of the stack.

To solve this problem, we represent each cube by a graph that tells us which pairs of colours appear on opposite faces. More precisely, we represent each cube by a graph with four vertices R, B, G, Y (corresponding to the four colours) in which two vertices are adjacent when the cube in question has the corresponding colours on opposite faces. For example, in cube 1, blue and yellow appear on opposite faces, and so the vertices B and Y are joined in the corresponding graph.

The graphs for the above set of cubes are:

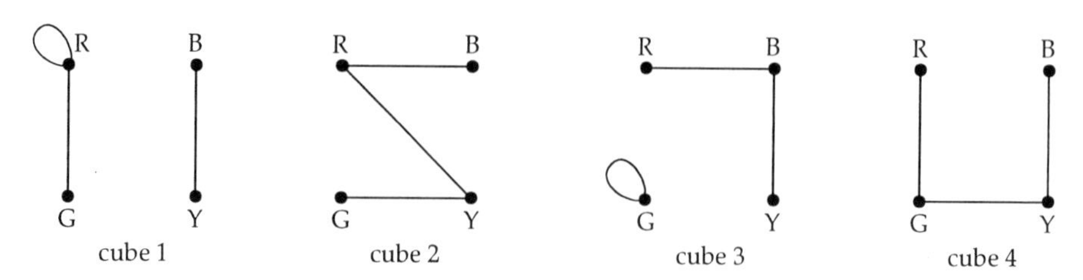

We now superimpose them to give a new graph G:

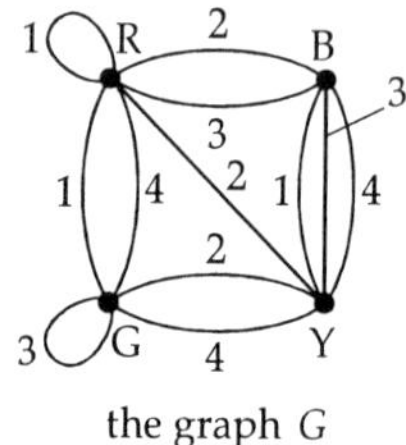

the graph G

A solution to the four cubes problem is obtained by finding two particular subgraphs H_1 and H_2 of G. The subgraph H_1 tells us which pair of colours appears on the front and back faces of each cube, and the subgraph H_2 tells us which pair of colours appears on the left-hand and right-hand faces of each cube. The subgraphs H_1 and H_2 must possess three properties:

(a) each subgraph contains exactly one edge from the graph of each cube;
(b) the subgraphs have no edges in common;
(c) each vertex is incident with two edges.

Property (a) tells us that each cube has a front and a back, and a left side and a right side, and the subgraphs H_1 and H_2 tell us which pairs of colours appear on these faces.
Property (b) tells us that the faces appearing on the front and back of a cube cannot be the same as those appearing on the sides.
Property (c) tells us that each colour appears exactly twice on the sides of the stack (once on each side), and exactly twice on the front and back (once on the front and once on the back).

A solution for the above set of cubes is shown below. In this solution, the subgraphs H_1 and H_2 tell us that cube 1 has yellow on the front and blue on the back (from H_1) and red on the left and green on the right (from H_2), and similarly for the other cubes.

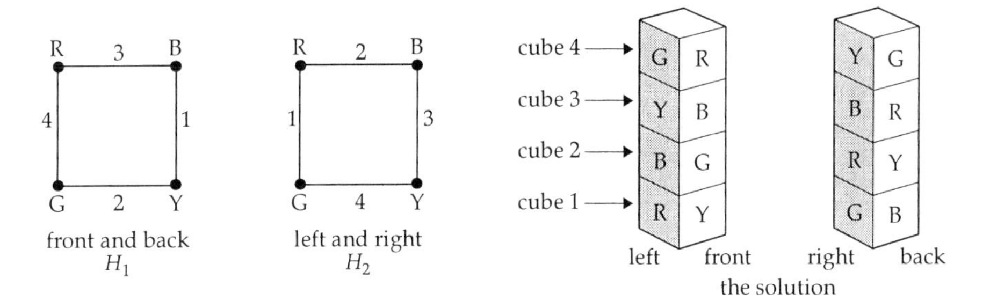

This solution is the only one possible in this instance (see Exercise 2.16). Other instances exist that have no solution or several solutions.

Problem 2.24

Use the above approach to find a solution to the four-cubes problem for the following set of cubes:

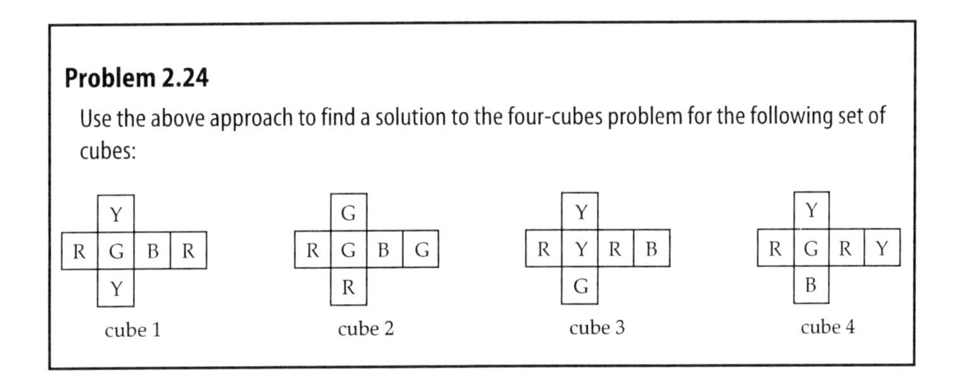

Social Networks

Graphs have been used extensively in the social sciences to represent *interpersonal relationships*. The vertices correspond to individuals in a group or society, and the edges join pairs of individuals that are related in some way – for example, x is joined to y if x likes, hates, agrees with, avoids, or communicates with y. Such representations have been extended to relationships between groups of individuals, and have proved useful in a number of contexts ranging from kinship relationships in certain primitive tribes to relationships between political parties. Graphs have also been used by political scientists to study international relations, where the vertices correspond to nations or groups of nations, and the edges join pairs of nations that are allied, maintain diplomatic relations, agree on a particular strategy, etc.

We can analyse the possible tension in such situations by using the concept of a **signed graph.** This is a graph with either + or – associated with each edge, indicating a positive relationship (likes, loves, agrees with, communicates with, etc.) or a negative one (dislikes, hates, disagrees with, avoids, etc.). For example, in the signed graph below, Jack likes Jill but not John, Jill likes Jack and Mary but not John, Mary likes John and Jill, and John likes Mary but not Jack or Jill; Jack and Mary have no strong feelings about each other, and are therefore not joined by an edge.

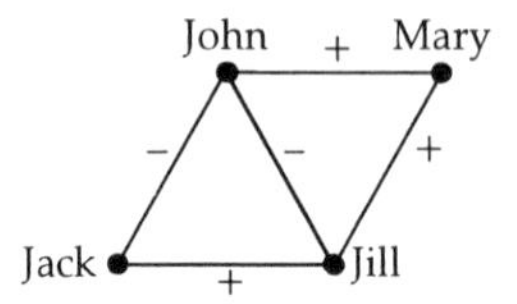

Consider now the following diagrams, which illustrate some of the situations that can occur when three people work together. Which of these situations is most likely to cause tension between John, Jack and Jill?

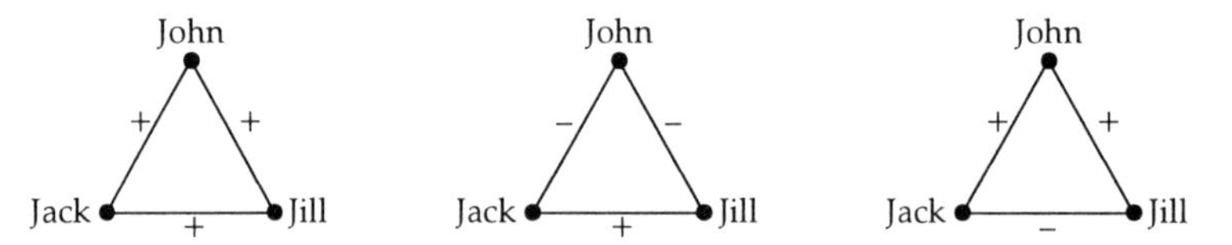

In the first case, all three get on well, and there is no tension. In the second case, Jack and Jill get on well and both dislike John; the result is that John works on his own, and again there is no tension. In the third case, John likes both Jack and Jill and would like to work with them, but Jack and Jill dislike each other and do not wish to work together; in this case, a suitable working arrangement cannot be found, and there is tension. We express this by saying that the first two situations are *balanced*, whereas the third is *unbalanced*.

More generally, we say that a signed graph is **balanced** if we can colour its vertices black or white in such a way that positive edges have ends of the same colour, and negative edges have a black end and a white end. Clearly, the first two of the above diagrams can be coloured in this way, whereas the third cannot:

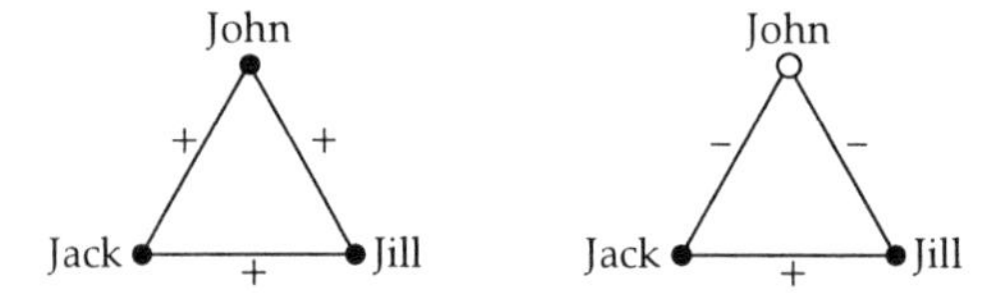

This definition resembles that of a bipartite graph. To see the connection, take a balanced signed graph and remove all the positive edges; this leaves a bipartite graph, as indicated by the following diagram:

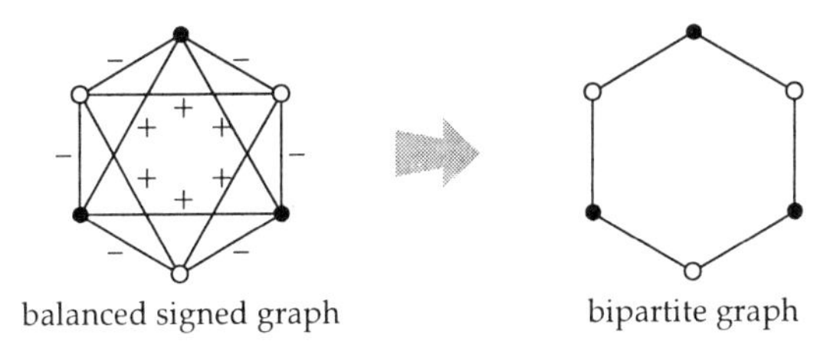

balanced signed graph bipartite graph

We can exploit this connection between balanced signed graphs and bipartite graphs a little further. Recall from Problem 2.20 that

> in a bipartite graph, every cycle has an even number of edges.

For *balanced* signed graphs the corresponding result is

> in a balanced signed graph, every cycle has an even number of *negative* edges.

Problem 2.25

Decide which of the following signed graphs are balanced, and find the corresponding bipartite graph in each case:

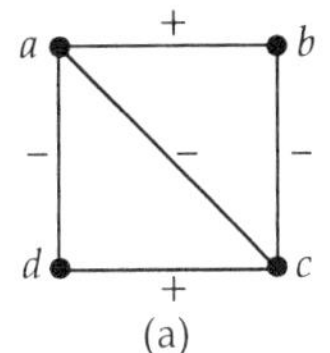

(a)

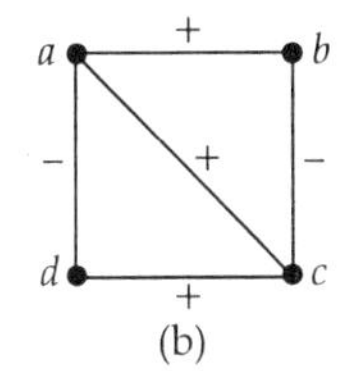

(b)

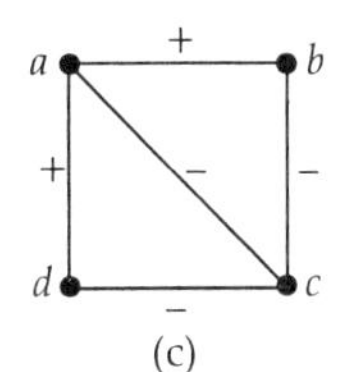

(c)

Before studying Chapter 3, you may find it helpful to read the Appendix: Methods of Proof.

Exercises 2

Graphs and Subgraphs

2.1 Consider the graph G shown on the right. Which of the following statements hold for G?

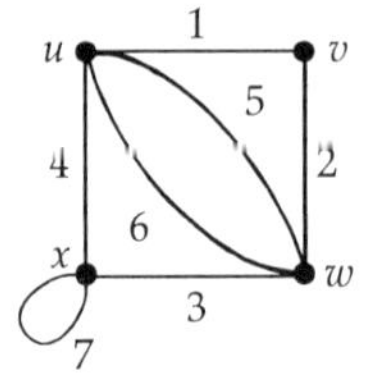

(a) vertices v and x are adjacent;

(b) edge 6 is incident with vertex w;

(c) vertex x is incident with edge 4;

(d) vertex w and edges 5 and 6 form a subgraph of G.

2.2 By suitably labelling the vertices, show that the following graphs are isomorphic:

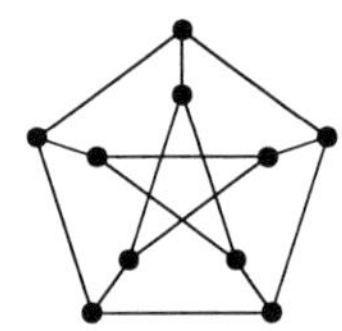

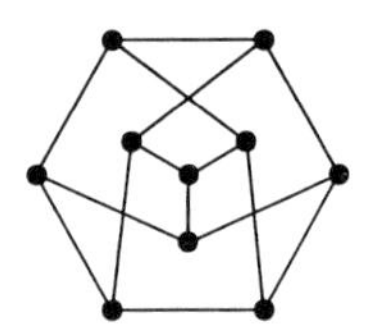

2.3 Draw the eleven unlabelled simple graphs with four vertices.

Vertex Degrees

2.4 (a) If two graphs have the same degree sequence, must they be isomorphic?

(b) If two graphs are isomorphic, must they have the same degree sequence?

2.5 Let G be a graph with degree sequence (1, 2, 3, 4). Write down the number of vertices and number of edges of G, and construct such a graph. Are there any *simple* graphs with degree sequence (1, 2, 3, 4)?

2.6 Prove that, if G is a simple graph with at least two vertices, then G has two or more vertices of the same degree.

Paths and Cycles

2.7 For the graph shown on the right, write down:

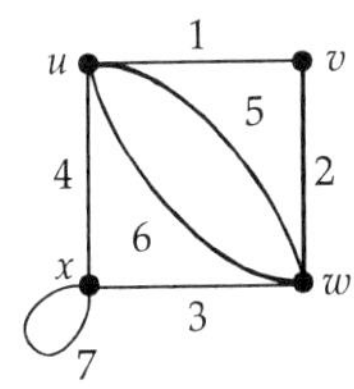

(a) a walk of length 7 between u and w;
(b) all the cycles of lengths 1, 2, 3 and 4;
(c) a path of maximum length.

2.8 Draw four connected graphs, G_1, G_2, G_3 and G_4, each with 5 vertices and 8 edges, satisfying the following conditions:

G_1 is a simple graph;
G_2 is a non-simple graph with no loops;
G_3 is a non-simple graph with no multiple edges;
G_4 is a graph with both loops and multiple edges.

2.9 (a) Draw a simple connected graph with degree sequence (1, 1, 2, 3, 3, 4, 4, 6).
(b) Draw a simple connected graph with degree sequence (3, 3, 3, 3, 3, 5, 5, 5).

Regular and Bipartite Graphs

2.10 Draw:
(a) two non-isomorphic regular graphs with 8 vertices and 12 edges;
(b) two non-isomorphic regular graphs with 10 vertices and 20 edges.

2.11 Determine the number of edges of each of the following graphs:
(a) C_{10}; (b) $K_{9,10}$; (c) K_{10}; (d) Q_5;
(e) the dodecahedron.

2.12 The **complement** of a simple graph G is obtained by taking the vertices of G and joining two of them whenever they are *not* joined in G.

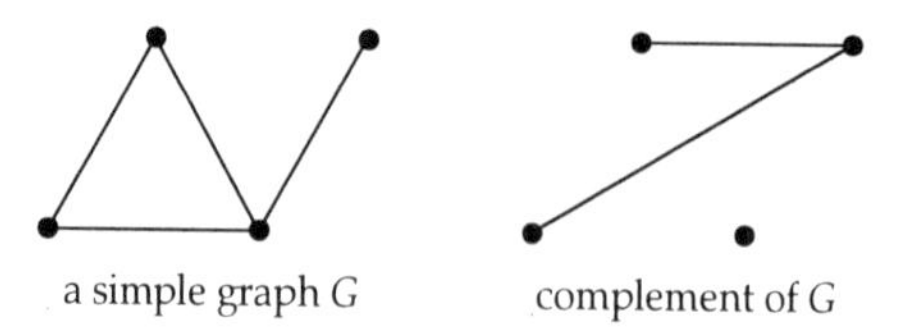

a simple graph G complement of G

(a) Verify that the complement of P_4 is P_4.
(b) What is the complement of K_4? of $K_{3,3}$? of C_5?
(c) Show that, if a simple graph G is isomorphic to its complement, then the number of vertices of G has the form $4k$ or $4k + 1$, for some positive integer k.
(d) Find all the simple graphs with four or five vertices that are isomorphic to their complements.
(e) Construct a graph with eight vertices that is isomorphic to its complement.

2.13 The **girth** of a graph G is the length of a *shortest* cycle in G, and the **circumference** of G is the length of a *longest* cycle in G. Find the girth and circumference of:

(a) the Petersen graph; (b) the 4-cube graph Q_4.

2.14 Prove that, if every cycle of a graph has an even number of edges, then the graph is bipartite. (This is the converse of Problem 2.20.)

Hint Consider a connected graph G. Choose a vertex v in G and consider those vertices whose minimum distance from v is even and those whose minimum distance from v is odd. To which vertices are the 'odd' vertices adjacent? To which vertices are the 'even' vertices adjacent?

2.15 The **line graph** $L(G)$ of a simple graph G is the graph obtained by taking the *edges* of G as *vertices*, and joining two of these vertices whenever the corresponding edges of G have a vertex in common. Find an expression for the number of edges of $L(G)$ in terms of the degrees of the vertices of G, and show that:

(a) $L(C_n)$ is isomorphic to C_n;
(b) $L(K_n)$ has $\frac{1}{2}n(n-1)$ vertices and is regular of degree $2n - 4$;
(c) L(tetrahedron graph) = octahedron graph;
(d) the complement of $L(K_5)$ is the Petersen graph.

Case Studies

Four Cubes Problem

2.16 Show that the subgraphs H_1 and H_2 of the graph G shown below are the only pair of subgraphs possessing the following properties for the given set of cubes:

(a) each contains exactly one edge from the graph for each cube;
(b) they have no edges in common;
(c) each vertex is incident with two edges.

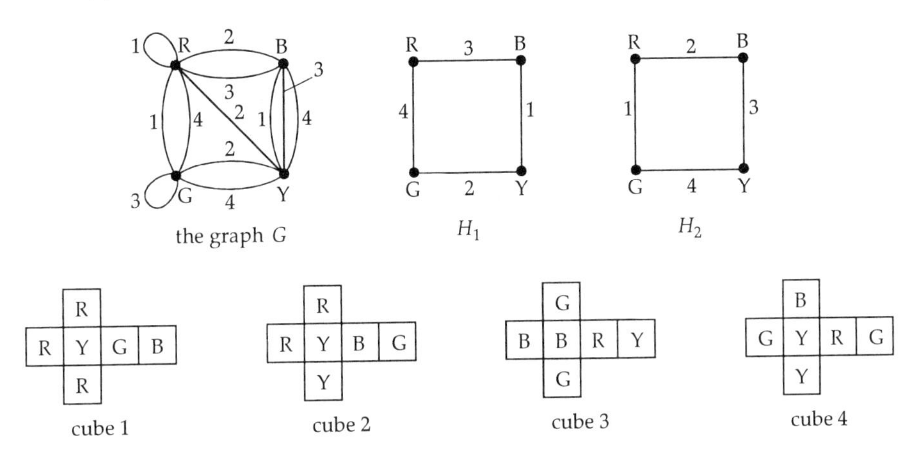

Hint First show that neither subgraph can contain the loop at R, then repeat the process for the loop at G, and then for the edge joining R and Y.

2.17 Show that there is *no* solution to the four cubes problem for the following set of cubes:

	G		
Y	G	R	B
	R		

cube 1

	B		
G	R	R	Y
	G		

cube 2

	Y		
R	Y	G	B
	Y		

cube 3

	B		
Y	B	G	R
	B		

cube 4

Social Networks

2.18 Draw the signed graph representing the following relationships, and determine whether it is balanced.

John likes Joan, Jean and Jane, but dislikes both Joe and Jill;
Jill and Joe like each other, but both dislike John, Joan, Jean and Jane;
Joan, Jean and Jane like each other and John, but each dislikes Joe and Jill.

2.19 Draw the signed graph representing the following relationships, and determine whether it is balanced. (Assume that all relationships are symmetric – that is, x likes y if and only if y likes x.)

Michael likes Ian and Leslie, but dislikes Jean and Kate;
Kate dislikes Ian and Leslie;
Jean likes Ian but dislikes Leslie.

2.20 Prove that in any balanced signed graph every cycle has an even number of edges.

Chapter 3
Eulerian and Hamiltonian Graphs

After studying this chapter, you should be able to:

- explain the terms *Eulerian graph* and *Eulerian trail*, and state a necessary and sufficient condition for a connected graph to be Eulerian;
- explain the terms *semi-Eulerian graph* and *semi-Eulerian trail*, and state a necessary and sufficient condition for a connected graph to be semi-Eulerian;
- explain the terms *Hamiltonian graph*, *Hamiltonian cycle*, *semi-Hamiltonian graph* and *semi-Hamiltonian path*, and state a sufficient condition for a simple connected graph to be Hamiltonian;
- explain the relevance of the above ideas to the Königsberg bridges problem, the icosian game, dominoes, diagram-tracing puzzles, the knight's tour problem and Gray codes.

In this chapter we introduce two important types of graph – *Eulerian* and *Hamiltonian* graphs, named after the mathematicians Leonhard Euler and William Rowan Hamilton. In particular, we give a necessary and sufficient condition for a connected graph to be Eulerian, and show the connection between Eulerian graphs and diagram-tracing puzzles. We also give sufficient conditions for a connected graph to be Hamiltonian, and show the connection between Hamiltonian graphs and the knight's tour problem and Gray codes. Because of the importance of Eulerian and Hamiltonian graphs in the development of graph theory, some of this chapter is presented from a historical point of view.

3.1 Exploring and Travelling

In this section, we consider two types of problem concerned with the routes joining a number of cities in a road map.

Explorer's Problem

An explorer wishes to find a tour that traverses each *road* exactly once and returns to the starting point.

Traveller's Problem

A traveller wishes to find a tour that visits each *city* exactly once and returns to the starting point.

To appreciate the difference between these two problems, consider the following road map.

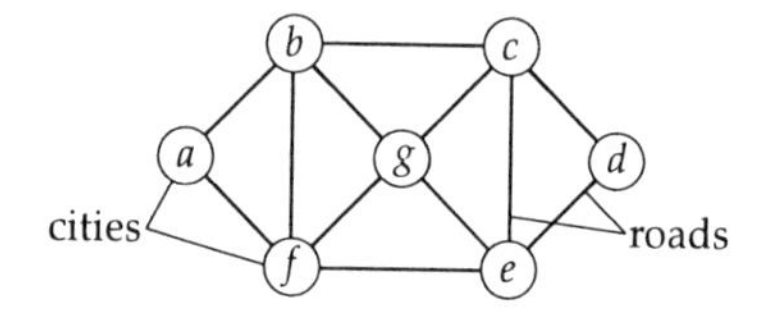

The explorer wishes to find a tour that starts at city a, goes along each road exactly once (in either direction) and ends back at a; two examples of such a tour are

$$a\,b\,c\,d\,e\,f\,b\,g\,c\,e\,g\,f\,a \quad \text{and} \quad a\,f\,g\,c\,d\,e\,g\,b\,c\,e\,f\,b\,a.$$

The traveller wishes to find a tour that starts at city a, goes to each city exactly once and ends back at a; two examples of such a tour are

$$a\,b\,c\,d\,e\,g\,f\,a \quad \text{and} \quad a\,f\,e\,d\,c\,g\,b\,a.$$

Note that the explorer travels along each road just once, but may visit a particular city several times, whereas the traveller visits each city just once, but may omit several roads.

Let us regard the road map as a connected graph whose vertices correspond to the cities and whose edges correspond to the roads. The explorer's problem is now to

find a closed trail that includes every edge of the graph,

whereas the traveller's problem is now to

find a cycle that includes every vertex of the graph.

With this in mind, we make the following definitions.

Definitions

A connected graph is **Eulerian** if it contains a closed trail that includes every edge; such a trail is an **Eulerian trail**.

A connected graph is **Hamiltonian** if it contains a cycle that includes every vertex; such a cycle is a **Hamiltonian cycle**.

Remark It is easy to remember which definition is which, since Eulerian graphs are defined in terms of Edges.

For example, consider the following four graphs:

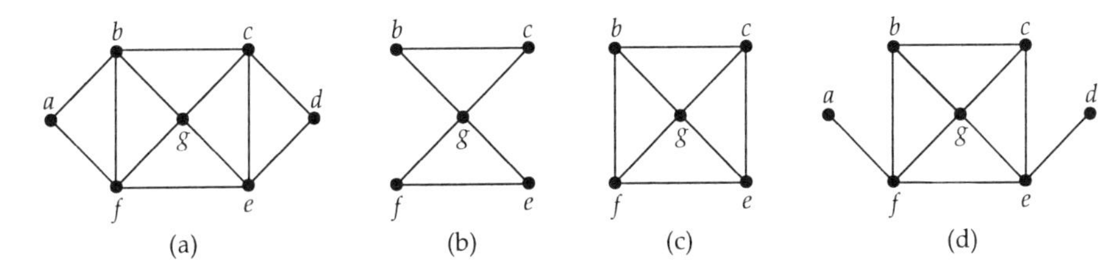

graph (a) is both Eulerian and Hamiltonian, as we saw above;
graph (b) is Eulerian – an Eulerian trail is $b\,c\,g\,f\,e\,g\,b$; it is not Hamiltonian;
graph (c) is Hamiltonian – a Hamiltonian cycle is $b\,c\,g\,e\,f\,b$; it is not Eulerian;
graph (d) is neither Eulerian nor Hamiltonian.

Problem 3.1

Decide which of the following graphs are Eulerian and/or Hamiltonian, and write down an Eulerian trail or Hamiltonian cycle where possible.

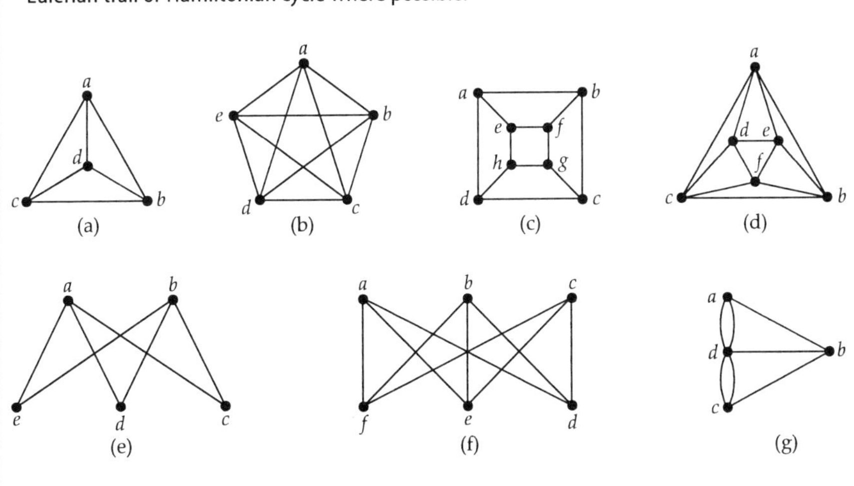

3.2 Eulerian Graphs

In Chapter 1, you met the *Königsberg bridges problem*: is it possible to find a route that crosses each of the seven bridges of Königsberg exactly once and returns to the starting point? It was not until Leonhard Euler tackled the problem in the 1730s that this was proved to be impossible.

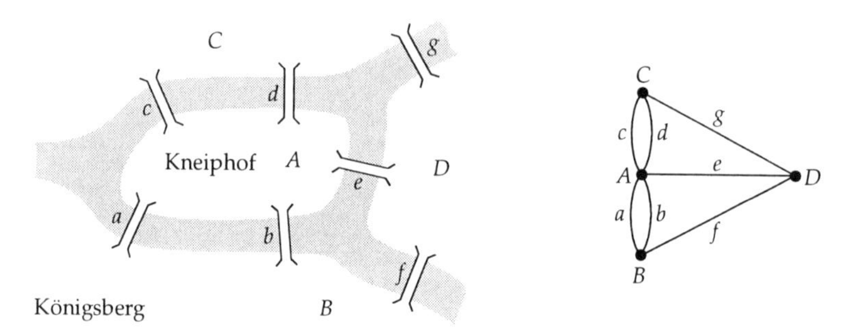

We represent the Königsberg bridges problem in terms of a graph by taking the four land areas as vertices and the seven bridges as edges joining the corresponding pairs of vertices. This gives the graph shown on the right above. The problem of finding a route crossing each bridge exactly once corresponds to that of finding an Eulerian trail in this graph, and you have already seen in Problem 3.1(g) that no such trail exists. It follows that there is no route of the desired kind crossing the seven bridges of Königsberg.

Euler also considered the corresponding problem of finding a route crossing all the bridges in a more general arrangement of bridges and land areas. This led him to present a rule that tells us when such a route is possible, and hence when a graph is Eulerian. In the following problem we ask you to try to formulate this rule.

Problem 3.2

(a) Finding a route crossing each bridge just once and returning to the starting point (that is, finding an Eulerian trail in the corresponding graph) is possible only when the following condition is satisfied:

> whenever you cross into a part of the city, you must be able to leave it by another bridge.

What does this tell you about the vertex degrees in an Eulerian graph?

(b) Using the result of part (a), guess a rule that tells you whether a given connected graph is Eulerian, and test your rule on the graphs of Problem 3.1.

Historical Note

Leonhard Euler (1707–1783), possibly the most prolific mathematician of all time, solved the Königsberg bridges problem in an important paper entitled *Solutio problematis ad geometriam situs pertinentis* (the solution of a problem relating to the geometry of position).

In the solution to Problem 3.2, we gave a criterion that enables us to tell whether a given connected graph is Eulerian – namely, *check whether all the vertex degrees are even*. Before stating this criterion formally, we state a result needed in the proof; its proof is given at the end of this section.

Theorem 3.1

Let G be a graph in which each vertex has even degree. Then G can be split into cycles, no two of which have an edge in common.

Problem 3.3

Show how the following graph can be split into cycles, no two of which have an edge in common.

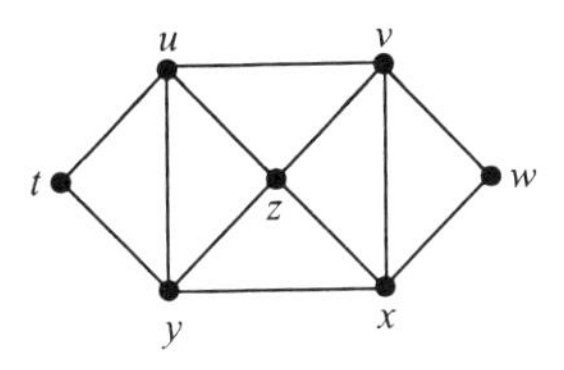

How can these cycles be combined to form an Eulerian trail?

We now state the above criterion formally; its proof is given at the end of this section.

Theorem 3.2

A connected graph is Eulerian if and only if each vertex has even degree.

Remark **E**ulerian graphs have **E**ven degrees.

This theorem gives a necessary and sufficient condition for a connected graph to be Eulerian. It is equivalent to the following two statements for a connected graph G:

(a) if G is Eulerian, then each vertex of G has even degree;
(b) if each vertex of G has even degree, then G is Eulerian.

Problem 3.4

Use Theorem 3.2 to determine which of the following graphs are Eulerian:

(a) the complete graph K_8;
(b) the complete bipartite graph $K_{8,8}$;
(c) the cycle graph C_8;
(d) the dodecahedron graph;
(e) the cube graph Q_8.

Combining the statements of Theorems 3.1 and 3.2, we obtain the following theorem.

Theorem 3.3

An Eulerian graph can be split into cycles, no two of which have an edge in common.

Semi-Eulerian Graphs

There are several simple modifications of the above ideas. The most important of these arises when we do not insist that the citizens of Königsberg return to their starting point.

Suppose that the citizens of Königsberg are keen to cross each of the seven bridges exactly once, but are content to start and finish their walk at different places. Is the walk possible under these conditions?

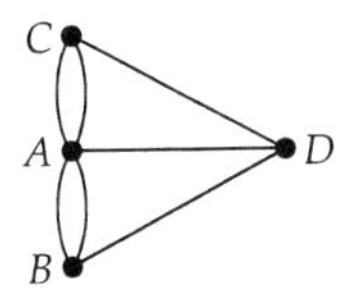

A little experimentation with the above graph should convince you that, even with this modification to the conditions, such a walk is not possible.

This idea leads to the following definition.

Definition

A connected graph is **semi-Eulerian** if there is an open trail that includes every edge; such a trail is a **semi-Eulerian trail.**

The following theorem is derived from Theorem 3.2; its proof is given at the end of this section.

Theorem 3.4

A connected graph is semi-Eulerian if and only if it has exactly two vertices of odd degree.

This theorem gives a necessary and sufficient condition for a connected graph to be semi-Eulerian. It is equivalent to the following two statements for a connected graph G:

(a) if G is semi-Eulerian, then G has exactly two vertices of odd degree;
(b) if G has exactly two vertices of odd degree, then G is semi-Eulerian.

It follows from the above discussion that, in a semi-Eulerian graph G, the starting and finishing vertices of an open trail that includes every edge of G must be the two vertices of odd degree.

Problem 3.5

Use Theorem 3.4 to determine which of the following graphs are semi-Eulerian, and write down a corresponding open trail where possible:

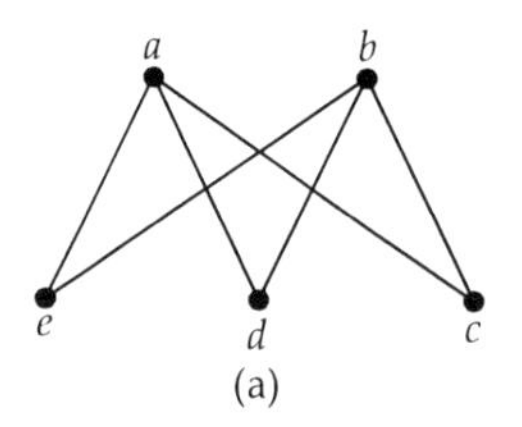

(a)

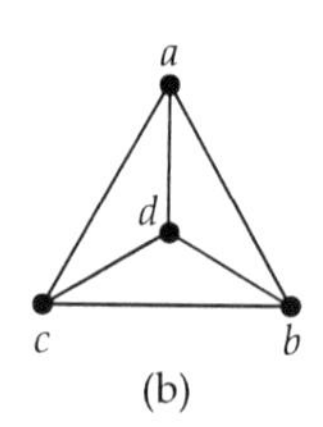

(b)

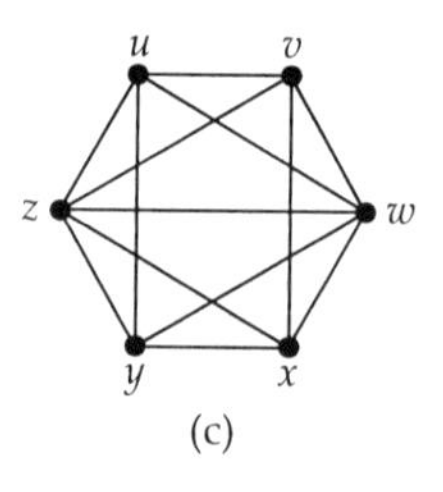

(c)

Proofs of Theorems

We now supply the three proofs omitted earlier.

Theorem 3.1

Let G be a graph in which each vertex has even degree. Then G can be split into cycles, no two of which have an edge in common.

Proof Let G be a graph in which each vertex has even degree. We obtain our first cycle in G by starting at any vertex u and traversing edges in an arbitrary manner, never repeating any edge. Because each vertex has even degree, we know that, whenever we enter a vertex, we must be able to leave it via a different edge. Since there is only a finite number of vertices, we must eventually reach a vertex v that we have met before. The edges of the trail between the two occurrences of the vertex v must therefore form a cycle, C_1.

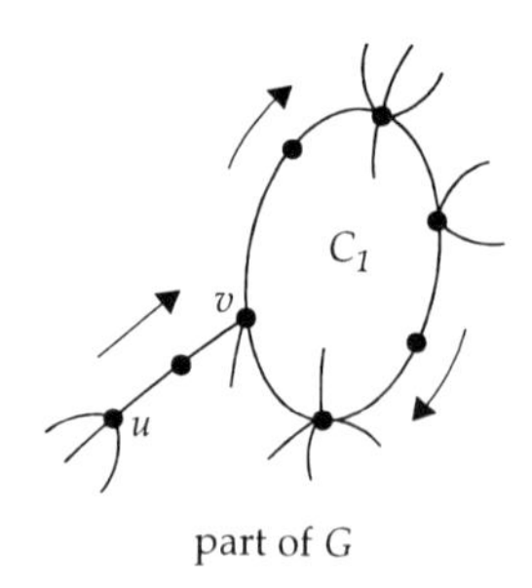

part of G

We now remove from G the edges of C_1. This leaves a graph H (possibly disconnected) in which each vertex has even degree. If H has any edges (that is, if G is not just C_1), we can repeat the procedure above to find a cycle C_2 in H, with no edges in common with C_1.

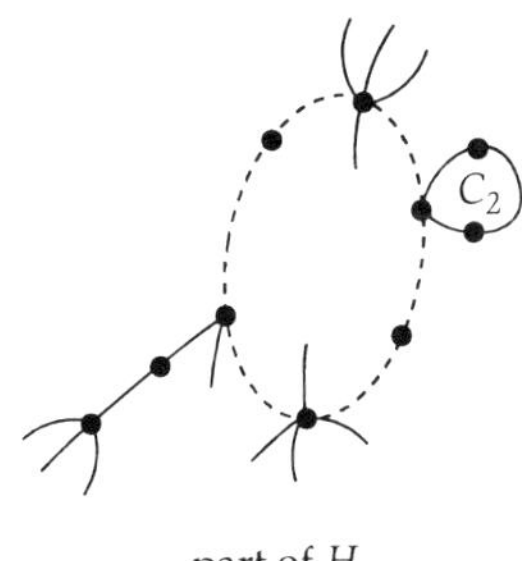

part of H

Removal of the edges of C_2 from H leaves yet another graph in which each vertex has even degree, and which therefore contains a cycle C_3. We continue in this way until there are no edges left, at which stage we have a number of cycles C_1, C_2, ..., C_k which together include every edge of G, and no two of which have any edges in common. ■

Theorem 3.2

A connected graph is Eulerian if and only if each vertex has even degree.

Proof There are two statements to prove.

(a) *If G is Eulerian, then each vertex of G has even degree.*
Let G be an Eulerian graph; then there is an Eulerian trail. Whenever this trail passes through a vertex, there is a contribution of 2 to the degree of that vertex. Since each edge is used just once, the degree of each vertex is a sum of 2s – that is, an even number.

(b) *If each vertex of a connected graph G has even degree, then G is Eulerian.*
Let G be a connected graph in which each vertex has even degree. We know from Theorem 3.1 that G can be split into cycles, no two of which have an edge in common.

We now fit these cycles together to make an Eulerian trail. We start at any vertex of a cycle C_1 and travel round C_1 until we meet a vertex of another cycle, C_2 say.

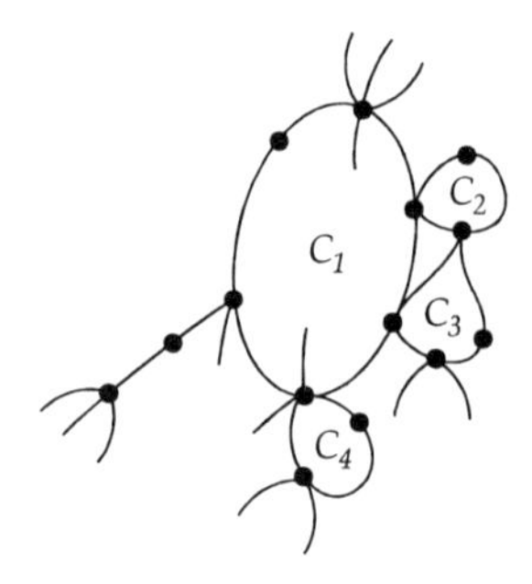

We traverse the edges of this cycle, and then resume travelling round C_1. This gives a closed trail that includes C_1 and C_2. If this trail includes all the edges in G, then we have the required Eulerian trail. If not, we travel round our new closed trail, and add a new cycle, C_3 say, when we come to it; since G is connected, there is always at least one cycle to add to our closed trail. We continue this process until all the cycles have been traversed, at which stage we have the required Eulerian trail. It follows that G is Eulerian. ■

The above proof is not the shortest possible; for example, there is a shorter proof that uses the method of mathematical induction. However, the advantage of the above proof is that it is constructive – it gives a method for *constructing* an Eulerian trail in a given graph.

Theorem 3.4

A connected graph is semi-Eulerian if and only if it has exactly two vertices of odd degree.

Proof There are two statements to prove.

(a) *If G is semi-Eulerian, then G has exactly two vertices of odd degree.*

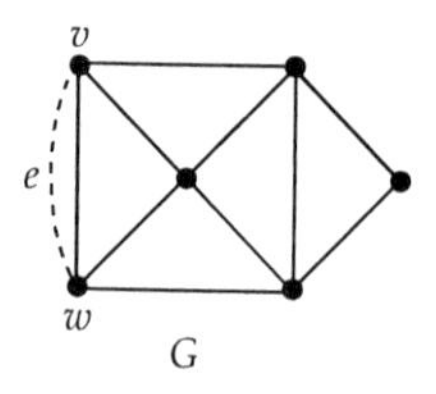

Let G be a semi-Eulerian graph, and let v and w be the starting and finishing vertices of an open trail. Let us add an edge e joining v and w; then we obtain an Eulerian graph in which each vertex has even degree, by Theorem 3.2. If we now recover G by removing the edge e, we see that v and w are the only vertices of odd degree.

(b) *If a connected graph G has exactly two vertices of odd degree, then G is semi-Eulerian.*

Let G be a graph with exactly two vertices of odd degree, v and w. Let us add an edge e joining v and w; then we obtain a connected graph in which each vertex has even degree. By Theorem 3.2, this graph is Eulerian, and so has an Eulerian trail. Removal of the edge e from this trail produces an open trail that includes every edge of G, so G is semi-Eulerian. ■

3.3 Hamiltonian Graphs

We now turn our attention to Hamiltonian graphs – graphs in which there is a cycle passing through every vertex.

The name *Hamiltonian* derives from a game invented by Sir William Rowan Hamilton (1805–1865), one of the leading mathematicians of his time. He did brilliant work in geometrical optics, dynamics and algebra. His *icosian calculus* can be expressed in terms of finding Hamiltonian cycles in the graph of the regular dodecahedron, shown below.

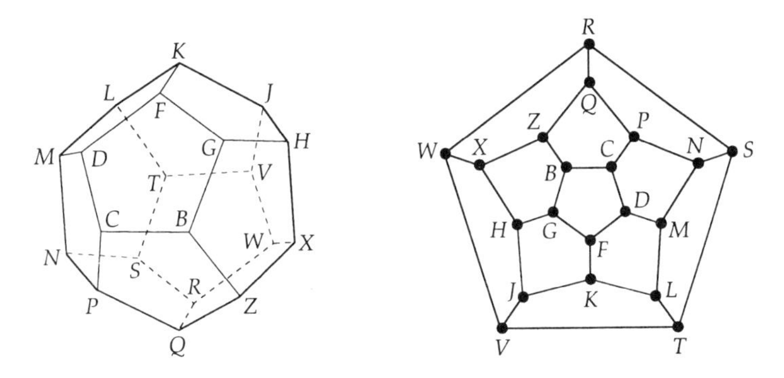

He also turned the problem into a game, the *icosian game*, in which the player has to find Hamiltonian cycles starting with five given letters. For example, given the initial letters *BCPNM*, the player can complete a Hamiltonian cycle in exactly two ways:

B C P N M D F K L T S R Q Z X W V J H G B;
B C P N M D F G H X W V J K L T S R Q Z B.

The game was marketed in 1859. It also appeared in a solid dodecahedron form under the title *A voyage round the world*, with the vertices representing places – Brussels, Canton, Delhi, ..., Zanzibar.

Problem 3.6

How many Hamiltonian cycles on the dodecahedron begin with *JVTSR*?

Problem 3.7

Find a path on the dodecahedron starting with *BCD*, ending with *T*, and including each vertex just once.

The name *Hamiltonian cycle* can be regarded as a misnomer, since Hamilton was not the first to look for cycles that pass through every vertex of a graph. An earlier example of a problem that can be expressed in terms of Hamiltonian cycles is the *knight's tour problem*, which we discuss later in this chapter. Yet another is the *travelling salesman problem*, introduced in Chapter 1. In this problem, a graph is given in which the vertices represent locations, and each edge has a *weight*, representing the distance between its endpoints. The problem is to find a route that visits each vertex just once and returns to the starting point, covering the shortest possible total distance – that is, to find a *minimum-weight Hamiltonian cycle* in the graph.

Properties of Hamiltonian Graphs

At first sight, the problem of deciding whether a given graph is Hamiltonian may seem similar to that of deciding whether it is Eulerian, and we might expect there to be a simple necessary and sufficient condition for a graph to be Hamiltonian, analogous to that of Theorem 3.2 for Eulerian graphs. However, no such condition is known, and the search for necessary or sufficient conditions for a graph to be Hamiltonian is a major area of study in graph theory.

Faced with this situation, the best we can do is to look for various classes of graphs that are Hamiltonian. For example, it is clear that the cycle graph C_n is Hamiltonian for all values of n. Also, the graph K_n is Hamiltonian for $n \geq 3$; if the vertices are denoted by 1, 2, ..., n, then a Hamiltonian cycle is 1 2 3 ... n 1.

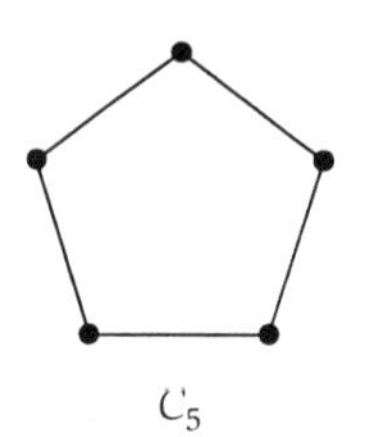

C_5

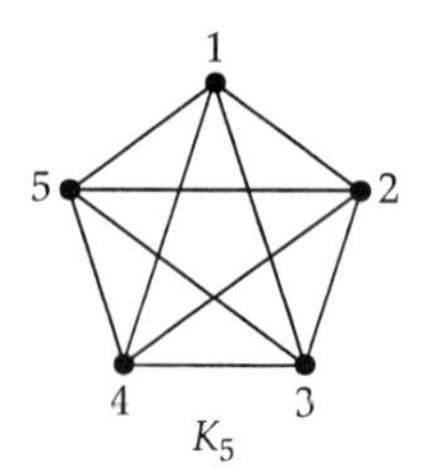

K_5

Problem 3.8

Which of the following graphs are Hamiltonian?

(a) the complete bipartite graph $K_{4,4}$;
(b) a tree.

Problem 3.9

(a) Prove that a bipartite graph with an odd number of vertices is not Hamiltonian.
(b) Use the result of part (a) to prove that the following graph is not Hamiltonian.

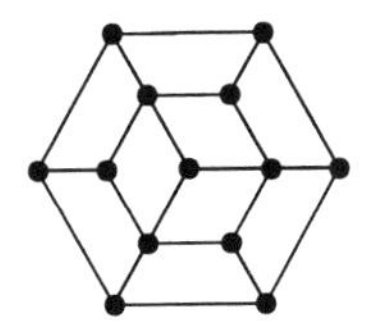

If we take a Hamiltonian graph and add an edge to it, then we obtain another Hamiltonian graph, since we can take the same Hamiltonian cycle as before. It follows that graphs with large vertex degrees, and hence many edges, are more likely to be Hamiltonian than graphs with small vertex degrees. We can make this idea precise in various ways. One of these is the following result of Oystein Ore, proved in 1960; the proof is omitted.

Theorem 3.5: Ore's Theorem

Let G be a simple connected graph with n vertices, where $n \geq 3$, and

$$\deg v + \deg w \geq n,$$

for each pair of non-adjacent vertices v and w. Then G is Hamiltonian.

For example, for the graph shown below, $\deg v + \deg w \geq 5$ for each pair of non-adjacent vertices v and w (in fact, for all pairs of vertices), so this graph is Hamiltonian, by Ore's theorem.

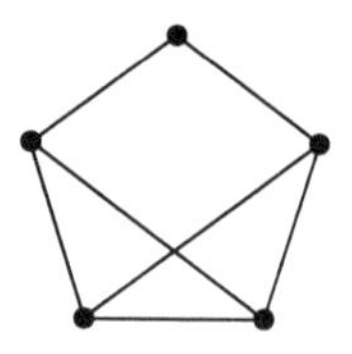

Problem 3.10

(a) Let G be a simple connected graph with n vertices, where $n \geq 3$ and $\deg v \geq n/2$ for each vertex v. Use Ore's theorem to show that G is Hamiltonian.
(This result is known as *Dirac's theorem*; it was proved in 1952.)

(b) Give an example of a Hamiltonian graph that does not satisfy the conditions of Ore's theorem.

Just as for Eulerian graphs, there are several variations of the above ideas and results. For example, we can define *semi-Hamiltonian graphs* – graphs in which it is possible to visit every vertex, but not return to the starting point.

Definition

A connected graph is **semi-Hamiltonian** if there is a path, but not a cycle, that includes every vertex; such a path is a **semi-Hamiltonian path.**

There is no known general criterion for testing whether a given graph is semi-Hamiltonian.

Problem 3.11

Determine which of the following graphs are semi-Hamiltonian, and write down a corresponding semi-Hamiltonian path where possible:

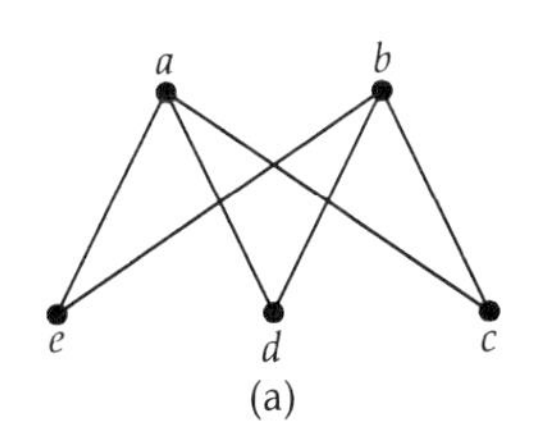

(a)

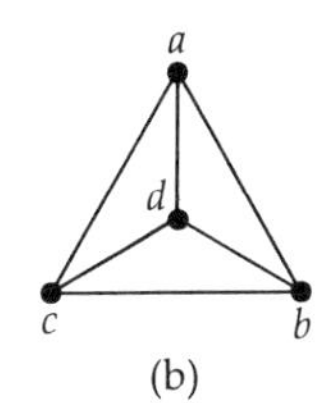

(b)

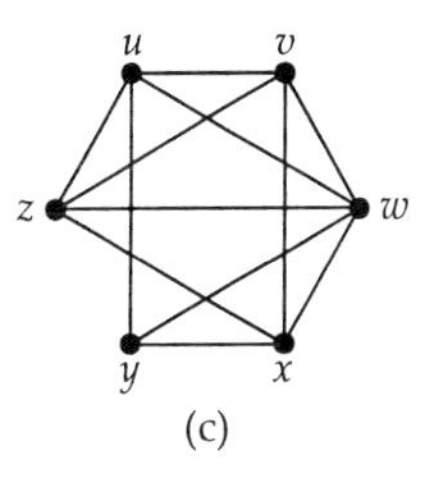

(c)

3.4 Case Studies

We conclude this chapter with four case studies – dominoes, diagram-tracing puzzles, the knight's tour problem and Gray codes.

Dominoes

An unusual application of Eulerian graphs is to the game of dominoes. We use the complete graph K_7, which is Eulerian since each vertex has degree 6.

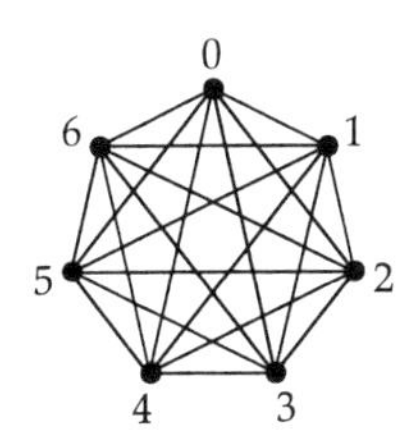

Let us label the vertices 0, 1, 2, 3, 4, 5, 6, consecutively. Then an Eulerian trail is obtained by tracing the edges in the following order:

01, 12, 23, 34, 45, 56, 60, 02, 24, 46, 61, 13, 35, 50, 03, 36, 62, 25, 51, 14, 40.

(There are many other Eulerian trails.)

We can regard each of these edges as a domino – for example, the edge 24 corresponds to the domino

It follows that the above Eulerian trail corresponds to an arrangement of the dominoes of a normal set (other than the doubles 0–0, 1–1, … , 6–6) in a continuous sequence. Once this basic sequence is found, we can insert the doubles at appropriate places, thus showing that a complete game of dominoes is possible. The following ring of dominoes corresponds to the above Eulerian trail:

Problem 3.12

By finding an Eulerian trail in K_5, arrange a set of fifteen dominoes (from 0–0 to 4–4) in a ring.

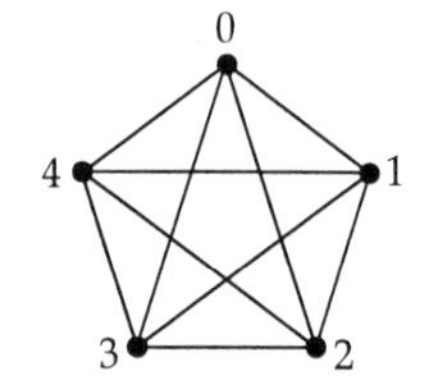

Diagram-Tracing Puzzles

A common type of recreational puzzle is that of drawing a given diagram with as few continuous pen-strokes as possible, without covering any part of the diagram twice. For example, it is easy to draw the following diagram with four continuous strokes, but can it be done with three?

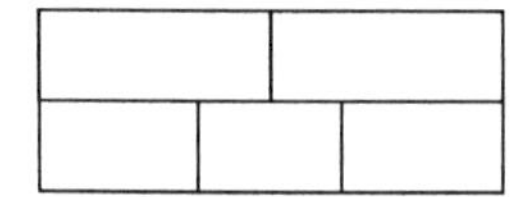

Such problems are equivalent to determining the minimum number of open trails with no edge in common that make up the corresponding graph.

In 1809 Louis Poinsot, unaware of Euler's solution to the Königsberg bridges problem, showed that diagrams consisting of n mutually connected points can be drawn in one continuous stroke if n is odd, but not if n is even:

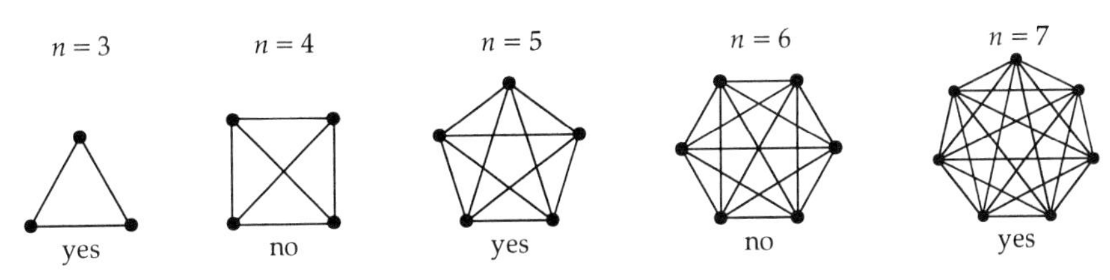

In the terminology of graph theory, this amounts to saying that the complete graph K_n is Eulerian only for odd values of n, since such graphs have even vertex degrees. Poinsot's account of the subject included an ingenious construction for finding an Eulerian trail when n is odd – no mean feat, as you will see if you try to describe a method for constructing an Eulerian trail in (say) K_{99}.

In 1847, Johann Listing wrote an important treatise entitled *Vorstudien zur Topologie* (Introductory studies in topology), which included a discussion of diagram-tracing puzzles. In particular, he observed that the following diagram has eight vertices of odd degree, and so cannot be drawn with fewer than four continuous strokes.

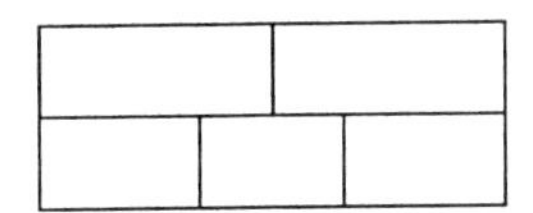

He also remarked that the following diagram can be drawn in one continuous stroke, starting at one end and ending at the other, since these are the only points that correspond to vertices of odd degree:

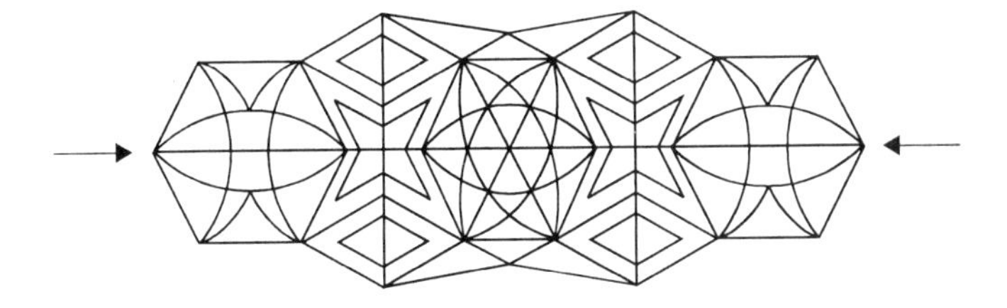

Problem 3.13

How many continuous pen-strokes are needed to draw the following diagram?

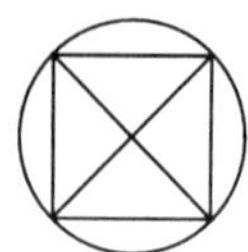

The following problem answers the question of how many continuous pen-strokes are needed to draw a given connected graph.

Problem 3.14

Prove that, if a connected graph G has k vertices of odd degree, then the smallest number of continuous pen-strokes needed to cover all the edges is $k/2$.

Hint Add $k/2$ edges to G, in a suitable manner, to obtain an Eulerian graph.

Knight's Tour Problem

On a chessboard, a knight always moves two squares in a horizontal or vertical direction and one square in a perpendicular direction, as illustrated below.

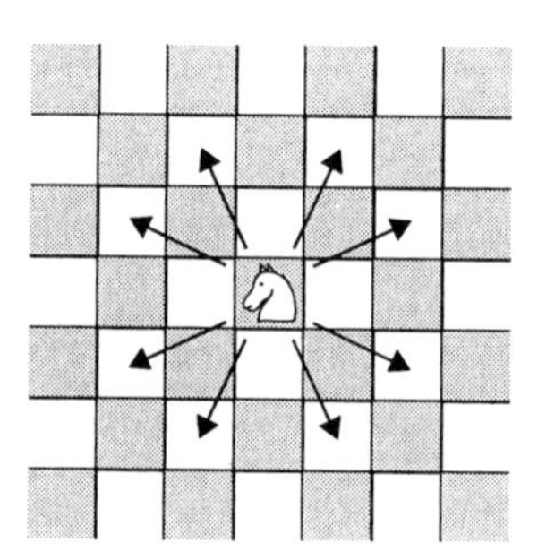

A celebrated recreational problem, which has been studied for many hundreds of years, is the following.

Knight's Tour Problem

Can a knight visit each square of a chessboard just once by a sequence of knight's moves, and finish on the same square as it began?

We represent the board as a graph in which each vertex corresponds to a square, and each edge corresponds to a pair of squares connected by a knight's move. We deduce that finding a knight's tour is equivalent to finding a Hamiltonian cycle in the associated graph of the chessboard. The following diagram shows a 4×4 chessboard and its associated graph.

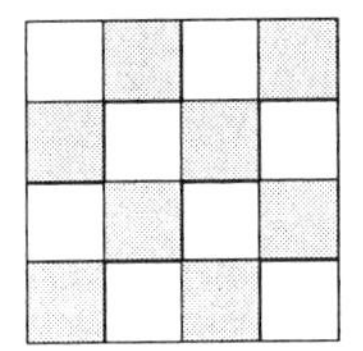

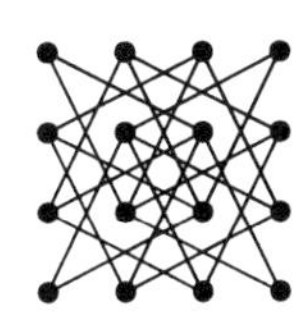

In fact, there is no knight's tour on a 4×4 chessboard. In order to see this, note that the only way that we can include the top-left square in the tour is to include the two moves shown in figure (a) below. Similarly, the only way that we can include the lower-right square is to include the two moves shown in figure (b). Combining these, we see that the tour has to include the four moves in figure (c). But these already form a cycle, so it is impossible to include them as part of a full tour. Thus, no knight's tour is possible on a 4×4 chessboard.

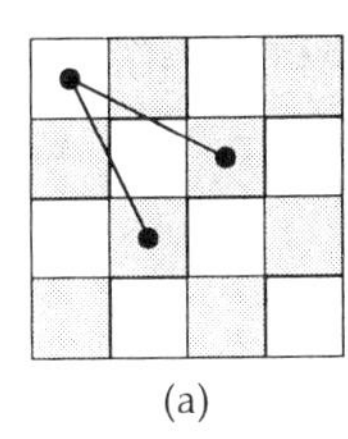

(a)

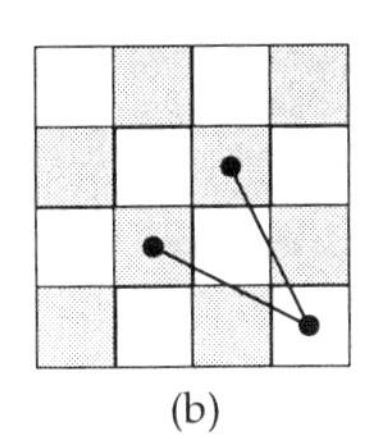

(b)

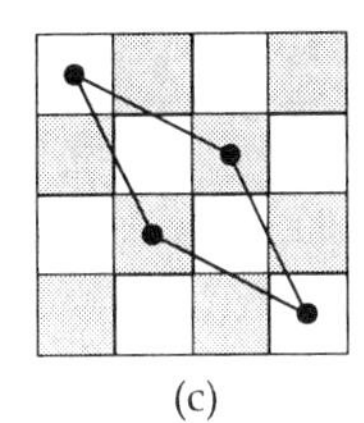

(c)

There is no knight's tour on a chessboard with an *odd* number of squares (such as a 5×5 chessboard), as you will see in the following problem. However, for certain other chessboards, a knight's tour is possible. The following diagram illustrates a knight's tour on an ordinary 8×8 chessboard, thus answering the original knight's tour problem in the affirmative.

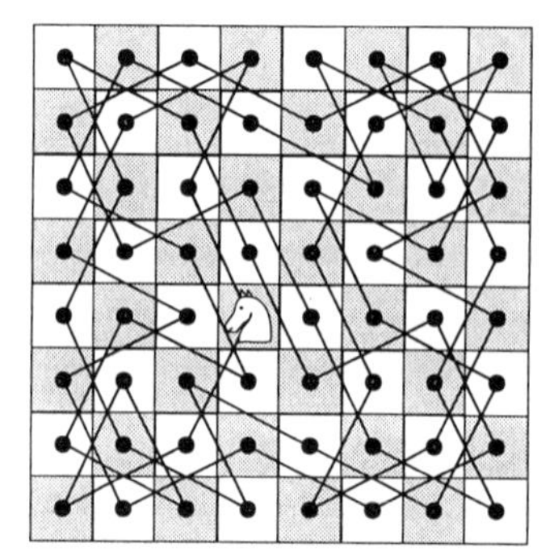

50	11	24	63	14	37	26	35
23	62	51	12	25	34	15	38
10	49	64	21	40	13	36	27
61	22	9	52	33	28	39	16
48	7	60	1	20	41	54	29
59	4	45	8	53	32	17	42
6	47	2	57	44	19	30	55
3	58	5	46	31	56	43	18

This solution is particularly interesting, because if we write down the order of the moves, as in the right-hand diagram, we get a *magic square*, in which the numbers in each row or column have the same total, 260.

Problem 3.15

Show that there is no knight's tour on a 5×5 or 7×7 chessboard.
Hint Use the result of Problem 3.9(a): a bipartite graph with an odd number of vertices is not Hamiltonian.

Gray Codes

Engineers sometimes wish to represent the angular position (in multiples of 45°) of a shaft that is rotating continuously. An arrangement of brushes on a commutator is used to read certain tracks inscribed on the shaft and convert the angle through which the shaft rotates into a 3-digit binary word (a string of 0s and 1s), as follows:

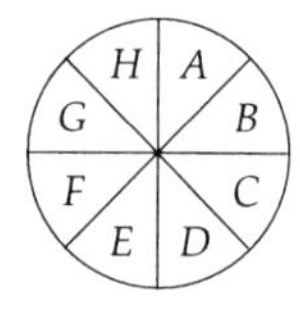

angle segment	*A*	*B*	*C*	*D*	*E*	*F*	*G*	*H*
binary word	000	001	011	010	110	111	101	100

Each 3-digit binary word identifies the angle segment in the position occupied by *A* in the diagram above.

As the shaft rotates, *the binary word changes by only one digit at a time as we progress from each word to the next in the sequence*. A sequence of binary words with this property is called a **Gray code**. The advantage of such a code is that it minimizes ambiguities that might be caused by misalignments of the brushes that read the tracks.

Gray codes can be found by tracing Hamiltonian cycles on the graph of a cube. For example, the above code,

000 → 001 → 011 → 010 → 110 → 111 → 101 → 100 (→ 000),

and the code

000 → 100 → 110 → 010 → 011 → 111 → 101 → 001 (→ 000),

both correspond to Hamiltonian cycles in the 3-cube, shown below.

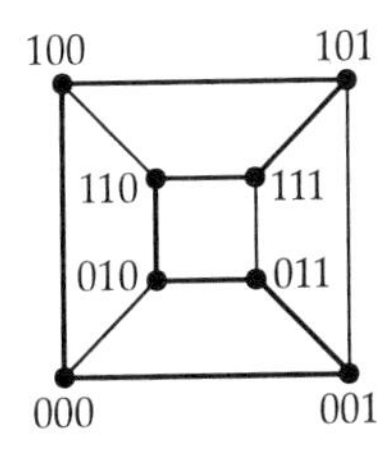

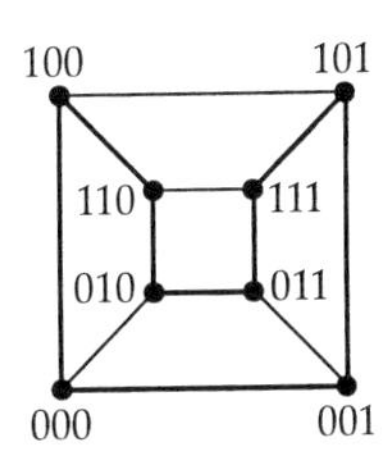

Similarly, to find a Gray code of 4-digit binary words, we trace a Hamiltonian cycle in the 4-cube. An example of such a code, illustrated below as a Hamiltonian cycle in the 4-cube, is

0000 → 0001 → 0011 → 0010 → 0110 → 0111 → 0101 → 0100 →
1100 → 1101 → 1111 → 1110 → 1010 → 1011 → 1001 → 1000 (→ 0000).

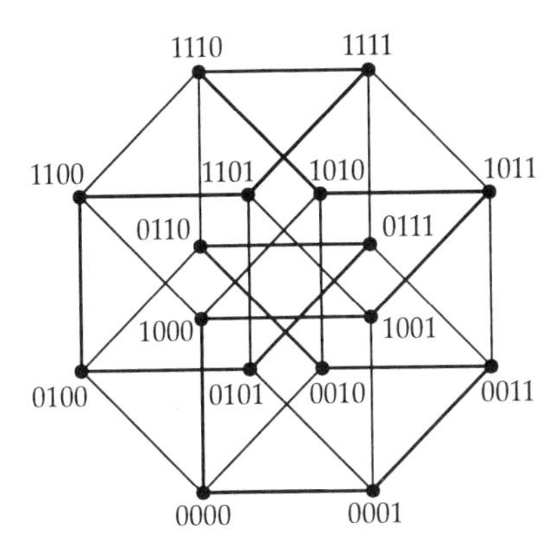

Problem 3.16

Find another Gray code of 4-digit binary words.

Exercises 3

Eulerian Graphs

3.1 For which values of n, r and s are the following graphs Eulerian? For which values are they semi-Eulerian?

(a) the complete graph K_n;

(b) the complete bipartite graph $K_{r,s}$;

(c) the n-cube Q_n.

3.2 Write down all the ways in which the following Eulerian graph can be split into cycles, no two of which have an edge in common.

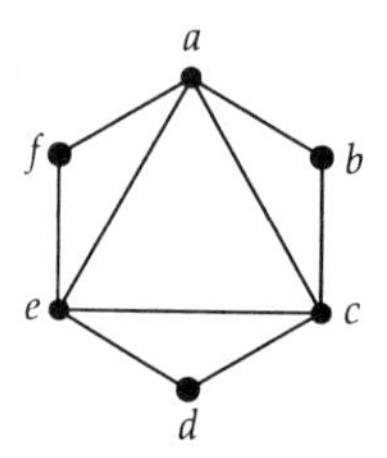

3.3 Theorems 3.2 and 3.4 tell us about the properties of connected graphs with zero or two vertices of odd degree. What can you say about connected graphs with exactly one vertex of odd degree?

Hamiltonian Graphs

3.4 For which values of n, r and s are the graphs in Exercise 3.1 Hamiltonian? For which values are they semi-Hamiltonian?

3.5 Draw two graphs each with 10 vertices and 13 edges: one that is Eulerian but not Hamiltonian and one that is Hamiltonian but not Eulerian.

3.6 Check whether the conditions of Ore's theorem hold for the following Hamiltonian graphs:

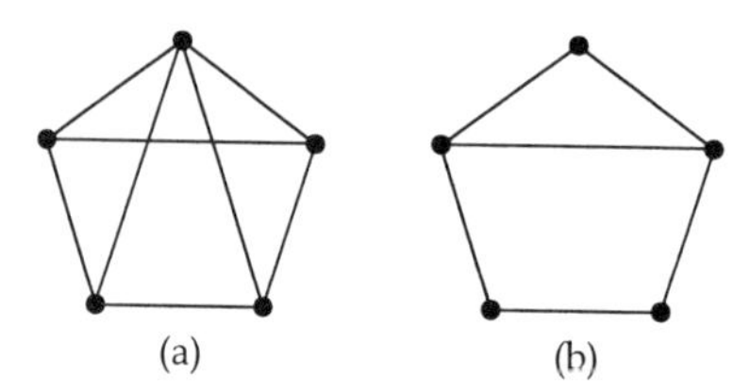

3.7 (a) Let G be a simple connected graph with n vertices and $\frac{1}{2}(n-1)(n-2)+2$ edges. Use Ore's theorem to prove that G is Hamiltonian.

(b) Give an example of a non-Hamiltonian simple connected graph with n vertices and $\frac{1}{2}(n-1)(n-2)+1$ edges.

Case Studies

Dominoes

3.8 Find two further ways of arranging:

(a) a full set of dominoes in a ring;

(b) fifteen dominoes from (0–0 to 4–4) in a ring.

Diagram-Tracing Puzzles

3.9 How many continuous pen-strokes are needed to draw each of the following diagrams without covering any part twice?

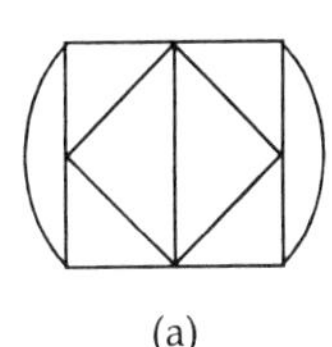

(a)

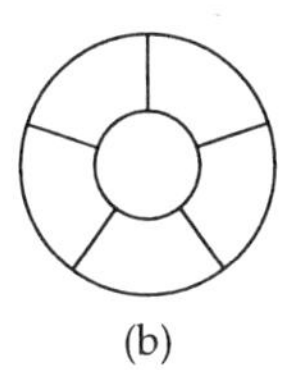

(b)

Knight's Tour Problem

3.10 Prove that there is no knight's tour on a 3×6 chessboard.

Gray Codes

3.11 Write down a Gray code of 5-digit binary words.

Chapter 4

Digraphs

After studying this chapter, you should be able to:

- explain the terms *digraph, labelled digraph, unlabelled digraph, vertex, arc, adjacent, incident, multiple arcs, loop, simple digraph, underlying graph* and *subdigraph;*
- determine whether two given digraphs are *isomorphic;*
- explain the terms *in-degree, out-degree, in-degree sequence* and *out-degree sequence;*
- state and use the *handshaking dilemma;*
- explain the terms *walk, trail, path, closed walk, closed trail, cycle, connected, disconnected* and *strongly connected* in the context of digraphs;
- explain the terms *Eulerian digraph* and *Eulerian trail*, and state a necessary and sufficient condition for a connected digraph to be Eulerian;
- explain the terms *Hamiltonian digraph* and *Hamiltonian cycle;*
- describe the use of digraphs in ecology, social networks, the rotating drum problem, and ranking in tournaments.

In this chapter we discuss digraphs and their properties. Our treatment of the subject is similar to that of Chapters 2 and 3 for graphs, except that we need to take account of the directions of the arcs.

4.1 Digraphs and Subdigraphs

We start by recalling the definition of a *digraph*.

Definitions

A **digraph** consists of a set of elements called **vertices** and a set of elements called **arcs**. Each arc **joins** two vertices in a specified direction.

For example, the digraph shown below has four vertices $\{u, v, w, x\}$ and six arcs $\{1, 2, 3, 4, 5, 6\}$. Arc 1 joins x to u, arc 2 joins u to w, arcs 3 and 4 join w to v, arc 5 joins x to w, and arc 6 joins the vertex x to itself.

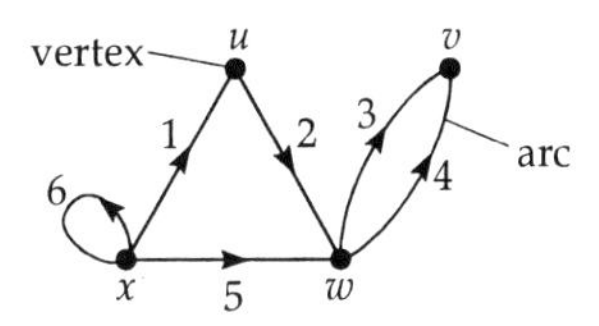

We often denote an arc by specifying its two vertices in order; for example, arc 1 is denoted by xu, arcs 3 and 4 are denoted by wv, and arc 6 is denoted by xx. Note that xu is not the same as ux.

The above digraph contains more than one arc joining w to v, and an arc joining the vertex x to itself. The following terminology is useful when discussing such digraphs.

Definitions

In a digraph, two or more arcs joining the same pair of vertices in the same direction are **multiple arcs**. An arc joining a vertex to itself is a **loop**.
A digraph with no multiple arcs or loops is a **simple digraph**.

For example, digraph (a) below has multiple arcs and digraph (b) has a loop, so neither is a simple digraph. Digraph (c) has no multiple arcs or loops, and is therefore a simple digraph.

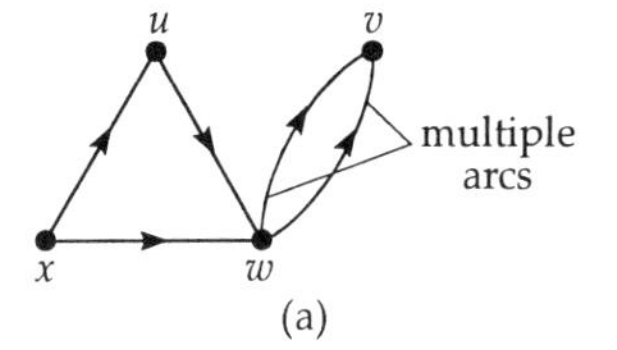

(a)

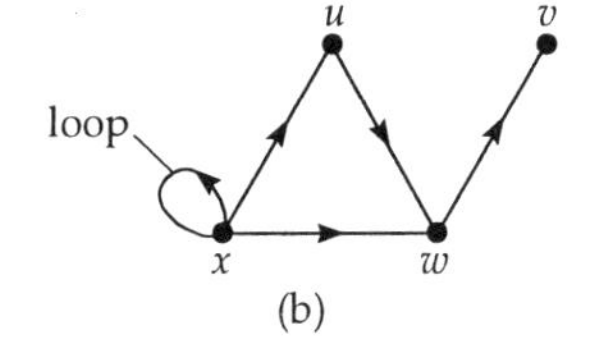

(b)

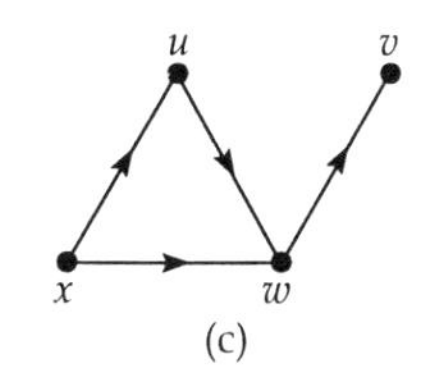

(c)

Problem 4.1

Write down the vertices and arcs of each of the following digraphs. Are these digraphs simple digraphs?

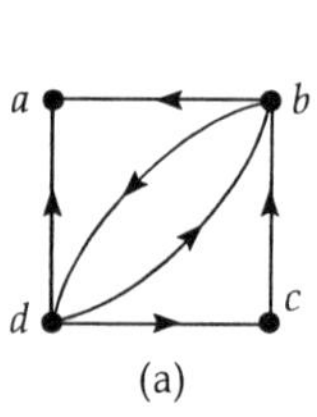

(a)

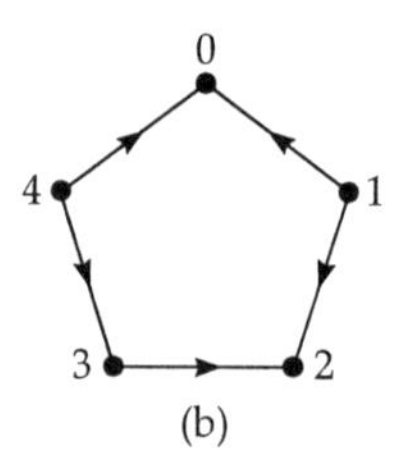

(b)

Problem 4.2

Draw the digraphs whose vertices and arcs are as follows. Are these digraphs simple digraphs?

(a) vertices: $\{u, v, w, x\}$ arcs: $\{vw, wu, wv, wx, xu\}$

(b) vertices: $\{1, 2, 3, 4, 5, 6, 7, 8\}$ arcs: $\{12, 22, 23, 34, 35, 67, 68, 78\}$

Adjacency and Incidence

The digraph analogues of *adjacency* and *incidence* are similar to the corresponding definitions for graphs, except that we take account of the directions of the arcs.

Definitions

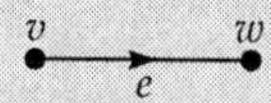

The vertices v and w of a digraph are **adjacent** vertices if they are joined (in either direction) by an arc e. An arc e that joins v to w is **incident from** v and **incident to** w; v is **incident to** e, and w is **incident from** e.

For example, in the digraph below, the vertices u and x are adjacent, vertex w is incident from arcs 2 and 5 and incident to arcs 3 and 4, and arc 6 is incident to (and from) the vertex x.

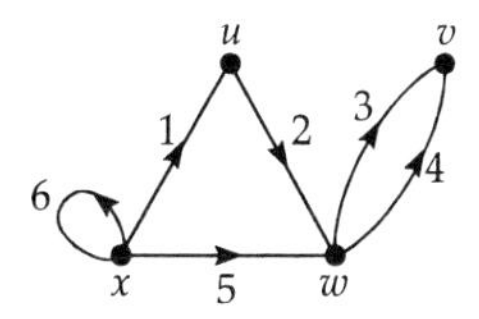

Problem 4.3

Which of the following statements hold for the digraph on the right?

(a) vertices v and w are adjacent;

(b) vertices v and x are adjacent;

(c) vertex u is incident to arc 2;

(d) arc 5 is incident from vertex v.

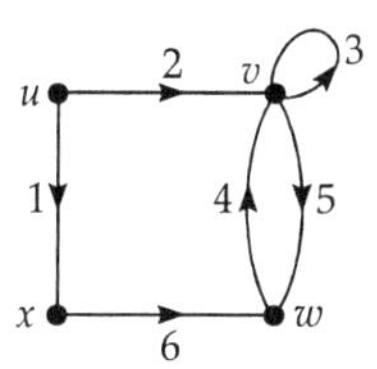

Isomorphism

It follows from the definition that a digraph is completely determined when we know its vertices and arcs, and that two digraphs are *the same* if they have the same vertices and arcs. Once we know the vertices and arcs, we can draw the digraph and, in principle, any picture we draw is as good as any other; the actual way in which the vertices and arcs are drawn is irrelevant – although some pictures are easier to use than others!

We extend the concept of isomorphism to digraphs, as follows.

Definition

Two digraphs C and D are **isomorphic** if D can be obtained by relabelling the vertices of C – that is, if there is a one–one correspondence between the vertices of C and those of D, such that the arcs joining each pair of vertices in C agree in both number and direction with the arcs joining the corresponding pair of vertices in D.

For example, the digraphs C and D represented by the diagrams

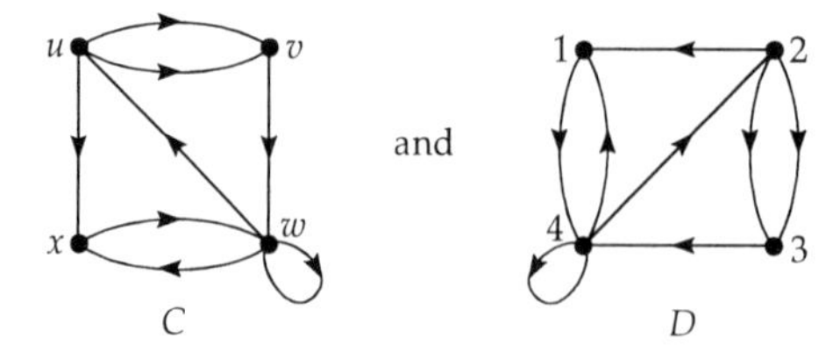

are not the same, but they are isomorphic, since we can relabel the vertices in the digraph C to get the digraph D, using the following one-one correspondence:

$$\begin{array}{ccc} C & \leftrightarrow & D \\ u & \leftrightarrow & 2 \\ v & \leftrightarrow & 3 \\ w & \leftrightarrow & 4 \\ x & \leftrightarrow & 1 \end{array}$$

Note that arcs in C correspond to arcs in D – for example:

the two arcs from u to v in C correspond to the two arcs from 2 to 3 in D;
the arcs wx and xw in C correspond to the arcs 41 and 14 in D;
the loop ww in C corresponds to the loop 44 in D.

Problem 4.4

By suitably relabelling the vertices, show that the following digraphs are isomorphic:

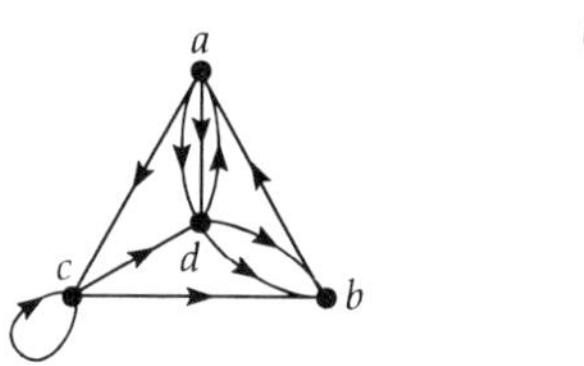

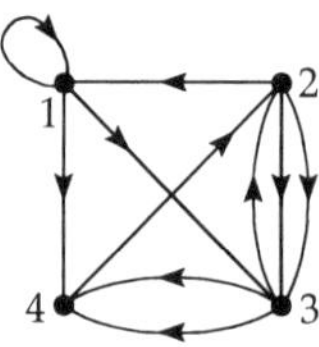

Problem 4.5

Are the following two digraphs isomorphic? If so, find a suitable one-one correspondence between the vertices of the first and those of the second; if not, explain why no such one-one correspondence exists.

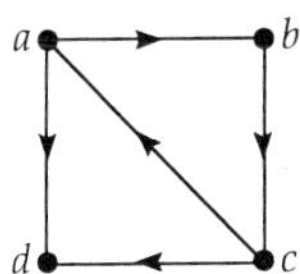

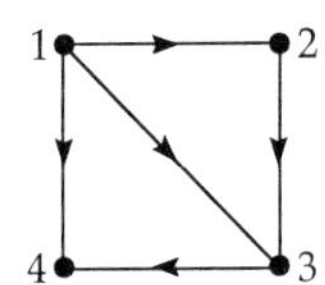

Sometimes it is unnecessary to have labels on the digraphs. In such cases, we omit the labels, and refer to the resulting object as an *unlabelled digraph*. For example, the unlabelled digraph

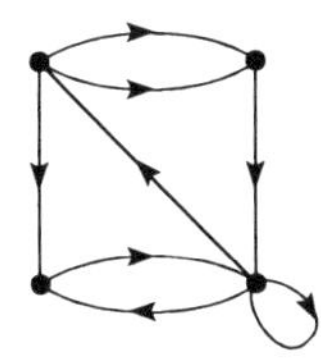

corresponds to any of the following isomorphic digraphs:

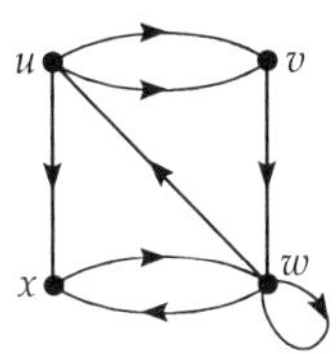

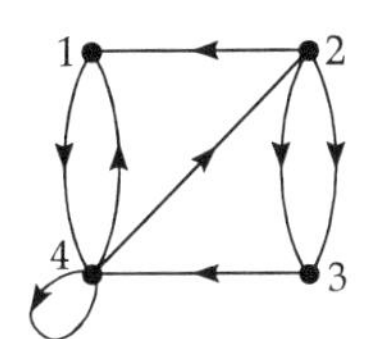

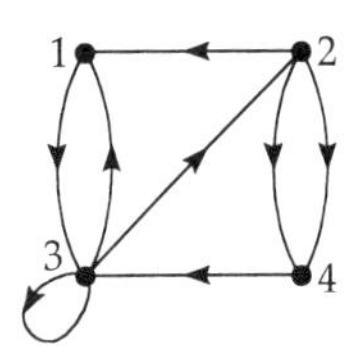

We say that two unlabelled digraphs are *isomorphic* if labels can be attached to their vertices so that they become the same digraph.

Problem 4.6

By suitably labelling the vertices, show that the following unlabelled digraphs are isomorphic:

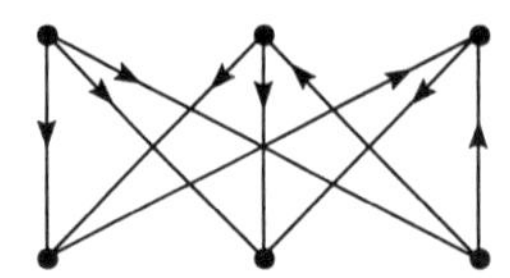

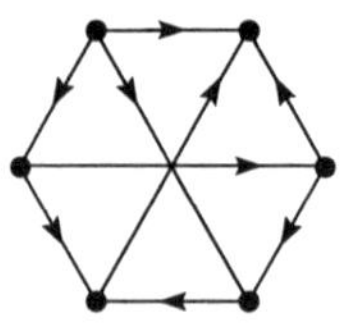

Subdigraphs and Underlying Graphs

It is convenient to define a concept analogous to that of a subgraph of a graph.

Definition

A **subdigraph** of a digraph D is a digraph all of whose vertices are vertices of D and all of whose arcs are arcs of D.

Remark Note that D is a subdigraph of itself.

For example, the following digraphs are all subdigraphs of the digraph D on the left, with vertices $\{u, v, w, x\}$ and arcs $\{1, 2, 3, 4, 5, 6\}$.

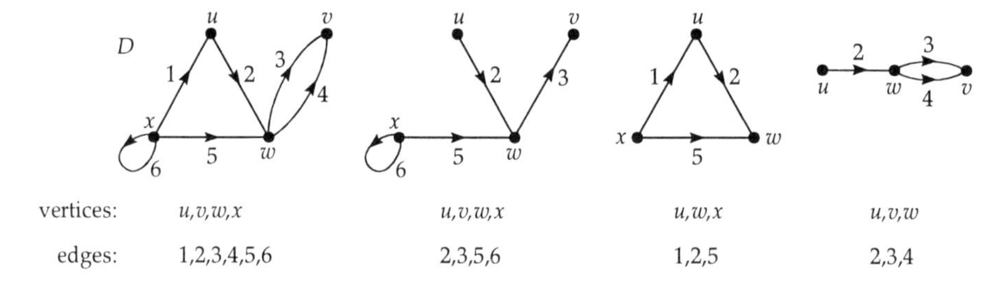

Problem 4.7

Which of the following digraphs are subdigraphs of the digraph D below?

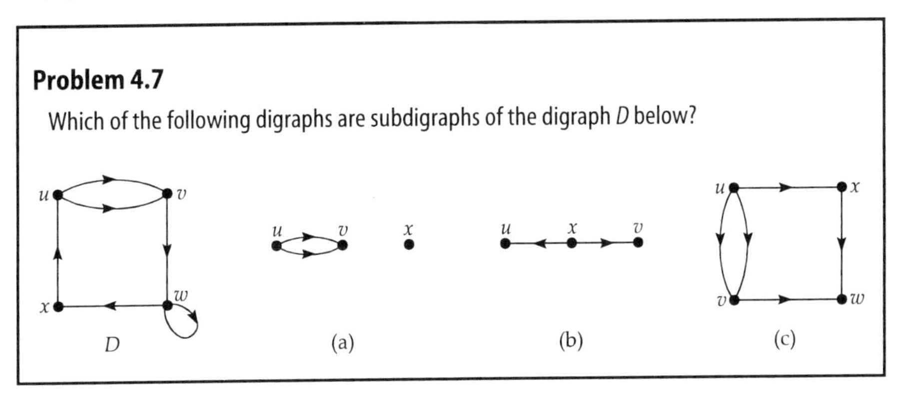

The idea of a subdigraph can be extended to unlabelled digraphs. For example, the following digraphs are all subdigraphs of the unlabelled digraph C on the left:

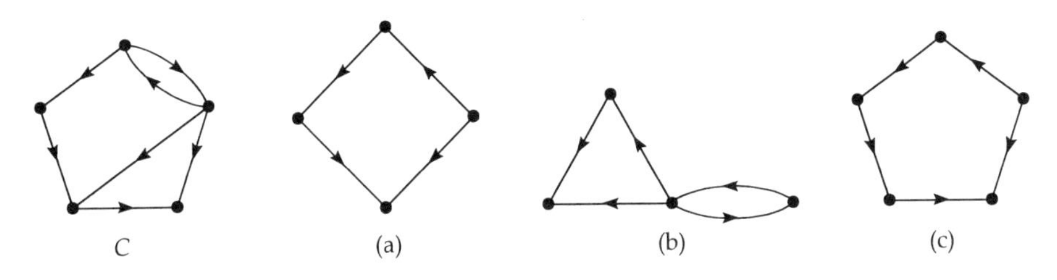

Problem 4.8

Which of the following digraphs are subdigraphs of the digraph C below?

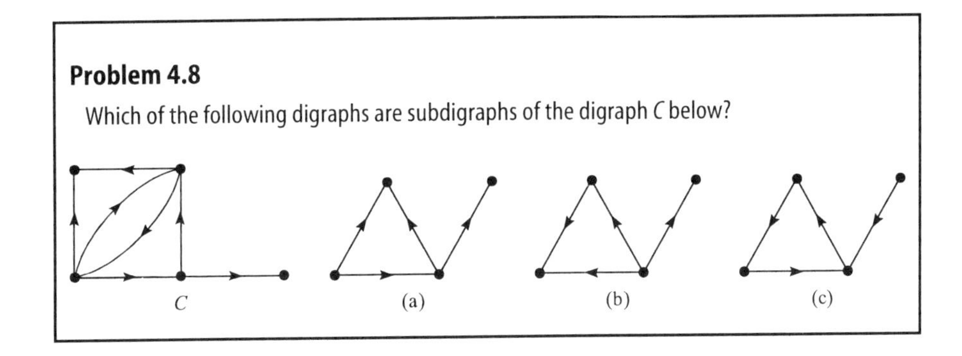

It is also convenient to introduce the idea of the *underlying graph* of a digraph.

Definition

The **underlying graph** of a digraph *D* is the graph obtained by replacing each arc of *D* by the corresponding undirected edge.

To obtain the underlying graph, we simply remove the arrows from the arcs; for example:

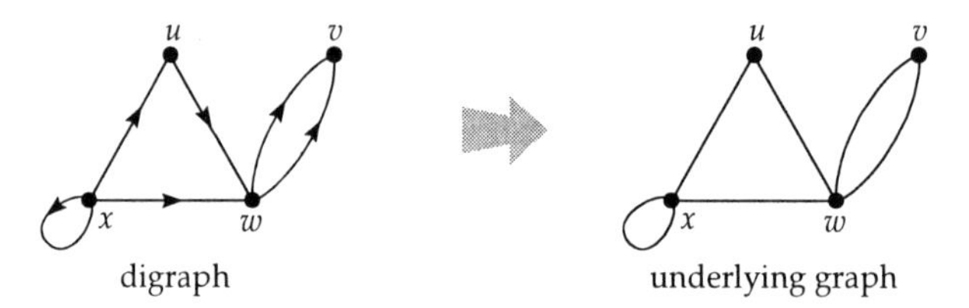

4.2 Vertex Degrees

We now give analogues of the degree of a vertex in a graph.

Definitions

In a digraph, the **out-degree** of a vertex *v* is the number of arcs incident from *v*, and is denoted by **outdeg *v***; the **in-degree** of *v* is the number of arcs incident to *v*, and is denoted by **indeg *v***.

Remark Each loop contributes 1 to both the in-degree and the out-degree of the corresponding vertex.

For example, the digraph below has the following out-degrees and in-degrees:

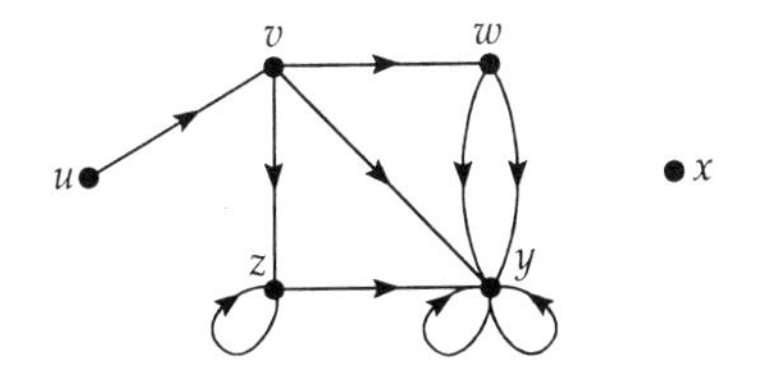

outdeg $u = 1$	outdeg $v = 3$	outdeg $w = 2$
indeg $u = 0$	indeg $v = 1$	indeg $w = 1$
outdeg $x = 0$	outdeg $y = 2$	outdeg $z = 2$
indeg $x = 0$	indeg $y = 6$	indeg $z = 2$

There are also analogues of the degree sequence of a graph, corresponding to the out-degree and in-degree of a vertex.

Definitions

The **out-degree sequence** of a digraph D is the sequence obtained by listing the out-degrees of D in increasing order, with repeats as necessary.

The **in-degree sequence** of D is defined analogously.

For example, the above digraph has out-degree sequence $(0, 1, 2, 2, 2, 3)$ and in-degree sequence $(0, 0, 1, 1, 2, 6)$.

Problem 4.9

Write down the out-degree and in-degree sequences of each of the following digraphs:

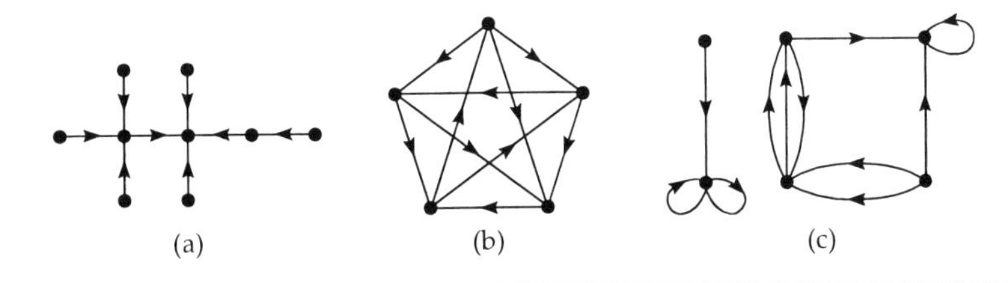

Problem 4.10

For each of the digraphs in Problem 4.9, write down:

the number of arcs;
the sum of the out-degrees of all the vertices;
the sum of the in-degrees of all the vertices.

What is the connection between your answers? Can you explain why this connection arises?

Handshaking Dilemma

In the solution to Problem 4.10, you should have noticed that the sum of the out-degrees and the sum of the in-degrees of each digraph are both equal to the number of arcs. A corresponding result holds for *any* digraph; we call it the *handshaking dilemma*!

Theorem 4.1: Handshaking Dilemma

In any digraph, the sum of all the out-degrees and the sum of all the in-degrees are both equal to the number of arcs.

Proof In any digraph, each arc has two ends, so it contributes exactly 1 to the sum of the out-degrees and exactly 1 to the sum of the in-degrees. The result follows immediately. ■

Problem 4.11

(a) Use the handshaking dilemma to prove that, in any digraph, if the number of vertices with odd out-degree is odd, then the number of vertices with odd in-degree is odd.
(b) Verify that the result of part (a) holds for the digraph in Problem 4.12.

4.3 Paths and Cycles

Just as you may be able to get from one vertex of a graph to another by tracing the edges of a walk, trail or path, so you may be able to get from one vertex of a digraph to another by tracing the arcs of a 'directed' walk, trail or path. This means that you have to follow the directions of the arcs as you go, just as if you were driving around a one-way street system in a town. We make this idea precise, as follows.

Definitions

A **walk of length *k*** in a digraph is a succession of k arcs of the form *uv*, *vw*, *wx*, ..., *yz*. This walk is denoted by *uvwx* ... *yz*, and is referred to as a **walk from *u* to *z***.

A **trail** is a walk in which all the arcs, but not necessarily all the vertices, are different.

A **path** is a walk in which all the arcs and all the vertices are different.

In the following diagram, the walk *vwxyvwyzzu* is a walk of length 9 from *v* to *u*, which includes the arc *vw* twice and the vertices *v*, *w*, *y* and *z* twice. The walk *uvwyvz* is a trail which is not a path, since the vertex *v* occurs twice, whereas the walk *vwxyz* has no repeated vertices and is therefore a path.

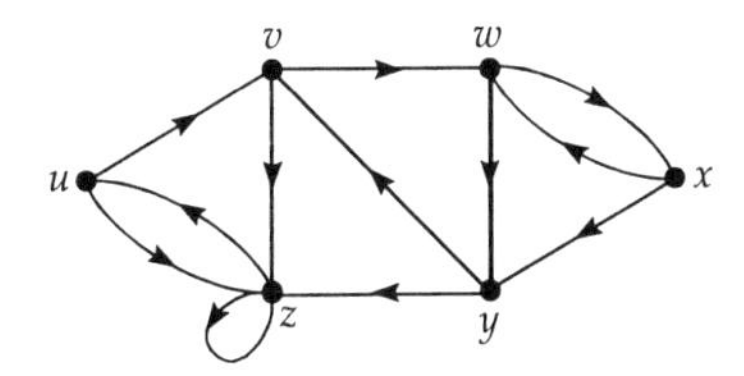

The terms *closed walk*, *closed trail* and *cycle* also apply to digraphs.

Definitions

A **closed walk** in a digraph is a succession of arcs of the form

uv, *vw*, *wx*, ..., *yz*, *zu*.

A **closed trail** is a closed walk in which all the arcs are different.

A **cycle** is a closed trail in which all the intermediate vertices are different.

In the digraph above, the closed walk *uvwyvzu* is a closed trail which is not a cycle (since the vertex *v* occurs twice), whereas the closed trails *zz*, *wxw*, *vwxyv* and *uvwxyzu* are all cycles. In describing closed walks, we can allow any vertex to be the starting vertex. For example, the triangle *vwyv* can also be written as *wyvw* or *yvwy*.

Problem 4.12

For the digraph on the right, write down:

(a) all the paths from *t* to *w*;

(b) all the paths from *w* to *t*;

(c) a closed trail of length 8 containing *t* and *z*.

(d) all the cycles containing both *t* and *w*.

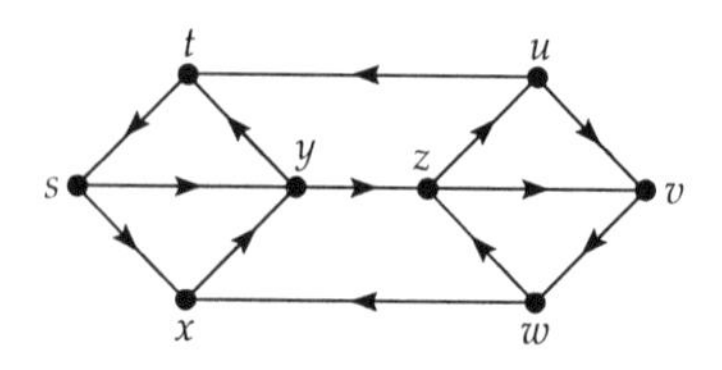

As with graphs, we can use the concept of a path to tell us whether or not a digraph is connected. Recall that a graph is connected if it is 'in one piece', and this means that there is a path between each pair of vertices. For digraphs *these two ideas are not the same,* and this leads to two different definitions of the word *connected* for digraphs.

Definitions

A digraph is **connected** if its underlying graph is a connected graph, and is **disconnected** otherwise.

A digraph is **strongly connected** if there is a path between each pair of vertices.

These three types of digraph are illustrated below:

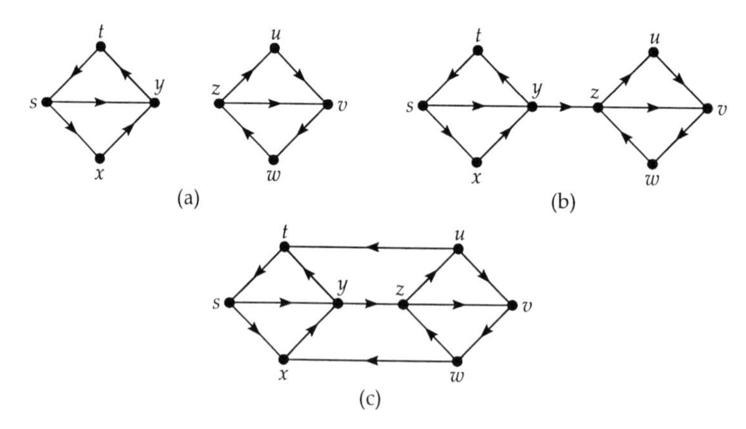

Digraph (a) is disconnected, since its underlying graph is a disconnected graph. Digraph (b) is connected but is not strongly connected since, for example, there is no path from z to y. Digraph (c) is strongly connected, since there are paths joining all pairs of vertices.

Alternatively, you can think of driving around a one-way street system in a town. If the town is strongly connected, then you can drive from any part of the town to any other, following the directions of the one-way streets as you go; if the town is merely connected, then you can still drive from any part of the town to any other, but you may have to ignore the directions of the one-way streets!

Problem 4.13

Classify each of the following digraphs as disconnected, connected but not strongly connected, or strongly connected:

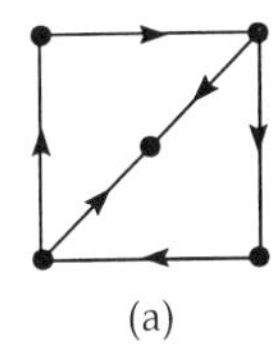

(a)

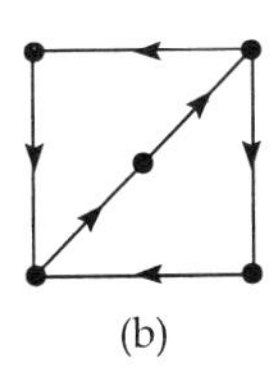

(b)

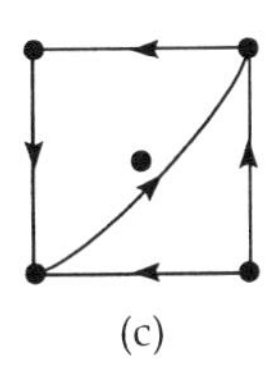

(c)

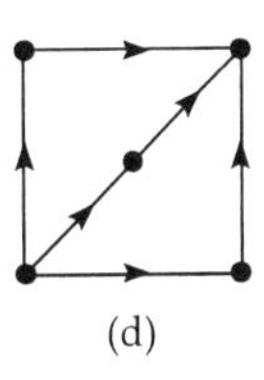

(d)

4.4 Eulerian and Hamiltonian Digraphs

In Chapter 3, we discussed the problem of finding a route that includes every edge or every vertex of a graph exactly once, and it is natural to consider the corresponding problem for digraphs. This leads to the following definitions.

Definitions

A connected digraph is **Eulerian** if it contains a closed trail that includes every arc; such a trail is an **Eulerian trail**.

A connected digraph is **Hamiltonian** if it contains a cycle that includes every vertex; such a cycle is a **Hamiltonian cycle**.

For example, consider the following four digraphs:

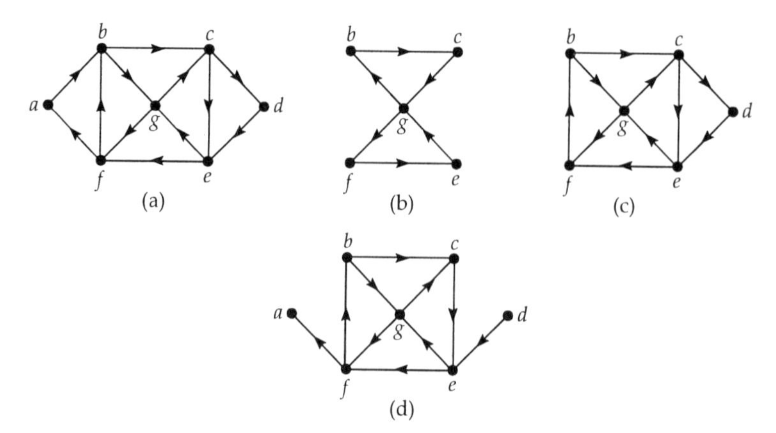

digraph (a) is Eulerian – an Eulerian trail is *a b c d e f b g c e g f a*
and Hamiltonian – a Hamiltonian cycle is *a b c d e g f a*;
digraph (b) is Eulerian – an Eulerian trail is *b c g f e g b*
it is not Hamiltonian;
digraph (c) is Hamiltonian – a Hamiltonian cycle is *b c d e g f b*
it is not Eulerian;
digraph (d) is neither Eulerian nor Hamiltonian.

Problem 4.14

Decide which of the following digraphs are Eulerian and/or Hamiltonian, and write down an Eulerian trail or Hamiltonian cycle where possible.

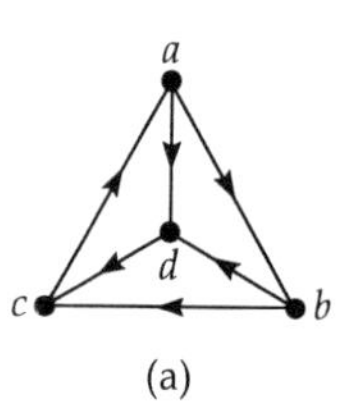

(a)

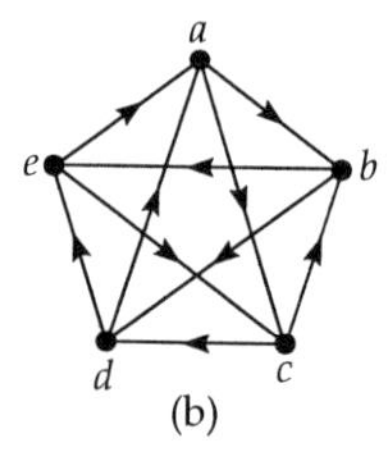

(b)

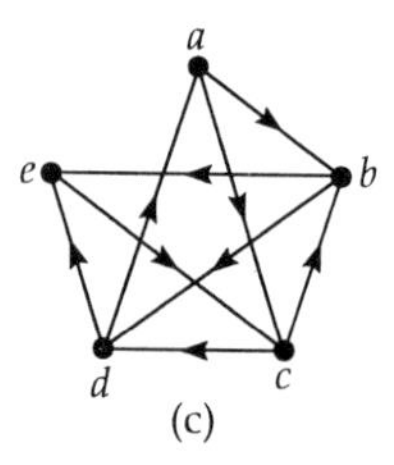

(c)

Much of the earlier discussion of Eulerian and Hamiltonian graphs can be adapted to Eulerian and Hamiltonian digraphs. In particular, there is an analogue of Theorem 3.2. We ask you to discover this analogue in the following problem.

Problem 4.15

(a) Guess a necessary and sufficient condition for a digraph to be Eulerian, involving the in-degree and out-degree of each vertex.
(b) Use the condition obtained in part (a) to check which of the digraphs in Problem 4.14 are Eulerian.

We now state the analogues of Theorem 3.2 and Theorem 3.3.

Theorem 4.2

A connected digraph is Eulerian if and only if, for each vertex, the out-degree equals the in-degree.

Theorem 4.3

An Eulerian digraph can be split into cycles, no two of which have an arc in common.

The proofs of these theorems are similar to those of Theorems 3.2 and 3.3. In the sufficiency part of the proof of Theorem 4.2, the basic idea is to show that the digraph contains a (directed) cycle, and then to build up the required Eulerian trail from cycles step by step, as in the proof of Theorem 3.2. We omit the details.

There is an analogue of Ore's theorem for Hamiltonian graphs, but it is harder to state and prove than the theorem for graphs, so we omit it.

4.5 Case Studies

We conclude this chapter with four case studies – ecology, social networks, the rotating drum problem (involving Eulerian digraphs) and tournaments (involving Hamiltonian digraphs).

Ecology

Snakes eat frogs, and birds eat spiders; birds and spiders both eat insects; frogs eat snails, spiders and insects. Given any such tangle of interrelationships between predator and prey, how do ecologists sort out the overall predatory behaviour of the various species they are investigating?

When studying relationships between animals and plants and their environment, ecologists sometimes use a digraph known as a *food web*. In such a digraph, the vertices correspond to the species under investigation, and there is an arc from a species A to a species B whenever A preys on B.

As an example of a food web, consider the following digraph, which represents the predatory habits of organisms in a Canadian willow forest.

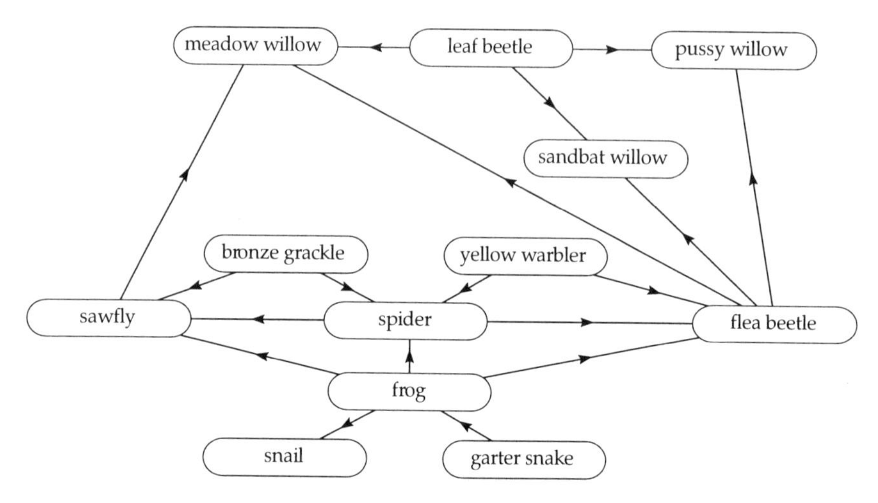

In untangling such food webs, ecologists introduce a graph that tells them which species compete for food. This graph is known as the **competition graph** or **niche overlap graph**, and its edges join pairs of vertices representing species that share a common prey. For example, in the above food web the bronze grackle and the yellow warbler both eat spiders, so the corresponding vertices are adjacent in the competition graph:

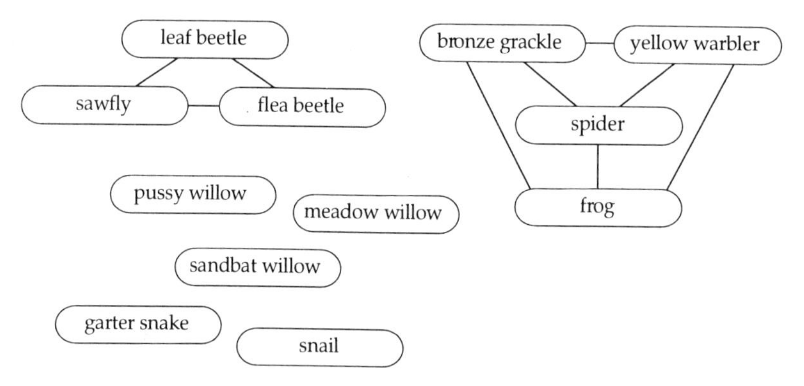

Such a representation has ecological significance in that adjacent vertices tend to correspond to species that react in the same way to particular environmental factors such as temperature, humidity or altitude. In the above example, the beetles and the sawfly have similar predatory behaviour, as do the birds, the spider and the frog.

Problem 4.16

Draw the competition graph of the following food web.

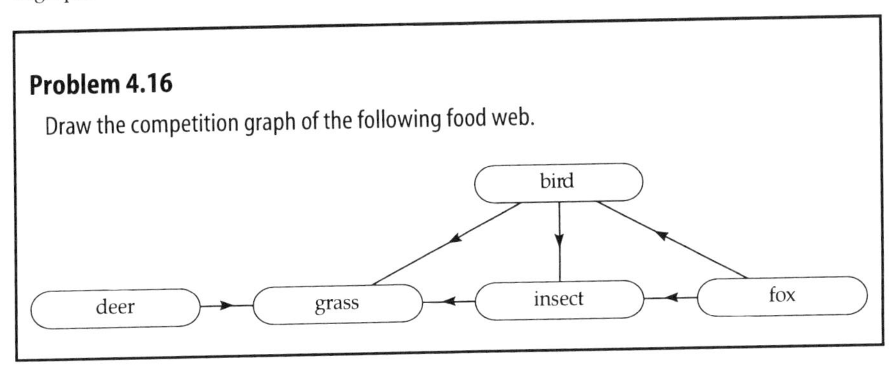

Social Networks

In Section 2.5 we described the use of signed graphs to represent *symmetric* relationships (x likes y if and only if y likes x). When some relationships are not symmetric (x likes y, but y dislikes x), we use a **signed digraph**. This is a digraph with either + or – associated with each arc, indicating a positive relationship (likes, supports, threatens, etc.) or a negative one (dislikes, is junior to, is afraid of, etc.). For example, in the signed digraph below, John and Jack like each other, Mary likes Jill but Jill dislikes Mary, John dislikes Jill but we have no information about Jill's feelings for John, and so on.

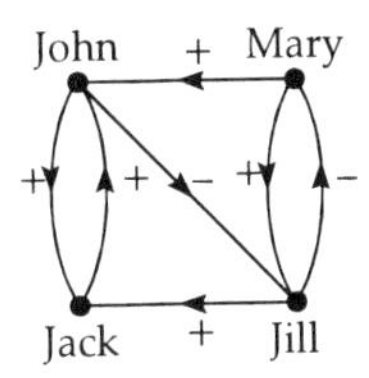

Note that a negative arc from x to y (Jill dislikes Mary) is *not* the same as a positive arc from y to x (Mary likes Jill).

Signed digraphs also have other uses. Many problems of modern society involve complex systems made up of a number of variables that constantly change and interact. Often we wish to predict the future development of the system when the amount of available information is minimal. For such situations, signed digraphs have proved to be convenient, and their use has led to precise and valid conclusions. In particular, they have successfully been applied to problems of waste disposal, energy planning, research funding, environmental contamination, allocation of medical resources, and so on. Although our discussion here is necessarily simplified, the ideas are also revelant to more complex examples.

The signed digraph below gives a simplified representation of the consequences of changes in energy use. The arc pu is marked positive, since an increase in population in a given area is likely to increase the amount of energy used; the arc ur is marked negative, since the more energy is used, the less it costs per unit. There is no arc from j to r, since an increase in the number of available jobs has no direct effect on the unit cost of electricity.

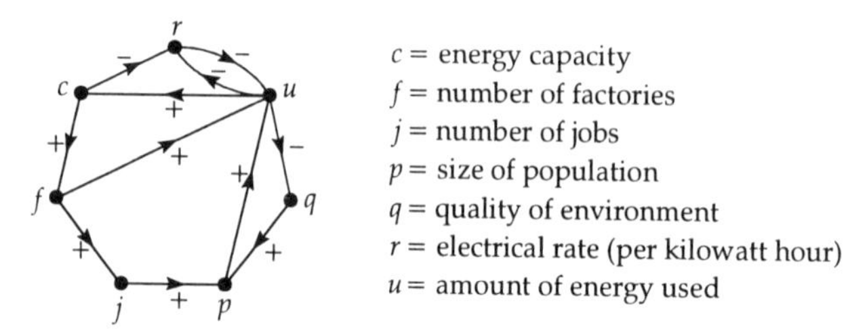

When following walks through such a signed digraph, we need to be careful. For example, the *negative* sign on the arc uq tells us that an *increase* in the amount of energy used leads to a *decrease* in the quality of environment, and the *positive* sign on the arc qp tells us that an *increase* in the quality of environment leads to an *increase* in the population. However, if we follow these arcs consecutively, from u to p, the first tells us that an *increase* in the amount of energy use leads to a *decrease* in the quality of environment, as before; but we now need to interpret the second arc differently – the *decrease* in the quality of environment together with the *positive* sign on the arc qp must be interpreted as leading to a *decrease* in the population. Similarly, following the *negative* arc cr and then the *negative* arc ru must be interpreted as an increase in the energy capacity leading to a *decrease* in the electrical rate, leading to an *increase* in the amount of electricity used.

Of particular interest in this digraph are the cycles. An *increase* in population (p) results in an *increase* in the amount of energy used (u), which in turn produces a *decrease* in the quality of environment (q), which then tends to *decrease* the population (p). A cycle of this kind, in which an *increase* in any variable (p) ultimately gives rise to a *decrease* in the same variable, is called a **negative feedback cycle**; thus the cycle $puqp$ is a negative feedback cycle. On the other hand, an *increase* in the energy capacity (c) tends to lead to an *increase* in the number of factories (f), leading to an *increase* in the amount of energy used (u), thereby *increasing* the energy capacity still further (c). A cycle of this kind, in which an *increase* in any variable (c) ultimately gives rise to a *further increase* in the same variable, is called a **positive feedback cycle**; thus the cycle $cfuc$ is a positive feedback cycle.

It is easy to see whether a given cycle is a positive or a negative feedback cycle, since

every *positive* feedback cycle has an *even* number of negative arcs,

whereas

every *negative* feedback cycle has an *odd* number of negative arcs.

In a positive feedback cycle, whenever an increase or decrease is counteracted by a negative arc, the counteraction is itself counteracted by the next negative arc. In a negative feedback cycle, one counteraction is never counteracted.

By counting the negative arcs in each cycle and using the above criterion, we see that the cycle *uqpu* is a negative feedback cycle, whereas the cycles *cruc*, *cfuc*, *rur* and *cfjpuc* are all positive feedback cycles. The existence of several positive feedback cycles containing the vertex *c* explains why the electrical energy demand system is so unstable, in the sense that initial increases in energy capacity lead eventually to further increases of the same kind. Although this had been observed empirically by environmentalists, the signed digraph representation tells us, from a structural point of view, why it occurs.

Even with such a simple model as this, we can make some remarkably accurate predictions. Some of the variables (such as 'quality of environment') may be difficult or impossible to measure, but this makes little difference to the conclusions we can draw.

Problem 4.17

The following signed digraph is adapted from a study by the Organization for Economic Co-operation and Development into the support that governments should provide for the funding of research projects in science and technology.

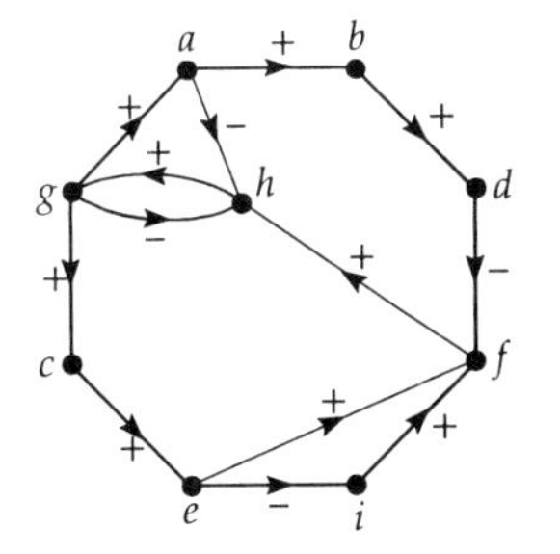

a = number of available jobs
b = number of poorly trained researchers
c = number of well trained researchers
d = amount of 'bad science' produced
e = amount of 'good science' produced
f = public opinion in favour of science
g = amount of available budget
h = pressure to increase budget
i = external or internal threats to society that call for science to alleviate them

List as many positive and negative feedback cycles as you can.

Rotating Drum Problem

A problem that has arisen in telecommunications (and also in cryptography and the design of washing machines) is the *rotating drum problem* or *teleprinter's problem*.

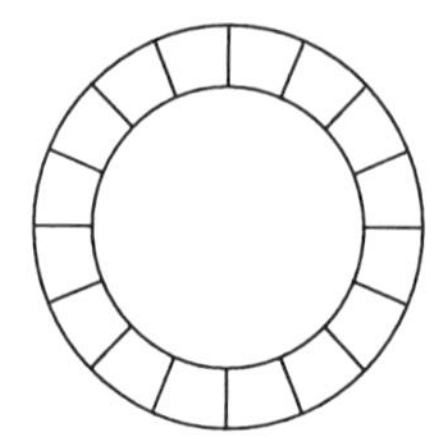

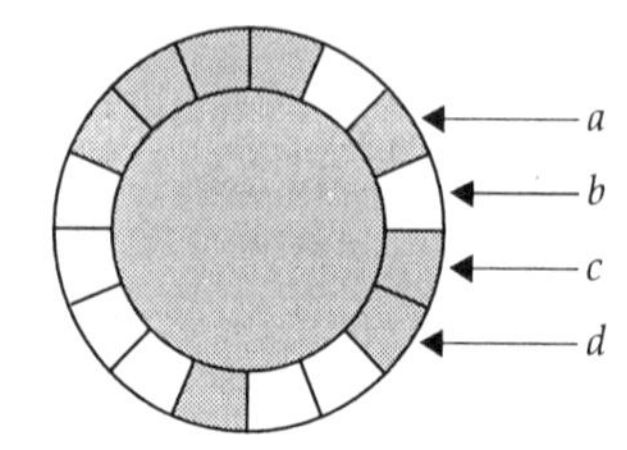

The surface of a rotating drum is divided into sixteen parts, as shown on the left; the shaded areas represent conducting materials and the unshaded areas represent non-conducting materials. We represent the position of the drum by four binary digits *a, b, c* and *d,* as indicated on the right. Depending on the position of the drum, the terminals represented by *a, b, c* and *d* are either earthed or insulated from the earth – for example, in the above diagram, the earthed terminals are *a, c* and *d.* The earthed terminals emit a signal represented by 1, and the insulated terminals emit a signal represented by 0.

In order that each of the sixteen positions of the drum may be represented *uniquely* by a four-digit binary word *abcd,* the conducting and non-conducting materials must be assigned to the sixteen positions in such a way that all sixteen possible four-digit binary words *abcd* occur. Can this be done? If so, how can it be arranged?

A solution is given in the right-hand diagram above. The position shown corresponds to the four-digit binary word 1011. Rotating the drum anticlockwise successively gives the binary words

0110, 1100, 1001, 0010, 0100, 1000, 0000, 0001,
0011, 0111, 1111, 1110, 1101, 1010, 0101, 1011.

These four-digit binary words are all different, and represent all sixteen positions of the drum.

In order to obtain this and other solutions, we construct a digraph with eight vertices, corresponding to the three-digit binary words

000, 001, 010, 011, 100, 101, 110, 111,

and with arcs from each vertex *abc* to the vertices *bc*0 and *bc*1.

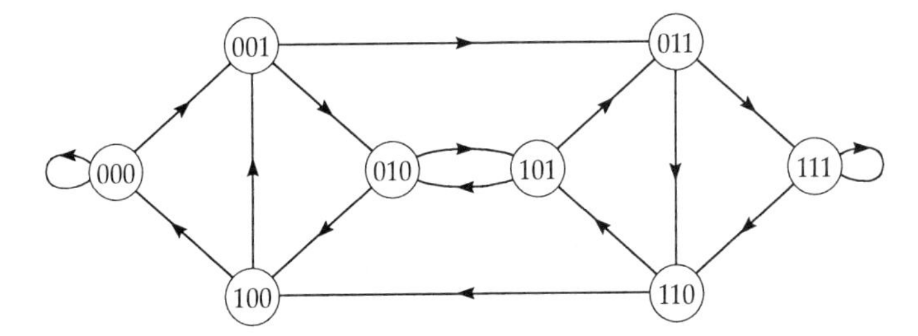

This digraph is clearly Eulerian, since the out-degree and in-degree of each vertex are both equal to 2. Any Eulerian trail gives a solution to the rotating drum problem. For example, if we take the Eulerian trail

101 → 011 → 110 → 100 → 001 → 010 → 100 → 000 →

000 → 001 → 011 → 111 → 111 → 110 → 101 → 010 → 101,

then we can 'compress' consecutive terms cumulatively (for example, the first three terms **101** → 01**1** → 11**0** compress to **10110**) to give the sequence

1011001000011110.

This gives the following circular arrangement of positions:

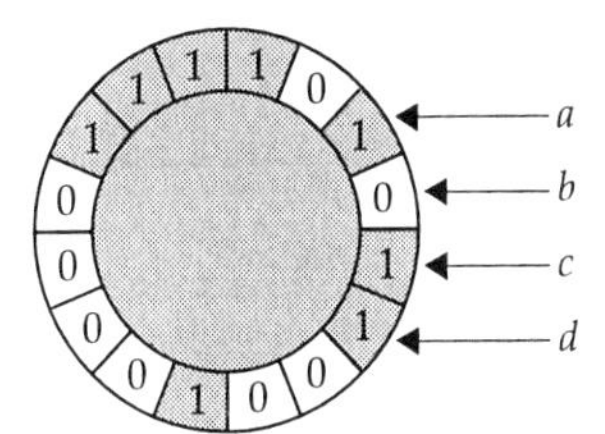

Using a similar argument, we can answer the corresponding question for rotating drums with 32, 64, ... divisions.

Problem 4.18

Find a different Eulerian trail in the above digraph, and hence construct a different solution to the rotating drum problem for sixteen divisions.

Ranking in Tournaments

We conclude this chapter with an application of Hamiltonian digraphs that arises in statistics.

A **tournament** is a digraph whose underlying graph is a complete graph. For example, the following diagram shows tournaments with 3 and 4 vertices:

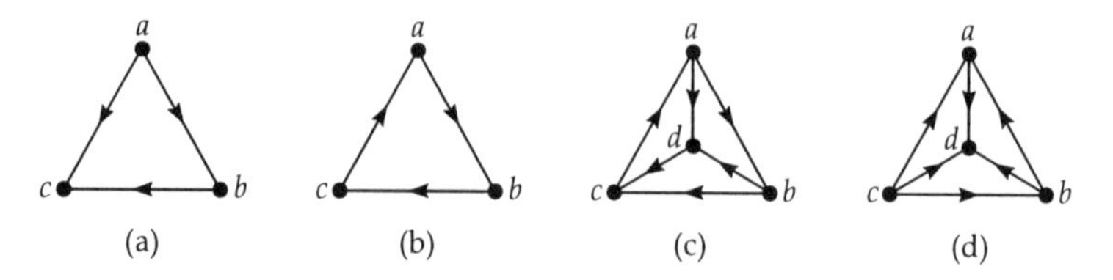

Such a digraph can be used to record the winners in a round-robin tournament in which each player plays each of the others. For example:

in tournament (a), a beats both b and c, and b beats c;
in tournament (d), c beats a, d and b; b beats a and d; and a beats d.

Tournaments also arise in other contexts, such as in the *method of paired comparisons*, in which we compare a number of commodities by testing them in pairs. For example, consider the following tournament, used for comparing six types of dog food. These delicacies were tested in pairs on a number of dogs, and the following preferences were recorded.

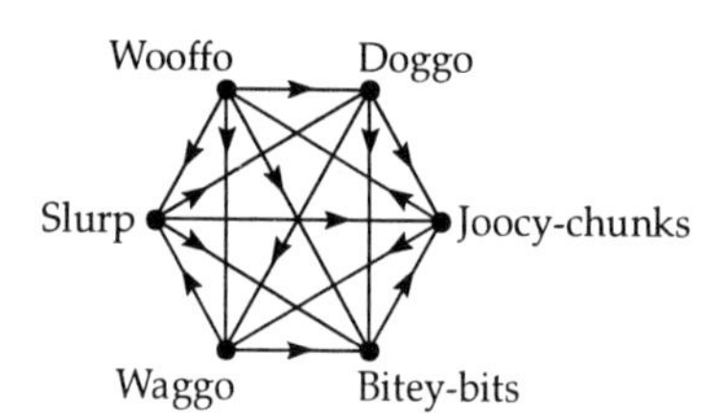

The problem now arises as to how we rank the various commodities in order of preference. For some tournaments there is no difficulty, since we can order them in such a way that each vertex 'beats' the others beneath it; for example, in tournaments (a) and (d) we can rank the participants in this way, as shown below.

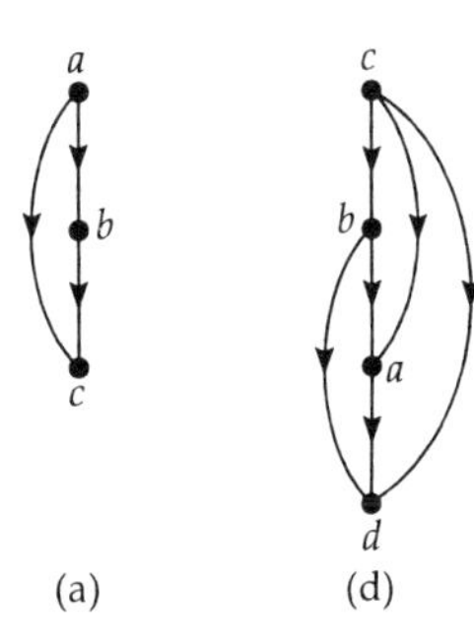

Unfortunately, in many practical examples a consistent ranking is impossible; for example, in tournament (b), a beats b, b beats c, and c beats a, so it is not possible to rank these three players directly. There is a similar inconsistency in the dog-food example, where Wooffo was preferred to Doggo, Doggo was preferred to Joocy-chunks, and Joocy-chunks was preferred to Wooffo. For such tournaments we must find alternative methods for ranking the participants or commodities.

In such circumstances, no method is entirely satisfactory, but a method that has been much used in practice is to look for paths containing each vertex – that is, semi-Hamiltonian paths. It can be proved that every tournament has at least one path of this kind, and each such path leads to a ranking.

For example, in tournament (c), possible rankings include a, b, d, c and b, c, a, d, whereas for the dog-food example, possible rankings include:

Wooffo, Doggo, Joocy-chunks, Waggo, Slurp, Bitey-bits

and

Bitey-bits, Joocy-chunks, Wooffo, Doggo, Waggo, Slurp.

Once we have listed all the possible rankings of this kind, we then take other considerations into account in deciding which ranking is best for our purposes.

Problem 4.19

How many rankings are possible in the following tournament?

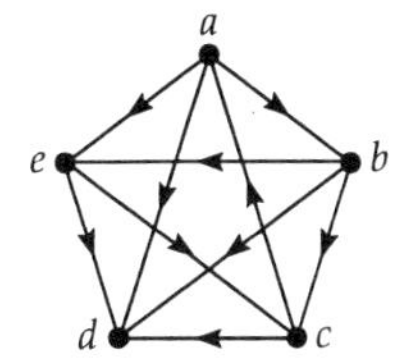

Exercises 4

Digraphs and Subdigraphs

4.1 Consider the digraph D shown on the right. Which of the following statements hold for D?

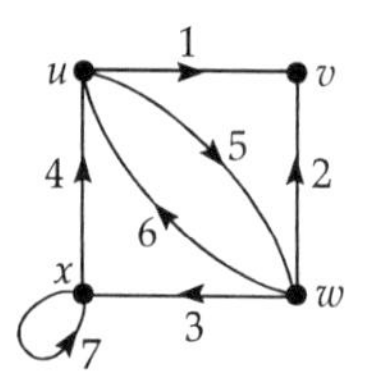

(a) vertices u and x are adjacent;
(b) arc 2 is incident to vertex w;
(c) vertex x is incident from arc 3;
(d) vertex x and arc 7 form a subdigraph of D.

4.2 Of the following four digraphs, which two are the same, which one is isomorphic to these two, and which is not isomorphic to any of the others?

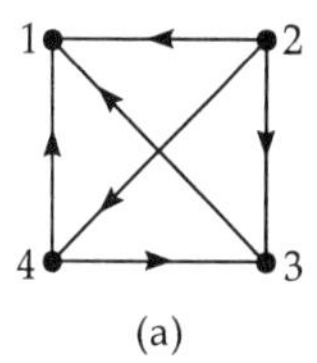

(a)

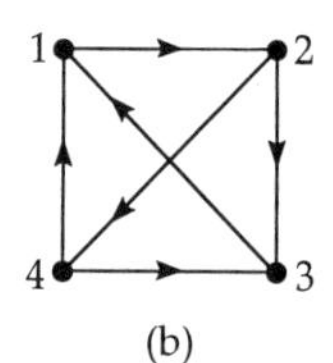

(b)

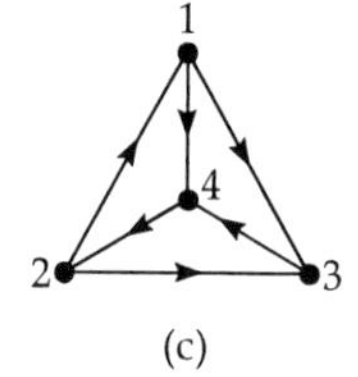

(c)

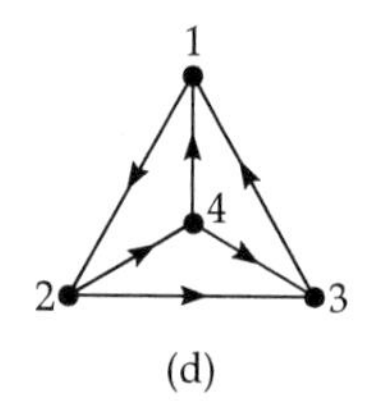

(d)

4.3 Draw two non-isomorphic non-simple digraphs, each with 4 vertices and 7 arcs. Explain why your digraphs are not isomorphic.

Vertex Degrees

4.4 Write down the out-degree sequence and the in-degree sequence for each of the digraphs in Exercise 4.2.

4.5 (a) If two digraphs have the same out-degree sequence and the same in-degree sequence, must they be isomorphic?
(b) If two digraphs are isomorphic, must they have the same out-degree sequence and the same in-degree sequence ?

4.6 Draw a digraph with 4 vertices and 7 arcs such that the number of vertices with odd out-degree is odd and the number of vertices with odd in-degree is odd.

Paths and Cycles

4.7 For the digraph shown on the right, write down (if possible):

(a) a walk of length 7 from u to w;

(b) cycles of lengths 1, 2, 3 and 4;

(c) a path of maximum length.

4.8 Draw four connected digraphs, D_1, D_2, D_3 and D_4, each with 5 vertices and 8 arcs, satisfying the following conditions:

D_1 is a simple digraph;

D_2 is a non-simple digraph with no loops;

D_3 is a digraph with both loops and multiple arcs;

D_4 is strongly connected.

4.9 Classify each of the following digraphs as disconnected, connected but not strongly connected, or strongly connected:

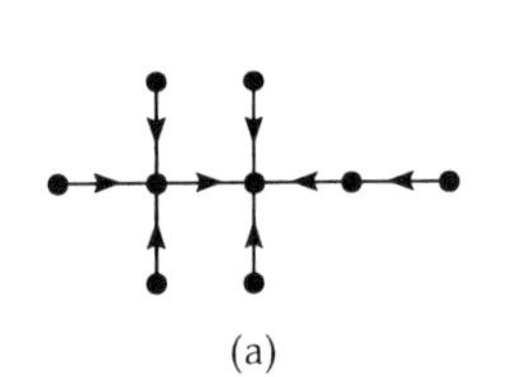

(a)

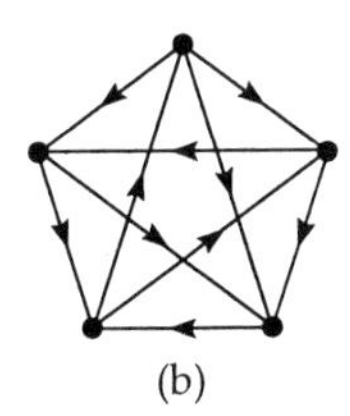

(b)

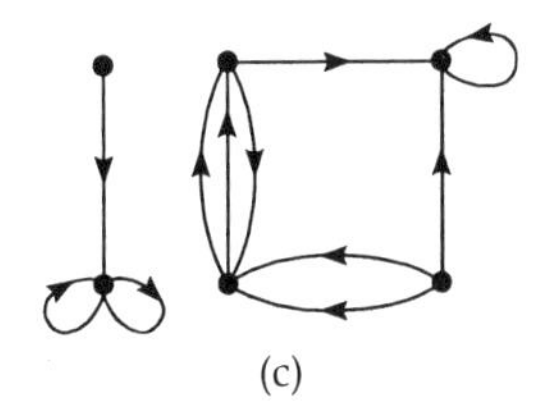

(c)

4.10 A graph is **orientable** if a direction can be assigned to each edge in such a way that the resulting digraph is strongly connected. Show that K_5 and the Petersen graph are orientable, and find a graph that is not.

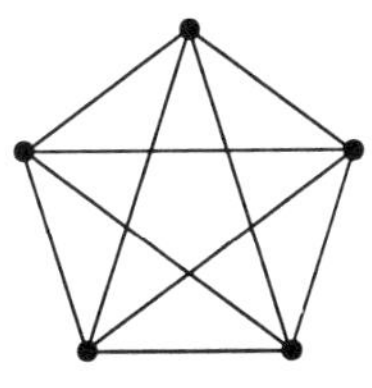

K_5

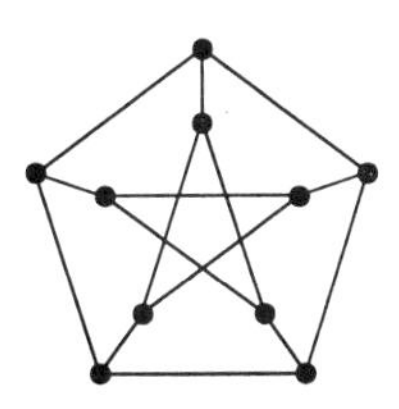

Petersen

Eulerian and Hamiltonian Digraphs

4.11 Are the following digraphs Eulerian? Hamiltonian?

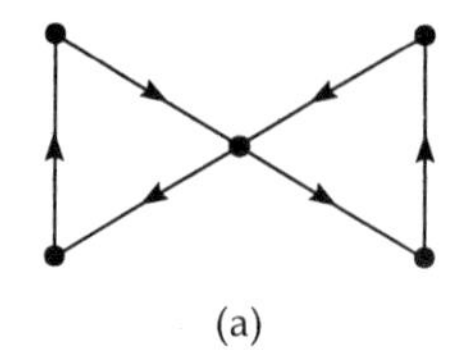

(a)

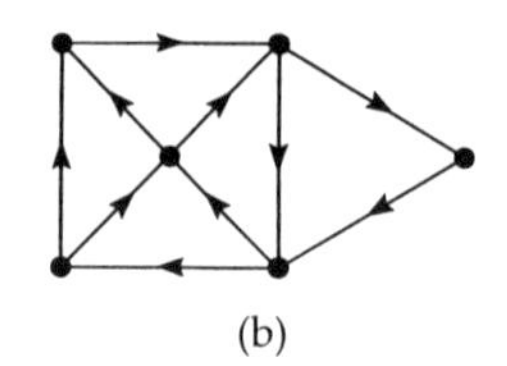

(b)

4.12 In the digraph on the right, find:

(a) all cycles of lengths 3, 4 and 5;
(b) an Eulerian trail;
(c) a Hamiltonian cycle.

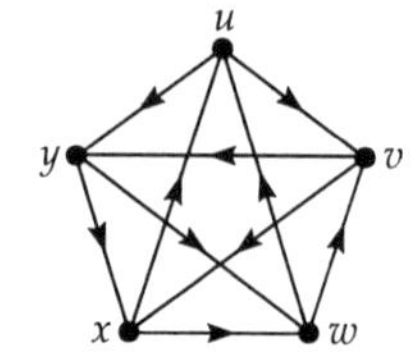

Case Studies

Ecology

4.13 A food web involves ten species, 1, ..., 10, with the following behaviour:

species 1 eats species 3, 7, 10;
species 2 eats species 7;
species 3 eats species 7;
species 4 eats species 2, 5, 6, 8, 10;
species 5 eats species 6;
species 6 eats species 7;
species 8 eats species 7;
species 9 eats species 2, 6.

(a) Draw the corresponding food web and competition graph.
(b) Draw the food web and competition graph that result if species 7 dies out.

Social Networks

4.14 The following signed digraph was used in a transport study in Vancouver to determine whether a large increase in the funding of public transport could make city travelling easier:

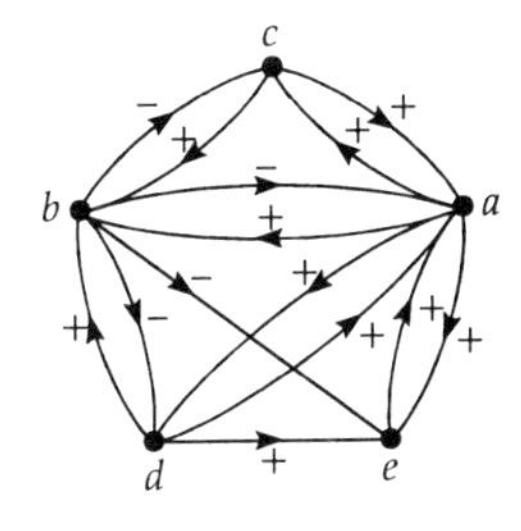

a = cost of an automobile
b = amount of automobile use
c = convenience of automobile use
d = freedom of choice in travel time
e = speed

Determine whether each of the following cycles is a positive or a negative feedback cycle:

(a) *abca*; (b) *beacb*; (c) *adea*.

Rotating Drum Problem

4.15 Solve the rotating drum problem for a drum with 32 divisions.

Ranking in Tournaments

4.16 Draw all the tournaments with four vertices.
In which of them can the participants be ranked in just one way?

4.17 How many rankings are possible in the following tournament?

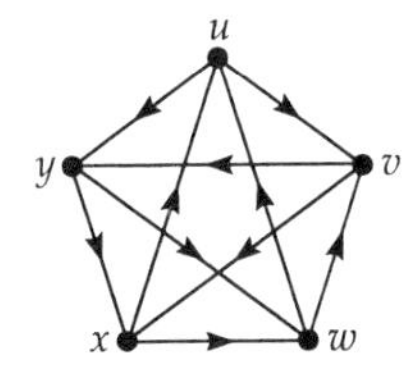

Chapter 5

Matrix Representations

After studying this chapter, you should be able to:

- write down the *adjacency matrix* and *incidence matrix* of a given labelled graph or digraph, and draw the graph or digraph with a given adjacency or incidence matrix;
- use an adjacency matrix to determine the number of walks between two given vertices in a graph or digraph;
- use an adjacency matrix to determine whether a given graph/digraph is connected/strongly connected;
- describe the connections between adjacency matrices and problems in archaeology and genetics;
- describe the connections between adjacency matrices and Markov chains.

Up to now, you have seen two ways of representing a graph or digraph – as a diagram of points joined by lines, and as a set of vertices and a set of edges or arcs. The pictorial representation is useful in many situations, especially when we wish to examine the structure of the graph or digraph as a whole, but its value diminishes as soon as we need to describe large or complicated graphs and digraphs. For example, if we need to store a large graph in a computer, then a pictorial representation is unsuitable and some other representation is necessary.

One possibility is to store the set of vertices and the set of edges or arcs. This method is often used, especially when the graph or digraph is 'sparse', with many vertices but relatively few edges or arcs. Another method is to take each vertex in turn and list those vertices adjacent to it; by joining each vertex to its neighbours, we can reconstruct the graph or digraph. Yet another method is to give a table indicating which pairs of vertices are adjacent, or indicating which vertices are incident to which edges or arcs.

Each of these methods has its advantages, but the last one is particularly useful. Using this method, we represent each graph or digraph by a rectangular array of numbers, called a *matrix*. Such matrices lend themselves to computational techniques, and are often the most natural way of formulating a problem. There are various types of matrix that we can use to specify a given graph or digraph. Here we describe the two simplest types – the *adjacency matrix* and the *incidence matrix*.

5.1 Adjacency Matrices

Consider the following example:

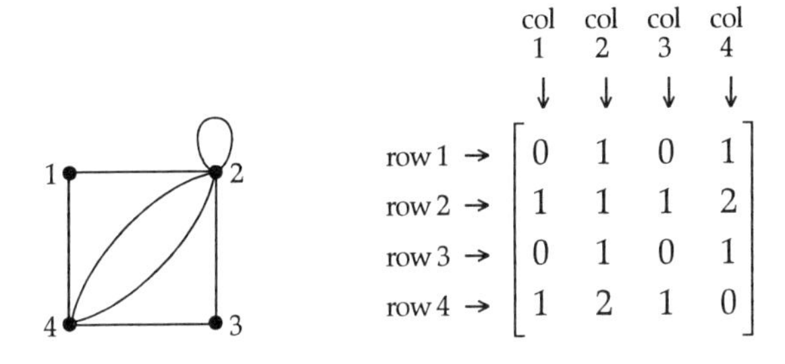

On the left we have a graph with *four* labelled vertices, and on the right we have a matrix with *four* rows and *four* columns – that is, a 4×4 matrix. The numbers appearing in the matrix refer to the number of edges joining the corresponding vertices in the graph. For example,

vertices 1 and 2 are joined by **1** edge,
so **1** appears in row 1 column 2, and in row 2 column 1;
vertices 2 and 4 are joined by **2** edges,
so **2** appears in row 2 column 4, and in row 4 column 2;
vertices 1 and 3 are joined by **0** edges,
so **0** appears in row 1 column 3, and in row 3 column 1;
vertex 2 is joined to itself by **1** edge,
so **1** appears in row 2 column 2.

We generalize this idea, as follows.

Definition

Let G be a graph with n vertices labelled $1, 2, 3, ..., n$.
The **adjacency matrix A**(G) of G is the $n \times n$ matrix in which the entry in row i and column j is the number of edges joining the vertices i and j.

Problem 5.1

Write down the adjacency matrix of each of the following graphs:

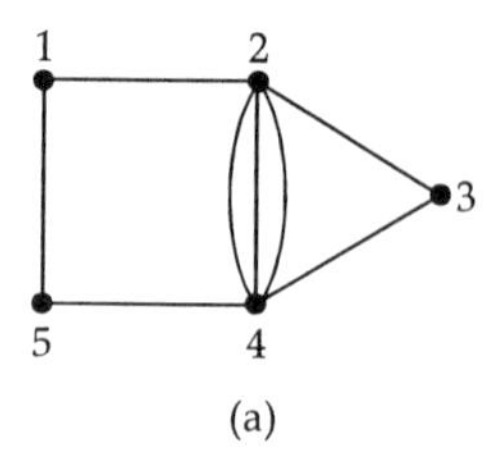

(a)

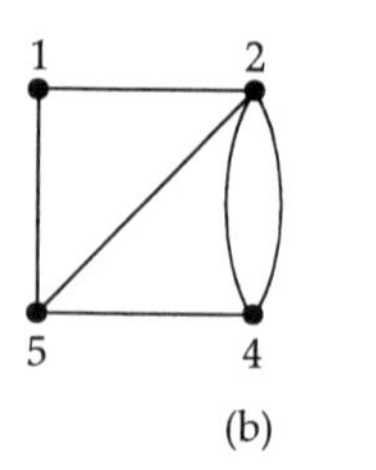

(b)

Problem 5.2

Draw the graph represented by each of the following adjacency matrices:

$$\begin{array}{c} \\ 1 \\ 2 \\ 3 \\ 4 \\ 5 \end{array}\begin{array}{c} \begin{array}{ccccc} 1 & 2 & 3 & 4 & 5 \end{array} \\ \begin{bmatrix} 0 & 2 & 0 & 1 & 1 \\ 2 & 0 & 0 & 1 & 1 \\ 0 & 0 & 0 & 0 & 0 \\ 1 & 1 & 0 & 0 & 2 \\ 1 & 1 & 0 & 2 & 0 \end{bmatrix} \end{array}$$

(a)

$$\begin{array}{c} \\ 1 \\ 2 \\ 3 \\ 4 \\ 5 \\ 6 \end{array}\begin{array}{c} \begin{array}{cccccc} 1 & 2 & 3 & 4 & 5 & 6 \end{array} \\ \begin{bmatrix} 0 & 1 & 1 & 1 & 0 & 0 \\ 1 & 0 & 0 & 1 & 0 & 0 \\ 1 & 0 & 0 & 1 & 0 & 0 \\ 1 & 1 & 1 & 0 & 0 & 0 \\ 0 & 0 & 0 & 0 & 0 & 1 \\ 0 & 0 & 0 & 0 & 1 & 0 \end{bmatrix} \end{array}$$

(b)

The adjacency matrix of a graph is symmetrical about the *main diagonal* (top-left to bottom-right). Also, for a graph without loops, each entry on the main diagonal is 0, and the sum of the entries in any row or column is the degree of the vertex corresponding to that row or column.

The representation of a graph by an adjacency matrix has a digraph analogue that is frequently used when storing large digraphs in a computer. When defining the adjacency matrix of a digraph, we have to take into account the directions of the arcs.

Consider the following example:

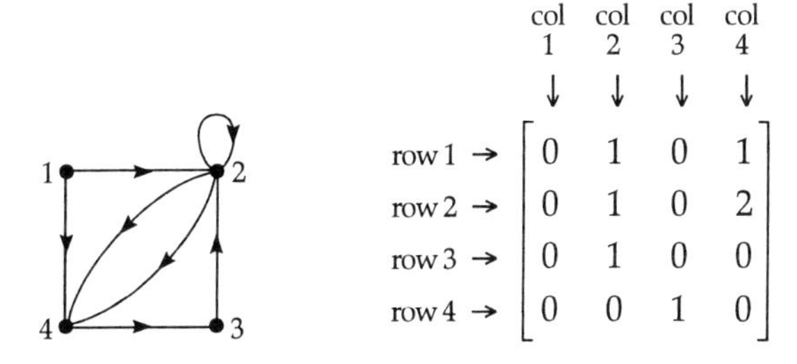

$$\begin{bmatrix} 0 & 1 & 0 & 1 \\ 0 & 1 & 0 & 2 \\ 0 & 1 & 0 & 0 \\ 0 & 0 & 1 & 0 \end{bmatrix}$$

On the left we have a digraph with *four labelled vertices,* and on the right we have a matrix with *four rows* and *four columns*. The numbers appearing in the matrix refer to the number of arcs joining the corresponding vertices in the digraph. For example,

vertices 1 and 2 are joined (in that order) by **1** arc,
so **1** appears in row 1 column 2;
vertices 2 and 4 are joined (in that order) by **2** arcs,
so **2** appears in row 2 column 4;
vertices 4 and 1 are joined (in that order) by **0** arcs,
so **0** appears in row 4 column 1;
vertex 2 is joined to itself by **1** arc,
so **1** appears in row 2 column 2.

We generalize this idea, as follows.

Definition

Let D be a digraph with n vertices labelled 1, 2, 3, ..., n.
The **adjacency matrix** $\boldsymbol{A}(D)$ of D is the $n \times n$ matrix in which the entry in row i and column j is the number of arcs from vertex i to vertex j.

Problem 5.3

Write down the adjacency matrix of each of the following digraphs:

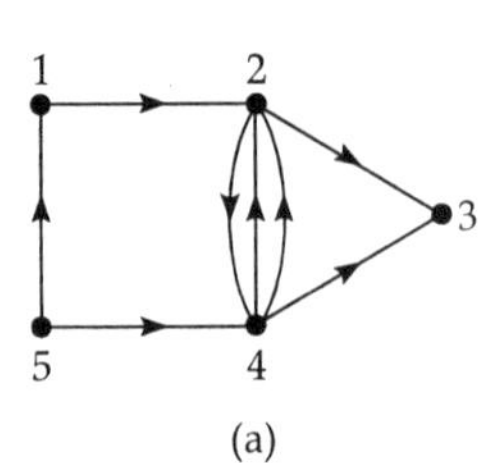

(a)

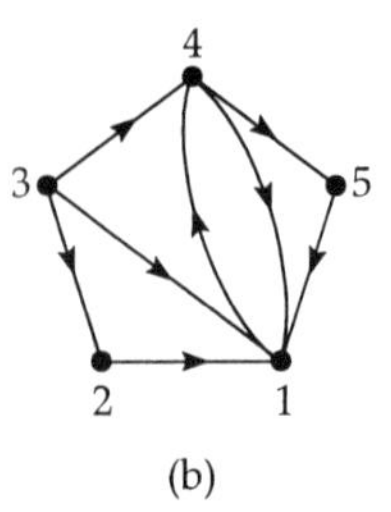

(b)

Problem 5.4

Draw the digraph represented by each of the following adjacency matrices:

$$\begin{array}{c} \\ 1 \\ 2 \\ 3 \\ 4 \\ 5 \end{array}\begin{array}{c} \begin{array}{ccccc} 1 & 2 & 3 & 4 & 5 \end{array} \\ \begin{bmatrix} 0 & 1 & 0 & 0 & 1 \\ 1 & 0 & 0 & 1 & 0 \\ 0 & 0 & 0 & 0 & 0 \\ 1 & 0 & 0 & 0 & 0 \\ 0 & 1 & 0 & 2 & 0 \end{bmatrix} \end{array}$$

(a)

$$\begin{array}{c} \\ 1 \\ 2 \\ 3 \\ 4 \\ 5 \end{array}\begin{array}{c} \begin{array}{ccccc} 1 & 2 & 3 & 4 & 5 \end{array} \\ \begin{bmatrix} 0 & 0 & 0 & 0 & 1 \\ 0 & 0 & 0 & 0 & 1 \\ 1 & 0 & 0 & 0 & 1 \\ 1 & 1 & 0 & 0 & 0 \\ 1 & 0 & 0 & 1 & 0 \end{bmatrix} \end{array}$$

(b)

The adjacency matrix of a digraph is not usually symmetrical about the main diagonal. Also, if the digraph has no loops, then each entry on the main diagonal is 0, the sum of the entries in any row is the out-degree of the vertex corresponding to that row, and the sum of the numbers in any column is the in-degree of the vertex corresponding to that column.

5.2 Walks in Graphs and Digraphs

We can establish the existence of walks in a graph or digraph by using the adjacency matrix. In the following, we restrict our attention to digraphs: similar results can be derived for graphs.

Consider the following digraph and table:

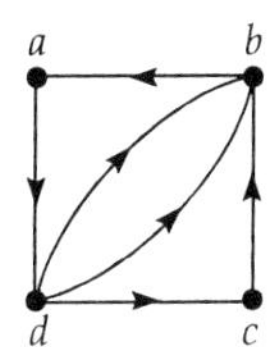

	a	*b*	*c*	*d*
a	0	0	0	1
b	1	0	0	0
c	0	1	0	0
d	0	2	1	0

The table shows the number of walks of length 1 between each pair of vertices. For example,

the number of walks of length 1 from a to c is **0**,
so **0** appears in row 1 column 3;
the number of walks of length 1 from b to a is **1**,
so **1** appears in row 2 column 1;
the number of walks of length 1 from d to b is **2**,
so **2** appears in row 4 column 2.

Now a walk of length 1 is an arc, so the table above is the *adjacency matrix* **A** of the digraph:

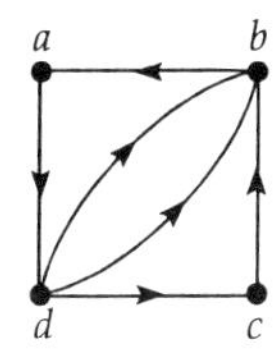

$$\begin{bmatrix} 0 & 0 & 0 & 1 \\ 1 & 0 & 0 & 0 \\ 0 & 1 & 0 & 0 \\ 0 & 2 & 1 & 0 \end{bmatrix}$$

adjacency matrix **A**

Next, we consider walks of lengths 2 and 3. For example, there are two different walks of length 2 from a to b, because there is one arc from a to d and two arcs from d to b. Similarly, there are two different walks of length 3 from d to d, since there are two arcs from d to b, and one walk of length 2 from b to d, namely, *bad*.

Problem 5.5

(a) Complete the following tables for the numbers of walks of lengths 2 and 3 in the above digraph.

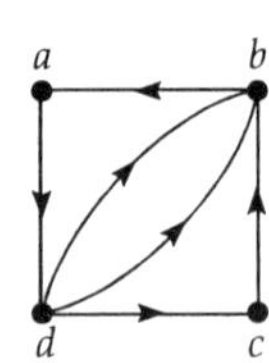

	a	b	c	d
a		2		
b				1
c				
d				

numbers of walks of length 2

	a	b	c	d
a				
b				
c				
d				2

numbers of walks of length 3

(b) Find the matrix products $\mathbf{A}^2$ and $\mathbf{A}^3$, where $\mathbf{A}$ is the adjacency matrix of the above digraph.

(c) Comment on your results.

The solution to the above problem illustrates the following theorem; the proof is given at the end of this section.

Theorem 5.1

Let D be a digraph with n vertices labelled $1, 2, \ldots, n$, let $\mathbf{A}$ be its adjacency matrix with respect to this listing of the vertices, and let k be any positive integer.
Then the number of walks of length k from vertex i to vertex j is equal to the entry in row i and column j of the matrix $\mathbf{A}^k$ (the kth power of the matrix $\mathbf{A}$).

Problem 5.6

Consider the following digraph:

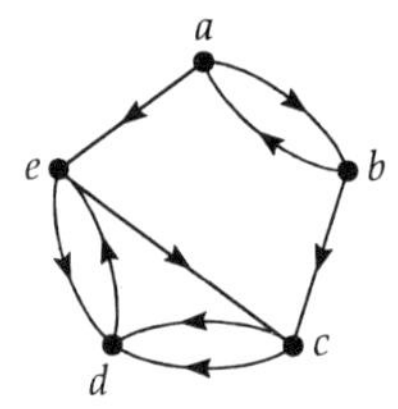

Write down the adjacency matrix $\mathbf{A}$, calculate the matrices $\mathbf{A}^2$, $\mathbf{A}^3$ and $\mathbf{A}^4$, and hence find the numbers of walks of lengths 1, 2, 3 and 4 from b to d. Are there walks of lengths 1, 2, 3 or 4 from d to b?

Theorem 5.1 also gives a method of determining whether a digraph is strongly connected, by working directly from its adjacency matrix.

Recall that a digraph is *strongly connected* if there is a path from vertex i to vertex j, for each pair of distinct vertices i and j, and that a path is a walk in which all the vertices are different. For example, in the digraph considered earlier, there are four vertices, so a path has length 1, 2 or 3. We have seen that the numbers of walks (including the paths) of lengths 1, 2 and 3 between pairs of distinct vertices are given by the non-diagonal entries in the matrices

$$\mathbf{A} = \begin{bmatrix} 0 & 0 & 0 & 1 \\ 1 & 0 & 0 & 0 \\ 0 & 1 & 0 & 0 \\ 0 & 2 & 1 & 0 \end{bmatrix}, \quad \mathbf{A}^2 = \begin{bmatrix} 0 & 2 & 1 & 0 \\ 0 & 0 & 0 & 1 \\ 1 & 0 & 0 & 0 \\ 2 & 1 & 0 & 0 \end{bmatrix}, \quad \mathbf{A}^3 = \begin{bmatrix} 2 & 1 & 0 & 0 \\ 0 & 2 & 1 & 0 \\ 0 & 0 & 0 & 1 \\ 1 & 0 & 0 & 2 \end{bmatrix}.$$

By examining these matrices, we can see that each pair of distinct vertices is indeed joined by at least one path of length 1, 2 or 3, so the digraph is strongly connected. However, we can check this more easily if we consider the matrix **B** obtained by adding the three matrices together:

$$\mathbf{B} = \mathbf{A} + \mathbf{A}^2 + \mathbf{A}^3 = \begin{bmatrix} 2 & 3 & 1 & 1 \\ 1 & 2 & 1 & 1 \\ 1 & 1 & 0 & 1 \\ 3 & 3 & 1 & 2 \end{bmatrix}.$$

Let b_{ij} denote the entry in row i and column j in the matrix **B**. Then each entry b_{ij} is the total number of walks of lengths 1, 2 and 3 from vertex i to vertex j. Since all the non-diagonal entries are positive, each pair of distinct vertices is connected by a path, so the digraph is strongly connected.

We generalize this result in the following theorem; the proof is given at the end of this section.

Theorem 5.2

Let D be a digraph with n vertices labelled $1, 2, ..., n$, let **A** be its adjacency matrix with respect to this listing of the vertices, and let **B** be the matrix

$$\mathbf{B} = \mathbf{A} + \mathbf{A}^2 + ... + \mathbf{A}^{n-1}.$$

Then D is strongly connected if and only if each non-diagonal entry in **B** is positive – that is, $b_{ij} > 0$ whenever $i \neq j$.

Problem 5.7

Find **B** for the digraph in Problem 5.6, and hence determine whether the digraph is strongly connected.

Problem 5.8

Determine whether the digraph with the following adjacency matrix is strongly connected:

$$\begin{bmatrix} 0 & 0 & 0 & 1 & 0 \\ 1 & 0 & 1 & 0 & 0 \\ 0 & 0 & 0 & 1 & 0 \\ 0 & 0 & 0 & 0 & 1 \\ 0 & 1 & 0 & 0 & 0 \end{bmatrix}.$$

Proofs of Theorems

We now supply the proofs of Theorems 5.1 and 5.2.

Theorem 5.1

Let D be a digraph with n vertices labelled 1, 2, ..., n, let **A** be its adjacency matrix with respect to this listing of the vertices, and let k be any positive integer.
Then the number of walks of length k from vertex i to vertex j is equal to the entry in row i and column j of the matrix $\mathbf{A}^k$ (the kth power of the matrix **A**).

Proof The proof is by mathematical induction on k, the length of the walk.

Step 1 The statement is true when $k = 1$, since the number of walks of length 1 from vertex i to vertex j is the number of arcs from vertex i to vertex j, and this is equal to a_{ij}, the entry in row i and column j of the adjacency matrix **A**.

Step 2 We assume that $k > 1$, and that the statement is true for all positive integers less than k. We wish to prove that the statement is true for the positive integer k.

Consider any walk of length k from vertex i to vertex j. Such a walk consists of a walk of length $k - 1$ from vertex i to some vertex r adjacent to vertex j, followed by a walk of length 1 from vertex r to vertex j.

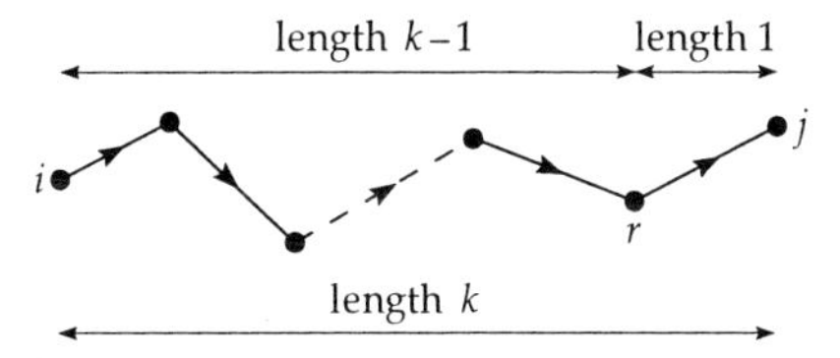

By our assumption, the number of walks of length $k - 1$ from vertex i to vertex r is the entry in row i and column r of the matrix $\mathbf{A}^{k-1}$, which we denote by $a_{ir}^{(k-1)}$. Since the number of walks of length 1 from vertex r to vertex j is a_{rj}, it follows that

the number of walks of length k from vertex i to vertex j via vertex r (at the previous step) is $a_{ir}^{(k-1)} a_{rj}$. $\quad (*)$

Now the total number of walks of length k from vertex i to vertex j equals

the number of such walks via vertex 1 (at the previous step)
\+ the number of such walks via vertex 2 (at the previous step)
...
\+ the number of such walks via vertex r (at the previous step)
...
\+ the number of such walks via vertex n (at the previous step).

By our previous result $(*)$, this is equal to

$$a_{i1}^{(k-1)} a_{1j} + a_{i2}^{(k-1)} a_{2j} + \ldots + a_{ir}^{(k-1)} a_{rj} + \ldots + a_{in}^{(k-1)} a_{nj}$$

By the rules for matrix multiplication, this is just the entry in row i and column j of the matrix $\mathbf{A}^{k-1}\mathbf{A} = \mathbf{A}^k$, as required.

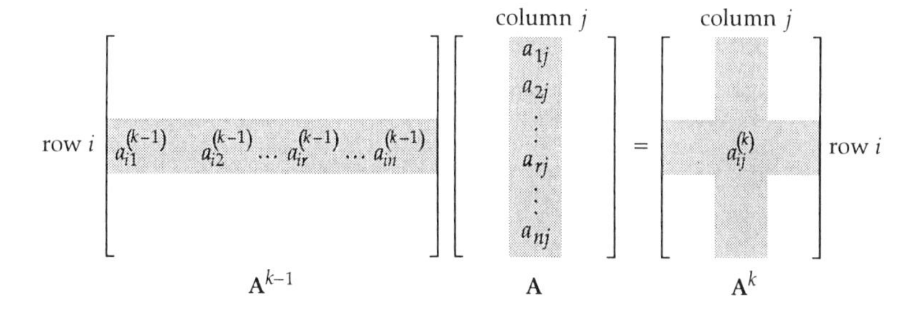

Thus, if the statement is true for all positive integers less than k, then it is true for the integer k. This completes Step 2.

Therefore, by the principle of mathematical induction, the statement is true for all positive integers k. ■

Theorem 5.2

Let D be a digraph with n vertices labelled $1, 2, \ldots, n$, let $\mathbf{A}$ be its adjacency matrix with respect to this listing of the vertices, and let $\mathbf{B}$ be the matrix

$$\mathbf{B} = \mathbf{A} + \mathbf{A}^2 + \ldots + \mathbf{A}^{n-1}.$$

Then D is strongly connected if and only if each non-diagonal entry in $\mathbf{B}$ is positive – that is, $b_{ij} > 0$ whenever $i \neq j$.

Proof There are two statements to prove.

(a) *If each non-diagonal entry in* **B** *is positive, then D is strongly connected.*

Let D be a digraph that satisfies the given conditions, and suppose that each non-diagonal entry in $\mathbf{B}$ is positive – that is, $b_{ij} > 0$ whenever $i \neq j$ – then $a_{ij}^{(k)} > 0$ for some $k \leq n-1$. Therefore there is a walk of length at most $n-1$ from vertex i to vertex j whenever $i \neq j$, so the digraph D is strongly connected.

(b) *If the digraph D is strongly connected, then each non-diagonal entry in* **B** *is positive.*

Let D be a strongly connected digraph that satisfies the given conditions; then there is a path from any vertex to any other. Since D has n vertices, such a path has length at most $n-1$. It follows that $a_{ij}^{(k)} > 0$ for at least one value of $k \leq n-1$, and hence that the entry in row i and column j of $\mathbf{B}$ is positive; that is, $b_{ij} > 0$ whenever $i \neq j$. ■

5.3 Incidence Matrices

For convenience, in this section we restrict our attention to graphs and digraphs without loops.

Whereas the adjacency matrix of a graph or digraph involves the adjacency of vertices, the incidence matrix involves the incidence of vertices and edges or arcs. To see what is involved, consider the following example:

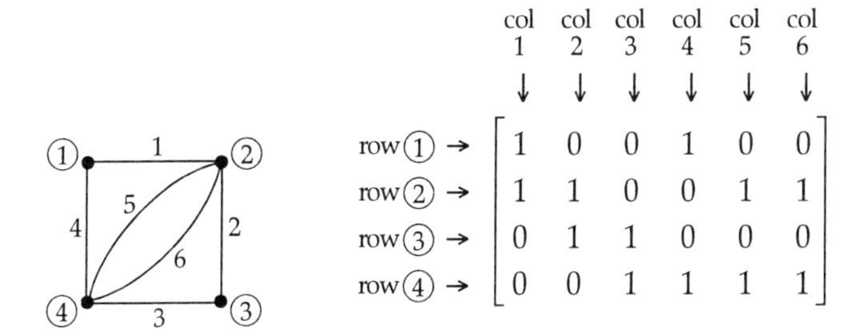

On the left we have a graph with *four* labelled vertices and *six* labelled edges, and on the right we have a matrix with *four* rows and *six* columns. Each of the numbers appearing in the matrix is 1 or 0, depending on whether the corresponding vertex and edge are incident with each other. For example,

vertex ① is incident with edge 4,
so **1** appears in row 1 column 4;
vertex ② is not incident with edge 4,
so **0** appears in row 2 column 4.

We generalize this idea, as follows.

Definition

Let G be a graph without loops, with n vertices labelled ①, ②, ..., ⓝ, and m edges labelled 1, 2, 3, ..., m.

The **incidence matrix I**(G) of G is the $n \times m$ matrix in which the entry in row i and column j is

$$\begin{cases} 1 & \text{if the vertex } i \text{ is incident with the edge } j, \\ 0 & \text{otherwise.} \end{cases}$$

Problem 5.9

Write down the incidence matrix of each of the following graphs:

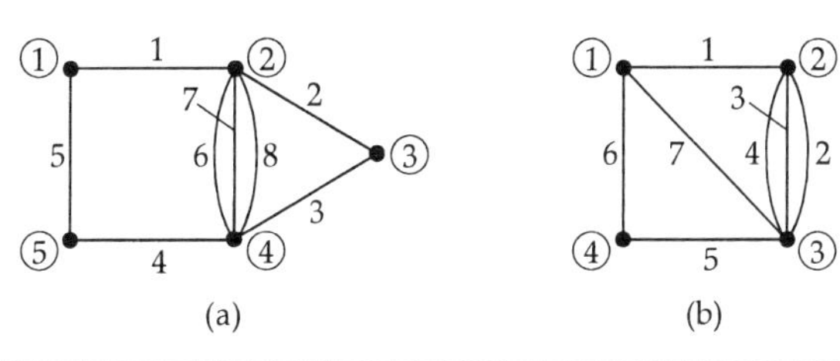

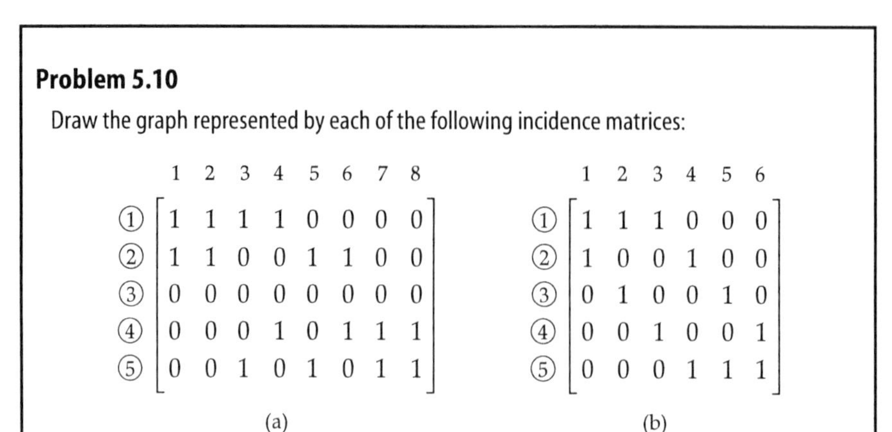

Problem 5.10

Draw the graph represented by each of the following incidence matrices:

(a)

	1	2	3	4	5	6	7	8
①	1	1	1	1	0	0	0	0
②	1	1	0	0	1	1	0	0
③	0	0	0	0	0	0	0	0
④	0	0	0	1	0	1	1	1
⑤	0	0	1	0	1	0	1	1

(b)

	1	2	3	4	5	6
①	1	1	1	0	0	0
②	1	0	0	1	0	0
③	0	1	0	0	1	0
④	0	0	1	0	0	1
⑤	0	0	0	1	1	1

In the incidence matrix of a graph without loops, each column contains exactly two 1s, as each edge is incident with just two vertices; the sum of the numbers in a row is the degree of the vertex corresponding to that row.

Whereas the adjacency matrix of a digraph involves the adjacency of vertices, the incidence matrix of a digraph involves the incidence of vertices and arcs. Since an arc can be incident from, incident to, or not incident with a vertex, we have to take account of this when defining the matrix. To see what is involved, consider the following example:

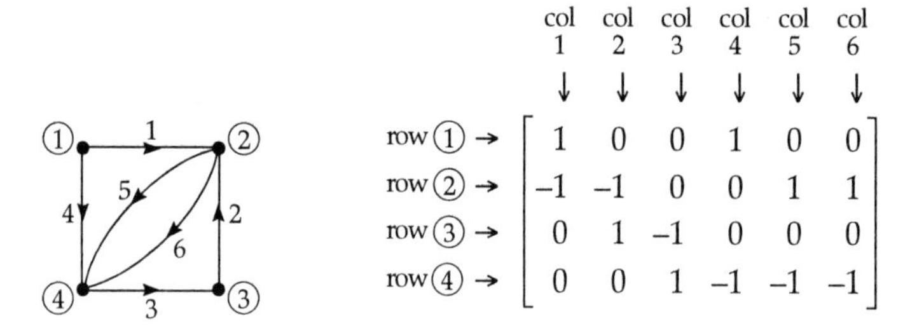

	col 1	col 2	col 3	col 4	col 5	col 6
row ① →	1	0	0	1	0	0
row ② →	–1	–1	0	0	1	1
row ③ →	0	1	–1	0	0	0
row ④ →	0	0	1	–1	–1	–1

On the left we have a digraph with *four* labelled vertices and *six* labelled arcs, and on the right we have a matrix with *four* rows and *six* columns. Each of the numbers appearing in the matrix is 1, –1 or 0, depending on whether the corresponding arc is incident from, incident to, or not incident with, the corresponding vertex. For example,

arc 4 is incident from vertex ①,
  so **1** appears in row 1 column 4;
arc 5 is incident to vertex ④,
  so **–1** appears in row 4 column 5;
arc 4 is not incident with vertex ②,
  so **0** appears in row 2 column 4.

We generalize this idea, as follows.

Definition

Let D be a digraph without loops, with n vertices labelled ①, ②, ..., ⓝ and m arcs labelled 1, 2, 3, ..., m.

The **incidence matrix I**(D) of D is the $n \times m$ matrix in which the entry in row i and column j is

$$\begin{cases} 1 & \text{if arc } j \text{ is incident from vertex } i, \\ -1 & \text{if arc } j \text{ is incident to vertex } i, \\ 0 & \text{otherwise.} \end{cases}$$

Problem 5.11

Write down the incidence matrix of each of the following digraphs:

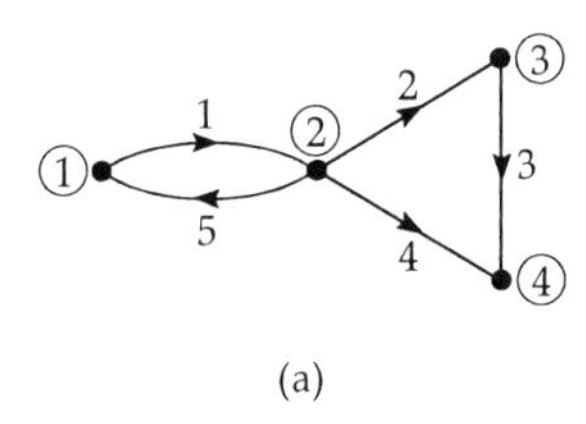

(a)

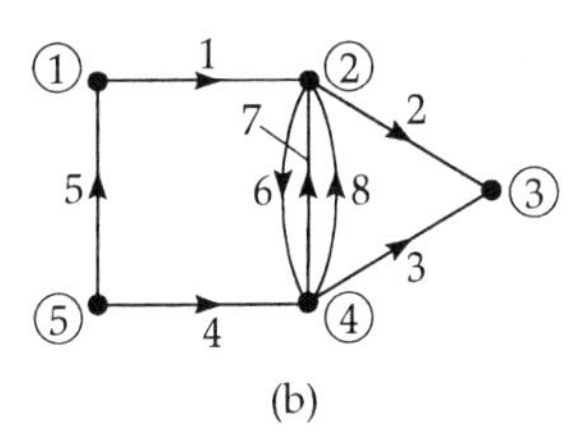

(b)

Problem 5.12

Draw the digraph represented by each of the following incidence matrices:

$$\begin{array}{c} \\ ① \\ ② \\ ③ \\ ④ \end{array}
\begin{array}{c} \begin{array}{ccccc} 1 & 2 & 3 & 4 & 5 \end{array} \\ \left[\begin{array}{ccccc} 1 & 1 & 0 & 0 & 0 \\ -1 & 0 & 0 & 1 & 1 \\ 0 & 0 & -1 & 0 & -1 \\ 0 & -1 & 1 & -1 & 0 \end{array}\right] \end{array}$$

(a)

$$\begin{array}{c} \\ ① \\ ② \\ ③ \\ ④ \\ ⑤ \end{array}
\begin{array}{c} \begin{array}{cccccccc} 1 & 2 & 3 & 4 & 5 & 6 & 7 & 8 \end{array} \\ \left[\begin{array}{cccccccc} 1 & -1 & 1 & -1 & 0 & 0 & 0 & 0 \\ -1 & 1 & 0 & 0 & -1 & 1 & 0 & 0 \\ 0 & 0 & 0 & 0 & 0 & 0 & 0 & 0 \\ 0 & 0 & 0 & 1 & 0 & -1 & -1 & -1 \\ 0 & 0 & -1 & 0 & 1 & 0 & 1 & 1 \end{array}\right] \end{array}$$

(b)

In the incidence matrix of a digraph without loops, each column has exactly one 1 and one –1, since each arc is incident from one vertex and incident to one vertex; the number of 1s in any row is the out-degree of the vertex corresponding to that row, and the number of –1s in any row is the in-degree of the vertex corresponding to that row.

5.4 Case Studies

We conclude this chapter with two case studies – interval graphs and Markov chains.

Interval Graphs

Interval graphs have been used extensively in situations involving the arrangement of data into chronological order. In such graphs, the vertices correspond to the objects being arranged and the edges correspond to pairs of objects that overlap in some way. Although interval graphs first arose in a genetic context, they have also been used in areas such as archaeology. We give a brief account of these applications, indicating how the relevant data can be represented by an interval graph.

Archaeology

At the end of the nineteenth century archaeologists were interested in the various types of pottery and other artefacts that had been found in several graves in predynastic Egypt (*c.* 4000–2400 BC). In particular, Sir Flinders Petrie used the data from nine hundred graves in the cemeteries of Naqada, Ballas, Abadiyeh and Hu in an attempt to arrange the graves chronologically and assign a time period to each artefact found in them – this process is known as *sequence dating* or *seriation*.

pottery found by Sir Flinders Petrie

In dating the graves, they assumed that if two different artefacts occurred together in the same grave, then their time periods must have overlapped. They also assumed, since the number of graves was large, that if the time periods of two artefacts overlapped, then the artefacts should appear together in some of the graves.

One of the most promising approaches to seriation problems in archaeology has been the representation of such data as a graph in which the vertices correspond to the artefacts, and the edges correspond to pairs of artefacts that have appeared together in the same grave. To see how this arises, suppose that there are just six artefacts *a, b, c, d, e, f*, and that the matrix on the left below tells us which pairs of artefacts occurred together in the same grave; for example, artefacts *a* and *b* occurred together in some grave, whereas *a* and *f* did not. We can regard such a matrix as the adjacency matrix of a graph, by replacing each ✓ by 1 and each × or – by 0; the adjacency matrix and the corresponding graph are shown below.

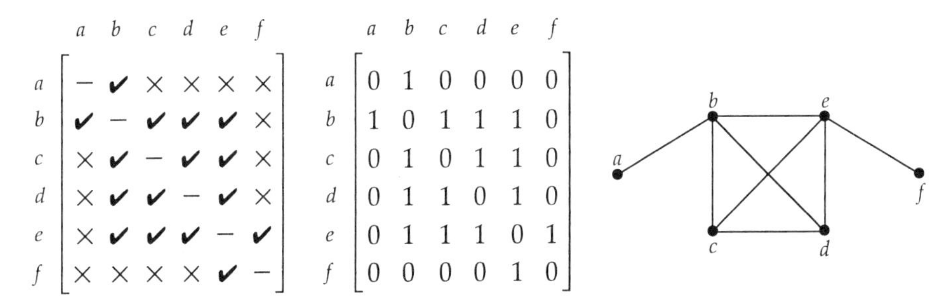

	a	*b*	*c*	*d*	*e*	*f*
a	–	✓	×	×	×	×
b	✓	–	✓	✓	✓	×
c	×	✓	–	✓	✓	×
d	×	✓	✓	–	✓	×
e	×	✓	✓	✓	–	✓
f	×	×	×	×	✓	–

	a	*b*	*c*	*d*	*e*	*f*
a	0	1	0	0	0	0
b	1	0	1	1	1	0
c	0	1	0	1	1	0
d	0	1	1	0	1	0
e	0	1	1	1	0	1
f	0	0	0	0	1	0

The problem is now to represent this information in chronological form. To do this, we construct a set of intervals on the real line corresponding to the time periods during which the artefacts were in use. Artefacts correspond to intervals, and pairs of artefacts that occurred together in the same grave correspond to overlapping intervals. This means that each vertex of the graph gives rise to an interval and each edge gives rise to overlapping intervals.

One way of doing this is shown below. Note, for example, that the vertices corresponding to artefacts *a* and *b* are adjacent, so their intervals overlap; however, the vertices corresponding to artefacts *a* and *f* are not adjacent, and so their intervals do not overlap. Any graph that gives rise to a set of intervals in this way is called an **interval graph**.

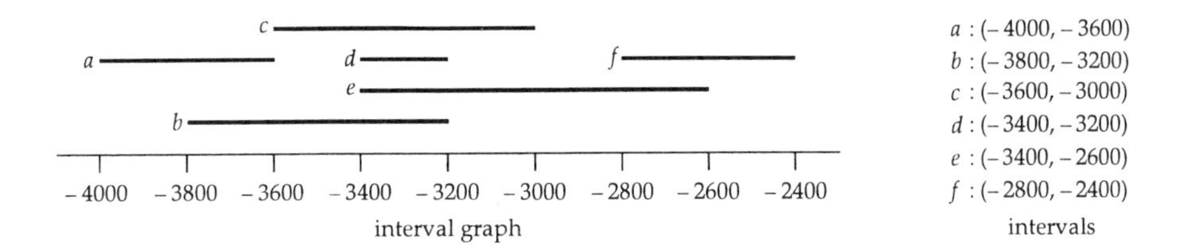

Unfortunately, the problem is not as simple in practice as this example may imply. In particular, several different arrangements of intervals may arise from the same graph – for example, we can simply reverse the entire pattern of intervals – and it is usually impossible to choose the correct arrangement unless some other information is available. In spite of this drawback, the interval graph approach has had some spectacular successes, and has led to the solution of many seriation problems, including the chronological ordering of bronze-age material in Central Europe, arrowheads in a palaeo-Indian site in Wyoming, and Greek inscriptions at Histria, in Romania.

Problem 5.13

Draw the graph that gives rise to the following set of intervals:

$$(1, 2), (3, 4), (5, 6), (7, 8), (1, 6), (2, 7), (3, 8).$$

Genetics

For some time, geneticists have regarded the chromosome as a linear arrangement of genes, and it is natural to ask whether the fine structure inside the gene is also arranged in a linear manner; this problem is called *Benzer's problem*. Unfortunately, this fine structure is too detailed to be observed directly, and so one has to study changes in the structure of the whole gene, known as *mutations*.

In analysing the genetic structure of a particular bacterial virus called phage T4, Seymour Benzer considered the mutations that result when part of the gene is missing. In particular, he studied mutations in which the missing segments overlap, and expressed his results in the form of an *overlap matrix*, part of which is shown as (a) opposite. This 19×19 matrix is the adjacency matrix of the graph (b), in which the vertices correspond to mutations, and the edges correspond to pairs of mutations whose missing segments overlap. In these terms, Benzer's problem is that of determining whether the matrix (a) represents the overlapping of a suitably chosen collection of intervals, or (equivalently) of determining whether the graph (b) is an interval graph. In (c) we see that this is indeed the case – there *are* intervals that arise from this adjacency matrix and graph. This interval graph is consistent with the parts of the gene corresponding to each mutation being lined up within the gene in a linear fashion in left-to-right order, as shown in (c).

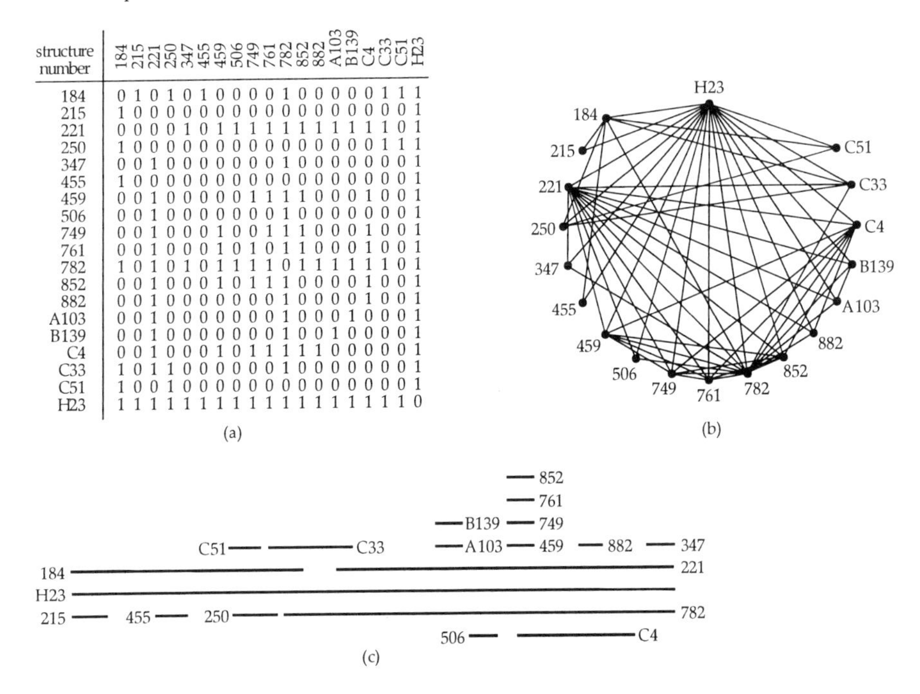

structure number	184	215	221	250	347	455	459	506	749	761	782	852	882	A103	B139	C4	C33	C51	H23
184	0	1	0	1	0	1	0	0	0	0	1	0	0	0	0	0	1	1	1
215	1	0	0	0	0	0	0	0	0	0	0	0	0	0	0	0	0	0	1
221	0	0	0	0	1	0	1	1	1	1	1	1	1	1	1	1	1	0	1
250	1	0	0	0	0	0	0	0	0	0	0	0	0	0	0	0	1	1	1
347	0	0	1	0	0	0	0	0	0	0	1	0	0	0	0	0	0	0	1
455	1	0	0	0	0	0	0	0	0	0	0	0	0	0	0	0	0	0	1
459	0	0	1	0	0	0	0	0	1	1	1	1	0	0	0	1	0	0	1
506	0	0	1	0	0	0	0	0	0	0	1	0	0	0	0	0	0	0	1
749	0	0	1	0	0	0	1	0	0	1	1	1	0	0	0	1	0	0	1
761	0	0	1	0	0	0	1	0	1	0	1	1	0	0	0	1	0	0	1
782	1	0	1	0	1	0	1	1	1	1	0	1	1	1	1	1	1	0	1
852	0	0	1	0	0	0	1	0	1	1	1	0	0	0	0	1	0	0	1
882	0	0	1	0	0	0	0	0	0	0	1	0	0	0	0	1	0	0	1
A103	0	0	1	0	0	0	0	0	0	0	1	0	0	0	1	0	0	0	1
B139	0	0	1	0	0	0	0	0	0	0	1	0	0	1	0	0	0	0	1
C4	0	0	1	0	0	0	1	0	1	1	1	1	1	0	0	0	0	0	1
C33	1	0	1	1	0	0	0	0	0	0	1	0	0	0	0	0	0	0	1
C51	1	0	0	1	0	0	0	0	0	0	0	0	0	0	0	0	0	0	1
H23	1	1	1	1	1	1	1	1	1	1	1	1	1	1	1	1	1	1	0

(a)

(b)

(c)

Although the representation of this data as an interval graph does not *prove* that the fine structure inside the gene is arranged linearly, it certainly provides support for such a hypothesis. In fact, Benzer extended his analysis to no fewer than 145 mutations and showed that, even with this number of rows, the resulting matrix can still be represented by an interval graph. By this means he was able to show that, for this virus at least, the evidence for a linear arrangement is overwhelming.

Markov Chains

The study of Markov chains has arisen in a wide variety of areas, ranging from genetics and statistics to computing and sociology. For ease of presentation we consider a rather trivial Markov chain problem, that of a drunkard standing directly between his two favourite pubs, *The Source and Sink* and *The Black Vertex*.

Every minute he behaves in one of three ways, each with a given probability:

he staggers ten metres towards the first pub with probability $\frac{1}{2}$,
or he staggers ten metres towards the second pub with probability $\frac{1}{3}$,
or he stays where he is with probability $\frac{1}{6}$.

Such a procedure is called a one-dimensional **random walk**. We assume that the two pubs are 'absorbing', in the sense that if he arrives at either of them he stays there. Given the distance between the two pubs and his initial position, there are several questions we can ask. For example, which pub is he more likely to reach first? How long is he likely to take getting there?

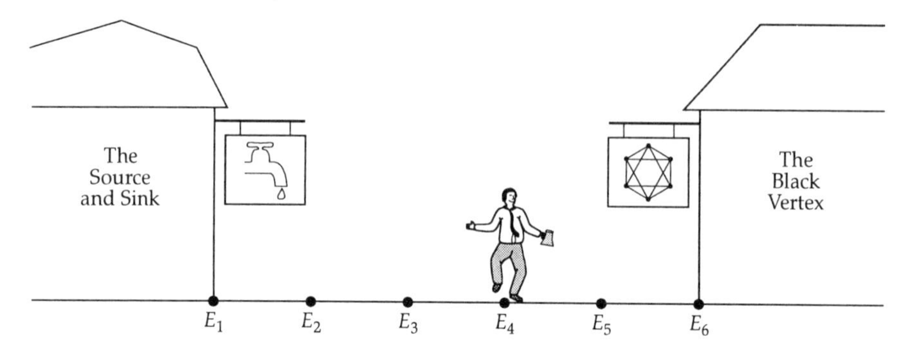

Let us suppose that the two pubs are fifty metres apart and that our friend is initially twenty metres from *The Black Vertex*. Let us denote the various places at which he can stop by $E_1, ..., E_6$, where E_1 and E_6 denote the two pubs. Then his initial position E_4 can be described by the vector $\mathbf{x} = [0, 0, 0, 1, 0, 0]$, in which the ith component is the probability that he is initially at E_i. Furthermore, the probabilities of his position after one minute are given by the vector $[0, 0, \frac{1}{2}, \frac{1}{6}, \frac{1}{3}, 0]$, and after two minutes by $[0, \frac{1}{4}, \frac{1}{6}, \frac{13}{36}, \frac{1}{9}, \frac{1}{9}]$; for example, the probability that he is at E_4 after two minutes is given by $(\frac{1}{2} \times \frac{1}{3}) + (\frac{1}{6} \times \frac{1}{6}) + (\frac{1}{3} \times \frac{1}{2}) = \frac{13}{36}$.

It is awkward to calculate *directly* the probability of his being at a given place after k minutes. Fortunately, there is a more convenient way of doing this, by introducing the idea of a *transition matrix*.

Let p_{ij} be the probability that he moves from E_i to E_j in one minute; for example, $p_{23} = \frac{1}{3}$, and $p_{24} = 0$. These probabilities p_{ij} are called the *transition probabilities*, and the 6×6 matrix $\mathbf{P}$ in which the entry in row i and column j is p_{ij} is known as the *transition matrix*, shown below. Note that each entry of $\mathbf{P}$ is non-negative and that the sum of the entries in any row is 1.

$$\mathbf{P} = \begin{bmatrix} 1 & 0 & 0 & 0 & 0 & 0 \\ \frac{1}{2} & \frac{1}{6} & \frac{1}{3} & 0 & 0 & 0 \\ 0 & \frac{1}{2} & \frac{1}{6} & \frac{1}{3} & 0 & 0 \\ 0 & 0 & \frac{1}{2} & \frac{1}{6} & \frac{1}{3} & 0 \\ 0 & 0 & 0 & \frac{1}{2} & \frac{1}{6} & \frac{1}{3} \\ 0 & 0 & 0 & 0 & 0 & 1 \end{bmatrix}$$

It now follows that if $\mathbf{x}$ is the initial row vector defined above, then the probabilities of his position after one minute are given by the row vector

$$\mathbf{xP} = [0 \quad 0 \quad 0 \quad 1 \quad 0 \quad 0]\begin{bmatrix} 1 & 0 & 0 & 0 & 0 & 0 \\ \frac{1}{2} & \frac{1}{6} & \frac{1}{3} & 0 & 0 & 0 \\ 0 & \frac{1}{2} & \frac{1}{6} & \frac{1}{3} & 0 & 0 \\ 0 & 0 & \frac{1}{2} & \frac{1}{6} & \frac{1}{3} & 0 \\ 0 & 0 & 0 & \frac{1}{2} & \frac{1}{6} & \frac{1}{3} \\ 0 & 0 & 0 & 0 & 0 & 1 \end{bmatrix} = [0 \quad 0 \quad \tfrac{1}{2} \quad \tfrac{1}{3} \quad \tfrac{1}{6} \quad 0]$$

and after k minutes by the vector $\mathbf{xP}^k$. In other words, the ith component of $\mathbf{xP}^k$ represents the probability that he is at position E_i after k minutes have elapsed.

In general, a **transition matrix** is a square matrix, each of whose rows contains non-negative numbers, called **transition probabilities**, with sum 1, and a **Markov chain** consists of an $n \times n$ transition matrix $\mathbf{P}$ and a $1 \times n$ row vector $\mathbf{x}$. The positions E_i are the **states** of the Markov chain, and our aim is to describe a way of classifying them.

We are mainly concerned with whether we can get from a given state to another state, and if so, how long it takes. For example, in the above problem, the drunkard can get from E_4 to E_1 in three minutes and from E_4 to E_6 in two minutes, but he can never get from E_1 to E_4 because of our assumption that the pubs are 'absorbing'. It follows that our main concern is not with the actual probabilities p_{ij}, but with when they are non-zero. To decide this, we represent the situation by a digraph whose vertices correspond to the states and whose arcs tell us whether we can go from one state to another in one minute. Thus, if each state E_i is represented by a corresponding vertex v_i, then the required digraph is obtained by joining v_i to v_j if and only if $p_{ij} \neq 0$. We refer to this digraph as the **associated digraph** of the Markov chain. The associated digraph of the above problem is as follows:

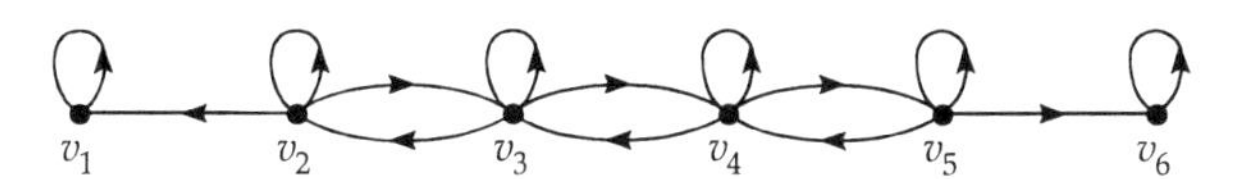

The adjacency matrix of the associated digraph of a Markov chain, known as the **associated adjacency matrix** of the Markov chain, is easily obtained from the transition matrix $\mathbf{P}$ by replacing each non-zero entry of $\mathbf{P}$ by 1. Thus the associated adjacency matrix for the above problem is as follows:

$$\begin{bmatrix} 1 & 0 & 0 & 0 & 0 & 0 \\ 1 & 1 & 1 & 0 & 0 & 0 \\ 0 & 1 & 1 & 1 & 0 & 0 \\ 0 & 0 & 1 & 1 & 1 & 0 \\ 0 & 0 & 0 & 1 & 1 & 1 \\ 0 & 0 & 0 & 0 & 0 & 1 \end{bmatrix}$$

As a further example, suppose that we are given a Markov chain whose transition matrix is as shown on the left below; then its associated adjacency matrix and digraph are as shown on the right below.

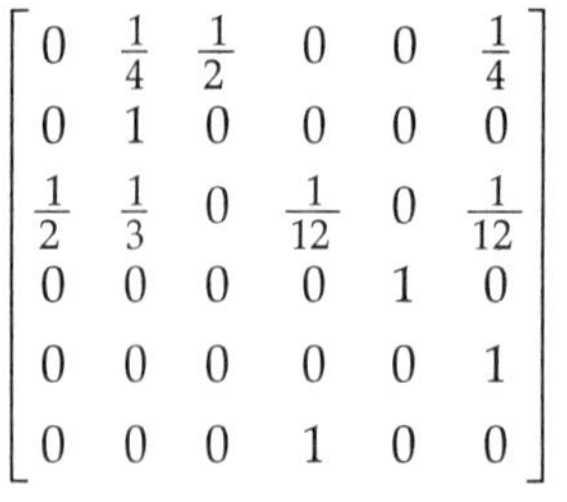

$$\begin{bmatrix} 0 & \frac{1}{4} & \frac{1}{2} & 0 & 0 & \frac{1}{4} \\ 0 & 1 & 0 & 0 & 0 & 0 \\ \frac{1}{2} & \frac{1}{3} & 0 & \frac{1}{12} & 0 & \frac{1}{12} \\ 0 & 0 & 0 & 0 & 1 & 0 \\ 0 & 0 & 0 & 0 & 0 & 1 \\ 0 & 0 & 0 & 1 & 0 & 0 \end{bmatrix}$$

transition matrix

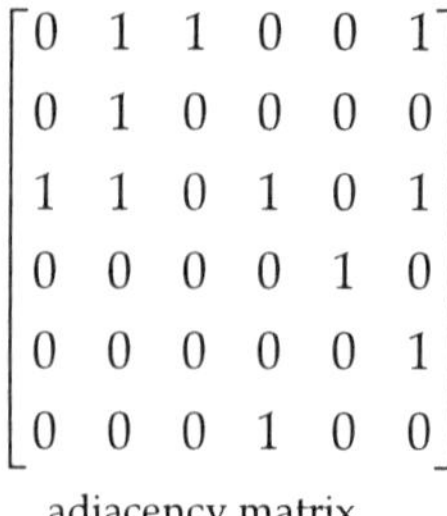

$$\begin{bmatrix} 0 & 1 & 1 & 0 & 0 & 1 \\ 0 & 1 & 0 & 0 & 0 & 0 \\ 1 & 1 & 0 & 1 & 0 & 1 \\ 0 & 0 & 0 & 0 & 1 & 0 \\ 0 & 0 & 0 & 0 & 0 & 1 \\ 0 & 0 & 0 & 1 & 0 & 0 \end{bmatrix}$$

adjacency matrix

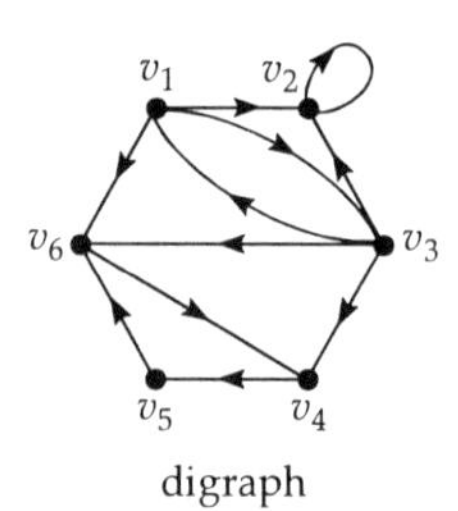

digraph

It should now be clear that we can get from a state E_i to a state E_j in a Markov chain if and only if there is a path from v_i to v_j in the associated digraph, and that the least possible time to do so is the length of the shortest such path.

A Markov chain in which we can get from any state to any other is called an **irreducible Markov chain**. Clearly a Markov chain is irreducible if and only if its associated digraph is strongly connected. Note that neither of the Markov chains described above is irreducible. For example, in the second Markov chain, there is no path from v_2 to any other vertex.

Problem 5.14

(a) Suppose that, in the problem of the drunkard, *The Black Vertex* ejects him as soon as he gets there. Write down the resulting transition matrix and its associated digraph, and decide whether the resulting Markov chain is irreducible.

(b) How would your answers to part (a) be changed if both pubs eject him?

Exercises 5

Adjacency Matrices

5.1 Write down the adjacency matrices of the following graph and digraph.

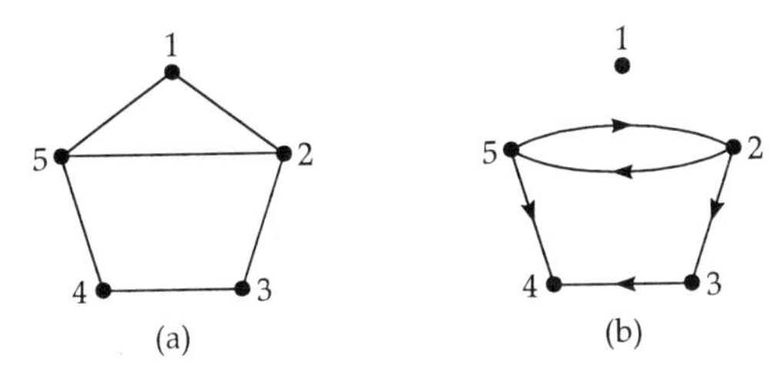

5.2 Draw the graph corresponding to adjacency matrix (a) and the digraph corresponding to adjacency matrix (b).

$$\begin{bmatrix} 0 & 1 & 1 & 1 & 0 \\ 1 & 0 & 0 & 0 & 1 \\ 1 & 0 & 0 & 0 & 1 \\ 1 & 0 & 0 & 0 & 1 \\ 0 & 1 & 1 & 1 & 0 \end{bmatrix} \qquad \begin{bmatrix} 0 & 1 & 0 & 0 & 1 \\ 1 & 0 & 0 & 1 & 0 \\ 0 & 0 & 0 & 0 & 0 \\ 1 & 0 & 1 & 0 & 0 \\ 0 & 1 & 0 & 2 & 0 \end{bmatrix}$$

(a) (b)

5.3 Consider the following adjacency matrix of a digraph D:

$$\begin{bmatrix} 0 & 1 & 0 & 0 & 0 \\ 0 & 0 & 1 & 0 & 0 \\ 1 & 0 & 0 & 0 & 1 \\ 0 & 0 & 1 & 0 & 0 \\ 0 & 0 & 0 & 1 & 0 \end{bmatrix}$$

Write down the out-degree and the in-degree of each vertex of the digraph D, and hence determine whether D is Eulerian. Check your answer by drawing the digraph D.
Hint Use Theorem 4.2.

5.4 The following matrix is the adjacency matrix of a graph G.

$$\begin{bmatrix} 0 & 0 & 0 & 1 & 1 & 0 & 0 \\ 0 & 0 & 0 & 1 & 1 & 1 & 1 \\ 0 & 0 & 0 & 0 & 0 & 1 & 1 \\ 1 & 1 & 0 & 0 & 0 & 0 & 0 \\ 1 & 1 & 0 & 0 & 0 & 0 & 0 \\ 0 & 1 & 1 & 0 & 0 & 0 & 0 \\ 0 & 1 & 1 & 0 & 0 & 0 & 0 \end{bmatrix}$$

Which three of the following statements are TRUE?

(a) G is connected; (b) G is regular;
(c) G is bipartite; (d) G is a tree;
(e) G is Eulerian; (f) G is Hamiltonian.

Walks in Graphs and Digraphs

5.5 Consider the following digraph:

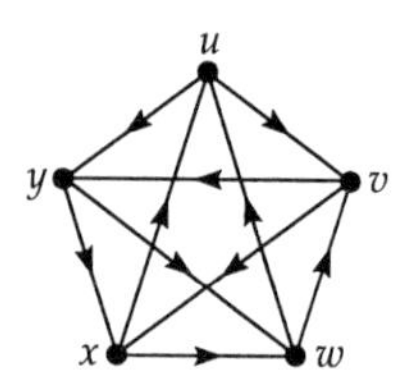

Write down the adjacency matrix $\mathbf{A}$, calculate the matrices $\mathbf{A}^2$, $\mathbf{A}^3$ and $\mathbf{A}^4$, and hence find the numbers of walks of lengths 1, 2, 3 and 4 from w to u. Is there a walk of length 1, 2, 3 or 4 from u to w?

5.6 Find the matrix $\mathbf{B}$ for the digraph in Exercise 5.5, and hence determine whether the digraph is strongly connected.

5.7 Use Theorem 5.2 to determine whether the digraphs with the following adjacency matrices are strongly connected:

$$\begin{bmatrix} 0 & 0 & 0 & 0 & 1 \\ 0 & 0 & 0 & 1 & 0 \\ 1 & 0 & 0 & 0 & 0 \\ 1 & 0 & 0 & 0 & 0 \\ 0 & 1 & 1 & 0 & 0 \end{bmatrix} \qquad \begin{bmatrix} 0 & 0 & 0 & 0 & 1 \\ 0 & 0 & 1 & 0 & 0 \\ 1 & 0 & 1 & 0 & 0 \\ 1 & 0 & 0 & 0 & 0 \\ 0 & 1 & 1 & 0 & 0 \end{bmatrix}$$

(a) (b)

5.8 Write an analogue for graphs of Theorem 5.2, and use it to determine whether the graphs with the following adjacency matrices are connected:

$$\begin{bmatrix} 0 & 0 & 0 & 1 & 0 \\ 0 & 0 & 1 & 0 & 0 \\ 0 & 1 & 0 & 0 & 0 \\ 1 & 0 & 0 & 0 & 1 \\ 0 & 0 & 0 & 1 & 0 \end{bmatrix} \qquad \begin{bmatrix} 0 & 0 & 0 & 1 & 1 \\ 0 & 0 & 1 & 0 & 0 \\ 0 & 1 & 0 & 0 & 1 \\ 1 & 0 & 0 & 0 & 0 \\ 1 & 0 & 1 & 0 & 0 \end{bmatrix}$$

(a) (b)

Incidence Matrices

5.9 Write down the incidence matrices of the following graph and digraph.

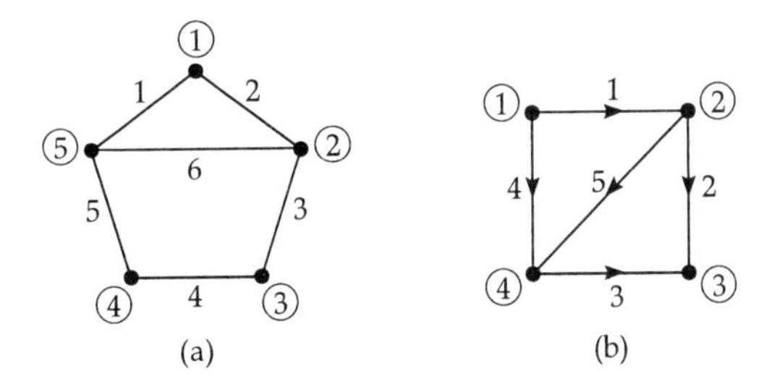

5.10 Draw the digraph whose incidence matrix is

$$\begin{bmatrix} 1 & -1 & 1 & -1 & 0 & 0 & 0 & 0 \\ -1 & 1 & 0 & 0 & -1 & 1 & 0 & 0 \\ 0 & 0 & 0 & 0 & 0 & 0 & 0 & 0 \\ 0 & 0 & 0 & 1 & 0 & -1 & -1 & -1 \\ 0 & 0 & -1 & 0 & 1 & 0 & 1 & 1 \end{bmatrix}$$

5.11 The following matrix is the incidence matrix of a graph G. What is the adjacency matrix of G with the same labelling?

$$\begin{bmatrix} 0 & 1 & 0 & 1 & 0 & 0 & 0 \\ 0 & 0 & 1 & 1 & 1 & 0 & 0 \\ 0 & 0 & 1 & 0 & 0 & 1 & 1 \\ 1 & 0 & 0 & 0 & 1 & 1 & 0 \\ 1 & 1 & 0 & 0 & 0 & 0 & 1 \end{bmatrix}$$

Case Studies

Interval Graphs

5.12 Show that the cycle graph C_4 is not an interval graph.

5.13 Seven types of jewellery were found in five ancient tombs 1, 2, 3, 4, 5, as shown in the following table:

jewellery

tombs	a	b	c	d	e	f	g
1	–	✓	–	✓	–	–	–
2	✓	–	–	–	–	–	✓
3	–	✓	–	–	✓	–	–
4	✓	–	✓	–	–	✓	–
5	–	✓	✓	–	✓	–	–

Draw the graph with vertex set $\{a, b, c, d, e, f, g\}$ in which two vertices are joined when the corresponding types of jewellery are found in the same tomb. Show how this graph gives rise to a set of overlapping intervals, and hence write down a possible chronological ordering of the seven types of jewellery.

Markov Chains

5.14 The transition matrix of a Markov chain is as follows.

$$\begin{bmatrix} 0 & 0.5 & 0.3 & 0 & 0.2 \\ 1 & 0 & 0 & 0 & 0 \\ 0.5 & 0.5 & 0 & 0 & 0 \\ 0.4 & 0.4 & 0 & 0 & 0.2 \\ 0 & 0 & 0 & 0.1 & 0.9 \end{bmatrix}$$

Draw the associated adjacency matrix and digraph. Is the Markov chain irreducible?

5.15 A game is played with a die by five people around a circular table:

a player who throws an odd number passes the die to the player immediately to his/her left;

a player who throws 2 or 4 passes the die to the player two places to his/her right;

a player who throws 6 keeps the die and throws it again.

Write down the corresponding transition matrix, draw its associated digraph, and determine whether the corresponding Markov chain is irreducible.

Chapter 6

Tree Structures

After working through this chapter, you should be able to:

- state several properties of a tree, and give several equivalent definitions of a tree;
- distinguish between physical and conceptual tree structures, and give examples of each type;
- appreciate the uses of rooted trees in different areas;
- construct the bipartite graph representation of a given braced rectangular framework and use it to determine whether the system is rigid; if so, determine whether the system is minimally braced.

In this chapter we focus our attention on one particularly important and useful type of graph – a *tree*. Although trees are relatively simple structures, they form the basis of many of the practical techniques used to model and to design large-scale systems.

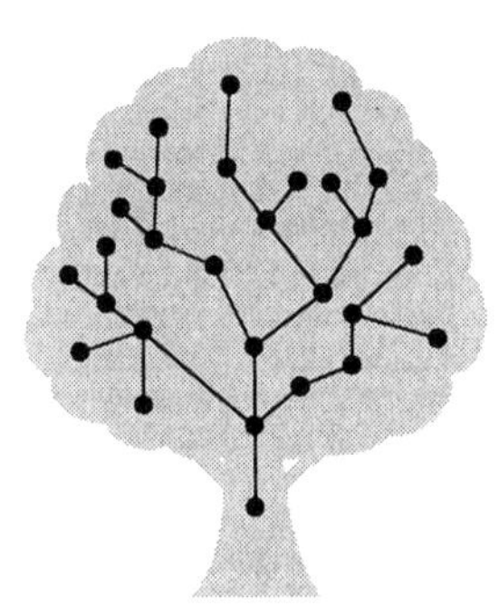

The concept of a *tree* is one of the most important and commonly used ideas in graph theory, especially in the applications of the subject. It arose in connection with the work of Gustav Kirchhoff on electrical networks in the 1840s, and later with Arthur Cayley's work on the enumeration of molecules in the 1870s. More recently, trees have proved to be of value in such areas as computer science, decision making, linguistics, and the design of gas pipeline systems.

Trees are often used to model situations involving various physical or conceptual tree-like structures. These structures are also commonly referred to as 'trees'. In the following examples, we classify such 'trees' in terms of the type of application in which they occur.

Many trees have a *physical* structure which may be either natural or artificial and either static or time-dependent. Two examples of natural trees are the biological variety with trunk, branches and leaves, and the drainage system of tributaries forming a river basin. Less obvious examples of tree structures are provided by the chemical structure of certain organic molecules.

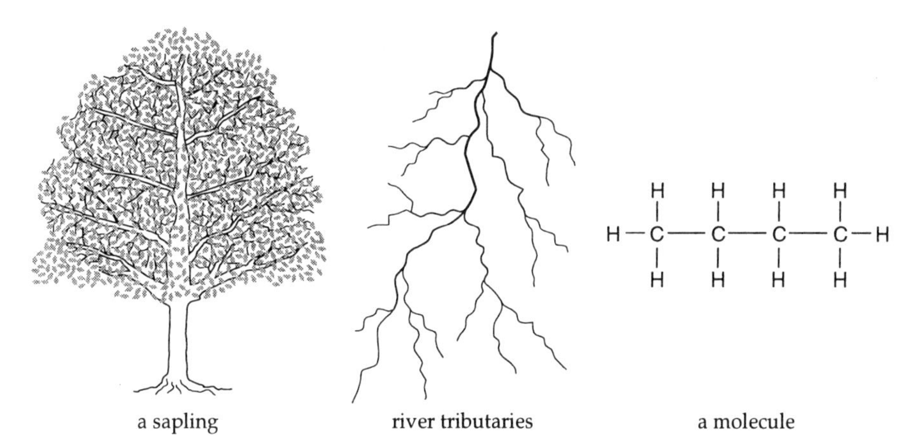

a sapling river tributaries a molecule

An example of the artificial variety of tree is an oil or gas pipeline distribution system, such as an undersea pipeline network; since the cost of constructing such a network may be very large, a tree structure with no unnecessary edges may be the most economical form for the network.

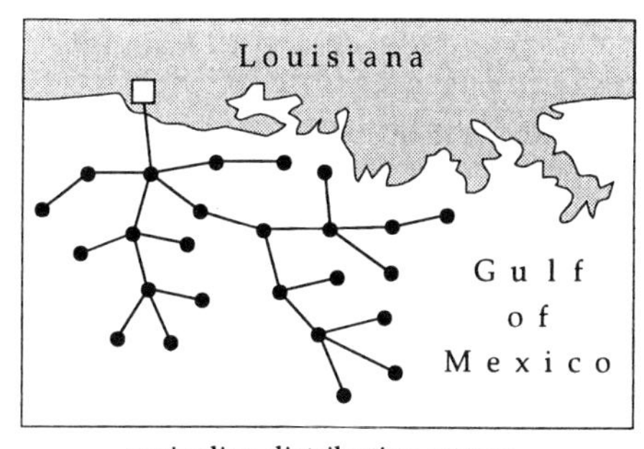

a pipeline distribution system

Many trees do not have a well-defined physical structure, but are *conceptual*. Probably the most familiar type of conceptual tree is a *family tree*, depicting ancestors and descendants. The following diagram illustrates part of the family tree of Saxon kings in the ninth century in England.

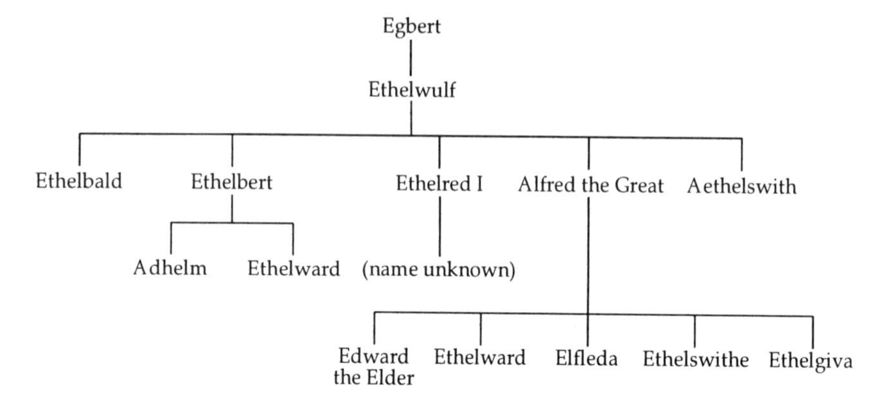

Another example of a conceptual tree is the following *hierarchical tree*, representing the lines of responsibility in a company.

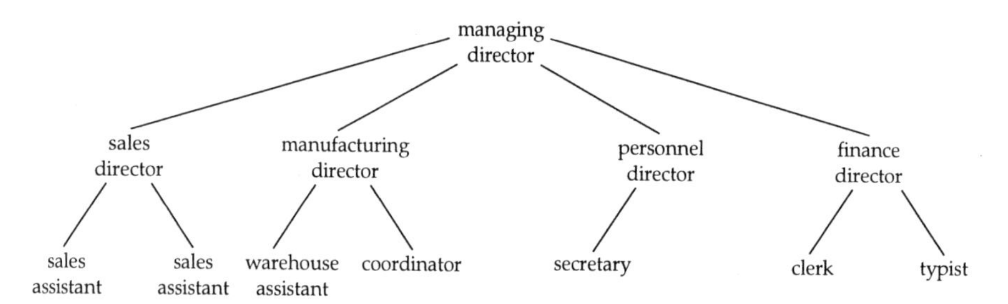

6.1 Mathematical Properties of Trees

We start by recalling the definition of a tree, given in Chapter 2.

> **Definition**
>
> A **tree** is a connected graph that has no cycles.

For example, the following diagram depicts all the unlabelled trees with at most five vertices.

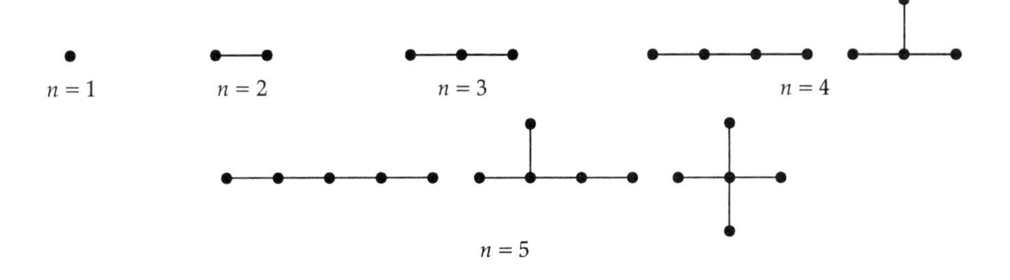

Problem 6.1

Draw the six unlabelled trees with six vertices.

Each unlabelled tree with six vertices can be obtained from an unlabelled tree with five vertices by adding an edge joining a new vertex to an existing vertex. This is a general procedure for increasing the size of a tree, since it creates no cycles and can be carried out systematically by adjoining the new edge to each vertex in turn. For example, consider the following tree with six vertices.

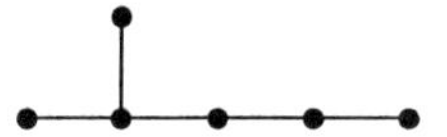

By adjoining a new edge to each vertex in turn, we obtain the following trees with seven vertices.

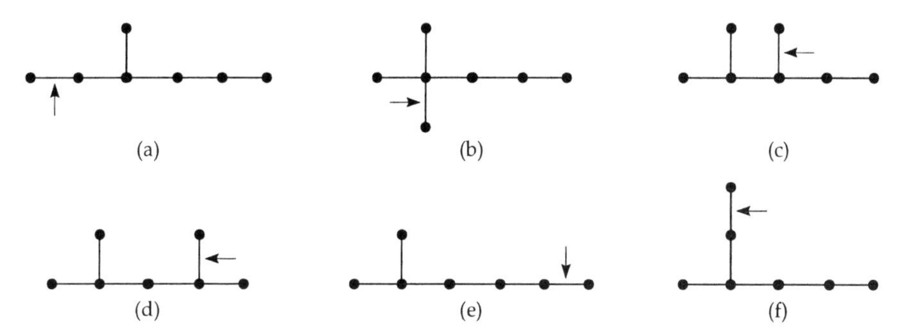

We can omit tree (f) from this list, since it is isomorphic to tree (a), so we obtain five trees with seven vertices from our original tree with six vertices. The difficulty of producing trees in this way lies in recognizing duplicates, but at least we can be sure that each tree with seven vertices can be thus obtained from some tree with six vertices.

Problem 6.2

By adding a new edge in all possible ways to each unlabelled tree with six vertices, draw the eleven unlabelled trees with seven vertices.

Starting with the tree with just one vertex, we can build up any tree we wish by successively adding a new edge and a new vertex. At each stage, the number of vertices exceeds the number of edges by 1, so

every tree with n vertices has exactly $n - 1$ edges.

At no stage is a cycle created, since each added edge joins an old vertex to a new vertex.

At each stage, the tree remains connected, so any two vertices must be connected by at least one path. However, they cannot be connected by more than one path, since any two such paths would contain a cycle (and possibly other edges as well).

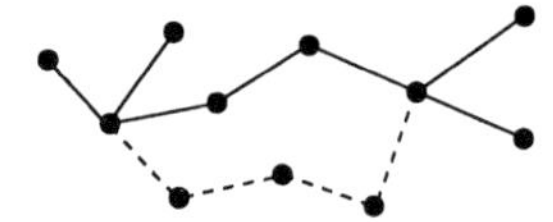

We therefore deduce that

any two vertices in a tree are connected by exactly one path.

In particular, any two *adjacent* vertices are connected by exactly one path – the edge joining them. If this edge is removed, then there is no path between the two vertices.

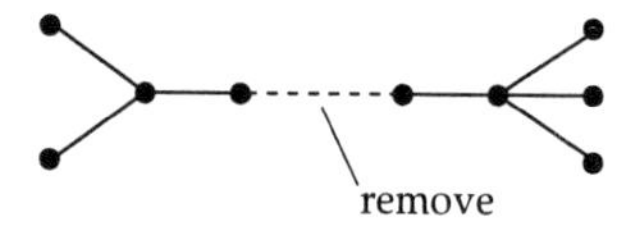

It follows that

the removal of any edge of a tree disconnects the tree.

Moreover, any two vertices v and w are connected by a path, and the addition of the edge vw produces a cycle – the cycle consisting of the path and the added edge vw.

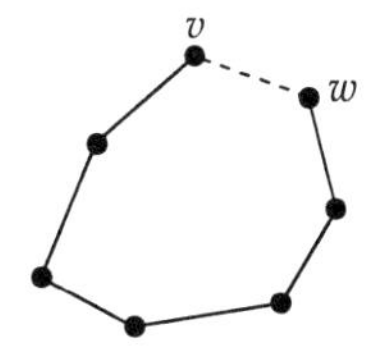

So

joining any two vertices of a tree by an edge creates a cycle.

Several of the above properties can be used as definitions of a tree. In the following theorem, we state six possible definitions. They are all *equivalent*: any one of them can be taken as the definition of a tree, and the other five can then be deduced. We omit the proof.

Theorem 6.1: Equivalent Definitions of a Tree

Let T be a graph with n vertices. Then the following statements are equivalent.

- T is connected and has no cycles.
- T has $n - 1$ edges and has no cycles.
- T is connected and has $n - 1$ edges.
- T is connected and the removal of any edge disconnects T.
- Any two vertices of T are connected by exactly one path.
- T contains no cycles, but the addition of any new edge creates a cycle.

Problem 6.3

(a) Use a proof by contradiction to show that the removal of an edge cannot disconnect a tree into more than two components.

(b) Use a proof by contradiction to show that the addition of a new edge to a tree cannot create more than one cycle.

Problem 6.4

(a) Give an example of a tree with seven vertices and
  (1) exactly two vertices of degree 1;
  (2) exactly four vertices of degree 1;
  (3) exactly six vertices of degree 1.

(b) Use the handshaking lemma to prove that every tree with n vertices, where $n \geq 2$, has at least two vertices of degree 1.

6.2 Spanning Trees

An important concept that we need later is that of a *spanning tree* in a graph.

Definition

Let G be a connected graph. Then a **spanning tree** in G is a subgraph of G that includes every vertex and is also a tree.

For example, the following diagram shows a graph and three of its spanning trees.

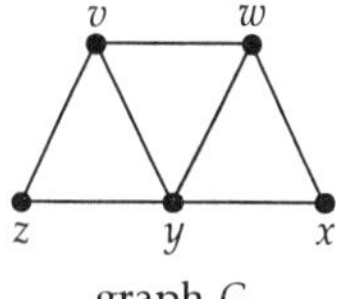

graph G

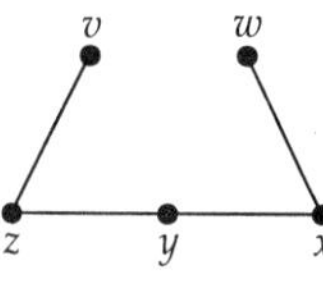

spanning tree

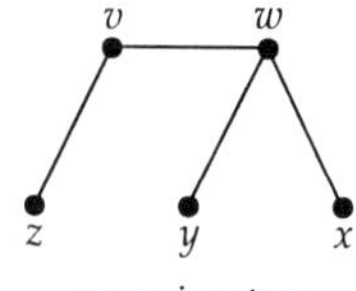

spanning tree

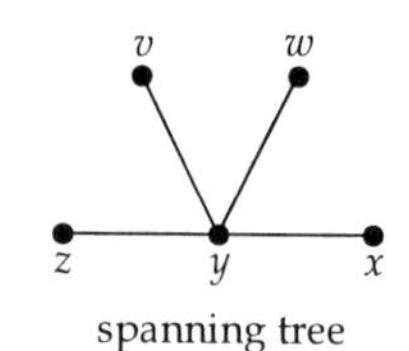

spanning tree

The number of spanning trees in a graph can be very large; for example, the Petersen graph has 2000 labelled spanning trees.

Given a connected graph, we can construct a spanning tree by using either of the following two methods. We illustrate these by applying them to the graph G above.

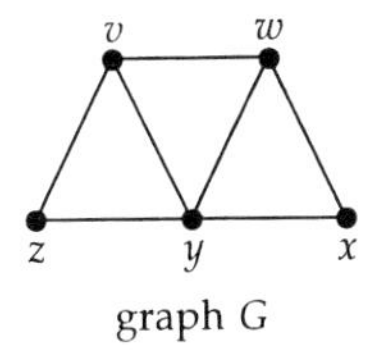

graph G

Building-up method
Select edges of the graph one at a time, in such a way that no cycles are created;
repeat this procedure until all vertices are included.

Example
In the above graph G, we select the edges
  vz, wx, xy, yz;
then no cycles are created.

We obtain the following spanning tree.

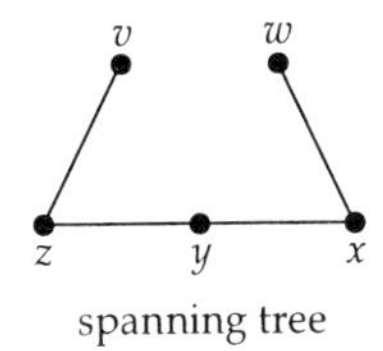

spanning tree

Cutting-down method
Choose any cycle and remove any one of its edges;
repeat this procedure until no cycles remain.

Example
From the above graph G, we remove the edges
  vy (destroying the cycle $vwyv$),
  yz (destroying the cycle $vwyzv$),
  xy (destroying the cycle $wxyw$).
We obtain the following spanning tree.

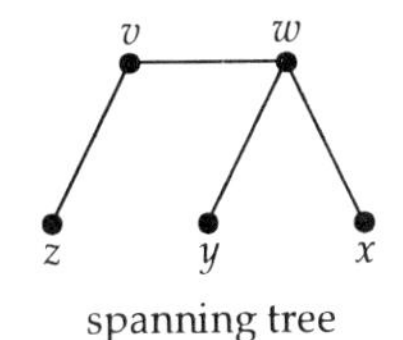

spanning tree

Problem 6.5

Use each method to construct a spanning tree in the complete graph K_5.

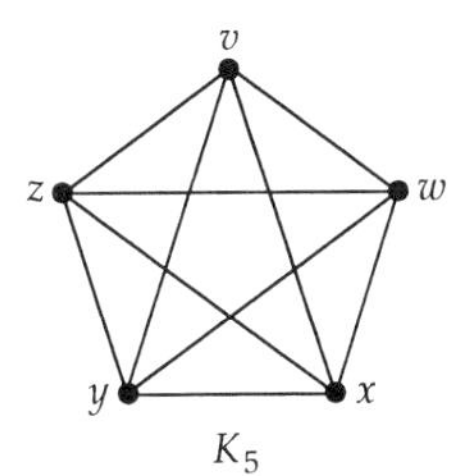

K_5

Problem 6.6

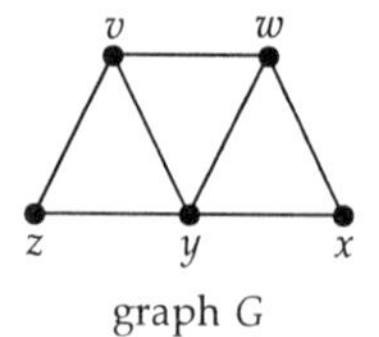

graph G

The above graph G has twenty-one spanning trees. Find as many of them as you can.

Problem 6.7

Find three spanning trees in the Petersen graph.

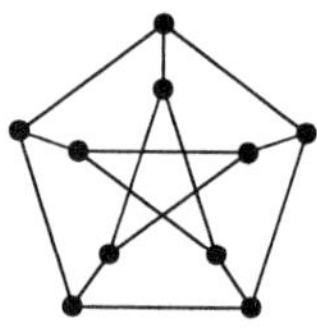

6.3 Rooted Trees

Among the examples of tree structures, one particular type of tree occurs repeatedly. This is the hierarchical structure in which one vertex is singled out as the starting point, and the branches fan out from this vertex. We call such trees **rooted trees**, and refer to the starting vertex as the **root**. For example, the tree representing the lines of responsibility of a company is a rooted tree, with the managing director as the root.

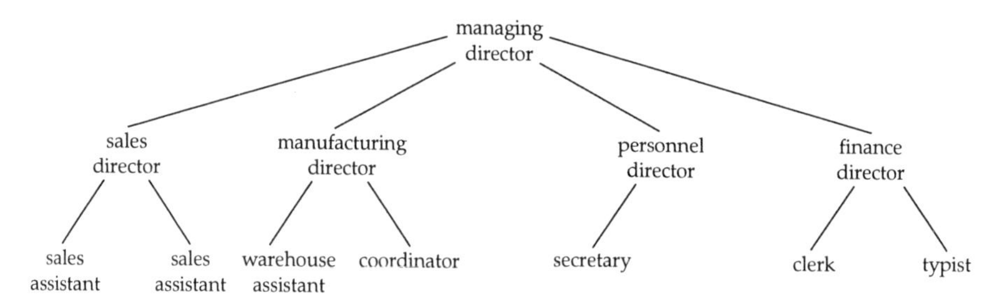

A rooted tree is often drawn as follows, with the root indicated by a small square at the top, and the various branches descending from it. When a path from the top reaches a vertex, it may split into several new branches. Although a top-to-bottom direction is often implied, we usually draw a rooted tree as a graph with undirected edges, rather than as a digraph with arcs directed downwards. A rooted tree in which there are at most *two* descending branches at any vertex is a *binary tree.*

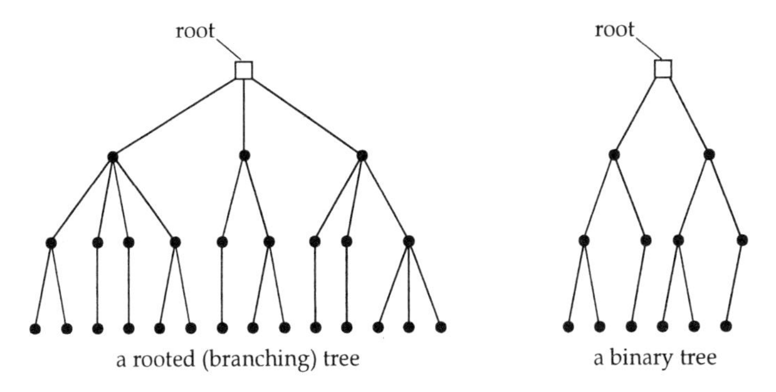

a rooted (branching) tree   a binary tree

Such trees are often called *branching trees*. We have already seen two instances of branching trees – the family tree and the hierarchical tree. There are many further examples, as we now show.

Outcomes of Experiments

If we toss a coin or throw a die several times, then the possible outcomes can be represented by a branching tree. In the case of tossing a coin, each possible outcome has two edges leading from it, since the next toss may be a head (H) or a tail (T), and we obtain a binary tree. For example, if we toss a coin three times, then there are eight possible outcomes, and we obtain the following branching tree.

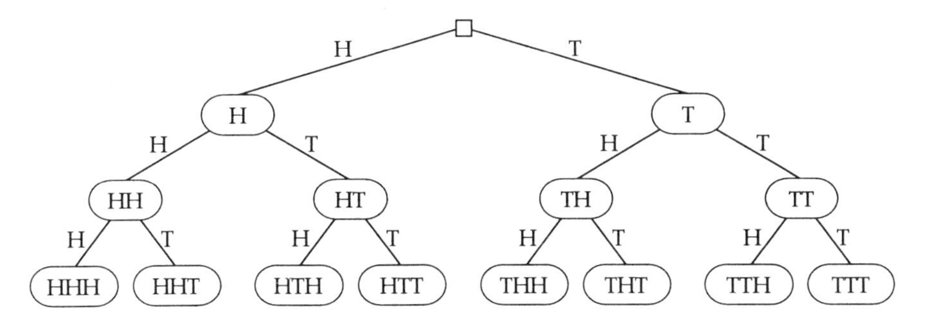

Problem 6.8

Draw the branching tree representing the outcomes of two throws of a six-sided die.

Games of Strategy

Branching trees arise in the analysis of games, particularly games of strategy such as chess and noughts-and-crosses (tic-tac-toe), and for strategic manoeuvres such as those arising in military situations. In such tree representations, a path from the top corresponds to a sequence of moves from each position to the next. The following diagram illustrates the branching tree representing the first three moves in a game of noughts-and-crosses.

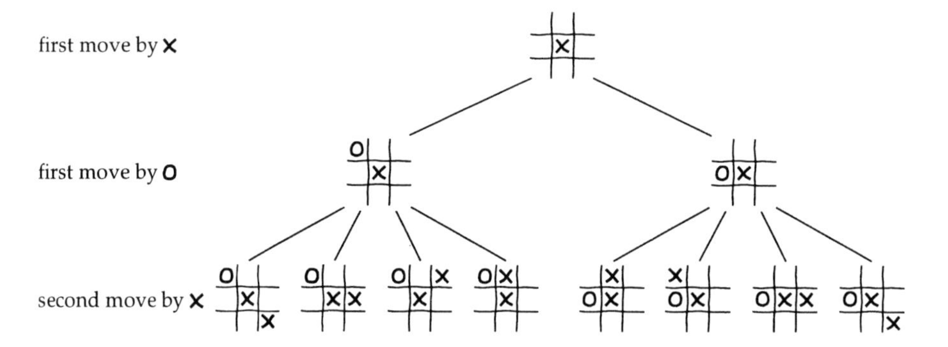

Equivalent moves such as

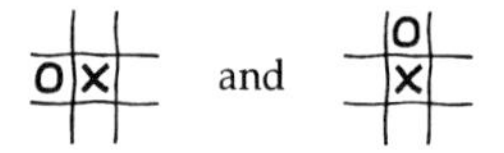

are considered to be the same.

Grammatical Trees

Branching trees occur in the parsing of a sentence in a natural language, such as English. The tree represents the interrelationships between the words and phrases of the sentence, and hence the underlying syntactic structure. Such a branching tree is obtained by splitting the sentence into noun phrases and verb phrases, then splitting these phrases into nouns, verbs, adjectives, and so on. For example, the structure of the sentence *Good students read books* can be represented by the following tree.

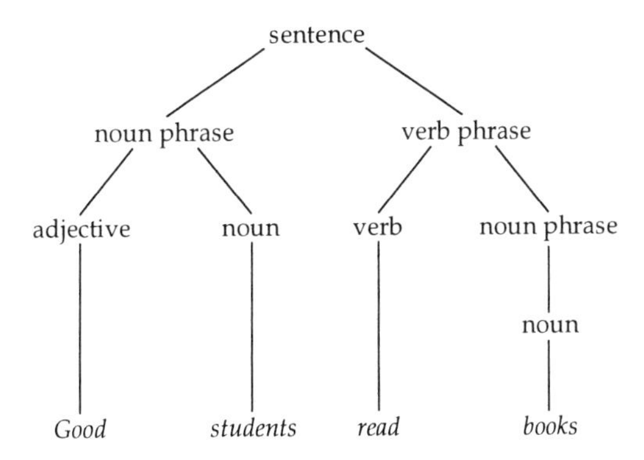

If a sentence is ambiguous, we can use branching trees to distinguish between the different sentence constructions. For example, the newspaper headline *Council rents rocket* can be interpreted in two ways, as illustrated by the following trees.

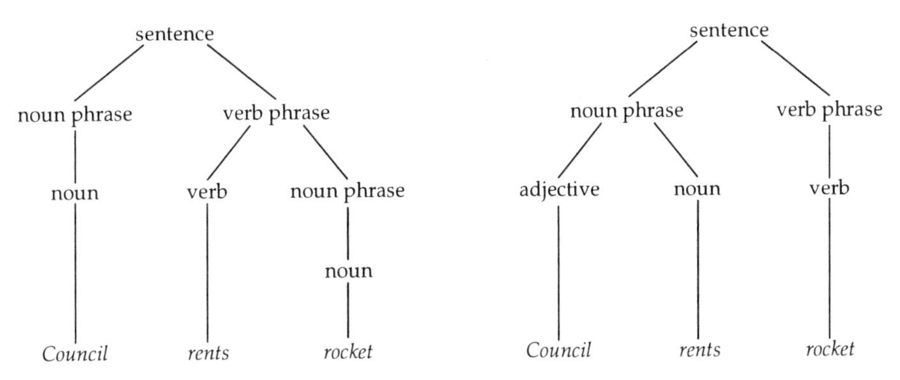

Computer Science

Rooted tree structures arise in computer science, where they are used to model and describe branching procedures in programming languages (the languages used to write algorithms to be interpreted by computers). In particular, they are used to store data in a computer's memory in many different ways. For example, consider the list of seven numbers 7, 5, 4, 2, 1, 6, 8. The following trees represent ways of storing this list in the memory – as a *stack* and as a *binary tree*. Each representation has its advantages, depending on how the data is to be manipulated, but in both representations it is important to distinguish where the data starts, so the trees are *rooted* trees.

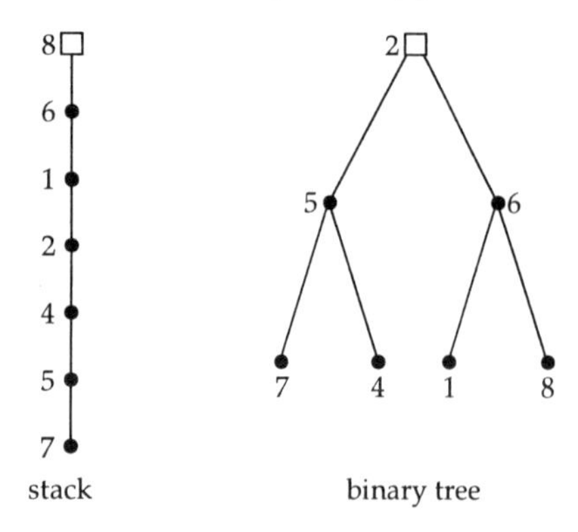

We obtain the tree by writing the numbers in a string 7542168, 'promoting' every second number (5, 2, 6) and then 'promoting' the new second number (2).

Sorting Trees

Sorting trees are branching trees that arise when we wish to make a succession of choices, each dependent on the previous one. For example, a sorting tree can be used to represent the sorting of mail according to postcode.

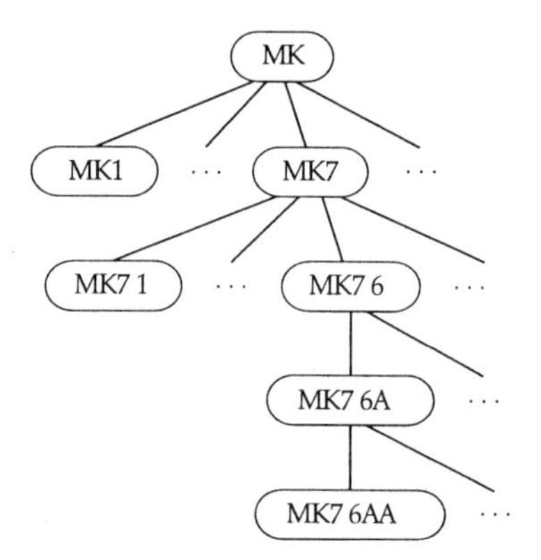

Equivalent Forms

We conclude this subsection by showing how rooted trees can be represented in several different ways. Because such trees are important and widespread, we need to be able to recognize these different forms.

The following diagrams illustrate four equivalent ways of representing the same rooted tree.

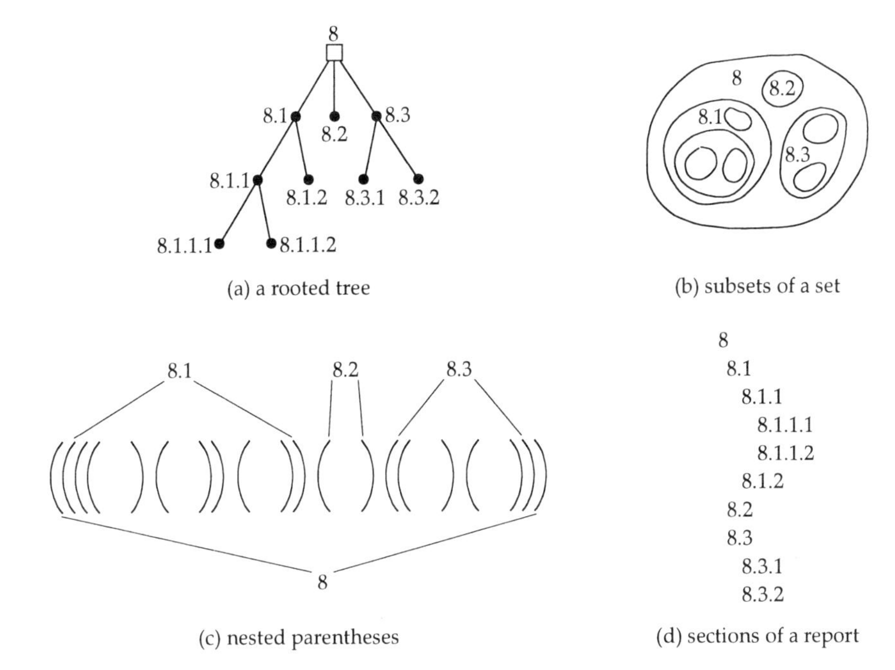

(a) a rooted tree

(b) subsets of a set

(c) nested parentheses

(d) sections of a report

Diagram (a) has the conventional appearance of a *rooted tree.*

Diagram (b) is a system of *subsets of a set* representing, say, the organization of subsystems within a complex machine system; this has the same tree structure as diagram (a), but the different levels are defined by the depth of nesting.

Diagram (c) is a system of *nested parentheses* as used in English text and mathematical equations; again, the level is defined by the depth of nesting.

Diagram (d) is provided by the organization of a report such as a government report or a legal contract; these are often arranged in *nested sections* (subsections, paragraphs, etc.), and the level of each section is indicated by indentation and by the length of the decimal number in the heading.

Problem 6.9

Draw the corresponding subsets of a set and nested parentheses for the following tree.

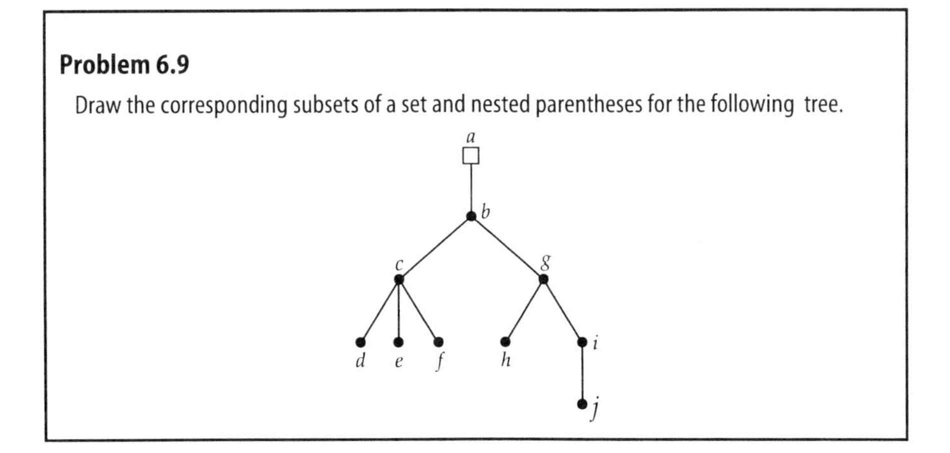

The advantage of tree structures is the ease with which they can be altered or updated. This is particularly important in computer applications, where we can insert or delete branches (such as subroutines) without having to change the whole system. On the other hand, a major drawback of tree structures is that they can be vulnerable to faults or damage. The removal of a single vertex or the breaking of a single edge is sufficient to disconnect or destroy the system, which can be disastrous for efficient operation. A striking example of this vulnerability was given by the collapse of the Inca civilization which virtually disintegrated overnight when the Spanish conquistador Pizarro captured the chief Inca, Emperor Atahuallpa, in 1532. The latter occupied the top position in a rigid hierarchical social pyramid, and his removal destroyed the root of the tree, thereby breaking the chain of command.

6.4 Case Study

Braced Rectangular Frameworks

Many buildings are supported by steel frameworks, which usually consist of rectangular arrangements of girder beams and welded or riveted joints. A simple rectangular framework by itself is inadequate as a stable support structure. A rectangular framework maintains its shape by having rigid joints at its corners and if these fail (often under relatively small loads) the rectangle deforms by bending at its corners to form a parallelogram, as we saw in Chapter 1.

Despite the inherent lack of stability in rectangular frameworks, they are used widely because they fit the natural bias towards the horizontal, the vertical and the perpendicular in buildings. Thus they provide strong vertical columns to support the weight of the building, and strong horizontal beams to support the concrete floors. For many purposes they may be regarded as planar structures (rather than spatial), with pin-joints holding them together (rather than rigid welds). Here we address the problem of making such rectangular frameworks rigid by adding braces in an optimal way.

Bracings

A rectangular framework consists of rectangles, or **bays**. A **brace** is a mechanical restriction on the motion of a bay. A **bracing** of a framework is a particular allocation of braces to bays of the framework. We indicate the insertion of a brace by shading the appropriate bay in the framework.

We simplify matters by adopting a standard rectangular framework in which all the bays have the same size and (undeformed) shape. As we saw in Chapter 1, is not necessary to brace every bay of the framework in order to make the framework rigid.

Consider the three frameworks shown below. Framework (a) has six bays braced, whereas each of the others has just five bays braced.

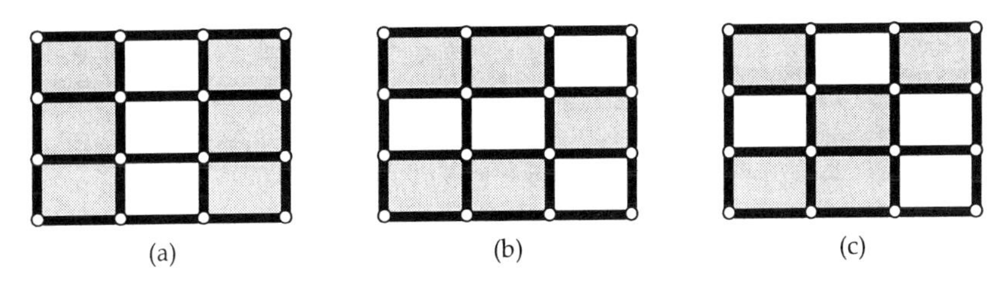

Only framework (c) is rigid, even though it has fewer braces than the first; frameworks (a) and (b) deform kinematically in the plane, as shown below.

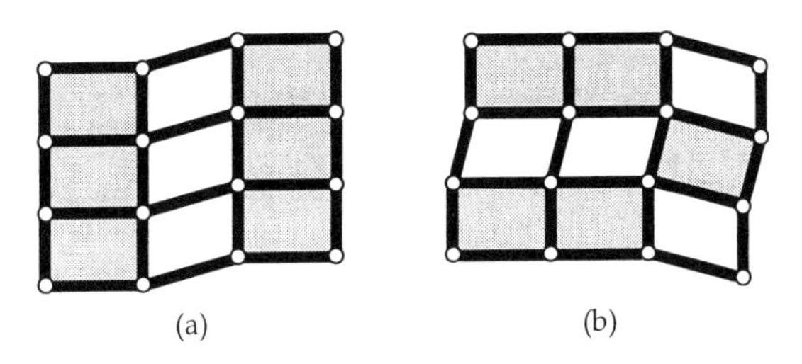

So, given a general framework, which bays need to be braced in order to make it rigid? To answer this question, we introduce some further terminology.

In a standard rectangular framework of congruent rectangles, a **row** is the set of all links forming the vertical sides of a horizontal string of bays, and a **column** is the set of all links forming the horizontal sides of a vertical string of bays. We number the rows $r_1, r_2, \ldots$ sequentially from top to bottom and the columns $c_1, c_2, \ldots$ sequentially from left to right. The bay in row i and column j is **bay(i, j)**.

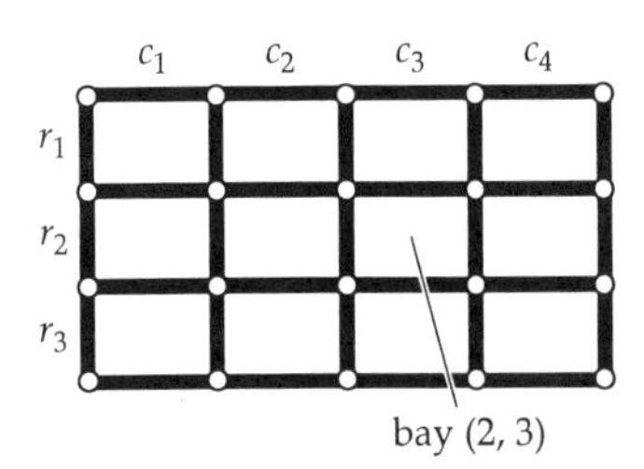

What happens to the links when a framework deforms? Under a deformation, all the links in any one row remain parallel. Similarly, all the links in any one column remain parallel. Also, if a bay in a framework is braced, then all the links in this bay's rows are perpendicular to all the links in this bay's columns; in general, the links in other rows and columns do not satisfy the parallelism or the perpendicularity relationships of the braced bay.

Now, if a framework is to be rigid, we must ensure that *all* the links in *all* the rows remain parallel and that *all* the links in *all* the columns remain parallel. For example, consider framework (c):

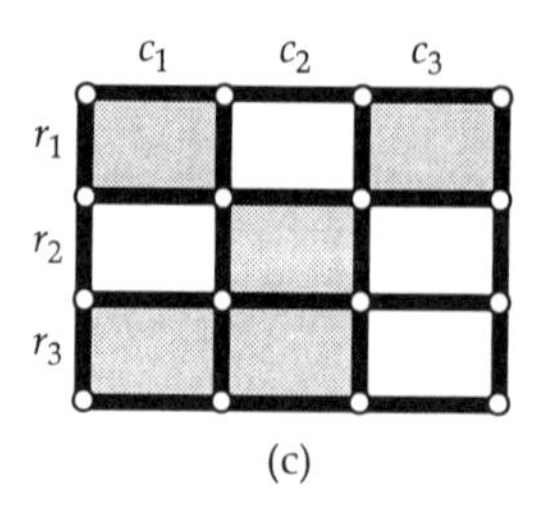

(c)

We deduce that

row r_1 must remain perpendicular to columns c_1 and c_3,
row r_2 must remain perpendicular to column c_2,
row r_3 must remain perpendicular to columns c_1 and c_2.

Now, since column c_1 must remain perpendicular to rows r_1 and r_3, rows r_1 and r_3 must remain parallel to each other. Similarly, since column c_2 must remain perpendicular to rows r_2 and r_3, rows r_2 and r_3 must remain parallel to each other. Hence *all three rows remain parallel to each other*. Consequently, since row r_1 must remain perpendicular to columns c_1 and c_3, and row r_2 must remain perpendicular to column c_2, *all three columns remain parallel to each other*. Furthermore, we can deduce that *each row remains perpendicular to each column*. The only possible conclusion is that the framework is rigid.

We now use these observations to deduce a criterion for a braced rectangular framework to be rigid.

Rigidity Criterion

If a braced rectangular framework, with rows $r_1, r_2, r_3, \ldots$, and columns $c_1, c_2, c_3, \ldots$, is rigid, then the braces must be located such that, under any attempted deformation of the framework,

- r_i remains parallel to r_j, for all r_i and r_j
- c_i remains parallel to c_j, for all c_i and c_j
- r_i remains perpendicular to c_j, for all r_i and c_j.

The task of checking the conditions above becomes tedious as the number of rows and columns in the framework increases significantly, so we look for a better approach.

Braced Frameworks and Bipartite Graphs

The rigidity conditions can be formulated in terms of a bipartite graph. This graph formulation provides a powerful technique for determining the rigidity of a given rectangular framework, and for deciding where to insert braces in a framework. It also provides a systematic method for obtaining a rigid framework which is *minimally* braced, in the sense that it has the fewest number of braces, so the approach leads to significant structural economy.

To see how it works, consider the 3×3 framework shown on the left below. We model this with a bipartite graph whose two sets of three vertices, labelled r_1, r_2, r_3 and c_1, c_2, c_3, represent the rows and columns of the framework, respectively. A vertex r_i is joined to a vertex c_j by an edge if and only if the bay(i, j) is braced.

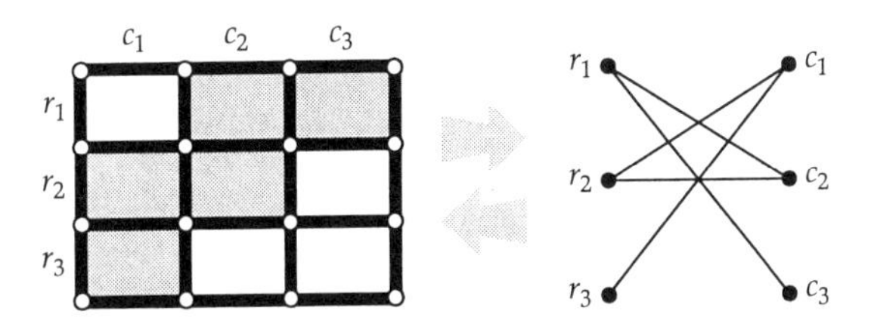

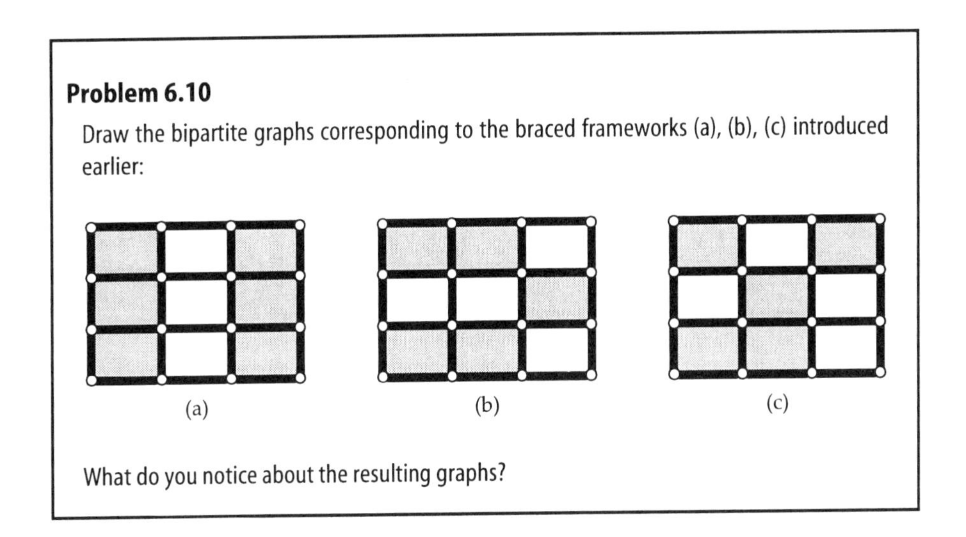

Problem 6.10

Draw the bipartite graphs corresponding to the braced frameworks (a), (b), (c) introduced earlier:

What do you notice about the resulting graphs?

Framework (c) in Problem 6.10 is rigid and its bipartite graph is connected. However, frameworks (a) and (b) in Problem 6.10 are not rigid and their bipartite graphs are not connected. So, does the connectedness of the bipartite graph correspond to the rigidity of the framework?

Consider the following framework and its bipartite graph.

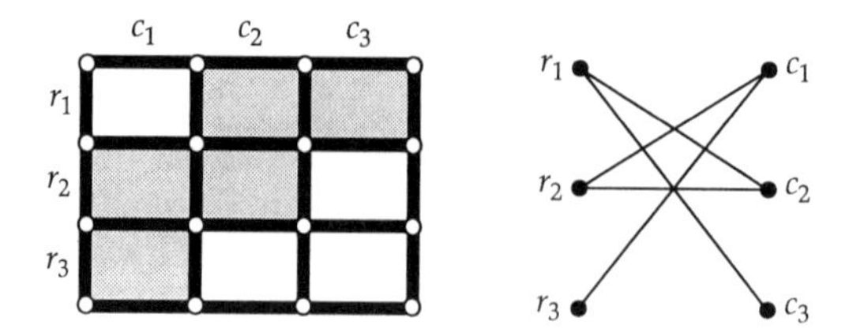

In the framework, bay(1, 2) and bay(2, 2) are braced, so

row r_1 remains perpendicular to column c_2,
row r_2 remains perpendicular to column c_2,

and hence

row r_1 remains parallel to row r_2.

Correspondingly, in the bipartite graph,

r_1 is joined to c_2 by an edge r_1c_2,
r_2 is joined to c_2 by an edge r_2c_2,

and hence

r_1 is joined to c_2 by a path $r_1c_2r_2$.

This correspondence between parallelism and perpendicularity in the framework and a path in the graph extends to all pairs of rows and/or columns that must remain parallel or perpendicular. Furthermore, the correspondence enables us to deduce the maintenance of parallelism or perpendicularity in the framework from any path in the graph.

For example, in the above bipartite graph,

r_3 is joined to r_2 by the path $r_3c_1r_2$,

so

r_3 is joined to c_1 by an edge,
r_2 is joined to c_1 by an edge.

It follows that, in the framework,

row r_3 remains perpendicular to column c_1,
row r_2 remains perpendicular to column c_1,

so

row r_3 remains parallel to row r_1.

In fact,

this framework is rigid and its bipartite graph is connected.

Now consider the following braced rectangular framework and its bipartite graph.

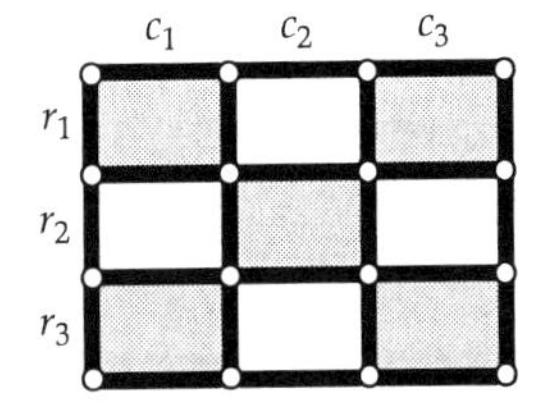

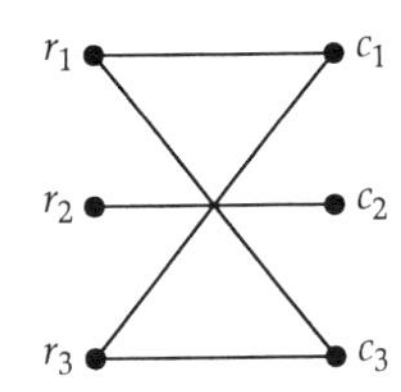

This time the situation is different. We can show that

row r_1 remains parallel to row r_3 (path $r_1c_1r_3$ in the graph),

and that

column c_1 remains parallel to column c_3 (path $c_1r_1c_3$ in the graph).

But we cannot show that

row r_2 remains parallel to row r_1,

nor that

row r_2 remains parallel to row r_3,

because there are no paths in the bipartite graph from r_2 to r_1 or r_3.
In this case,

the framework is not rigid and its bipartite graph is disconnected.

We summarize the result of our discussion as follows.

> **A braced rectangular framework is rigid if and only if its associated bipartite graph is connected.**

Problem 6.11

By constructing the associated bipartite graph, determine whether each of the following braced rectangular frameworks is rigid.

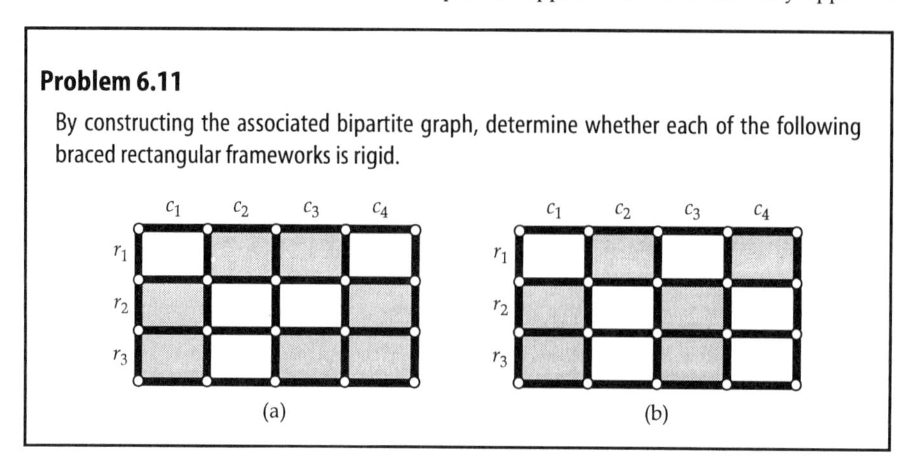

(a) (b)

Minimum Bracings

The associated bipartite graph of a rigid braced rectangular framework is connected. If the graph has a cycle, then the removal of any edge of this cycle does not disconnect the graph. Hence the new graph (with one less edge) represents another rigid bracing.

For example, in the following graph, we remove the edge r_3c_3 of the cycle $r_1c_1r_3c_3r_1$.

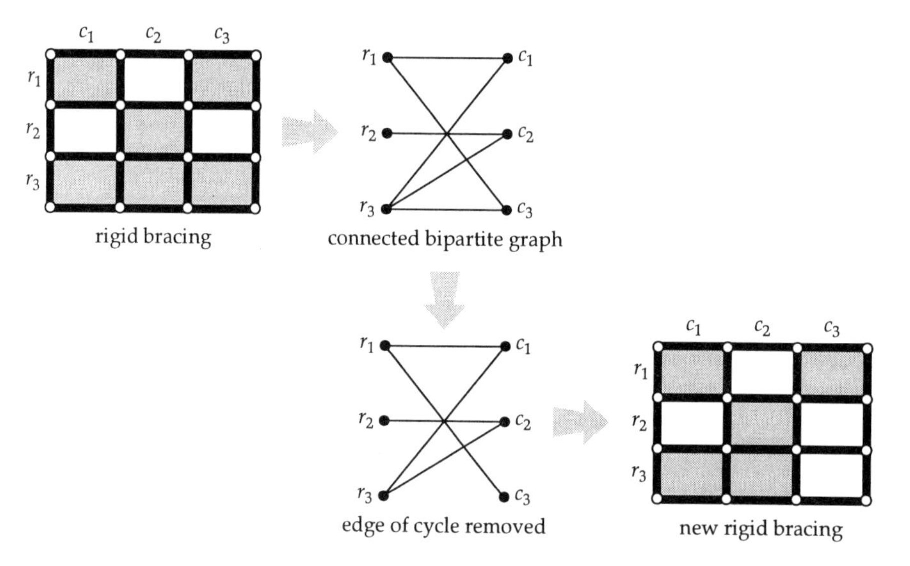

If we repeat this procedure and continue to remove edges from cycles in the bipartite graph of a rigid framework, we eventually obtain a tree. Moreover, since we destroy all the cycles in this process, we produce a *spanning* tree.

A braced rectangular framework whose associated bipartite graph is a spanning tree is **minimally braced**; the bracing is a **minimum bracing**.

Using the familiar properties of a tree, we state these results as follows.

A given bracing of a rectangular framework is not a minimum bracing if the associated bipartite graph has either of the following properties:

- the graph has n vertices and more than $n - 1$ edges;
- the graph contains a cycle.

Problem 6.12

(a) By constructing the associated bipartite graph, show that the following braced rectangular framework is minimally braced.

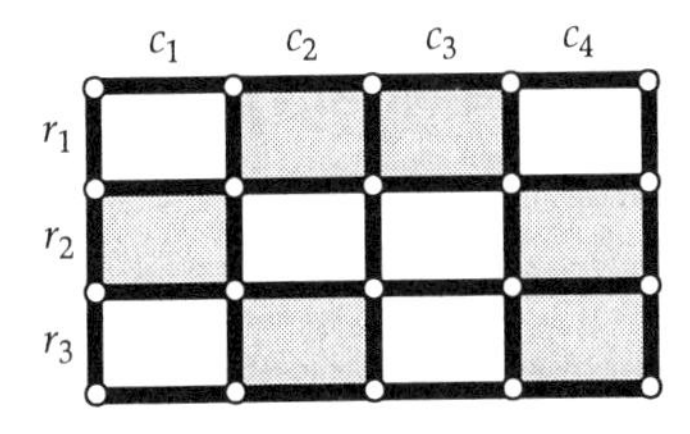

(b) Construct another minimum bracing for a 3×4 rectangular framework.

New Minimum Bracings From Old

We can derive further minimum bracings from a given minimum bracing by considering various alterations to the bipartite graph which maintain the rigidity. If we insert an extra brace into an unbraced bay in a minimum bracing, then we create a cycle in the corresponding bipartite graph, and of course the new framework remains rigid.

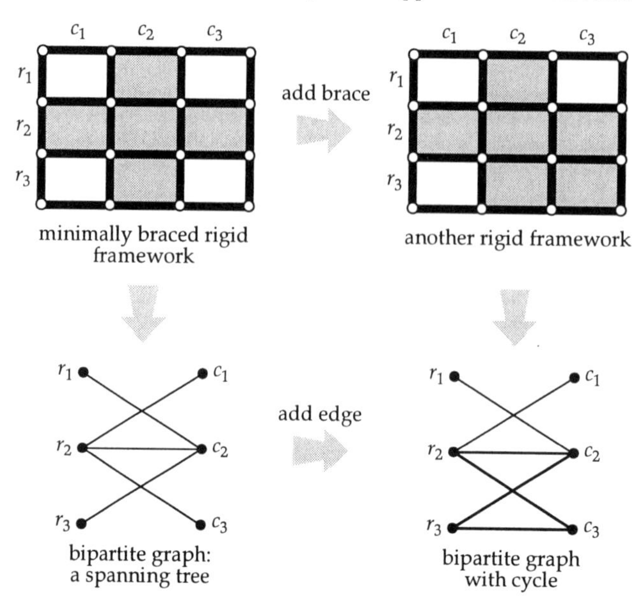

The new bipartite graph created by adding an extra brace in the above framework is no longer a spanning tree since it contains the cycle $r_2c_2r_3c_3r_2$. But if we now remove a *different* edge (that is, not the one we added) from this cycle, we obtain a spanning tree different from the original. Since the cycle we created has length 4, we can remove any one of its four edges. This leads to four different spanning trees: the original tree and three others, which give rise to three new minimum bracings for the framework.

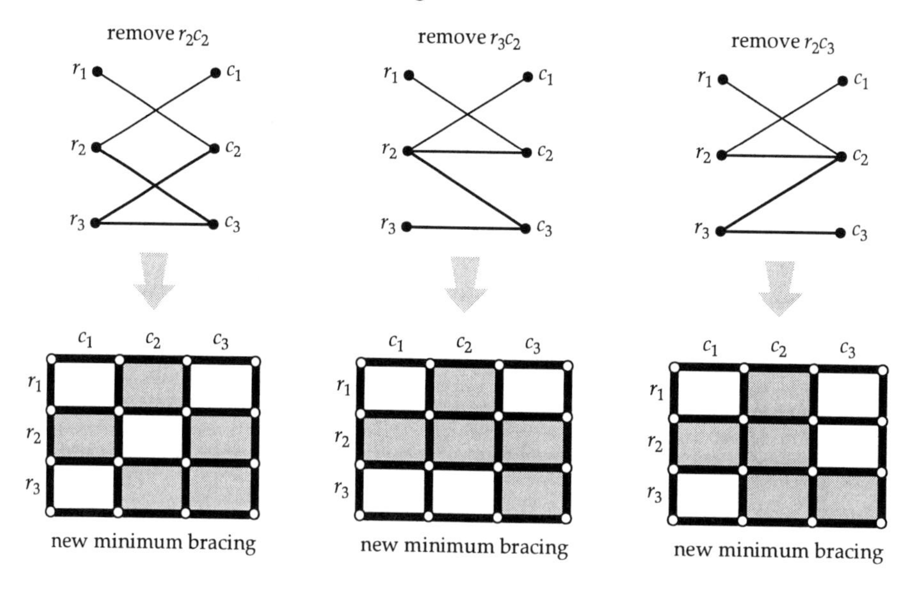

Exercises 6

Mathematical Properties of Trees

6.1 By adding a new edge in all possible ways to each unlabelled tree with seven vertices, draw the twenty-three unlabelled trees with eight vertices.

6.2 A **forest** is a graph (not necessarily connected), each of whose components is a tree.

(a) Let G be a forest with n vertices and k components. How many edges does G have?

(b) Construct a forest with 12 vertices and 9 edges.

(c) Is it true that every forest with k components has at least $2k$ vertices of degree 1?

Spanning Trees

6.3 Find all the spanning trees in each of the following graphs.

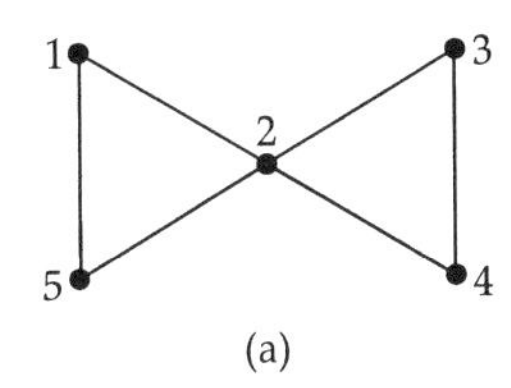

(a)

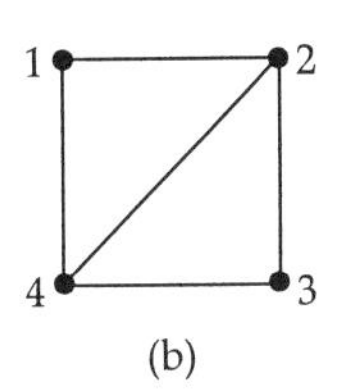

(b)

6.4 How many spanning trees has $K_{2,3}$? $K_{2,100}$?

6.5 A **spanning forest** in a graph G (not necessarily connected) is obtained by constructing a spanning tree for each component of G.

(a) Find a spanning forest for the following graph.

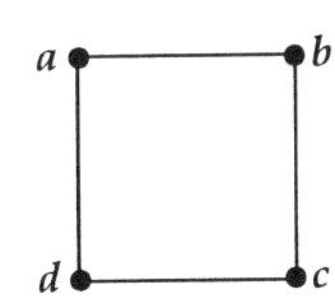

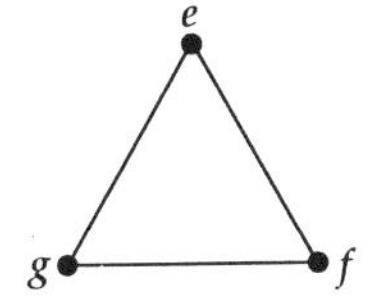

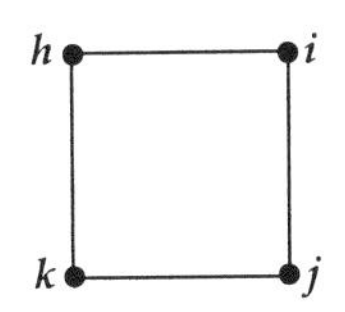

(b) Let G be a graph, and let F be a subgraph of G. If F is a forest which includes all vertices of G, is F necessarily a spanning forest of G?

Rooted Trees

6.6 The ambiguous sentence *Help rape victims* appeared as a newspaper headline, and can be interpreted in two ways. Draw two tree structures that correspond to this sentence.

6.7 Draw a rooted tree and a system of subsets of a set with essentially the same structure as the following nested parentheses:

(((( ))((( ))((( )))))( )).

Case Study

Braced Rectangular Frameworks

6.8 Consider the following braced rectangular framework:

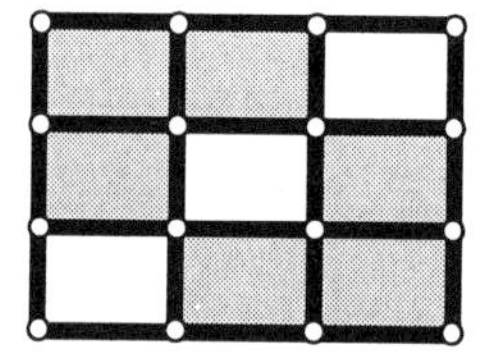

(a) Explain why the framework is rigid.

(b) Determine the braces that may be removed without destroying the rigidity.

6.9 Show that:

(a) if we permute the rows (or columns) of a rigid bracing, then we obtain another rigid bracing;

(b) if we permute the rows (or columns) of a minimum bracing, then we obtain another minimum bracing.

(This exercise indicates another method of obtaining new minimum bracings from old.)

6.10 Use one of the minimum bracings in a 3×3 rectangular framework given earlier to determine two further minimum bracings.

6.11 Determine the number of braced cells in a minimum bracing of an $r \times s$ rectangular framework.

Chapter 7
Counting Trees

After studying this chapter, you should be able to:

- find the *Prüfer sequence* associated with a given labelled tree, and *vice versa*;
- state and use Cayley's theorem for labelled trees;
- understand the method for counting binary trees;
- distinguish between *central* and *bicentral* trees, and describe their use in the counting of alkanes.

Much of the interest and importance of trees arises from the fact that, in many ways, a tree is the simplest non-trivial type of graph. Consequently, when investigating a problem in graph theory, it is sometimes convenient to start by investigating the corresponding problem for trees.

In this chapter we turn our attention to the problem of counting trees with particular properties.

Two typical enumeration problems are given below.

How many irrigation canal systems are there linking eight locations with seven canals?

How many molecules are there with the formula C_6H_{14}?

We can reduce such problems to that of determining the number of trees with a particular property. For example, the first problem amounts to counting labelled trees with eight vertices; the second reduces to that of determining the number of unlabelled trees with six vertices, each of degree 4 or less.

7.1 Counting Labelled Trees

As we saw in Chapter 2, counting problems for labelled graphs are usually much easier to solve than their analogues for unlabelled graphs; in fact, there are certain types of graph for which the former problem has been solved while the latter problem remains unsolved. However, the problems of counting labelled and unlabelled trees have both been solved, although the former problem is much easier to solve than the latter.

The following table lists the numbers of unlabelled and labelled trees with n vertices, where $n \leq 9$. The labelled case illustrates the *combinatorial explosion*, mentioned in Chapter 1.

n	1	2	3	4	5	6	7	8	9
unlabelled trees	1	1	1	2	3	6	11	23	47
labelled trees	1	1	3	16	125	1296	16807	262144	4782969

The labelled trees with at most three vertices are as follows.

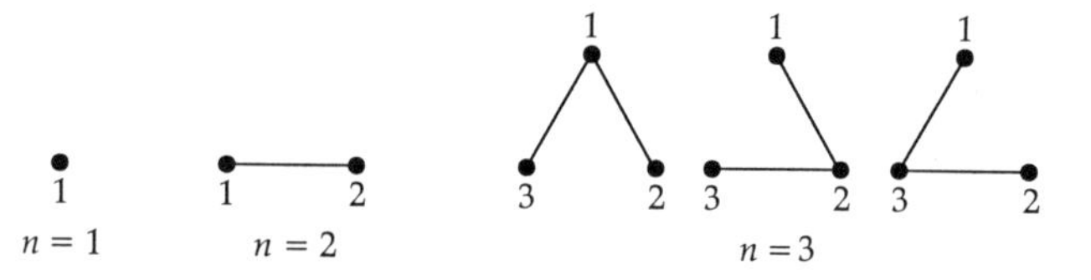

Problem 7.1

Draw the sixteen labelled trees with four vertices.

Hint Draw the two unlabelled trees with four vertices ($K_{1,3}$ and P_4) and label them in all possible ways.

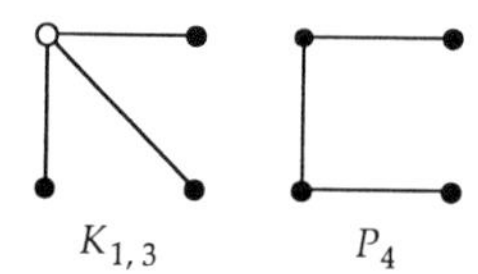

Problem 7.2

Using the above table, try to guess a simple formula for the number of labelled trees with n vertices.

In fact, there are exactly n^{n-2} labelled trees with n vertices; this is known as *Cayley's theorem*. To prove this result, we construct a one–one correspondence between labelled trees with n vertices and sequences of $n - 2$ numbers, called **Prüfer sequences**. (This construction is due to H. Prüfer.) The construction of the one–one correspondence between labelled trees and Prüfer sequences is in two parts.

In the first part, given a labelled tree with n vertices, we construct a Prüfer sequence (denoted by bold type)

$$(\mathbf{a}_1, \mathbf{a}_2, \mathbf{a}_3, \ldots, \mathbf{a}_{n-2}),$$

where each $\mathbf{a}_i$ is one of the integers 1, 2, 3, ..., n (allowing repetition); for example, for $n = 7$, two possible sequences are (**1**, **2**, **3**, **4**, **5**) and (**7**, **7**, **7**, **1**, **2**). This construction is given below and illustrated in Example 7.1A.

Construction A: To Construct a Prüfer Sequence from a Given Labelled Tree

Step 1 Find the vertices of degree 1 and choose the one with the smallest label.

Step 2 Look at the vertex adjacent to the one just chosen and place its label in the first available position in the Prüfer sequence.

Step 3 Remove the vertex chosen in Step 1 and its incident edge, leaving a smaller tree.

Repeat Steps 1–3 for the remaining tree, continuing until there are only two vertices left, then STOP: the required Prüfer sequence has been constructed.

Example 7.1A

Consider the following labelled tree.

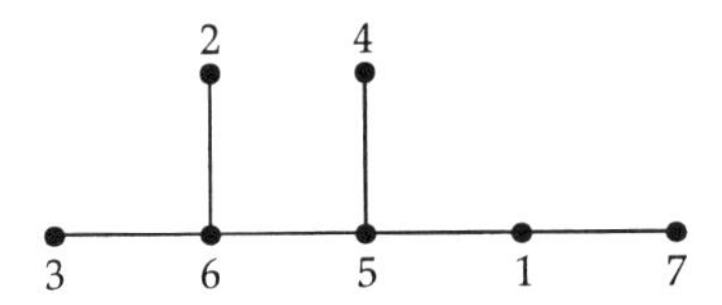

This labelled tree has 7 vertices, so the corresponding Prüfer sequence has 5 numbers.

First term		*Prüfer sequence*
Step 1	The vertices of degree 1 are vertices 3, 2, 4 and 7; the one with the smallest label is vertex 2.	
Step 2	The vertex adjacent to vertex 2 is vertex 6, so the first term in the sequence is **6**.	(**6**, ?, ?, ?, ?)

Step 3 Removal of vertex 2 and edge 2**6** leaves the following tree.

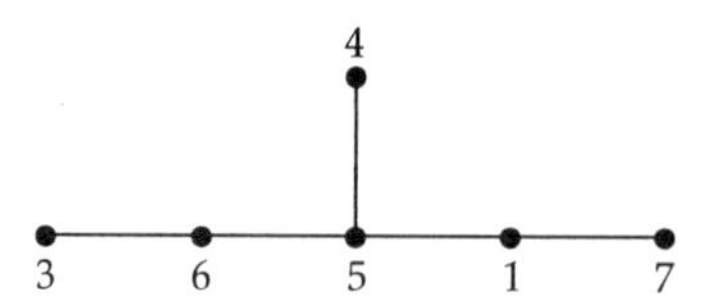

Second term

Step 1 The vertices of degree 1 are vertices 3, 4 and 7; the one with the smallest label is vertex 3.

Step 2 The vertex adjacent to vertex 3 is vertex 6, so the second term in the sequence is **6**.

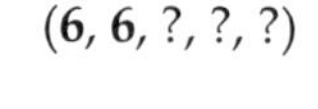

Step 3 Removal of vertex 3 and edge 3**6** leaves the following tree.

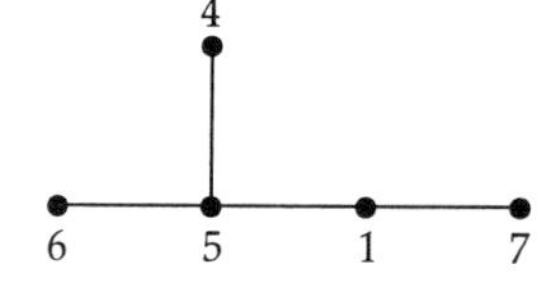

Third term

Step 1 The vertices of degree 1 are vertices 4, 6 and 7; the one with the smallest label is vertex 4.

Step 2 The vertex adjacent to vertex 4 is vertex 5, so the third term in the sequence is **5**.

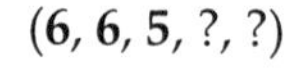

Step 3 Removal of vertex 4 and edge 4**5** leaves the following tree.

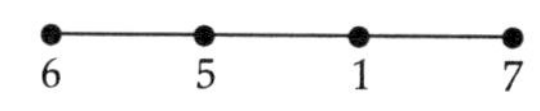

Fourth term

Step 1 The vertices of degree 1 are vertices 6 and 7; the one with the smaller label is vertex 6.

Step 2 The vertex adjacent to vertex 6 is vertex 5, so the fourth term in the sequence is **5**.

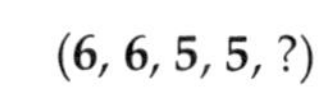

Step 3 Removal of vertex 6 and edge 6**5** leaves the following tree.

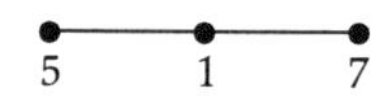

Fifth term

Step 1 The vertices of degree 1 are vertices 5 and 7; the one with the smaller label is vertex 5.

Step 2 The vertex adjacent to vertex 5 is vertex 1, so the next term in the sequence is **1**. **(6, 6, 5, 5, 1)**

Step 3 Removal of vertex 5 and edge **51** leaves a tree with only two vertices.

STOP.

We thus obtain the Prüfer sequence **(6, 6, 5, 5, 1)**. ❐

Problem 7.3

Find the Prüfer sequence corresponding to each of the following labelled trees.

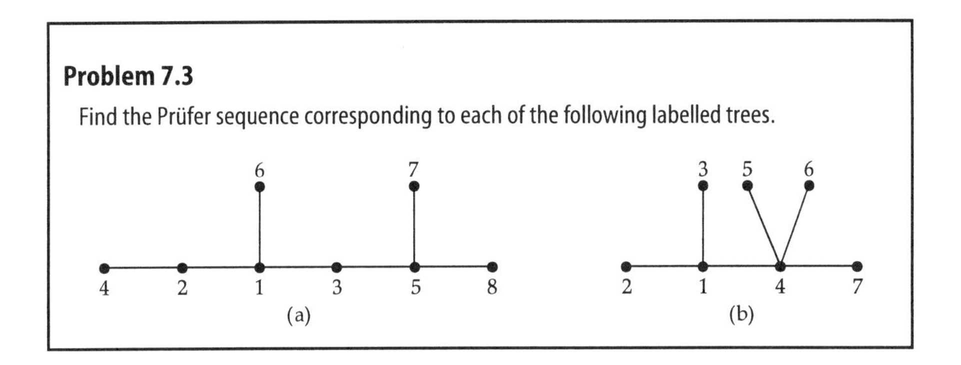

(a) (b)

In the second part of the construction of the one–one correspondence between labelled trees and Prüfer sequences, we give the converse of Construction A.

In this second part, given a sequence $(\mathbf{a}_1, \mathbf{a}_2, \mathbf{a}_3, \ldots, \mathbf{a}_{n-2})$, where each $\mathbf{a}_i$ is one of the integers 1, 2, 3, ..., n (allowing repetition), we construct a labelled tree with n vertices. This construction is given below and illustrated in Example 7.1B.

Construction B: To Construct a Labelled Tree from a Given Prüfer Sequence

Step 1 Draw the n vertices, labelling them from 1 to n, and list the integers from 1 to n.

Step 2 Find the smallest number that is in the list but not in the Prüfer sequence, and also find the first number in the sequence; then add an edge joining the vertices with these labels.

Step 3 Remove the first number found in Step 2 from the list and the second number found in Step 2 from the sequence, leaving a smaller list and a smaller sequence.

Repeat Steps 2 and 3 for the remaining list and sequence, continuing until there are only two terms left in the list. Join the vertices with these labels and STOP: the required labelled tree has been constructed.

Example 7.1B

Consider the Prüfer sequence (**6**, **6**, **5**, **5**, **1**). This sequence has five numbers, so the corresponding labelled tree has 7 vertices.

No edges *Labelled tree*

Step 1 Draw the 7 vertices, labelling them 1 to 7, and list the integers from 1 to 7.
The list is (1, 2, 3, 4, 5, 6, 7) and the sequence is (**6**, **6**, **5**, **5**, **1**).

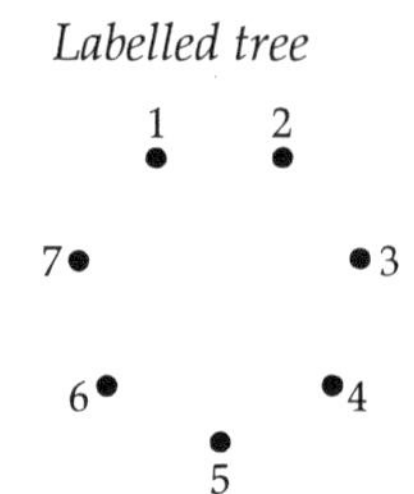

First edge

Step 2 The smallest number in the list but not in the sequence is 2, and the first number in the sequence is **6**, so we add an edge joining vertices 2 and **6**.

Step 3 We remove the number 2 from the list and the number **6** from the sequence.
This leaves the list (1, 3, 4, 5, 6, 7) and the sequence (**6**, **5**, **5**, **1**).

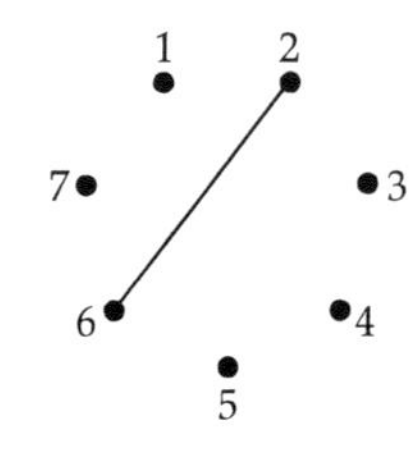

Second edge

Step 2 The smallest number in the new list but not in the new sequence is 3, and the first number in the new sequence is **6**, so we add an edge joining vertices 3 and **6**.

Step 3 We remove the number 3 from the list and the number **6** from the sequence.
This leaves the list (1, 4, 5, 6, 7) and the sequence (**5**, **5**, **1**).

Third edge

Step 2 The smallest number in the new list but not in the new sequence is 4, and the first number in the new sequence is **5**, so we add an edge joining vertices 4 and **5**.

Step 3 We remove the number 4 from the list and the number **5** from the sequence.
This leaves the list (1, 5, 6, 7) and the sequence (**5**, **1**).

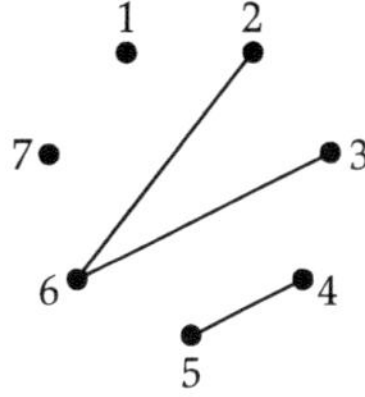

Fourth edge

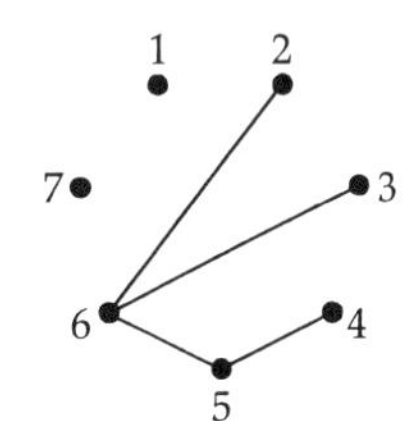

Step 2 The smallest number in the new list but not in the new sequence is 6, and the first number in the new sequence is **5**, so we add an edge joining vertices 6 and **5**.

Step 3 We remove the number 6 from the list and the number **5** from the sequence.
This leaves the list (1, 5, 7) and the sequence (**1**).

Fifth edge

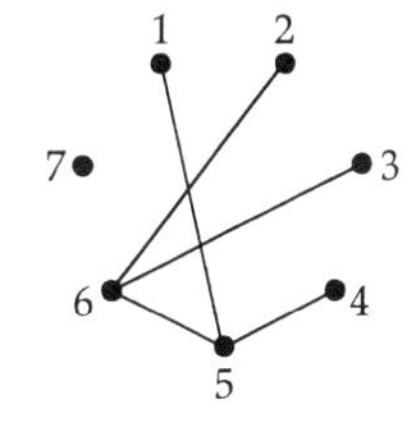

Step 2 The smallest number in the new list but not in the new sequence is 5, and the only number in the new sequence is **1**, so we add an edge joining vertices 5 and **1**.

Step 3 We remove the number 5 from the list and the number **1** from the sequence.
This leaves the list (1, 7) and an empty sequence.

Sixth edge

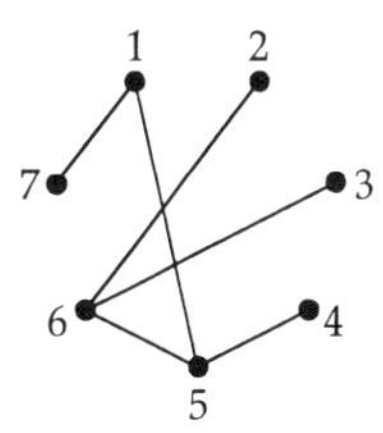

The labels 1 and 7 are the only two terms left in the list, so we join vertices 1 and 7.

STOP

We thus obtain the required labelled tree.

Redrawing the tree without edges crossing, we obtain the following.

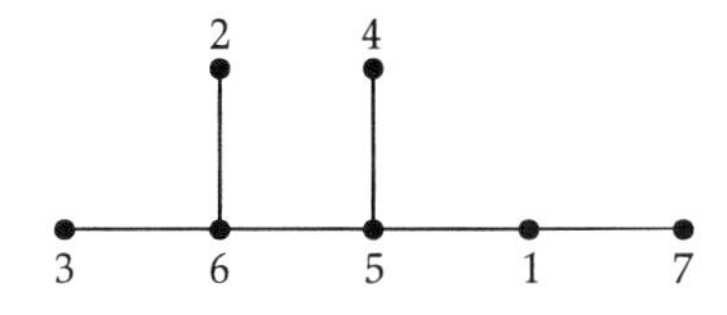

❒

Problem 7.4

Find the labelled tree corresponding to each of the following Prüfer sequences.

(a) **(2, 1, 1, 3, 5, 5)** (b) **(1, 1, 4, 4, 4)**

What do you notice about your results?

Notice that, both in the above examples and in Problems 7.3 and 7.4, the Prüfer sequence arising from a particular labelled tree in Construction A gives rise to the same labelled tree in Construction B; for example, the Prüfer sequence **(6, 6, 5, 5, 1)** arising from the labelled tree in Example 7.1A gives rise to the same labelled tree in Example 7.1B.

This happens in general – if you start with any labelled tree, find the corresponding Prüfer sequence, and then find the labelled tree corresponding to this sequence, you always get back to the original labelled tree. The two constructions above do indeed give us the required one–one correspondence between labelled trees and Prüfer sequences.

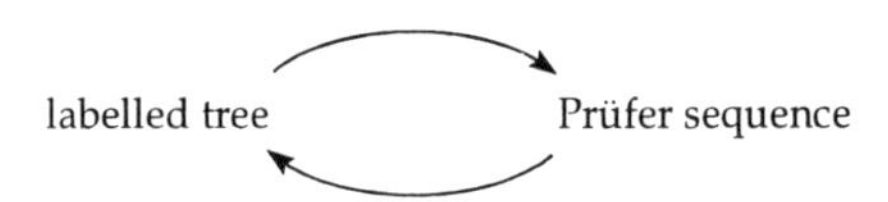

This one–one correspondence can be used to prove Cayley's theorem.

Theorem 7.1: Cayley's Theorem

The number of labelled trees with n vertices is n^{n-2}.

Proof We assume that $n \geq 3$, since the result is clearly true for $n = 1$ and 2.

We construct the above one–one correspondence between the set of labelled trees with n vertices and the set of all Prüfer sequences of the form $(\mathbf{a}_1, \mathbf{a}_2, \mathbf{a}_3, \ldots, \mathbf{a}_{n-2})$, where each $\mathbf{a}_i$ is one of the integers 1, 2, 3, ..., n (allowing repetition). Since there are exactly n possible values for each integer $\mathbf{a}_i$, the total number of such sequences is

$$\underbrace{n \times n \times \ldots \times n}_{(n-2 \text{ terms})} = n^{n-2}$$

so, by the one–one correspondence, the number of labelled trees with n vertices is also n^{n-2}. ■

Problem 7.5

Construct explicitly the one–one correspondence between the sixteen labelled trees with four vertices obtained in Problem 7.1 and the sixteen Prüfer sequences $(\mathbf{a}_1, \mathbf{a}_2)$, where each term is 1, 2, 3 or 4.

Problem 7.6

How many irrigation canal systems are there linking eight locations with seven canals?

Historical Note

The earliest statement of Cayley's theorem occurred in his article *A theorem on trees*, written in 1889, although related results had been described earlier. However, Cayley's 'proof' was unsatisfactory, since he discussed only the case $n = 6$ and his argument cannot easily be generalized to larger values of n. Since then, several proofs have appeared, of which Prüfer's, given in 1918, is probably the best known.

7.2 Counting Binary Trees

In this section we illustrate a different technique – this time for counting certain unlabelled trees – by deriving an equation expressing the number of trees with a given number of vertices in terms of the numbers of trees with fewer vertices; such an equation is called a *recurrence relation*. Most problems of this kind are too complicated for us to consider here, but the following discussion of *binary trees* illustrates some of the techniques involved.

Definition

A **binary tree** is a rooted tree in which the number of descending edges at each vertex is at most 2, and a distinction is made between left-hand and right-hand branches.

The binary trees with at most three vertices (including the root) are as follows; as before, we represent the root by a small square.

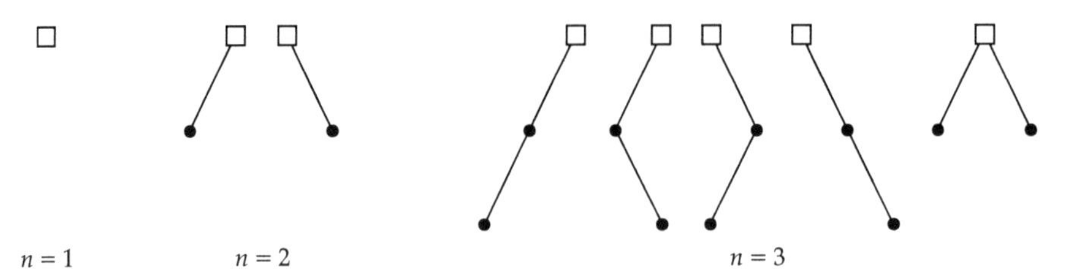

Note that branching to the left and branching to the right at each stage give rise to *different* binary trees. Thus there are five *binary* trees with three vertices, whereas there are only two *rooted* trees with three vertices, as shown below.

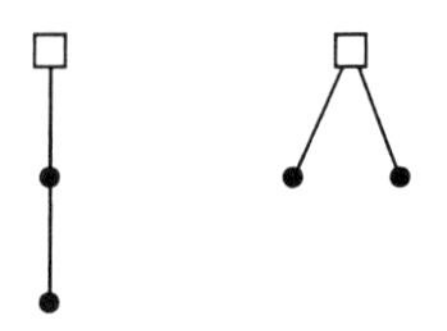

Problem 7.7

Draw the fourteen binary trees with four vertices.

We now consider the question: how many different binary trees are there with a given number of vertices?

Let u_n denote the number of binary trees with n vertices. Then, from the above diagrams and Problem 7.7, we have

$$u_1 = 1, \quad u_2 = 2, \quad u_3 = 5, \quad u_4 = 14.$$

In order to find u_n, for a general value of n, we distinguish between those binary trees with one edge emerging from the root to the left, those with one edge emerging from the root to the right, and those with two edges emerging from the root. For example, for $n = 3$, the first two binary trees depicted above have a 'root edge' to the left, the next two have a 'root edge' to the right, and the last one has both.

Let a_n denote the number of binary trees with n vertices and a root edge to the left, let b_n denote the number of binary trees with n vertices and a root edge to the right, and let c_n denote the number of binary trees with n vertices and two root edges. Then, from the above diagrams and Problem 7.7, we have

for the binary tree with 1 vertex, $a_1 = 0,\ b_1 = 0,\ c_1 = 0,\ u_1 = 1;$
for the binary trees with 2 vertices, $a_2 = 1,\ b_2 = 1,\ c_2 = 0,\ u_2 = 2;$
for the binary trees with 3 vertices, $a_3 = 2,\ b_3 = 2,\ c_3 = 1,\ u_3 = 5;$
for the binary trees with 4 vertices, $a_4 = 5,\ b_4 = 5,\ c_4 = 4,\ u_4 = 14;$

in fact, for $n \geq 2$, we have

$$u_n = a_n + b_n + c_n.$$

Now consider a_n and b_n. Any binary tree with n vertices and a single root edge has one of the forms shown below, and can be obtained by taking a binary tree with $n - 1$ vertices rooted at Q and joining it to the root R by the root edge RQ.

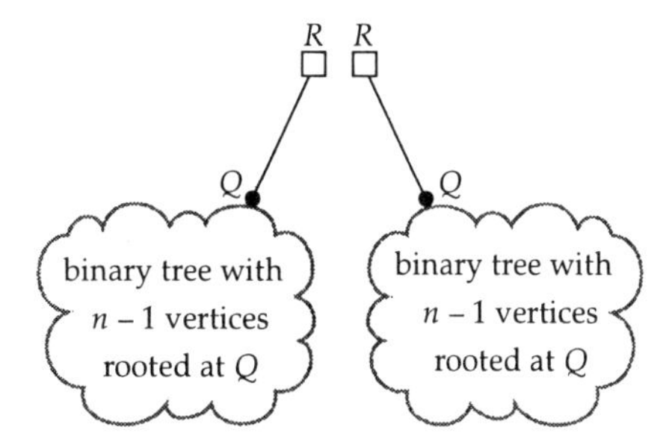

Since the number of binary trees rooted at Q is u_{n-1}, we deduce that

$$a_n = u_{n-1} \quad \text{and} \quad b_n = u_{n-1}, \quad \text{for } n \geq 2.$$

You can check from the above lists of numbers that this is correct for $n = 2, 3$ and 4; for example, $a_4 = b_4 = u_3 = 5$.

Next, consider c_n. Any binary tree with n vertices and two root edges has the form shown below, and can be obtained by taking a binary tree with k vertices rooted at P, and a binary tree with $(n - 1) - k$ vertices rooted at Q, and joining them both to the root R by the root edges RP and RQ.

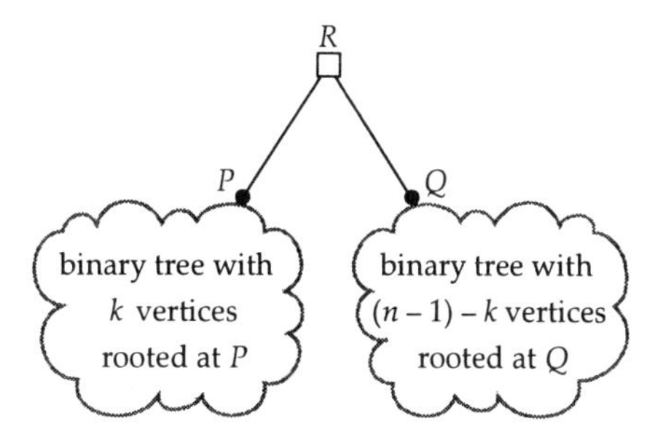

Since there are u_k such binary trees rooted at P, and u_{n-k-1} such binary trees rooted at Q, and as k can be any of the numbers 1, 2, 3, ..., $n-2$, we deduce that, for $n \geq 3$,

$$c_n = u_1 u_{n-2} + u_2 u_{n-3} + u_3 u_{n-4} + \ldots + u_{n-2} u_1$$

Again, you can check from the above lists of numbers that this is correct for $n = 3$ and 4; for example,

$$c_4 = u_1 u_2 + u_2 u_1 = (1 \times 2) + (2 \times 1) = 4.$$

If we now substitute these expressions for a_n, b_n and c_n into the equation $u_n = a_n + b_n + c_n$, we obtain:

$$u_n = 2u_{n-1} + (u_1 u_{n-2} + u_2 u_{n-3} + u_3 u_{n-4} + \ldots + u_{n-2} u_1).$$

Using this recurrence relation with $n = 5, 6, \ldots$, we can find the values of u_5, u_6, and so on, as far as we wish. For example,

$$\begin{aligned} u_5 &= 2u_4 + (u_1 u_3 + u_2 u_2 + u_3 u_1) \\ &= (2 \times 14) + (1 \times 5) + (2 \times 2) + (5 \times 1) = 42. \end{aligned}$$

Problem 7.8

Use the above recurrence relation to determine the number of binary trees with six vertices.

7.3 Counting Chemical Trees

We saw in Chapter 1 that a molecule can be represented as a graph whose vertices correspond to the atoms and whose edges correspond to the bonds connecting them. For example, the molecule *ethanol*, with formula C_2H_5OH, can be represented by a graph as follows:

In such a graph, the degree of each vertex is simply the valency of the corresponding atom – the carbon vertices have degree 4, the oxygen vertex has degree 2, and the hydrogen vertices have degree 1.

Diagrams of the above type can be used to represent the arrangement of atoms in a molecule. They explain the existence of *isomers* – molecules with the same formula but different properties. For example, the molecules *n*-butane and 2-methyl propane (formerly called butane and isobutane) both have the formula C_4H_{10}, but the atoms inside the molecule are arranged differently:

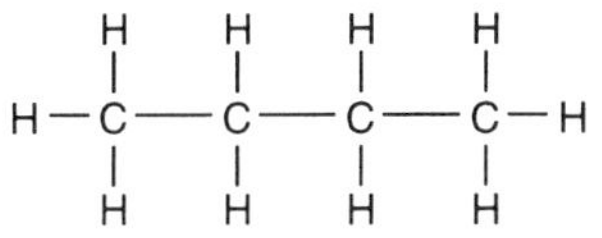

n-butane

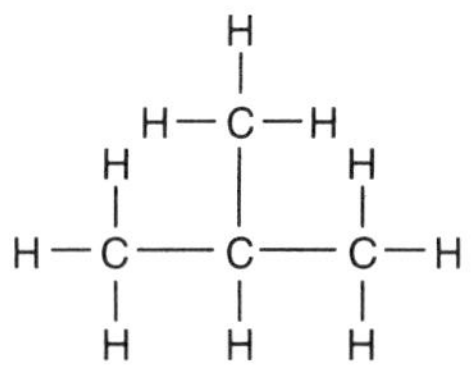

2-methyl propane

It is natural to ask whether there are any other molecules with the formula C_4H_{10}, and this leads us directly to the problem of *isomer enumeration* – the determination of the number of non-isomeric molecules with a given formula. The most celebrated problem of this kind is that of counting the alkanes (paraffins) C_nH_{2n+2}. For small values of n, we can construct a table, as on page 176, where for clarity the carbon vertices are drawn white.

Such diagrams become very complicated as n increases. We can simplify them considerably by removing all the hydrogen atoms, as follows:

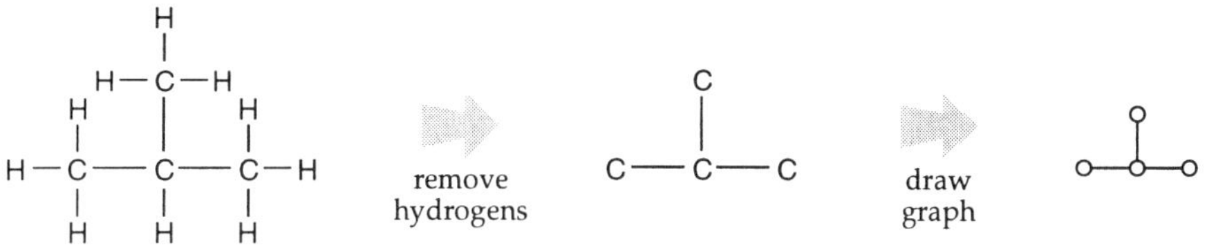

remove hydrogens

draw graph

This leaves the following non-isomorphic *carbon graphs* with up to five carbon atoms:

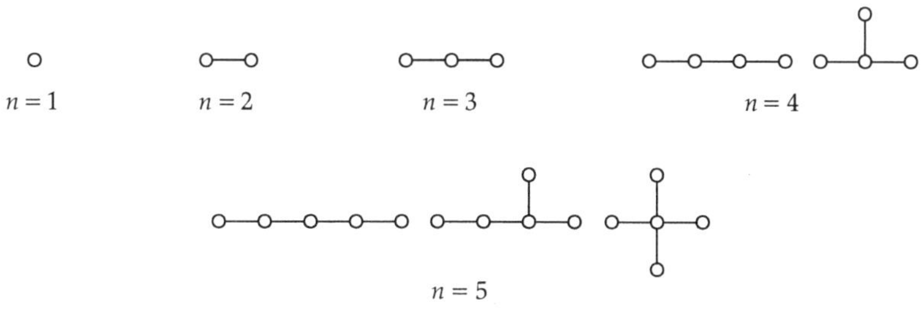

$n = 1$ $n = 2$ $n = 3$ $n = 4$

$n = 5$

Each of these carbon graphs has a tree-like structure in which each vertex has degree 4 or less.

Table of alkanes C_nH_{2n+2} *for* $n \leq 5$

n	chemical formula	name	molecule	graph
1	CH_4	methane	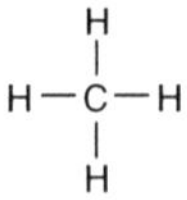	
2	C_2H_6	ethane	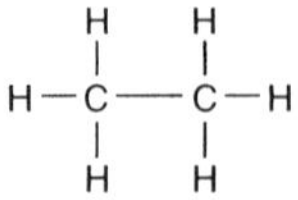	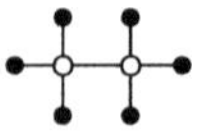
3	C_3H_8	propane	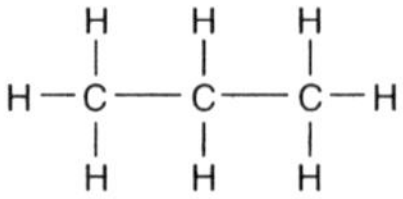	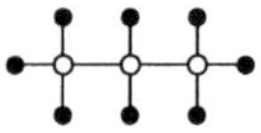
4	C_4H_{10}	n-butane	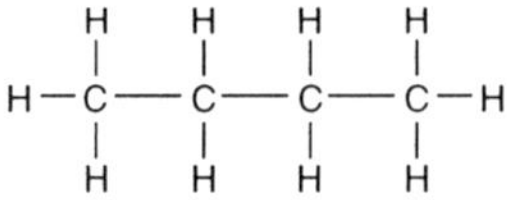	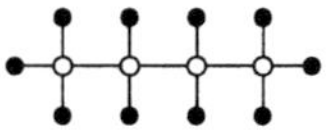
		2-methyl propane	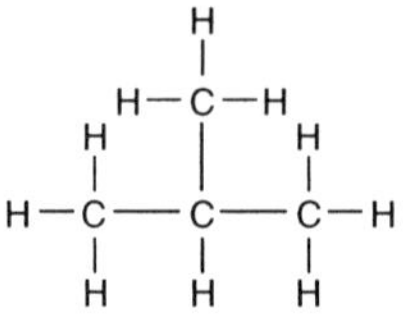	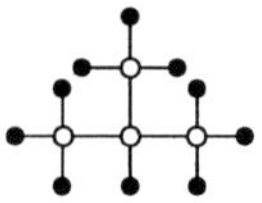
5	C_5H_{12}	n-pentane	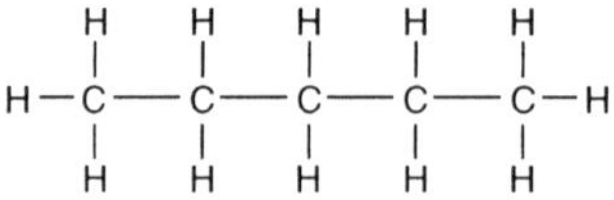	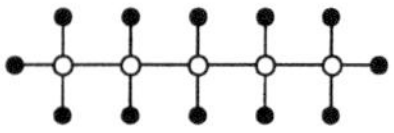
		2-methyl butane	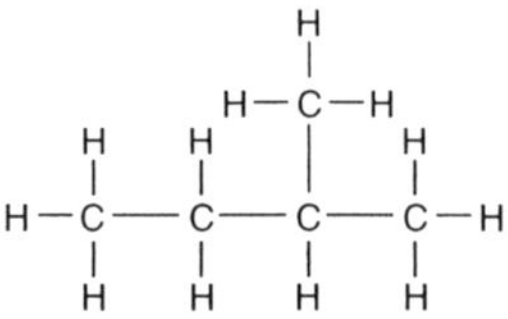	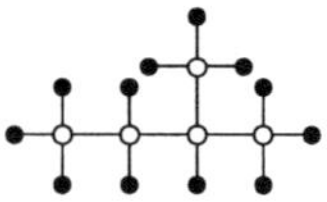
		2,2-dimethyl propane	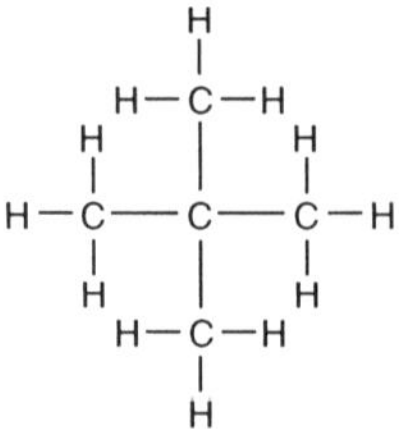	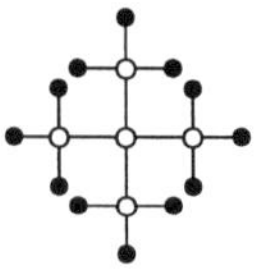

Conversely, from any tree in which each vertex has degree 4 or less, we can construct an alkane by adding enough hydrogen atoms to bring the degree of each carbon vertex up to 4, as follows:

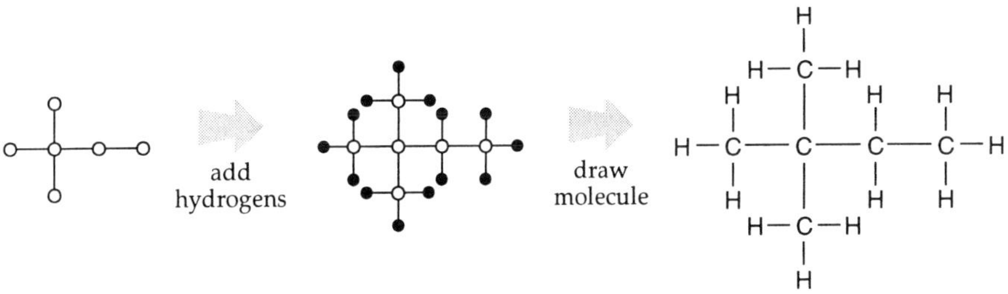

add hydrogens

draw molecule

Problem 7.9

(a) Draw the carbon graph of the molecule

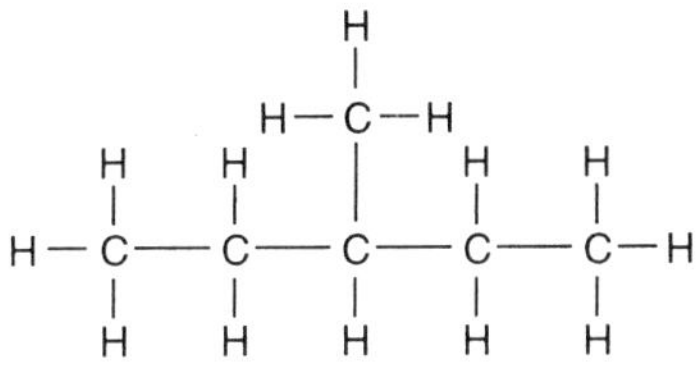

(b) Draw the molecule whose carbon graph is

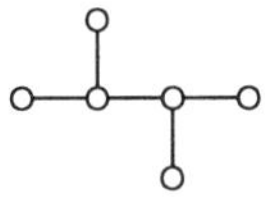

Problem 7.10

Determine the numbers of vertices and edges in the graph of a molecule with formula C_6H_{14}. Deduce that such a graph is a tree.

Using the list of unlabelled trees with six vertices (given in Solution 6.1), we can find all the alkanes with formula C_6H_{14}. There are six such trees:

The first five trees are the carbon graphs of the alkanes: hexane, 2-methyl pentane, 3-methyl pentane, 2,3-dimethyl butane and 2,2-dimethyl butane; the sixth tree cannot be the carbon graph of a molecule, as it has a vertex of degree 5.

We now ask you to show that the graph of *any* alkane is a tree. It follows that the problem of counting alkanes is essentially a tree-counting problem.

Problem 7.11

Determine the numbers of vertices and edges in the graph of any alkane with formula C_nH_{2n+2}. Deduce that such a graph is a tree.

The general problem of determining the number of alkanes with formula C_nH_{2n+2} was solved in the 1870s by the English mathematician Arthur Cayley. In order to outline his method, we introduce the concept of the 'middle' of a tree. For some trees this is easy to define:

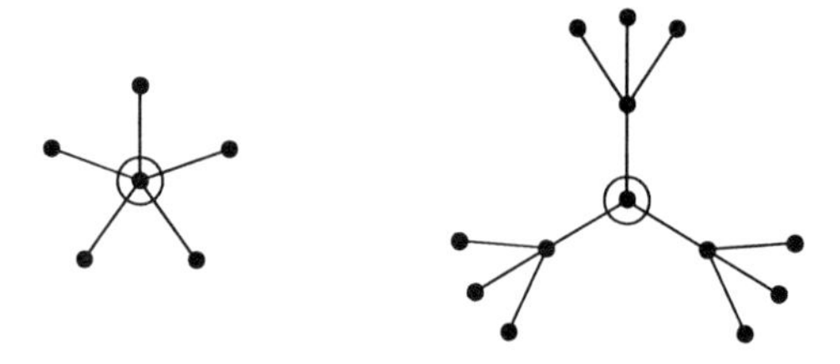

But how do we define the 'middle' of the following trees?

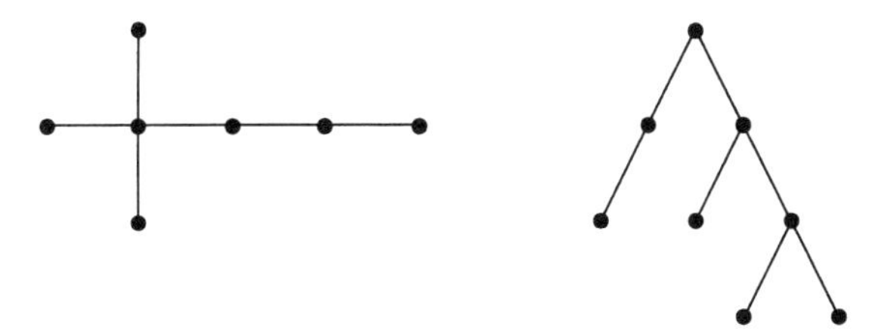

To answer this question, we take our tree and remove all the vertices of degree 1, together with their incident edges; we then repeat this process until we obtain either a single vertex, called the **centre**, or two adjacent vertices, called the **bicentre**. A tree with a centre is called a **central tree**, and a tree with a bicentre is called a **bicentral tree**. Every tree is either central or bicentral, but not both.

For example, the following tree is a *central tree* with centre e.

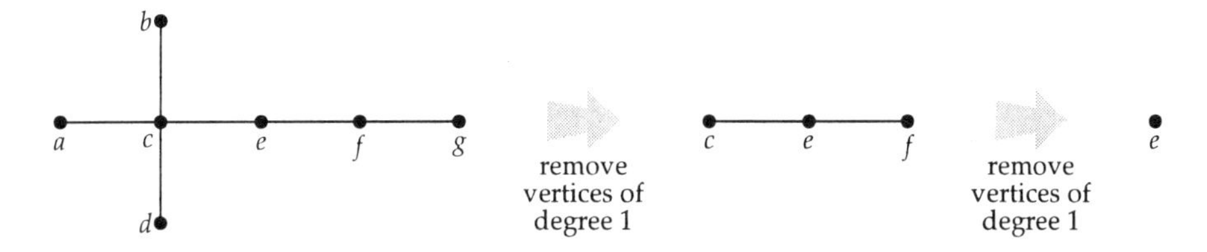

The following tree is a *bicentral tree* with bicentre cd.

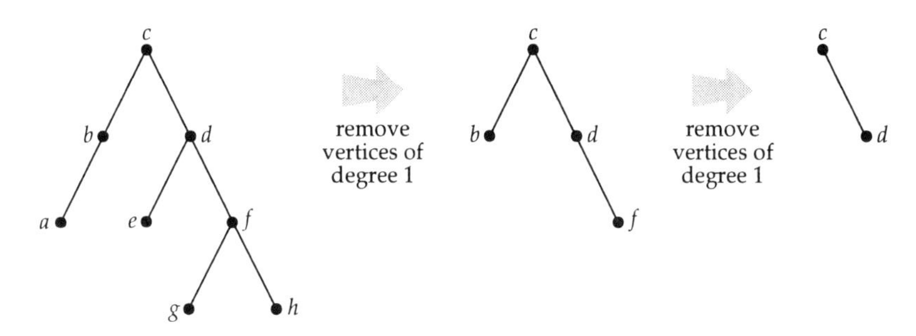

Problem 7.12

Classify each of the trees with five and six vertices as central or bicentral, and locate the centre or bicentre in each case.

Cayley's approach to the problem of finding the number of alkanes was to regard each molecule as a tree with either a centre (a carbon vertex of degree 4) or a bicentre (two such vertices, joined by an edge). By removing the centre or bicentre, he produced a number of smaller trees; thus he obtained a complicated recurrence relation that successively gives the number of alkanes with formula C_nH_{2n+2}, for increasing values of n. Using this, Cayley correctly calculated the number of alkanes with up to eleven carbon atoms.

The following table lists the number of different alkanes C_nH_{2n+2} with n carbon atoms, for $n = 1, \ldots, 15$.

n	1	2	3	4	5	6	7	8	9	10	11	12	13	14	15
number of alkanes	1	1	1	2	3	5	9	18	35	75	159	355	802	1858	4347

Historical Note

Although graph-like diagrams had been used as far back as 1789 to represent molecules, it was not until the 1850s that the way atoms combine was sufficiently well understood for meaningful diagrams to be drawn. This occurred when August Kekulé and others put forward ideas that led to the theory of valency. In 1864 Alexander Crum Brown introduced structural diagrams to represent this theory and explain the nature of isomerism. Meanwhile, the mathematician Arthur Cayley (1821–95) had spent some time studying and counting trees, and in 1875 showed how to calculate the number of alkanes with a given number of carbon atoms. Arthur Cayley had been interested in trees for almost twenty years before solving the alkane-counting problem. His first two papers on trees, *On the theory of the analytical forms called trees*, appeared in 1857 and 1859, and were concerned with counting rooted trees in connection with a problem in the differential calculus.

Exercises 7

Counting Labelled Trees

7.1 Use Cayley's theorem to write down the number of labelled spanning trees in the complete graph K_n.

7.2 Show that there are 125 labelled trees with five vertices.

7.3 Write down the Prüfer sequence associated with each of the following labelled trees.

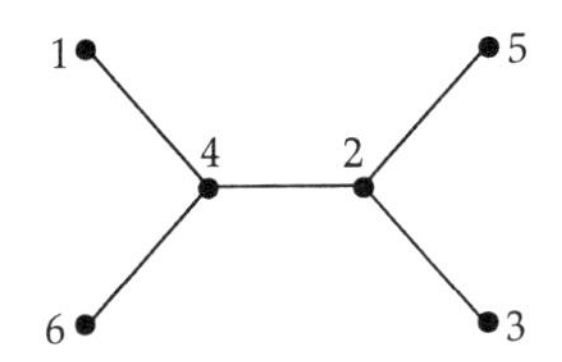

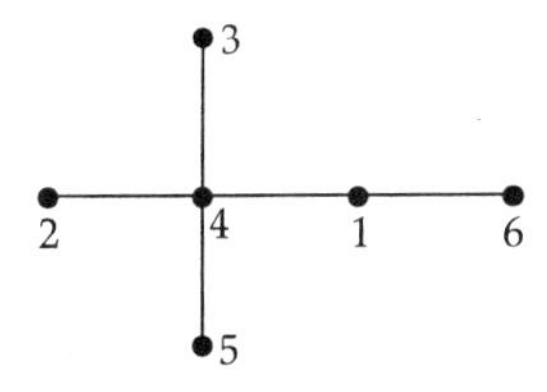

7.4 Draw the labelled tree associated with each of the Prüfer sequences $(\mathbf{1}, \mathbf{2}, \mathbf{3}, \mathbf{4})$ and $(\mathbf{3}, \mathbf{3}, \mathbf{3}, \mathbf{3})$.

7.5 Draw the labelled tree associated with the Prüfer sequence $(\mathbf{5}, \mathbf{2}, \mathbf{6}, \mathbf{2}, \mathbf{5}, \mathbf{1})$.

Counting Binary Trees

7.6 Determine the number of binary trees with:
(a) 7 vertices; (b) 8 vertices.

Counting Chemical Trees

7.7 Classify all the trees with seven vertices as *central* or *bicentral*. (These trees are shown in Solution 6.2.)

7.8 Draw a central tree and a bicentral tree, each with:
(a) 8 vertices; (b) 9 vertices.

7.9 A carbon atom has valency 4, an oxygen atom has valency 2, and a hydrogen atom has valency 1.

(a) There are two different molecules with formula C_3H_7OH. Draw the graphs representing these molecules.

(b) Prove that the graph of any alcohol $C_nH_{2n+1}OH$ is a tree.

(c) Is the graph of a molecule with formula $C_8H_{18}O_2$ a tree?

Chapter 8

Greedy Algorithms

After studying this chapter, you should be able to:

- define the terms *minimum spanning tree* and *minimum connector*;
- use Kruskal's and Prim's algorithms for solving the minimum connector problem;
- find upper bounds for the solution to the travelling salesman problem by using a heuristic algorithm;
- find lower bounds for the solution to the travelling salesman problem by using a minimum connector.

In this chapter we turn our attention to some common algorithms associated with the construction of trees. In particular, we introduce the *minimum connector problem*, and show how its solution can give information about the solution to the *travelling salesman problem*. Both these topics involve the idea of a *spanning tree* in a connected graph.

Recall that a *spanning tree* in a connected graph G is a subgraph of G that includes every vertex and is also a tree.

8.1 Minimum Connector Problem

Suppose that an irrigation canal system is to be built, interconnecting a number of given locations. The cost of digging and maintaining each canal in the system is known. Some pairs of locations cannot be joined by a canal for geographical reasons (for example, there is a gorge). How do we design a canal system that interconnects all the locations at minimum total cost?

This problem can be interpreted in two ways, depending on whether or not we allow extra 'locations' where canals may intersect. For example, in the case of the canal system shown below, we may be able to reduce the total cost by creating an extra location at the point E and linking it to A.

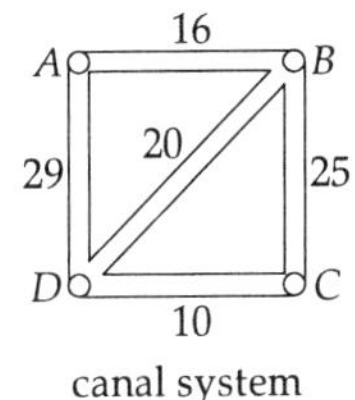

canal system

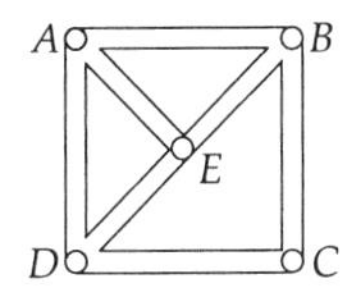

system with extra location E

Unfortunately, for many such problems, the additional cost of inserting an extra location (which may be a telephone exchange or power station) can exceed the possible saving in cost from a shorter connection system, and the mathematical analysis becomes complicated. In view of this, we adopt the second interpretation of the problem and assume that each connection joins two existing locations; no new locations are allowed.

We can model this problem graphically by representing the locations by vertices and the canals by edges – that is, by a weighted graph in which the weights are the costs. The problem is now to find a connected subgraph of minimum total weight, passing through each vertex. Such a subgraph is necessarily a spanning tree, because, if there is a cycle, we can lower the total cost by removing any one of its edges.

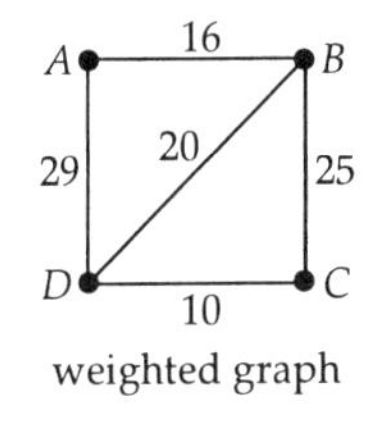

weighted graph

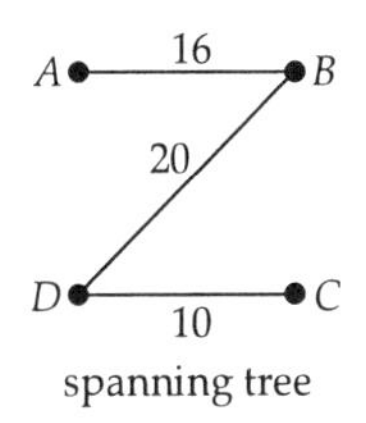

spanning tree

In our example, the graph has total weight 100. The removal of an edge from the cycle $ABDA$ and an edge from the remaining cycle – $ABCDA$ or $BCDB$ – lowers the total weight, and gives a spanning tree. The spanning tree of minimum total weight is obtained by removing the two edges of maximum weight – AD and BC. The minimum total cost is thus $16 + 20 + 10 = 46$. We call this spanning tree a *minimum connector* for the locations A, B, C, D.

Definition

Let T be a spanning tree of minimum total weight in a connected weighted graph G. Then T is a **minimum spanning tree** or a **minimum connector** in G.

We now state the minimum connector problem in graph-theoretical terms.

Minimum Connector Problem

Given a weighted graph, find a minimum spanning tree.

We present two algorithms for solving the minimum connector problem. Both belong to a class of algorithms known as *greedy algorithms*. The name arises from the fact that at each stage we make the 'greediest', or best, choice available, regardless of the subsequent effect of that choice. Such 'local' algorithms do not succeed for most combinatorial problems, but this is one case where they do. We present each algorithm formally as a sequence of steps, and then summarize it informally in a form suitable for small examples to be tackled without the use of a computer.

The first algorithm we describe is *Kruskal's algorithm*. (This algorithm was derived by Joseph Kruskal in 1956, but had been stated thirty years earlier by O. Borůvka.) In applying this algorithm, at each stage we choose the edge of least weight, provided that it does not create a cycle. We need to avoid cycles because, if there is a cycle, then we can decrease the total distance by removing any one of its edges.

Kruskal's Algorithm

START with a finite set of vertices, where each pair of vertices is joined by a weighted edge.

Step 1 List all the weights in ascending order.

Step 2 Draw the vertices and weighted edge corresponding to the first weight in the list, provided that, in so doing, no cycle is formed. Delete the weight from the list.

Repeat Step 2 until all the vertices are connected, then STOP.

The weighted graph obtained is a minimum connector, and the sum of the weights on its edges is the total weight of the minimum connector.

Remarks

1. When two or more weights are equal, they may appear in the list in any order. Strictly speaking, the weights are in non-descending order, rather than ascending order.

2. We do not necessarily obtain a *connected* graph at each intermediate stage.

When the number of vertices is small, we can solve the problem by hand, using the following simplified version of the algorithm.

Summary of Kruskal's Algorithm

To construct a minimum connector in a connected weighted graph G, successively choose edges of G of minimum weight in such a way that no cycles are created, until a spanning tree is obtained.

We illustrate the use of this form of the algorithm in the following example.

Example 8.1

Consider the given connected weighted graph.

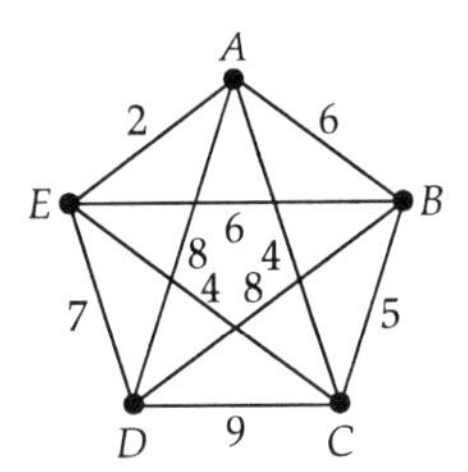

Note that, in this example, some of the weights are the same, so we may have a choice of edges at some stages, and there may be more than one minimum spanning tree.

First edge

We choose an edge of minimum weight – the only possibility is AE, with weight 2.

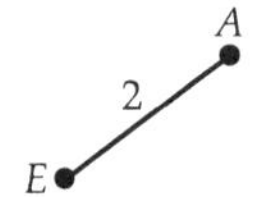

Second edge

We choose an edge of next smallest weight – we can choose either AC or CE, with weight 4; let us choose CE.

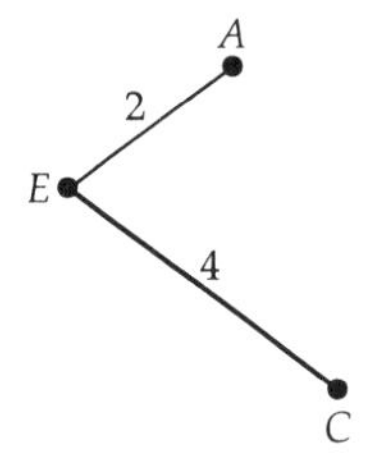

Third edge

We cannot now choose AC, also with weight 4, as this would create a cycle ($ACEA$), so we choose an edge of next smallest weight – the only possibility is BC, with weight 5.

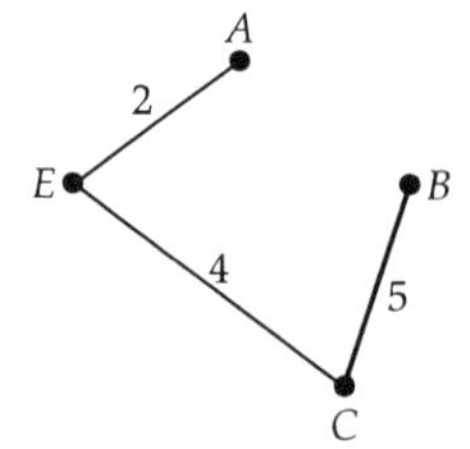

Fourth edge

The edges of next smallest weight are AB and BE, with weight 6. We cannot choose either of these, as each would create a cycle ($ABCEA$ or $BCEB$), so we choose an edge of next smallest weight – the only possibility is DE, with weight 7.

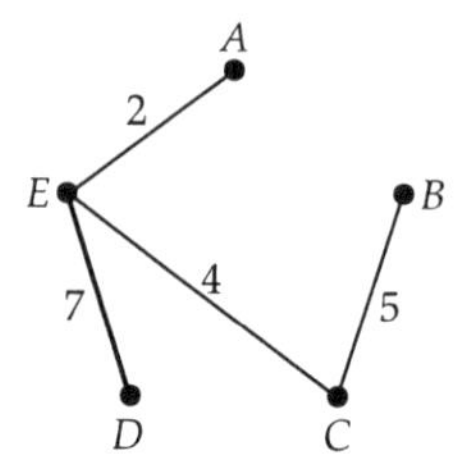

This completes a spanning tree, which is a minimum spanning tree of total weight $2 + 4 + 5 + 7 = 18$. ❐

Problem 8.1

In Example 8.1, which minimum spanning tree would we have obtained if at the second stage we had chosen the edge AC, instead of the edge CE?

When there are more than five vertices in a weighted graph, the weights are usually given in tabular form, as in the following example.

Example 8.2

The following table gives the distances (in hundreds of miles) between six cities. We use Kruskal's algorithm to find a minimum spanning tree connecting these cities.

	Berlin	London	Moscow	Paris	Rome	Seville
Berlin	—	7	11	7	10	15
London	7	—	18	3	12	11
Moscow	11	18	—	18	20	27
Paris	7	3	18	—	9	8
Rome	10	12	20	9	—	13
Seville	15	11	27	8	13	—

For a small table such as this, we can select the distances in ascending order directly from the table. However, you may prefer to begin by drawing up a table of the distances in ascending order.

First edge

We choose an edge of minimum weight – the only possibility is London–Paris, with weight 3.

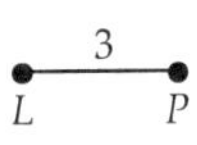

Second edge

We choose an edge of next smallest weight – we can choose either Berlin–London or Berlin–Paris, with weight 7; let us choose Berlin–London.

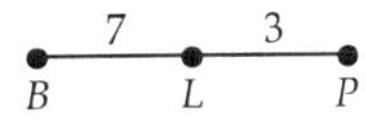

Third edge

We cannot now choose Berlin–Paris, also with weight 7, as this would create a cycle, so we choose an edge of next smallest weight – the only possibility is Paris–Seville, with weight 8.

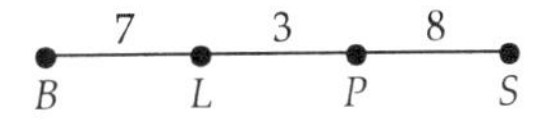

Fourth edge

We choose an edge of next smallest weight – the only possibility is Paris–Rome, with weight 9.

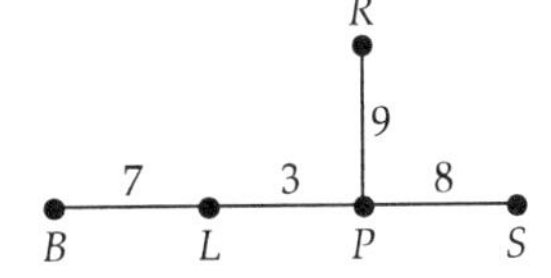

Fifth edge

We cannot now choose Berlin–Rome (weight 10) or London–Seville (weight 11), as each would create a cycle, so we choose Berlin–Moscow, with weight 11.

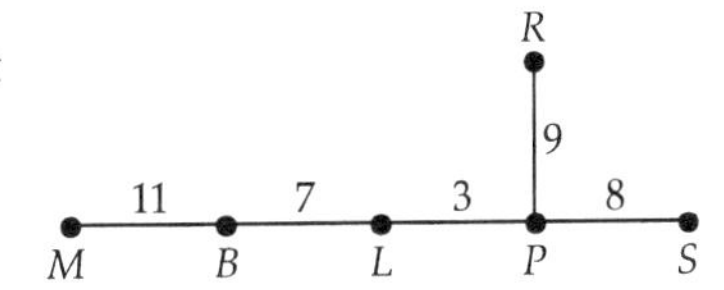

This completes a spanning tree, which is a minimum spanning tree of total weight $3 + 7 + 8 + 9 + 11 = 38$. ❐

Problem 8.2

The following table gives the distances (in miles) between six places in Ireland. Use Kruskal's algorithm to find a minimum spanning tree connecting these places.

	Athlone	Dublin	Galway	Limerick	Sligo	Wexford
Athlone	—	78	56	73	71	114
Dublin	78	—	132	121	135	96
Galway	56	132	—	64	85	154
Limerick	73	121	64	—	144	116
Sligo	71	135	85	144	—	185
Wexford	114	96	154	116	185	—

Although Kruskal's algorithm can be applied easily when the number of cities is small, it is not well suited for efficient computer implementation, due to the need to arrange the edges in order of ascending weight, and the need to recognize cycles as they are created.

Both these difficulties are overcome by a slight modification of the algorithm; the result is called *Prim's algorithm*. In applying this algorithm, we start with any vertex and build up the required spanning tree edge by edge.

Prim's Algorithm

START with a finite set of vertices, where each pair of vertices is joined by a weighted edge.

Step 1 Choose and draw any vertex.

Step 2 Find the edge of least weight joining a drawn vertex to a vertex *not* currently drawn. Draw this weighted edge and the corresponding new vertex.

Repeat Step 2 until all the vertices are connected, then STOP.

Remarks

1. When there are two or more edges with the same weight, choose any of them.
2. With this construction, we obtain a *connected* graph at each stage.

We can summarize Prim's algorithm as follows.

Summary of Prim's Algorithm

To construct a minimum connector T in a connected weighted graph G, build up T step by step as follows:

- put an arbitrary vertex of G into T;
- successively add edges of minimum weight joining a vertex already in T to a vertex not in T, until a spanning tree is obtained.

Example 8.3

The following table gives the distances (in hundreds of miles) between six cities. We use Prim's algorithm to find a minimum spanning tree T connecting these cities.

	Berlin	London	Moscow	Paris	Rome	Seville
Berlin	—	7	11	7	10	15
London	7	—	18	3	12	11
Moscow	11	18	—	18	20	27
Paris	7	3	18	—	9	8
Rome	10	12	20	9	—	13
Seville	15	11	27	8	13	—

We start by choosing any vertex and putting it in T. Let us choose Berlin.

First edge

We choose an edge of minimum weight joining Berlin to another vertex; we can choose either Berlin–London or Berlin–Paris, with weight 7; let us choose Berlin–London.

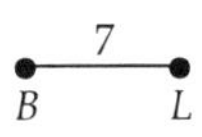

Second edge

We choose an edge of minimum weight joining Berlin or London to another vertex; the only possibility is London–Paris, with weight 3.

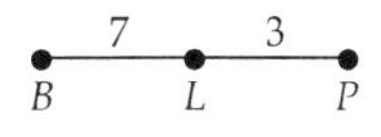

Third edge

We choose an edge of minimum weight joining Berlin, London or Paris to another vertex; the only possibility is Paris–Seville, with weight 8.

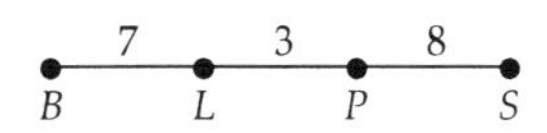

Fourth edge

We choose an edge of minimum weight joining Berlin, London, Paris or Seville to another vertex; the only possibility is Paris–Rome, with weight 9.

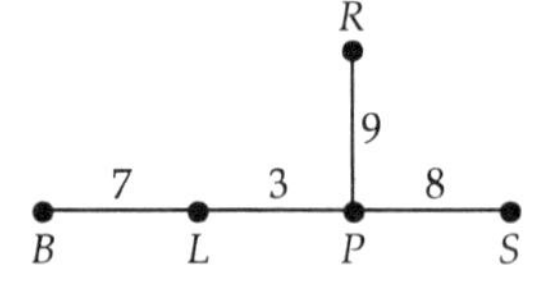

Fifth edge

We choose an edge of minimum weight joining Berlin, London, Paris, Rome or Seville to the remaining vertex; the only possibility is Berlin–Moscow, with weight 11.

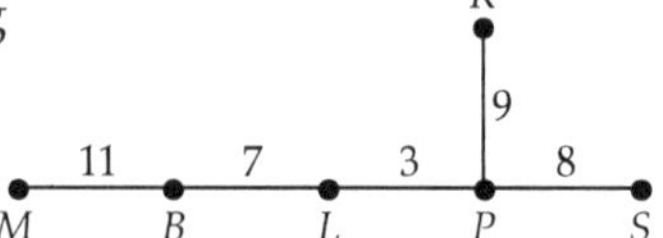

This completes a spanning tree, which is a minimum spanning tree of weight $7 + 3 + 8 + 9 + 11 = 38$. (It is the tree we obtained in Example 8.2.) ❐

Problem 8.3

In Example 8.3, which minimum spanning tree would we have obtained

(a) if at the first stage we had chosen Berlin–Paris, instead of Berlin–London?
(b) if we had started by choosing Rome as our first vertex, instead of Berlin?

We close this section by outlining a proof of the fact that the algorithms given above always produce a minimum connector.

Theorem 8.1

Prim's and Kruskal's algorithms always produce a spanning tree of minimum weight.

Outline of proof

The proof is by contradiction.

Suppose that the algorithm produces a tree T, and that there exists a spanning tree S with smaller total weight than T. Let e be an edge of smallest weight lying in T but not in S.

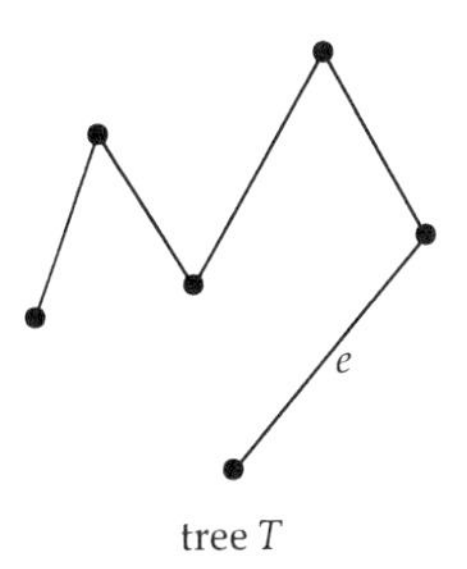

tree T

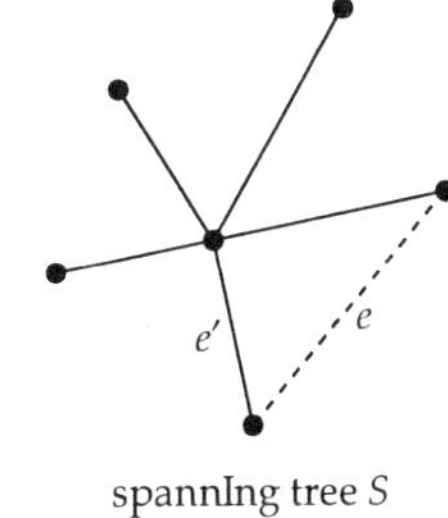

spanning tree S
(2 edges in common with T)

spanning tree S'
(3 edges in common with T)

If we add the edge e to S we create a cycle containing e. Since this cycle must contain an edge e' not contained in T, the subgraph obtained from S on replacing e' by e is also a spanning tree, S' say. It follows from the construction of T that the weight of e cannot exceed the weight of e'. So the total weight of S' does not exceed the total weight of S, and S' has one more edge in common with T than has S.

It follows that, by repeating this procedure, we can change S into T, one edge at a time, with the weight not increasing at each stage. This shows that the total weight of T does not exceed the total weight of S, contradicting the definition of S. This contradiction establishes the result. ■

8.2 Travelling Salesman Problem

In Chapter 1 we introduced the *travelling salesman problem*, in which a salesman wishes to visit a number of cities and return to the starting point, covering the *minimum* possible total distance. This is an important problem in practice, and can appear in a number of different guises; for example, it arises in situations such as drilling holes in a printed circuit board involving thousands of holes.

We can express the problem in graphical language. In such terms, we are given a weighted graph, usually a complete graph, and our aim is to find a cycle of minimum total weight passing through every vertex – in other words, a *minimum-weight Hamiltonian cycle.*

Travelling Salesman Problem

Given a weighted complete graph, find a minimum-weight Hamiltonian cycle.

In view of the simple nature of Kruskal's and Prim's algorithms for solving the minimum connector problem, we might hope that there is a simple algorithm here as well. Unfortunately, no such algorithm is known. We could, of course, try all possible Hamiltonian cycles, but this is a hopeless task, even on a computer, unless the number of vertices is small. For a travelling salesman problem involving 100 cities, there would be $\frac{1}{2}(99!) \approx 4.65 \times 10^{155}$ cycles to be considered, which is way beyond present computer capacities. In fact, there is no known efficient algorithm for the travelling salesman problem.

In view of this, we are forced to look for *approximate solutions* to the problem – that is, we look for *upper* and *lower bounds* for the length of a minimum-weight Hamiltonian cycle.

Historical Note

The travelling salesman problem appeared in rudimentary form in a practical German book of 1831 for the *Handlungsreisende* (travelling salesman). Its first appearance in mathematical circles was at Princeton University in the 1930s.

Upper Bound for the Solution to the Travelling Salesman Problem

We now describe a *heuristic* algorithm for the travelling salesman problem – that is, an algorithm that does not necessarily give the correct answer, but gives a good approximation to it.

The idea is to build up the required cycle step by step, starting with a single vertex. It is similar to Prim's algorithm, except that we build up a cycle rather than a tree.

Method for Finding An Upper Bound for the Solution to the Travelling Salesman Problem

START with a finite set of vertices, where each pair of vertices is joined by a weighted edge.

Step 1 Choose any vertex and find a vertex joined to it by an edge of minimum weight. Draw these two vertices and join them with two edges to form a 'cycle'; give the cycle a clockwise orientation.

Step 2 Find a vertex *not* currently drawn joined by an edge of least weight to a vertex already drawn.
Insert this new vertex into the cycle in front of the 'nearest' already-connected vertex.

Repeat Step 2 until all the vertices are joined by a cycle, then STOP.

The weighted cycle obtained is a Hamiltonian cycle, and its total weight – given by the sum of the weights on its edges – is an upper bound for the solution to the travelling salesman problem.

We can summarize this algorithm as follows.

Summary of Algorithm

To construct a cycle C that gives an approximate solution to the travelling salesman problem for a connected weighted graph G, build up C step by step as follows:

- choose an arbitrary vertex of G and its 'nearest neighbour' and put them into C;
- successively insert vertices joined by edges of minimum weight to a vertex already in C to a vertex not in C, until a Hamiltonian cycle is obtained.

Remark *Different* choices of the initial vertex may give *different* upper bounds. The *best* upper bound is the *smallest*: it gives the most information about the actual solution. (If you are saving up for something, then it is more helpful to know that you need at most £100 than to know that you need at most £120.)

Example 8.4

We find an upper bound for the solution to the travelling salesman problem for the six European cities.

	Berlin	London	Moscow	Paris	Rome	Seville
Berlin	—	7	11	7	10	15
London	7	—	18	3	12	11
Moscow	11	18	—	18	20	27
Paris	7	3	18	—	9	8
Rome	10	12	20	9	—	13
Seville	15	11	27	8	13	—

First vertex

We start by choosing any vertex. Let us choose Berlin.

Second vertex

The city nearest to Berlin is London or Paris (distance 7). Let us choose London.
We draw two vertices and two edges joining them, and give the 'cycle' a clockwise orientation.

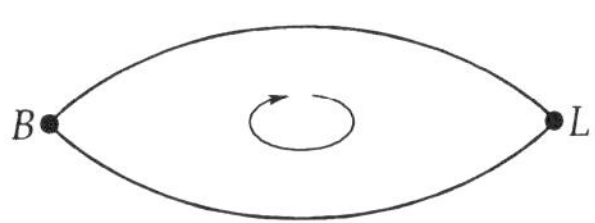

Third vertex

The city nearest to Berlin and London is Paris (distance 3 from London).
We insert Paris in front of London in the cycle.

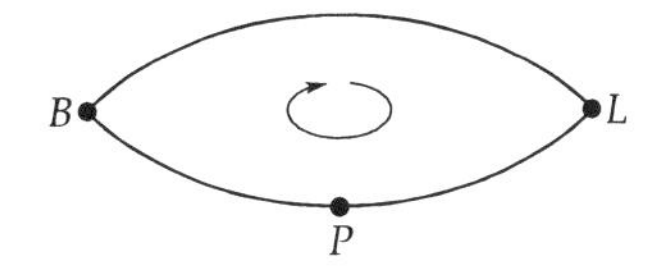

Fourth vertex

The city nearest to Berlin, London and Paris is Seville (distance 8 from Paris).
We insert Seville in front of Paris in the cycle.

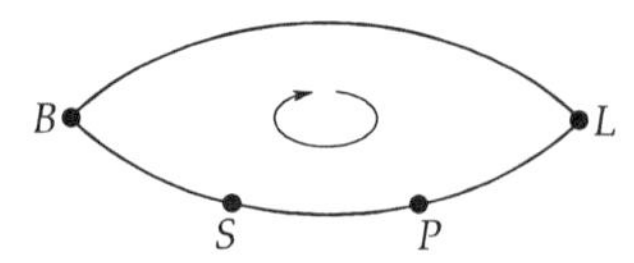

Fifth vertex

The city nearest to Berlin, London, Paris and Seville is Rome (distance 9 from Paris).
We insert Rome in front of Paris in the cycle.

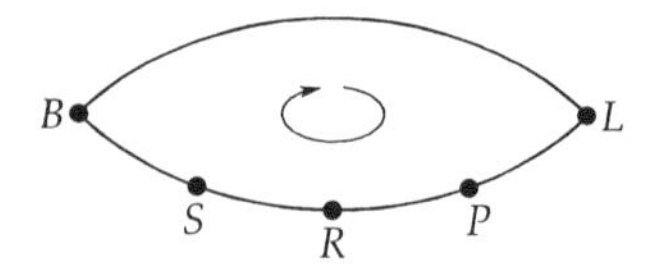

Sixth vertex

The final city is Moscow. It is closest to Berlin (distance 11).
We insert Moscow in front of Berlin in the cycle.

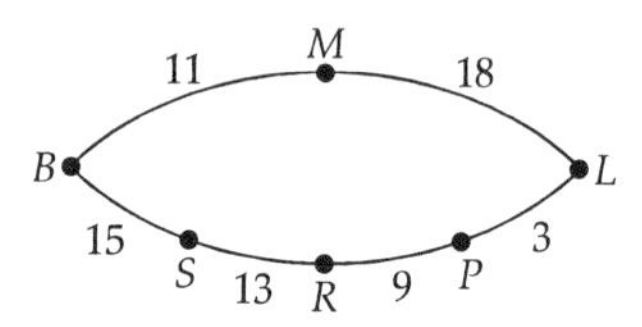

All the cities are now joined by the cycle

Berlin–Moscow–London–Paris–Rome–Seville–Berlin.

An upper bound for the solution is therefore

$$11 + 18 + 3 + 9 + 13 + 15 = 69.$$

❐

Problem 8.4

In Example 8.4, which upper bound would we have obtained if we had started with Rome?

Lower Bound for the Solution to the Travelling Salesman Problem

Another method for finding an approximate solution to the travelling salesman problem, which often works well in practice, is to find a *lower bound* for the total weight of a minimum-weight Hamiltonian cycle by solving a related minimum connector problem instead. We then know that the length of the cycle that is the correct solution must exceed this lower bound, although the method does not yield an approximate route. We illustrate the method by the following graph.

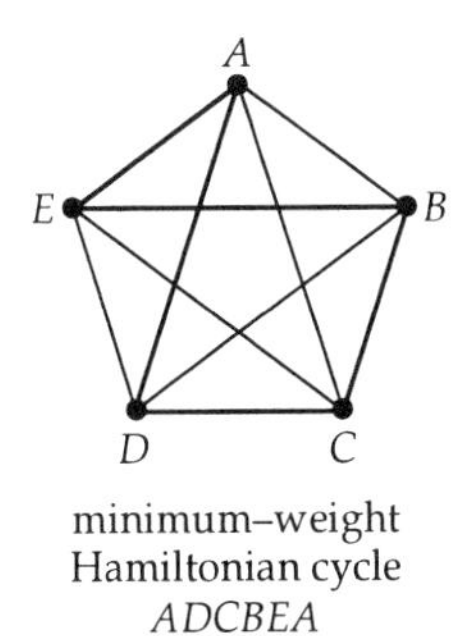

minimum-weight
Hamiltonian cycle
ADCBEA

(The weights are irrelevant here, so they are omitted.)

If we take a minimum-weight Hamiltonian cycle in this (weighted) complete graph and remove the vertex A and its incident edges, we get a path passing through the remaining vertices.

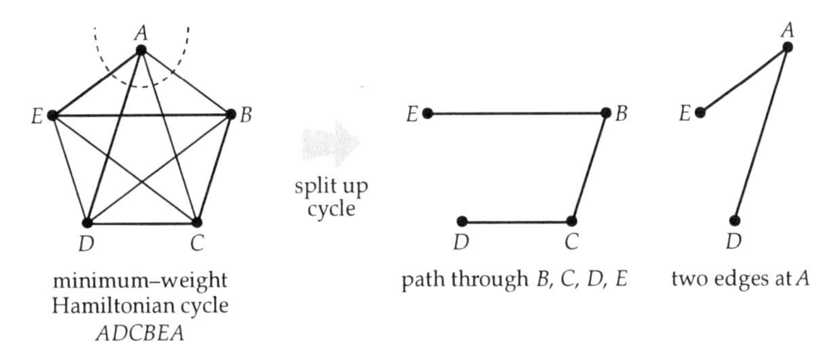

minimum-weight Hamiltonian cycle *ADCBEA* — path through B, C, D, E — two edges at A

Such a path is a spanning tree for the complete graph formed by these remaining vertices, and the weight of the Hamiltonian cycle is obtained by adding the weight of this spanning tree to the weights of the two edges incident with A:

total weight of minimum-weight Hamiltonian cycle	=	total weight of spanning tree connecting B, C, D, E	+	sum of weights of two edges incident with A

It follows that:

total weight of minimum-weight Hamiltonian cycle	$\geq$	total weight of *minimum* spanning tree connecting B, C, D, E	+	sum of two *smallest* weights of edges incident with A

So the expression on the right-hand side gives a *lower bound* for the solution to the travelling salesman problem in this case.

In general, we can use the generalized form of this inequality to obtain a lower bound to any travelling salesman problem, as outlined below.

Method for Finding a Lower Bound for the Solution to the Travelling Salesman Problem

Step 1 Choose a vertex V and remove it from the graph.
Step 2 Find a minimum spanning tree connecting the remaining vertices, and calculate its total weight w.
Step 3 Find the two smallest weights, w_1 and w_2, of the edges incident with V.
Step 4 Calculate the lower bound $w + w_1 + w_2$.

Remark *Different* choices of the initial vertex V may give *different* lower bounds. The *best* lower bound is the *largest*: it gives the most information about the actual solution. (If you are saving up for something, then it is more helpful to know that you need at least £100 than to know that you need at least £80.)

Example 8.5

We find a lower bound for the solution to the travelling salesman problem for the following weighted graph.

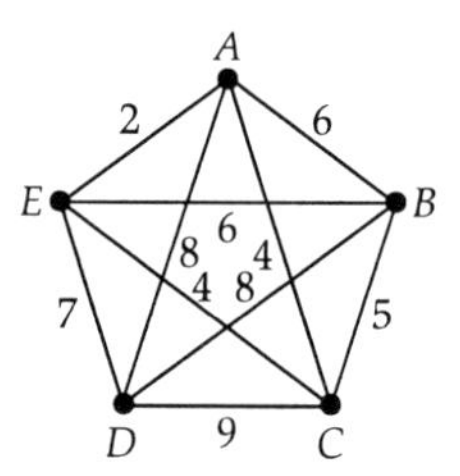

We start by removing any vertex. Let us choose the vertex A. Then the remaining weighted graph has the four vertices B, C, D, E.

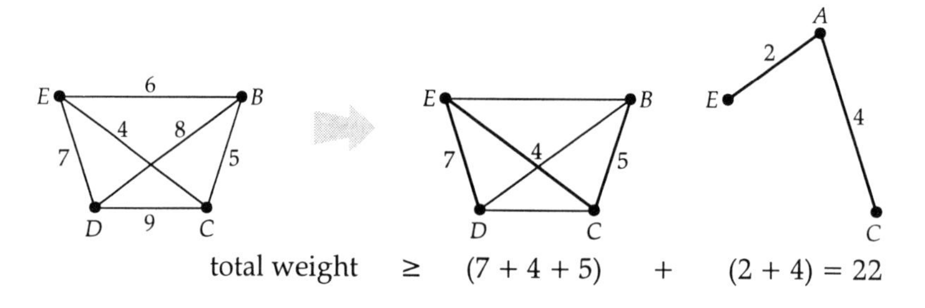

The minimum spanning tree joining these vertices is the tree whose edges are ED, CE, BC, with total weight $7 + 4 + 5 = 16$.

The two edges of smallest weight incident with A are AE and AC, with weights 2 and 4.

A lower bound for the solution is therefore

$$16 + 2 + 4 = 22.$$

A little experimentation shows that this lower bound is not very good: the actual solution to this problem is the cycle *ACBDEA* with total weight 26, so our lower bound is not the correct solution.

A better lower bound (that is, a greater one, giving more information) can be obtained by removing the vertex *D* instead of *A*. In this case, the remaining weighted graph has the four vertices *A, B, C, E*, and there are two minimum spanning trees joining these vertices, each with total weight $2 + 4 + 5 = 11$ (only one is highlighted below).

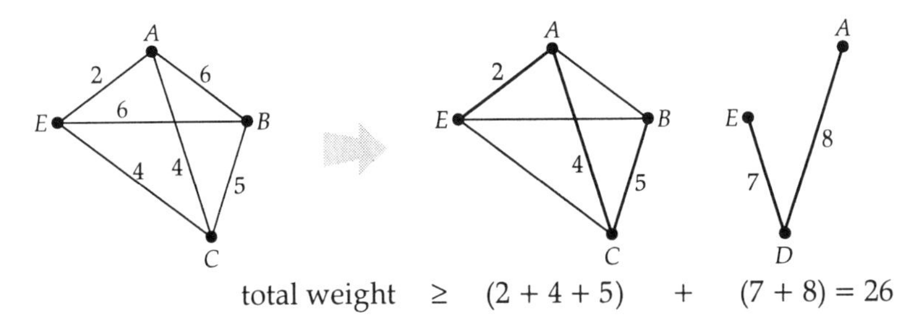

The two edges of smallest weight incident with *D* are *DE* and *DA*, or *DE* and *DB*, with weights 7 and 8.

A better lower bound for the solution is therefore $11 + 7 + 8 = 26$, which is the correct solution. ❒

In the above example, we found a minimum spanning tree by inspection of the diagram (with the given vertex removed). However, for larger examples, a diagram becomes congested and it is easier to find a minimum spanning tree using Kruskal's or Prim's algorithm.

Problem 8.5

In Example 8.5, which lower bounds would we have obtained if we had started by removing:

(a) vertex *B*? (b) vertex *E*?

Which of these lower bounds is the better one?

Exercises 8

Minimum connector problem

8.1 The following graph shows the distances (in miles) between five towns.

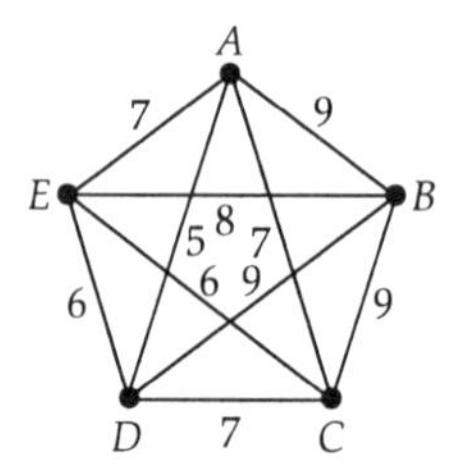

Find a minimum connector for this graph by using Kruskal's algorithm.

8.2 Find a minimum connector for the graph of Exercise 8.1 by using Prim's algorithm, starting at:

(a) vertex A; (b) vertex C.

8.3 The following table gives the distances (in miles) between six cities in England.

	Bristol	Exeter	Hull	Leeds	Oxford	York
Bristol	—	84	231	220	74	225
Exeter	84	—	305	271	154	280
Hull	231	305	—	61	189	37
Leeds	220	271	61	—	169	24
Oxford	74	154	189	169	—	183
York	225	280	37	24	183	—

Find a minimum connector for these six cities by using Kruskal's algorithm.

8.4 Find a minimum connector for the six cities of Exercise 8.3 by using Prim's algorithm, starting at:

(a) Oxford; (b) Hull.

8.5 The following table gives the distances (in miles) between six cities in Canada.

	Edmonton	London	Montreal	Ottawa	Regina	Toronto
Edmonton	—	2144	2301	2173	482	2107
London	2144	—	425	357	1696	122
Montreal	2301	425	—	127	1853	344
Ottawa	2173	357	127	—	1725	248
Regina	482	1696	1853	1725	—	1659
Toronto	2107	122	344	248	1659	—

Find a minimum connector for these six cities by using Kruskal's algorithm.

8.6 Find a minimum connector for the six cities of Exercise 8.5 by using Prim's algorithm, starting at:

(a) London; (b) Montreal.

Travelling Salesman Problem

8.7 Find an upper bound for the solution to the travelling salesman problem for the five towns of Exercise 8.1 by using the heuristic algorithm, starting with:

(a) town B; (b) town E.

Which of your upper bounds is the better one?
Find the correct solution by inspection.

8.8 Find an upper bound for the solution to the travelling salesman problem for the six European cities, by using the heuristic algorithm, starting with:

(a) Moscow; (b) Seville.

Which of your upper bounds is the better one?

	Berlin	London	Moscow	Paris	Rome	Seville
Berlin	—	7	11	7	10	15
London	7	—	18	3	12	11
Moscow	11	18	—	18	20	27
Paris	7	3	18	—	9	8
Rome	10	12	20	9	—	13
Seville	15	11	27	8	13	—

8.9 Find an upper bound for the solution to the travelling salesman problem for the six places in Ireland, by using the heuristic algorithm, starting with:

(a) Galway; (b) Sligo.

Which of your upper bounds is the better one?

	Athlone	Dublin	Galway	Limerick	Sligo	Wexford
Athlone	—	78	56	73	71	114
Dublin	78	—	132	121	135	96
Galway	56	132	—	64	85	154
Limerick	73	121	64	—	144	116
Sligo	71	135	85	144	—	185
Wexford	114	96	154	116	185	—

8.10 Find an upper bound for the solution to the travelling salesman problem for the six cities of Exercise 8.3 by using the heuristic algorithm, starting with:

(a) Bristol; (b) Exeter.

Which of your upper bounds is the better one?

8.11 Find an upper bound for the solution to the travelling salesman problem for the six cities of Exercise 8.5 by using the heuristic algorithm, starting with:

(a) Regina; (b) Toronto.

Which of your upper bounds is the better one?

8.12 Find a lower bound for the solution to the travelling salesman problem for the towns in Exercise 8.1 by removing:

(a) town B; (b) town E.

Which of your lower bounds is the better one?

8.13 Find a lower bound for the solution to the travelling salesman problem for the six cities of Exercise 8.3 by removing:

(a) Hull; (b) York.

Which of your lower bounds is the better one?

8.14 Find a lower bound for the solution to the travelling salesman problem for the six cities of Exercise 8.5 by removing:

(a) Edmonton; (b) Ottawa.

Which of your lower bounds is the better one?

8.15 Find a lower bound for the solution to the travelling salesman problem for the six places of Exercise 8.9 by removing

(a) Athlone; (b) Dublin.

Which of your lower bounds is the better one?

Chapter 9

Path Algorithms

After reading this chapter, you should be able to:

- use Fleury's algorithm to find an Eulerian trail in an Eulerian graph;
- use the shortest path algorithm to find the shortest path between two vertices of a weighted graph or digraph;
- solve simple instances of the Chinese postman problem.

In this chapter we consider some problems that involve finding a path with some special property in a given graph or digraph. First, we describe Fleury's algorithm for finding an Eulerian trail in a given Eulerian graph. We then turn our attention to weighted graphs and digraphs and describe an algorithm for finding a shortest path from any given vertex to another. Finally, we present a case study concerned with finding optimal routes for a postman.

9.1 Fleury's Algorithm

In Chapter 3 we described how to construct an Eulerian trail in a given Eulerian graph by combining cycles. Another way of constructing an Eulerian trail is to use the following algorithm.

Fleury's Algorithm

START with an Eulerian graph G.

Step 1 Choose a starting vertex.
Step 2 Starting from the current vertex, traverse any available edge, choosing a bridge only if there is no alternative. Then erase that edge and any isolated vertex.

Repeat Step 2 until there are no more edges, then STOP.

Recall that a *bridge* is an edge in a connected graph whose removal disconnects the graph.

Example 9.1

We illustrate the use of Fleury's algorithm by applying it to the following graph.

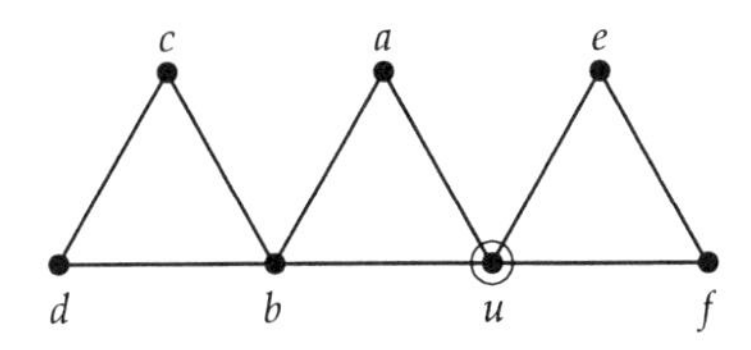

Step 1 We start at the vertex u.
Step 2 We traverse the edge ua and erase it.
Step 2 We traverse the edge ab and erase it, together with vertex a.

We now have the following graph:

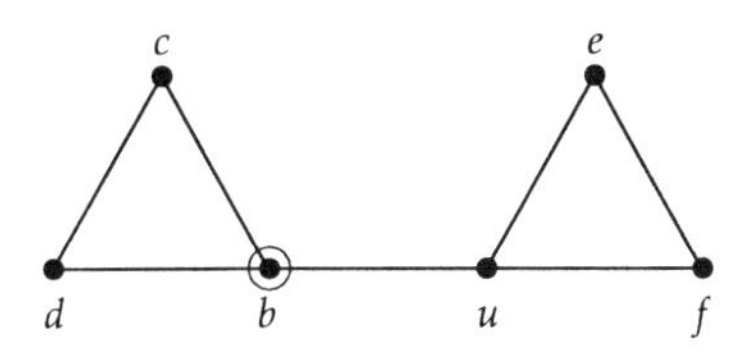

Step 2 We cannot use the edge bu because it is a bridge, so we choose the edge bc and erase it.
Step 2 We traverse the edge cd and erase it, together with vertex c.
Step 2 We traverse the edge db and erase it, together with vertex d.

We now have the following graph:

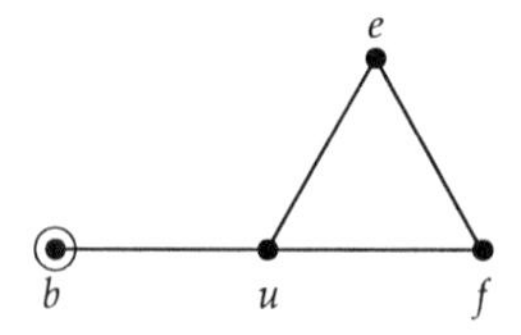

Step 2 We traverse the bridge *bu* and erase it, together with vertex *b*.
Step 2 We traverse the edge *ue* and erase it.
Step 2 We traverse the bridge *ef* and erase it, together with vertex *e*.
Step 2 We traverse the bridge *fu* and erase it, together with vertex *f*.

There are no more edges, so we STOP.
Thus we obtain the Eulerian trail *u a b c d b u e f u*. ❐

Problem 9.1

Use Fleury's algorithm to obtain an Eulerian trail in the following graph, starting with the edges *uv*, *vz*.

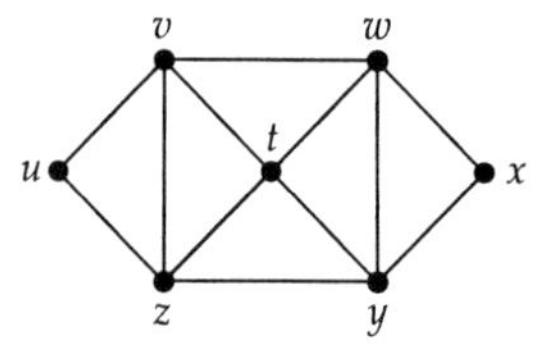

9.2 Shortest Path Algorithm

In this section we present the shortest path algorithm for finding a shortest path from a given vertex *S* in a weighted graph or digraph to another vertex *T*.

We start from *S* and calculate the shortest distance from *S* to each intermediate vertex as we go. At each stage of the algorithm, we look at all vertices reached by an edge or arc from the current vertex and assign to each such vertex a temporary label representing the shortest distance from *S* to that vertex by all paths so far considered. Eventually each vertex acquires a permanent label, called a **potential**, that represents the shortest distance from *S* to that vertex. Once *T* has been assigned a potential, we find the shortest path(s) from *S* to *T* by tracing back through the labels.

The algorithm may be used for graphs or digraphs; here we illustrate its use by applying it to digraphs. To see how it works, consider the following example.

Example 9.2

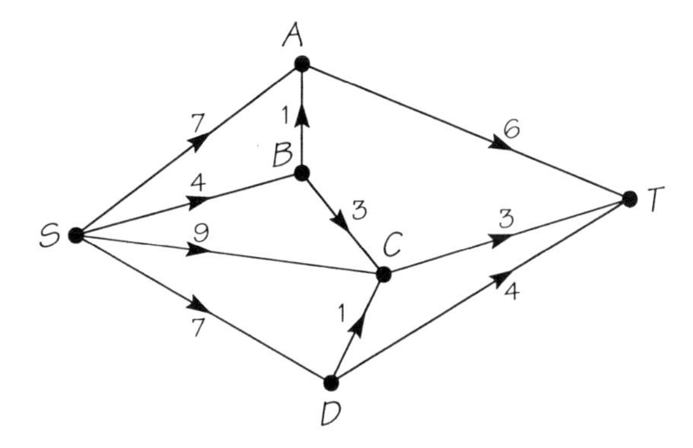

The problem is to find the shortest path from S to T, where the length of each arc is represented by the number next to the arc. We wish to assign a potential to each vertex; this represents the shortest distance along any path from S to that vertex. We denote a potential by a number inside a square. We start by assigning zero potential to S.

We illustrate the algorithm by drawing a sequence of diagrams. However, it is sometimes convenient to use a tabular method, rather than annotate a diagram. The tabular method enables the steps of the algorithm to be clearly displayed, and this makes for more accurate working and easier checking, particularly for large-scale examples. We illustrate the tabular method by giving the corresponding table for each iteration.

First Iteration: Vertices Reached Directly from S

We start from vertex S and consider each vertex that can be reached from S by a single arc – vertices A, B, C, D. We label each of these vertices with the distance along the corresponding arc from S, as shown below.

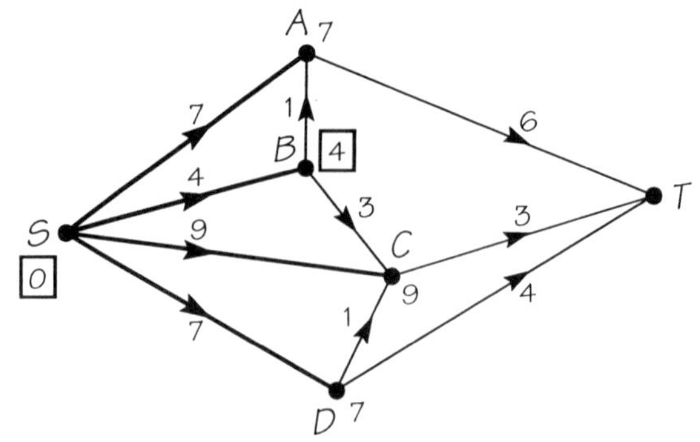

B has potential 4 assigned from S.

Now we take the *lowest* number on these vertices (that is, the shortest distance from *S*) and mark it as a potential. The shortest of the four distances *SA*, *SB*, *SC*, *SD* is *SB*, which is 4, so this is marked as a potential by putting a square round it, as shown above. This is the shortest distance to *B* along any path from *S*, since any other path is via *A*, *C* or *D*, and the first stage alone of such a path is longer than the direct path.

Using the tabular method, we draw up a table and record the information obtained in the first iteration in the first line, as follows:

iteration	origin vertex	vertices assigned labels				
		A	B	C	D	T
1	S	7	[4]	9	7	

Second Iteration: Vertices Reached Directly from B

We start from vertex *B*, the vertex just assigned a potential, and repeat the procedure with the vertices that can be reached from *B* by a single arc – *A* and *C*. Their distances from *S* via *B* are 5 and 7, respectively; these are labelled on the following figure.

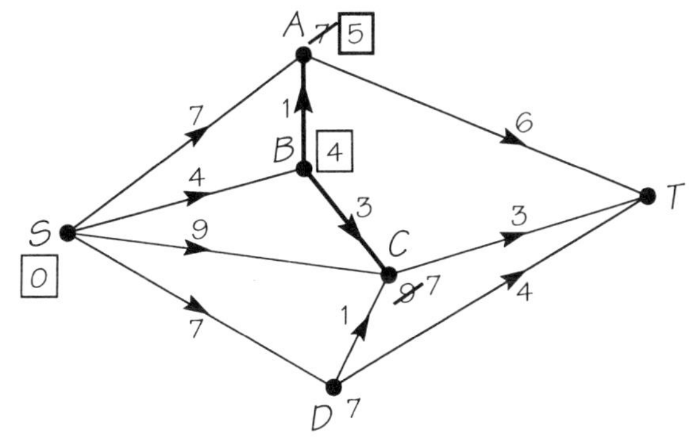

A has potential 5 assigned from B.

The distance to A via B is 5, which is less than the previously marked label 7 for B, so we cross out the 7 and replace it by 5. The same applies to vertex C. If, however, the existing label were lower than the new distance, then we would leave it in place, since we are assigning potentials equal to the shortest distances to vertices.

The shortest marked distance that is not a potential is 5, at A, so we make this a potential, as shown above.

Using the tabular method, we record this information in the second line of the table, as follows:

iteration	origin vertex	vertices assigned labels				
		A	B	C	D	T
1	S	~~7~~	[4]	~~9~~	7	
2	**B**	**[5]**		**7**		

Third Iteration: Vertices Reached Directly from A

We start from vertex A, the vertex just assigned a potential. Only vertex T can be reached from A by a single arc, so we label T with the distance from S via A. This is obtained by adding the potential at A to the distance along arc AT, that is, $5 + 6 = 11$.

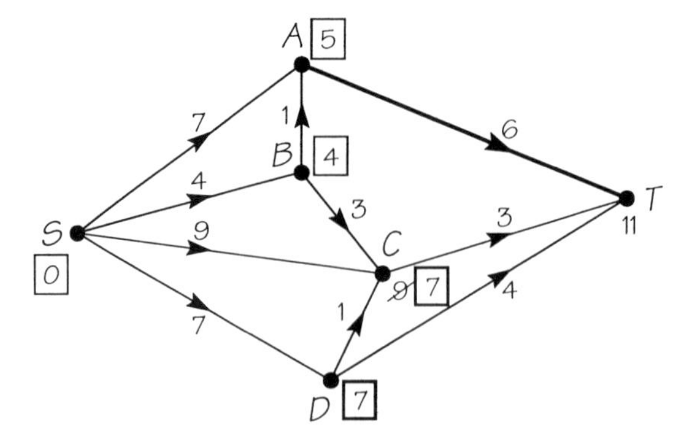

C has potential 7 assigned from *B*.
D has potential 7 assigned from *S*.

There are now two vertices – *C* and *D* – with the lowest labelled distance 7, so these distances become potentials, as shown above.

Using the tabular method, we record this information in the third line of the table. This time we add arrows to the boxes round the two new potentials, as shown below, to indicate that we are making potentials out of the label assigned to C at iteration 2 and the label assigned to *D* at iteration 1.

iteration	origin vertex	vertices assigned labels				
		A	*B*	*C*	*D*	*T*
1	*S*	~~7~~	[4]	~~9~~	7	
2	*B*	[5]		7		
3	**A**			**[7]**	**[7]**	**11**

Fourth Iteration: Vertices Reached Directly from C or D

We start from the vertices *C* and *D* and look at all the vertices that can be reached from either of these vertices by a single arc. There is only one such vertex – the vertex *T*. The path via *D* has total distance 7 + 4 = 11, which equals the current label on *T*, corresponding to the path via *A*. However, the path via *C* has distance 10, and is clearly the shortest of the three paths, so the label at *T*, and hence its potential, becomes 10, as shown below.

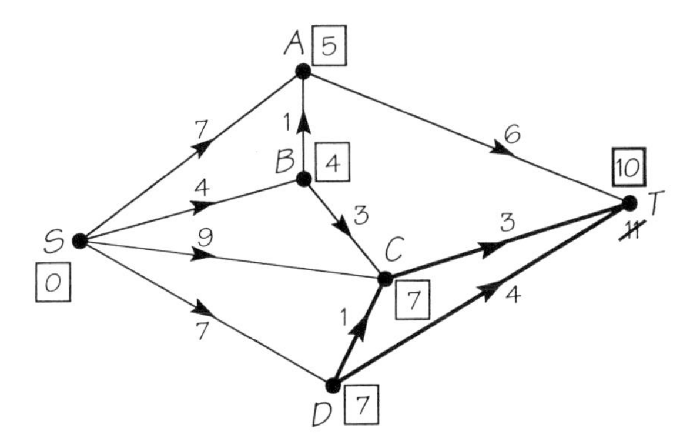

T has potential 10 assigned from *C*.

Using the tabular method, we record this information in two rows, one for each of the vertices *C* and *D*:

iteration	origin vertex	vertices assigned labels				
		A	B	C	D	T
1	S	~~7~~	[4]	~~9~~	7	
2	B	[5]		7		
3	A			[7]	[7]	11
4	**C**					**[10]**
	D					**11**

We have found that the shortest distance from *S* to *T* is 10, and it remains only to identify the corresponding shortest path. We do this by backtracking from *T*, choosing at each stage the vertex whose potential equals the potential at the previous vertex minus the distance along the arc joining them. Thus we backtrack from *T* to *C*, since 7 (the potential of *C*) is 10 (the potential of *T*) minus 3 (the length of the arc *CT*); then from *C* to *B*, and then from *B* to *S*. Hence the shortest path is *SBCT*.

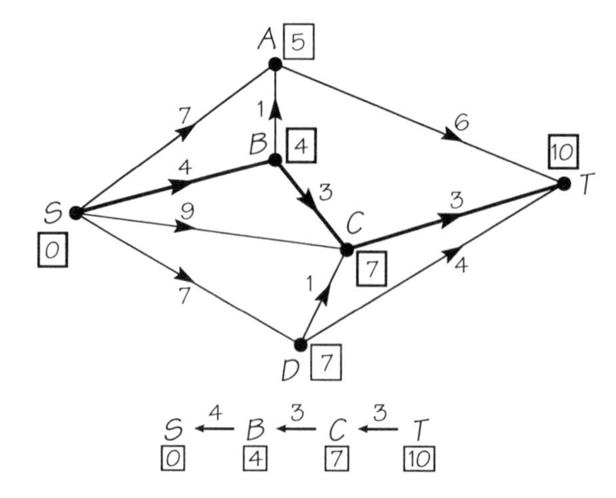

Using the tabular method, we can read off the shortest path(s) from the table. We find that

T has potential 10 assigned from *C*,
C has potential 7 assigned from *B*,
B has potential 4 assigned from *S*,

so

the shortest path is *SBCT* with length 10.

We trace the path back through the table as follows:

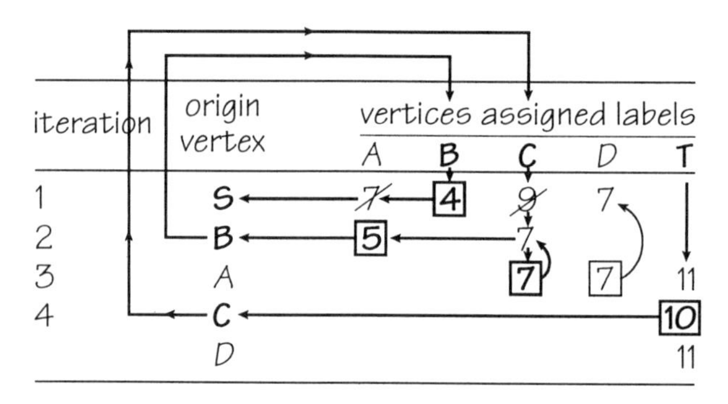

iteration	origin vertex	A	B	C	D	T
1	S	~~7~~	[4]	~~9~~	7	
2	B	[5]		7		
3	A			[7]	[7]	11
4	C					[10]
	D					11

❐

We now give a formal description of the algorithm.

Shortest Path Algorithm

START Assign potential 0 to S.

GENERAL STEP Consider the vertex (or vertices) just assigned a potential. For each such vertex V, consider each vertex W that can be reached from V along an arc VW, and assign W the label

$$(\text{potential of } V) + (\text{distance } VW)$$

unless W already has a *smaller* label assigned from an earlier iteration.
When all such vertices W have been labelled, choose the smallest vertex label that is not already a potential, and make it a potential at each vertex where it occurs.

REPEAT the general step with the new potential(s).

STOP when T has been assigned a potential.
The shortest distance from S to T is the potential of T.

To find a shortest path, trace backwards from T and include an arc VW whenever

$$(\text{potential of } W) - (\text{potential of } V) = \text{distance } VW$$

until S is reached.

Problem 9.2

Use the algorithm to find the shortest path from S to T in the following digraph, and record the results in a table.

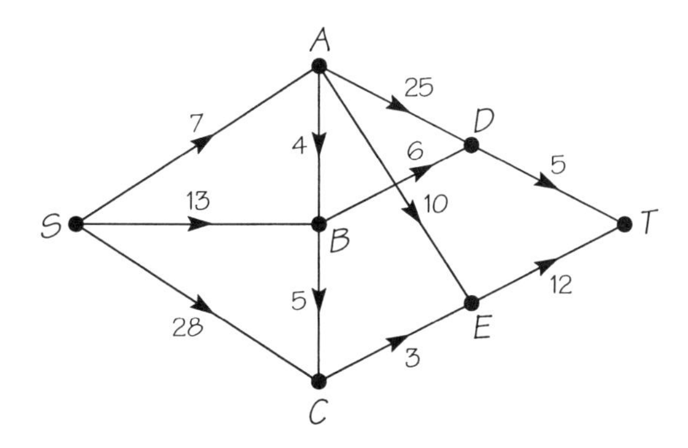

9.3 Case Study

Chinese Postman Problem

An important problem that has appeared in various guises is the so-called *Chinese postman problem,* which may be stated as follows.

Chinese Postman Problem

A postman wishes to deliver mail along all the streets in his area, and then return to the depot. How can he plan his route so as to cover the smallest total distance?

Remark The word *Chinese* refers to the problem, not the postman! The problem was considered in 1962 by Mei-ko Kwan.

If the map of his area happens to correspond to an Eulerian graph, there is no difficulty with this problem – he simply finds an Eulerian trail (using Fleury's algorithm, if necessary) and such a trail involves the smallest total distance; this situation occurs when all the vertices in the corresponding graph have even degree. What usually happens in practice, of course, is that the postman needs to visit some parts of the route more than once, and wants to minimize the amount of retracing. We assume that we know the length of each part of the route.

Similar problems have arisen in other contexts. For example, there was a major study of snow-clearing routes in Zürich some years ago. Snow-ploughs are expensive to operate, so it was necessary to arrange a route that involved revisiting streets as little as possible. Other cities have initiated similar investigations into the sweeping or cleaning of streets.

We can reformulate this problem in terms of weighted graphs as follows.

Chinese Postman Problem

Find a closed walk of minimum total weight that includes every edge at least once.

This problem has been solved in general, using an algorithm that combines features of Fleury's algorithm and the shortest path algorithm. If the graph is not Eulerian, then there exist vertices of odd degree. However, it follows from the handshaking lemma that the number of such vertices is even (see Problem 2.11), so is equal to $2k$, say. We can make the graph Eulerian by doubling the edges along k paths that join these vertices in pairs; we ensure that we obtain a closed walk of minimum weight by choosing the pairs so that the sum of the lengths of these k paths is as small as possible.

The details of the algorithm are too complicated to be given here, but we can get an idea of what is involved by considering the particular case of a graph with just two vertices v and w of odd degree, such as graph (a) below:

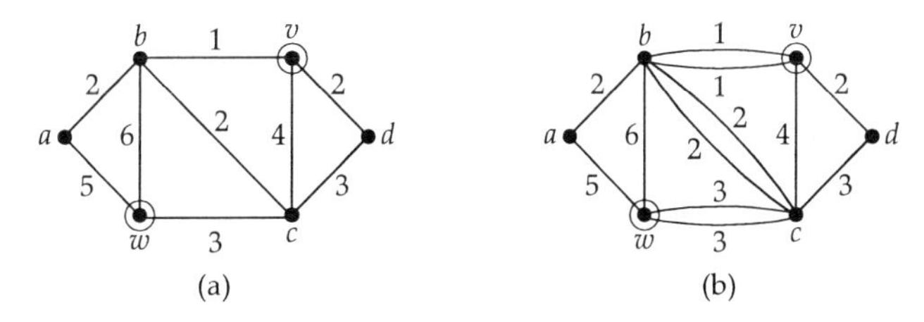

(a) (b)

The shortest path from v to w is $vbcw$, of total weight $1 + 2 + 3 = 6$. If we double up each of the edges in this path, we get the Eulerian graph (b). The required closed walk of minimum total weight is obtained by finding an Eulerian trail in this graph (using Fleury's algorithm if necessary), such as $a\,b\,v\,d\,c\,v\,b\,c\,b\,w\,c\,w\,a$. The only edges that need to be retraced are vb, bc and cw.

Alternatively, we find a semi-Eulerian path from v to w (using a modification of Fleury's algorithm) and then find the shortest path from w back to v. The combination of these paths is then a trail of minimum weight. For the above graph we obtain the route $v\,d\,c\,v\,b\,c\,w\,b\,a\,w\,c\,b\,v$.

Problem 9.3

Solve the Chinese postman problem for the following weighted graph:

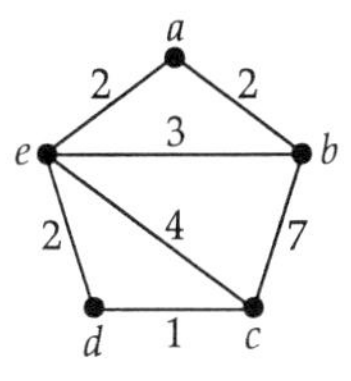

Now suppose that the problem is one of street-cleaning or snow-ploughing, and that the vehicle needs to travel along *both* sides of each road, once in each direction. We replace each edge of the corresponding graph by two arcs, one in each direction. We thus obtain a strongly connected digraph in which, for each vertex, the out-degree equals the in-degree. It follows from Theorem 4.2 that the digraph is Eulerian, so there exists a minimum-weight Eulerian trail; it can be found by a modification of Fleury's algorithm.

Exercises 9

Fleury's Algorithm

9.1 Use Fleury's algorithm to obtain an Eulerian trail in the following graph:

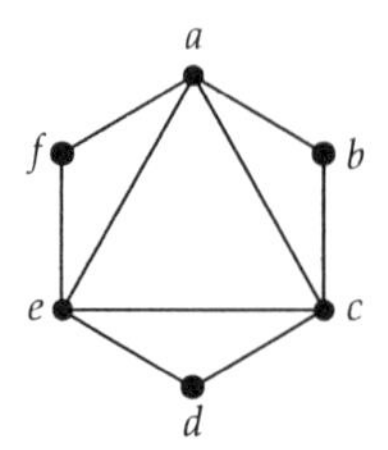

9.2 Use Fleury's algorithm to obtain an Eulerian trail in the following graph:

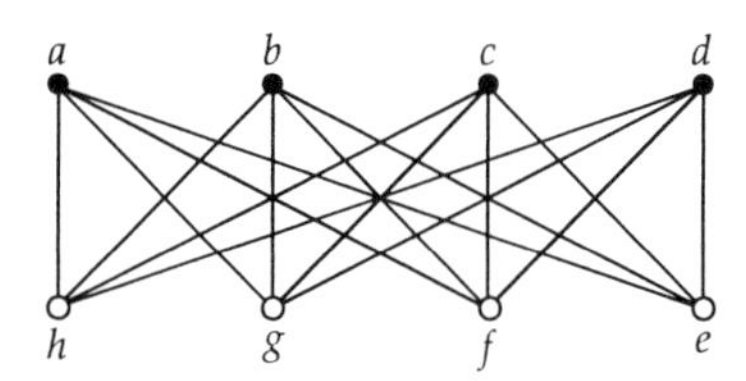

Shortest Path Algorithm

9.3 Find the shortest path from S to T and the shortest distance from S to each of the other vertices in the following weighted digraph:

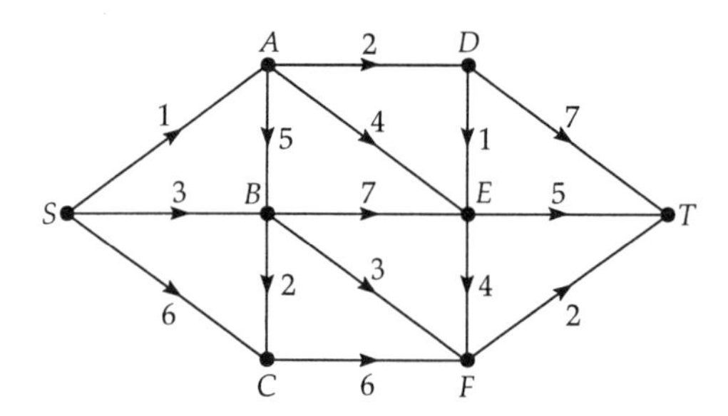

9.4 Find the shortest path(s) from S to T in the following weighted digraph:

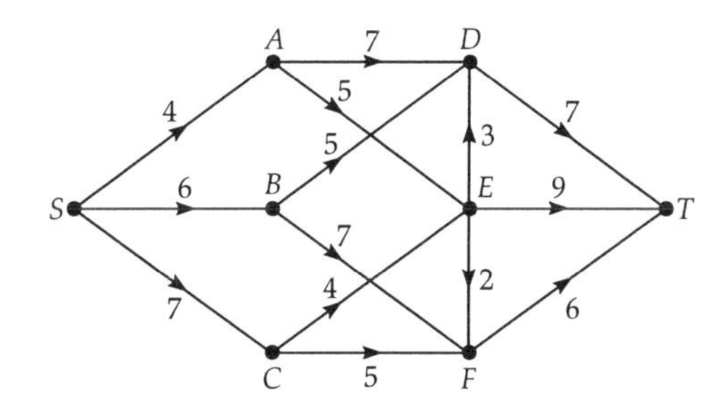

9.5 Find the shortest path(s) from S to T in the following weighted digraph:

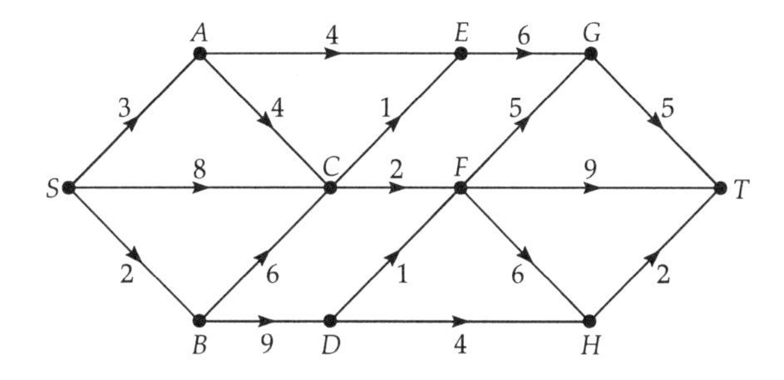

Chinese Postman Problem

9.6 Solve the Chinese postman problem for the following weighted graph:

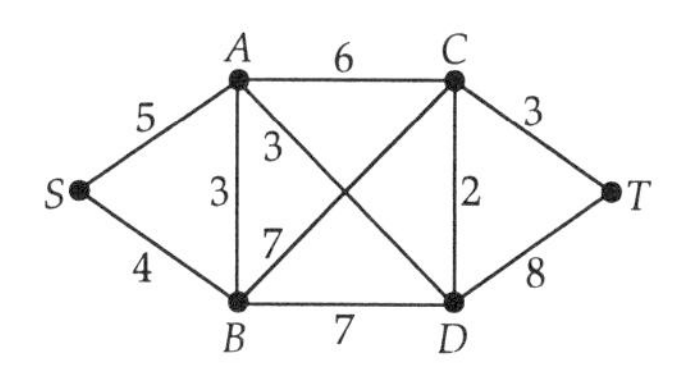

9.7 Solve the Chinese postman problem for the following weighted graph:

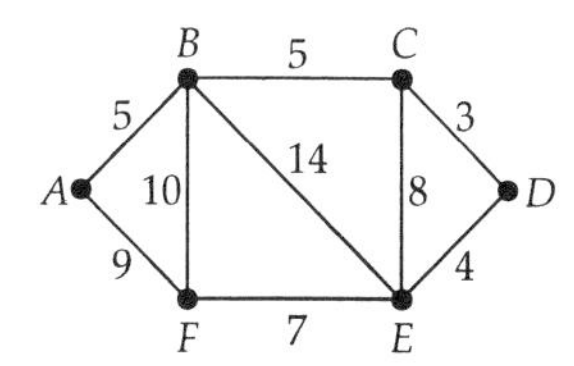

Chapter 10

Paths and Connectivity

After reading this chapter, you should be able to:

- explain the terms *edge connectivity*, *vertex connectivity*, *cutset* and *vertex cutset*;
- understand and use the various forms of Menger's theorem;
- discuss the reliability of a telecommunication network in terms of the concepts introduced in this chapter.

In this chapter we investigate the extent to which a given graph or digraph is connected. In particular, we discuss the following question:

> how many edges (or vertices) do we need to remove from a given connected graph so that it becomes disconnected?

This, and similar questions related to connectivity, are important ones to consider when designing telecommunication networks, road systems and other networks – for example, in a telecommunication network it is essential that the network should still be operable if some of the links between the exchanges become damaged, or are blocked by other calls. After discussing the theory of connectivity in graphs and digraphs in general, we describe its relevance to such networks.

10.1 Connected Graphs and Digraphs

In Chapter 2 we introduced the idea of a connected graph – a graph that is 'in one piece'. We noted that in any connected graph there is a path between each pair of vertices, and this led to the following definitions.

Definitions

A graph is **connected** if there is a path between each pair of vertices, and is **disconnected** otherwise.

Every disconnected graph can be split up into a number of connected subgraphs, called **components**.

For example:

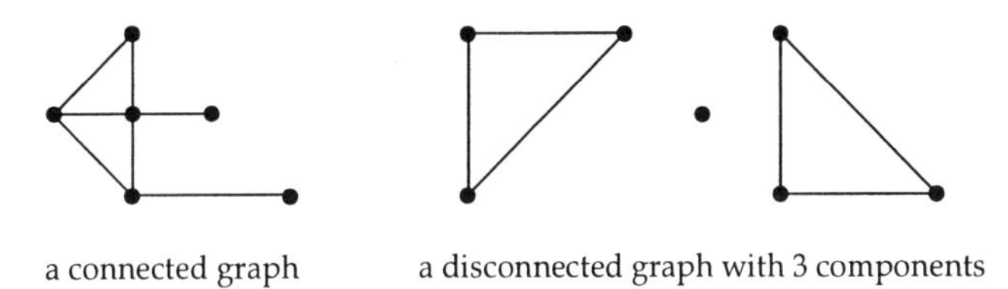

a connected graph a disconnected graph with 3 components

In the case of digraphs, it is not true in general that a digraph that is 'in one piece' has a (directed) path between each pair of vertices, and this observation led us to define two types of connected digraph, as follows.

Definitions

A digraph is **connected** if its underlying graph is a connected graph, and is **disconnected** otherwise.

A digraph is **strongly connected** if there is a path between each pair of vertices.

For example:

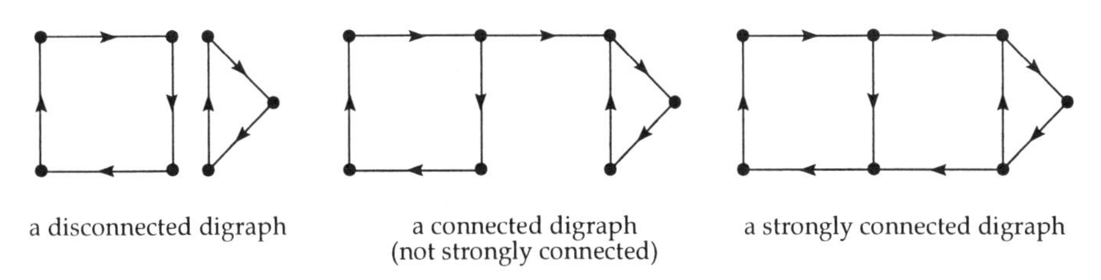

a disconnected digraph a connected digraph (not strongly connected) a strongly connected digraph

Edge Connectivity

For many applications it is necessary to know more about a graph than just whether it is connected. For example, in telecommunication networks there are usually several paths between a given pair of subscribers (vertices), and it is important to know how many links (edges) can be broken without preventing a call being made between the two subscribers. In order to answer this and similar questions, we investigate connected graphs in more detail.

Consider the following connected graphs:

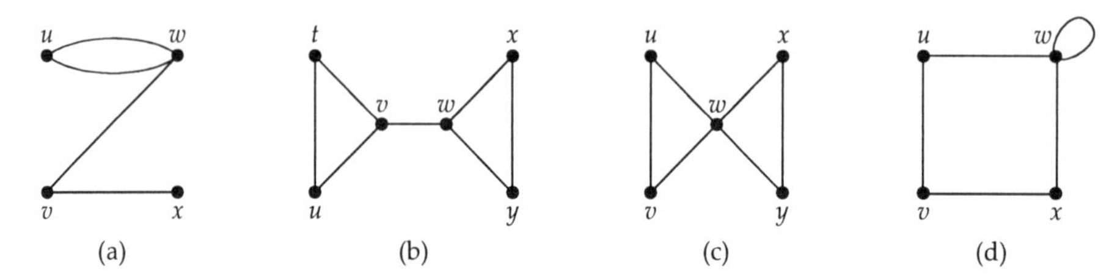

Graph (a) can be split into two components by removing one of the edges vw and vx. We say that the removal of either of these edges *disconnects* the graph.

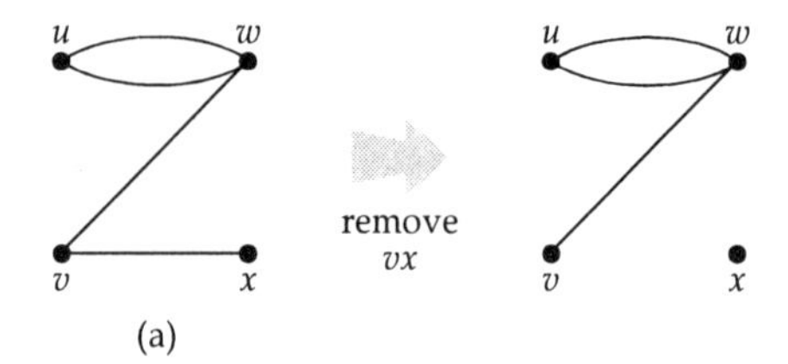

Graph (b) can also be disconnected by the removal of a single edge – vw.

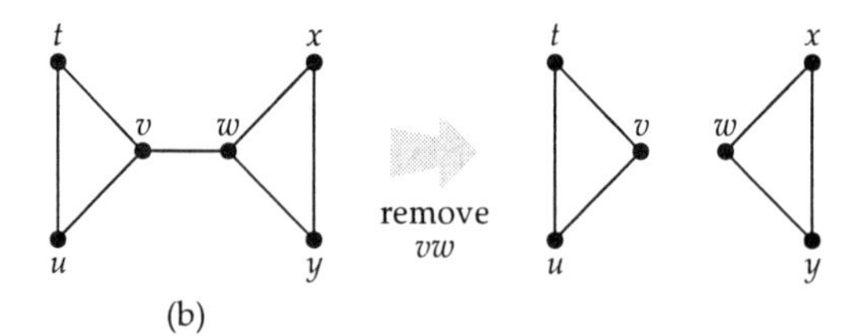

Graph (c) cannot be disconnected by removing a single edge, but the removal of two edges – for example, uw and vw – disconnects it.

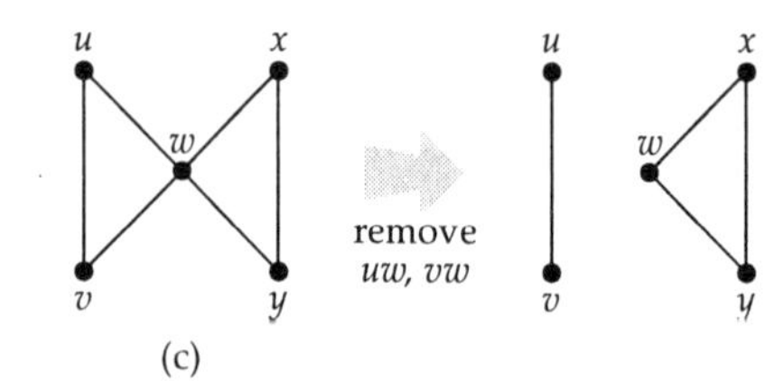

Similarly, graph (d) can be disconnected by removing two edges – for example, *uw* and *wx*.

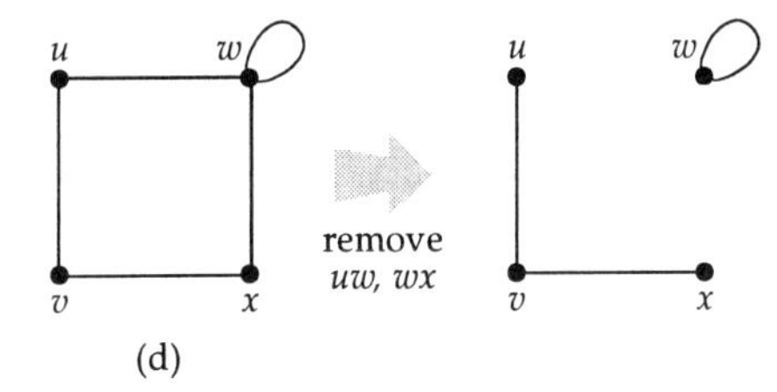

(d)

With these examples in mind, we define the *edge connectivity* of a graph as follows.

Definition

The **edge connectivity** $\lambda(G)$ of a connected graph *G* is the *smallest* number of edges whose removal disconnects *G*.

For example, graphs (a) and (b) have edge connectivity 1, and graphs (c) and (d) have edge connectivity 2.

If we wish to disconnect a graph by removing edges, we often have a choice of edges to delete. It seems natural to consider ways of disconnecting a graph that do not involve the removal of 'redundant' edges.

Consider the following graph *G*.

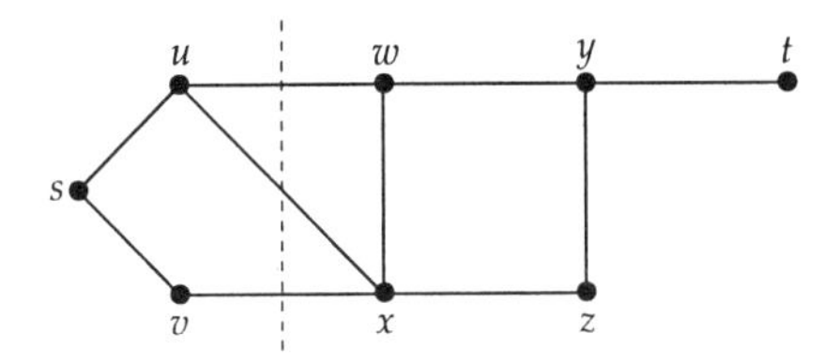

We can disconnect *G* by removing the three edges *uw*, *ux*, *vx*, but we cannot disconnect it by removing just two of these edges. We can also disconnect *G* by removing the four edges *uw*, *wx*, *xz*, *yz*, but the edge *yz* is redundant here, since we need remove only the edges *uw*, *wx*, *xz* to disconnect *G*. A set of such edges in which no edge is redundant – such as $\{uw, ux, vx\}$, $\{wy, xz\}$ or $\{yt\}$ – is called a *cutset*.

Definition

A **cutset** of a connected graph G is a set S of edges with the following two properties:

- removal of all the edges in S disconnects G;
- removal of some (but not all) of the edges in S does not disconnect G.

Remarks

1. Two cutsets of a graph need not necessarily have the same number of edges. For example, in the above graph, the sets $\{uw, ux, vx\}$, $\{wy, xz\}$ and $\{yt\}$ are all cutsets.
2. The edge connectivity $\lambda(G)$ of a graph G is the size of the *smallest* cutset of G. For example, for the above graph, $\lambda(G) = 1$.

Problem 10.1

Which of the following sets of edges are cutsets of the graph below?

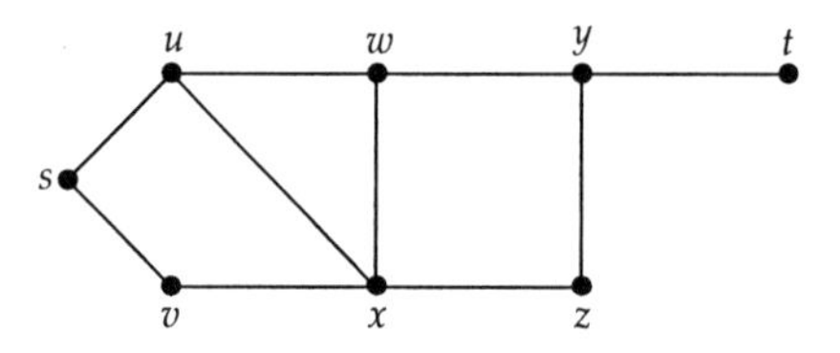

(a) $\{su, sv\}$; (b) $\{ux, wx, yz\}$; (c) $\{ux, vx, wx, yz\}$;
(d) $\{yt, yz\}$; (e) $\{wx, xz, yz\}$; (f) $\{uw, wx, wy\}$.

Problem 10.2

Write down the value of $\lambda(G)$ for each of the following graphs G:

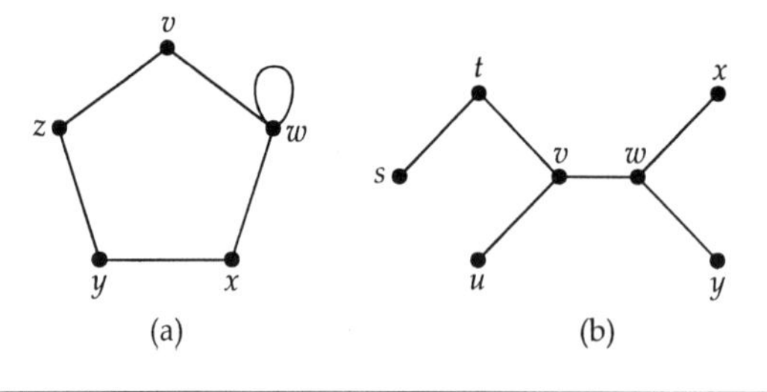

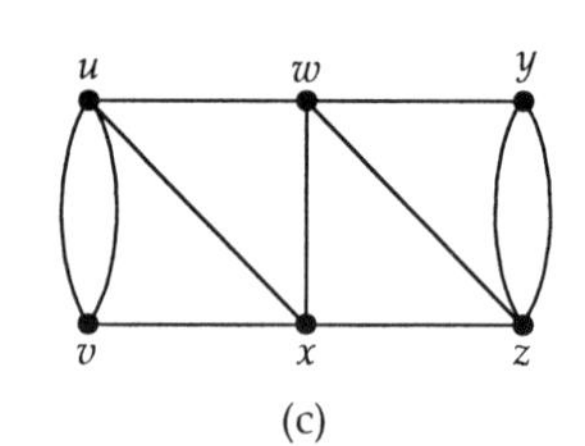

Vertex Connectivity

We can also think of connectivity in terms of the minimum number of *vertices* that need to be removed in order to disconnect a graph. When we remove a vertex, we must also remove the edges incident with it:

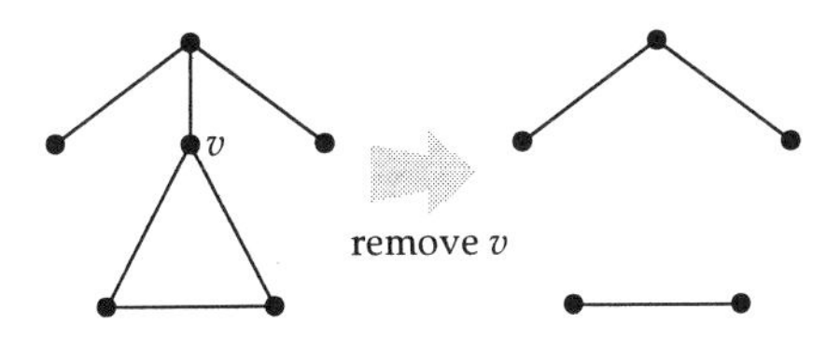

Consider again the connected graphs (a)–(d):

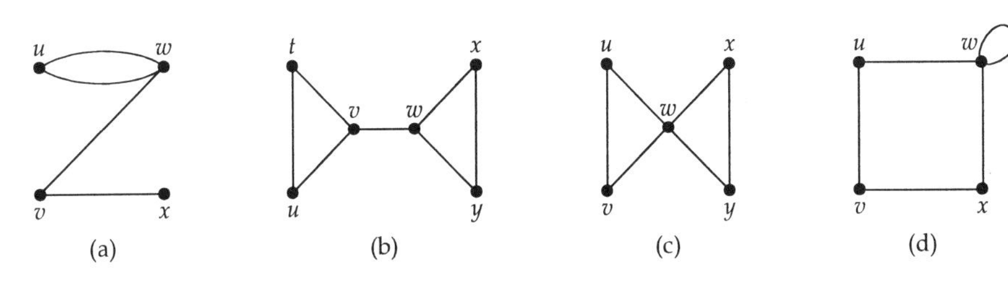

Graphs (a) and (b) can be disconnected by the removal of a single vertex, either v or w.

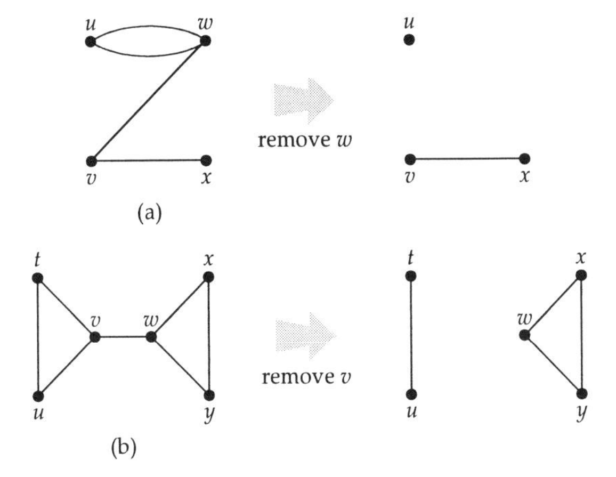

Graph (c) can also be disconnected by removing just one vertex, w.

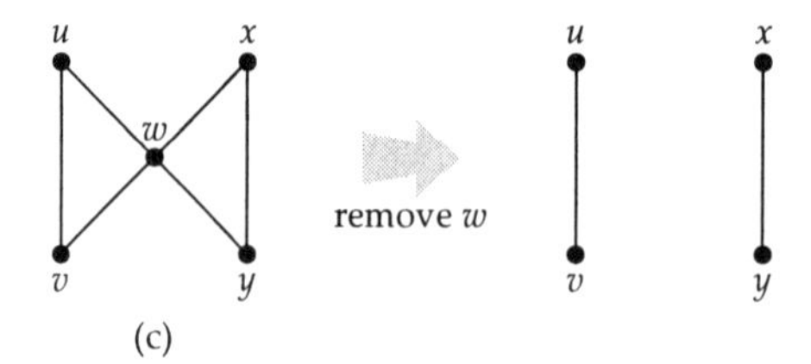

(c)

Graph (d) cannot be disconnected by removing a single vertex, but the removal of two non-adjacent vertices – for example, v and w – disconnects it.

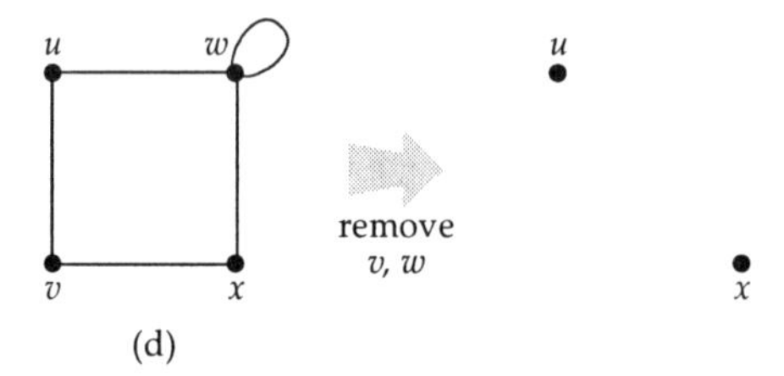

(d)

A single vertex whose removal disconnects a graph, such as v or w in graph (b), or w in graph (c), is called a **cut vertex**.

With these examples in mind, we define the *connectivity* (or *vertex connectivity*) of a graph as follows; we use the simpler term *connectivity* when there is no possibility of confusion with *edge connectivity*.

Definition

The **connectivity** (or **vertex connectivity**) $\kappa(G)$ of a connected graph G (other than a complete graph) is the *smallest* number of vertices whose removal disconnects G.

For example, graphs (a), (b) and (c) have connectivity 1, and graph (d) has connectivity 2.

The above definition breaks down when G is a complete graph, since we cannot then disconnect G by removing vertices. We therefore make the following definition.

> **Definition**
>
> The **connectivity $\kappa(K_n)$** of the complete graph K_n ($n \geq 3$) is $n - 1$.

There is also a 'vertex analogue' of the concept of a cutset.
Consider the following graph G.

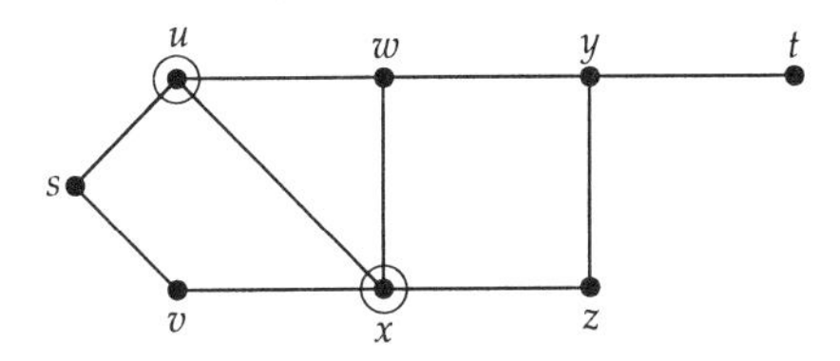

We can disconnect G by removing the two vertices u and x, but we cannot disconnect it by removing just one of these vertices. We can also disconnect G by removing the two vertices y and z, but the vertex z is redundant here, since we need remove only the vertex y to disconnect G. A set of such vertices in which no vertex is redundant, such as $\{u, x\}$ or $\{y\}$, is called a *vertex cutset*.

> **Definition**
>
> A **vertex cutset** of a connected graph G is a set S of vertices with the following two properties:
>
> - removal of all the vertices in S disconnects G;
> - removal of some (but not all) of the vertices in S does not disconnect G.

Remarks

1. Two vertex cutsets of a graph need not necessarily have the same number of vertices. For example, in the above graph, the sets $\{u, x\}$ and $\{y\}$ are both vertex cutsets.
2. The connectivity $\kappa(G)$ of a graph G is the size of the *smallest* vertex cutset of G. For example, for the above graph, $\kappa(G) = 1$.

Problem 10.3

Which of the following sets of vertices are vertex cutsets of the graph below?

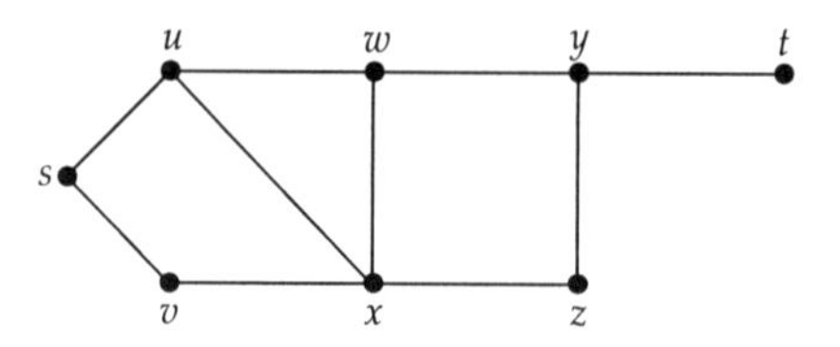

(a) $\{u, v\}$; (b) $\{v, w\}$; (c) $\{u, x, y\}$; (d) $\{w, z\}$.

Problem 10.4

Write down the value of $\kappa(G)$ for each of the following graphs G:

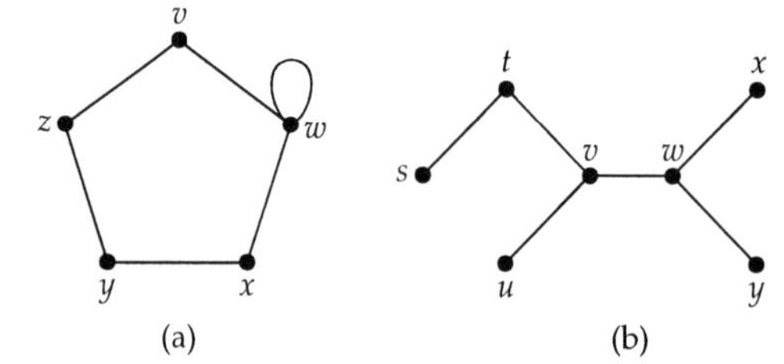

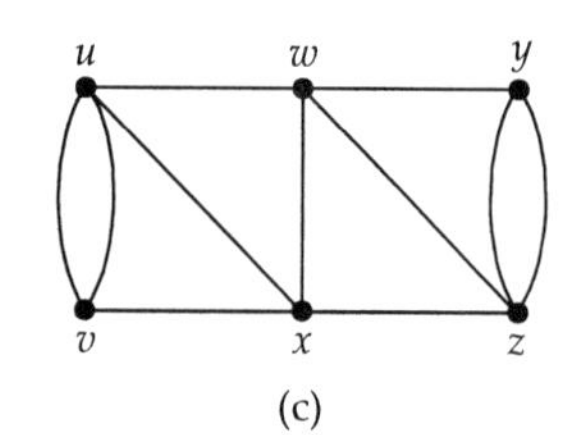

For each graph, list the values of $\kappa(G)$, $\lambda(G)$ (found in Problem 10.2), and the smallest vertex degree, $\delta(G)$.

In each part of Problem 10.4, you may have noticed that the vertex connectivity $\kappa(G)$ does not exceed the edge connectivity $\lambda(G)$, which does not exceed $\delta(G)$. These inequalities hold for all connected graphs.

Theorem 10.1

Let G be a connected graph with smallest vertex degree $\delta(G)$. Then

$$\kappa(G) \leq \lambda(G) \leq \delta(G).$$

Outline of Proof

If G is the complete graph K_n, then $\kappa(G) = \lambda(G) = \delta(G) = n - 1$.

If G is a connected graph that is not K_n, and if v is a vertex of degree $\delta(G)$, then G can be disconnected by removing all the $\delta(G)$ edges incident with v. It follows that $\lambda(G)$, the *minimum* number of edges whose removal disconnects G, cannot exceed $\delta(G)$. So

$$\lambda(G) \leq \delta(G). \qquad (*)$$

It remains to show that $\kappa(G) \leq \lambda(G)$ whenever G is not a complete graph.

Let G be a simple connected graph with n vertices and edge connectivity $\lambda(G)$. Since G is not K_n, the smallest vertex degree is at most $n - 2$, so, by equation $(*)$, $\lambda(G) \leq n - 2$. There is at least one set of $\lambda(G)$ edges whose removal disconnects G into two components G_1 and G_2, as illustrated below.

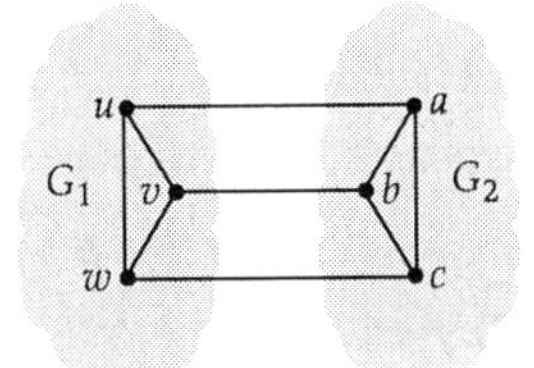

(For example, in the above case, $\lambda(G) = 3$ and we can disconnect the graph by removing the three edges ua, vb and wc.)

But we can also remove these edges by removing at most $\lambda(G)$ vertices, since we have only to remove one suitably chosen end-vertex from each of these $\lambda(G)$ edges. (For example, in the above case, we can remove the vertices u, v and c.) Let us remove these $\lambda(G)$ edges one at a time. Since there are at most n – 2 edges to be removed, and since at each step we can remove an edge either from G_1 or from G_2, we can achieve this by removing a set of vertices that leaves neither G_1 nor G_2 empty. It follows that the minimum number of vertices whose removal disconnects G cannot exceed $\lambda(G)$; that is, $\kappa(G) \leq \lambda(G)$.

Finally, let G be a non-simple connected graph, and let G' be the graph obtained by deleting loops and changing multiple edges to single edges. If $\kappa(G') \leq \lambda(G')$ for the simple graph G', then $\kappa(G) \leq \lambda(G)$, since the changes cannot affect $\kappa(G)$, and can only decrease $\lambda(G)$; that is,

$$\kappa(G) = \kappa(G') \leq \lambda(G') = \lambda(G).$$

Thus, for any connected graph G,

$$\kappa(G) \leq \lambda(G) \leq \delta(G). \qquad \blacksquare$$

Remark It is possible for both inequalities in Theorem 10.1 to be strict inequalities: $\kappa(G) < \lambda(G) < \delta(G)$. For example, for the graph shown below, $\kappa(G) = 1$, $\lambda(G) = 2$ and $\delta(G) = 3$:

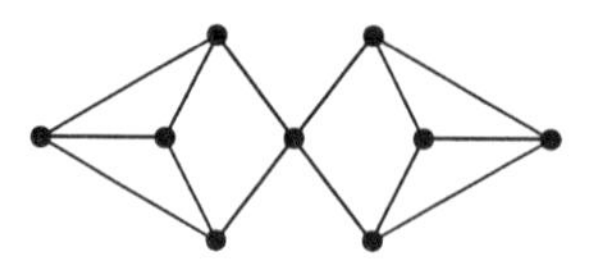

10.2 Menger's Theorem for Graphs

In this section we discuss an important result which relates the above ideas to the number of 'disjoint paths' between two vertices in a graph. This result is known as *Menger's theorem.*

We start by defining *disjoint paths* in a graph.

Definitions

Let G be a connected graph, and let s and t be vertices of G.

A path between s and t is an ***st*-path**. Two or more st-paths are **edge-disjoint** if they have no edges in common, and **vertex-disjoint** if they have no vertices in common, apart from s and t.

For example, in the following graph,

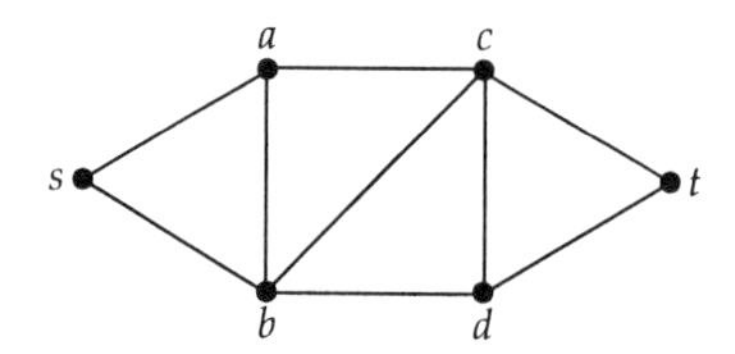

the paths *sact* and *sbdt* are both edge-disjoint and vertex-disjoint *st*-paths;

the paths *sact* and *sbct* are neither edge-disjoint nor vertex-disjoint, since they have the edge *ct* in common;

the paths *sact* and *sbcdt* are edge-disjoint, but not vertex-disjoint, since they have the vertex *c* in common.

Remark For convenience, we sometimes abuse the terminology and say 'there is at most *one* vertex-disjoint/edge-disjoint *st*-path' when there do not exist two vertex-disjoint/edge-disjoint *st*-paths (see Example 10.1).

Problem 10.5

Consider the following graph.

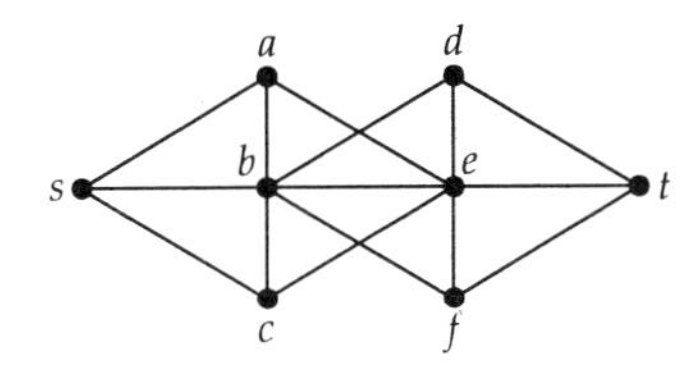

Write down

(a) three edge-disjoint st-paths;
(b) two st-paths that are edge-disjoint, but not vertex-disjoint;
(c) two vertex-disjoint st-paths.

Does this graph contain three vertex-disjoint st-paths? Justify your answer.

Problem 10.6

(a) Prove that if two st-paths in a graph are vertex-disjoint, then they are also edge-disjoint.
(b) Give an example of a graph in which no two edge-disjoint st-paths are vertex-disjoint.

We also need the following definitions.

Definitions

Let G be a connected graph, and let s and t be vertices of G.

Certain *edges* **separate s from t** if the removal of these edges destroys all paths between s and t.

Certain *vertices* **separate s from t** if the removal of these vertices destroys all paths between s and t.

For example, in the following graph,

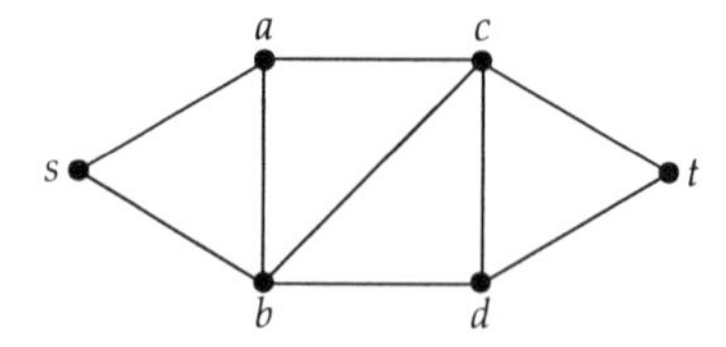

the edges *ac*, *bc*, *bd* separate *s* from *t*, as do the edges *sa*, *ac*, *bc*, *bd*, *dt*; the vertices *b* and *c* separate *s* from *t*, as do the vertices *a*, *b*, *d*.

We now show how these ideas are related to that of edge-disjoint *st*-paths. We motivate our discussion with three examples.

Example 10.1

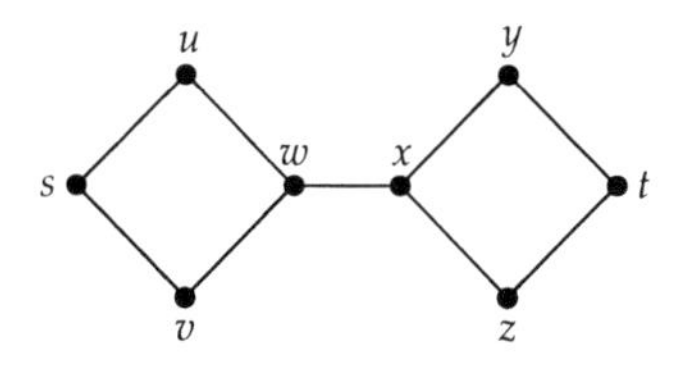

In this graph, the single edge *wx* separates *s* from *t*. It follows that *there cannot be two edge-disjoint st-paths*, since each *st*-path must include the edge *wx*, so there is at most *one* edge-disjoint *st*-path. ❐

Example 10.2

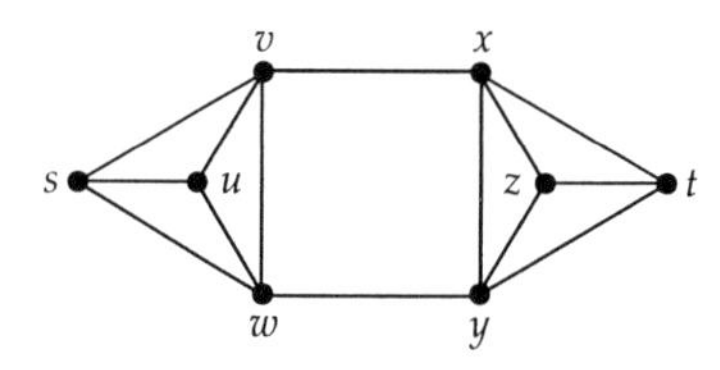

In this graph, the two edges *vx* and *wy* separate *s* from *t*. It follows that *there are at most two edge-disjoint st-paths*, since each *st*-path must include one of these two edges. ❐

Example 10.3

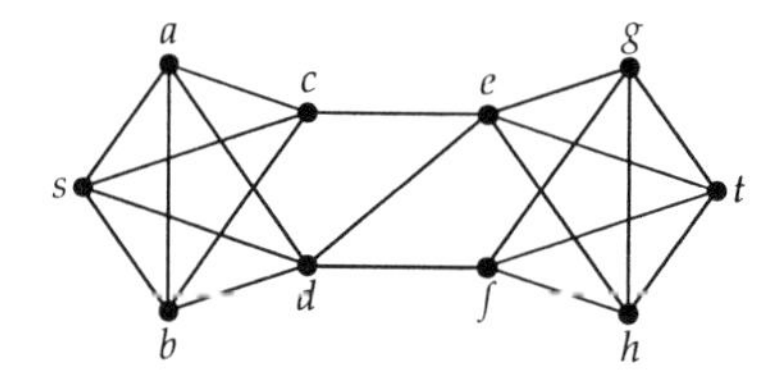

In this graph, the three edges *ce*, *de* and *df* separate *s* from *t*. It follows that *there are at most three edge-disjoint st-paths*, since each *st*-path must include one of these three edges. ❐

More generally, consider a set of edges separating *s* from *t* in an arbitrary connected graph.

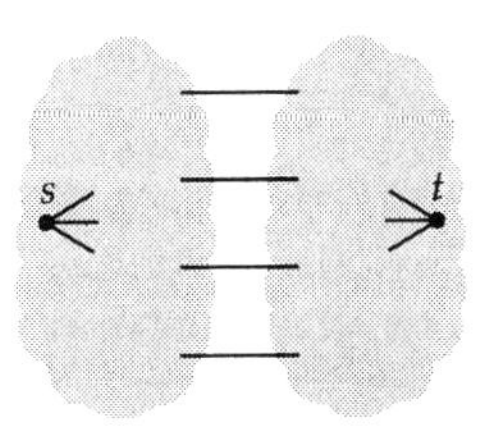

Since the removal of these edges destroys all paths between *s* and *t*, each *st*-path includes at least one of them. It follows that *the maximum number of edge-disjoint st-paths cannot exceed the number of edges in this set*. Since this applies to *any* set of edges separating *s* from *t*,

the *maximum* number of edge-disjoint *st*-paths	$\leq$	the number of edges in *any* set of edges separating *s* from *t*.

But this is true for *any* set of edges separating *s* from *t*, so it must be true for *a set with the smallest possible number of edges*. Thus

the *maximum* number of edge-disjoint *st*-paths	$\leq$	the *minimum* number of edges separating *s* from *t*.

The two numbers in the above inequality are, in fact, equal. This is the *edge form* of Menger's theorem for graphs, which we state formally as follows.

Theorem 10.2: Menger's Theorem for Graphs (Edge Form)

Let *G* be a connected graph, and let *s* and *t* be vertices of *G*.
Then the maximum number of edge-disjoint *st*-paths is equal to the minimum number of edges separating *s* from *t*.

Remark It follows that, if we can find *k* edge-disjoint *st*-paths and *k* edges separating *s* from *t* (for the same value of *k*), then *k* is the *maximum* number of edge-disjoint *st*-paths and the *minimum* number of edges separating *s* from *t*. These *k* edges separating *s* from *t* necessarily form a cutset. It follows that, when looking for them, we need consider only cutsets whose removal disconnects *G* into two components, one containing *s* and the other containing *t*.

Problem 10.7

By finding k edge-disjoint st-paths, and k edges separating s from t (for the same value of k), and using the edge form of Menger's theorem, find the maximum number of edge-disjoint st-paths for each of the following graphs:

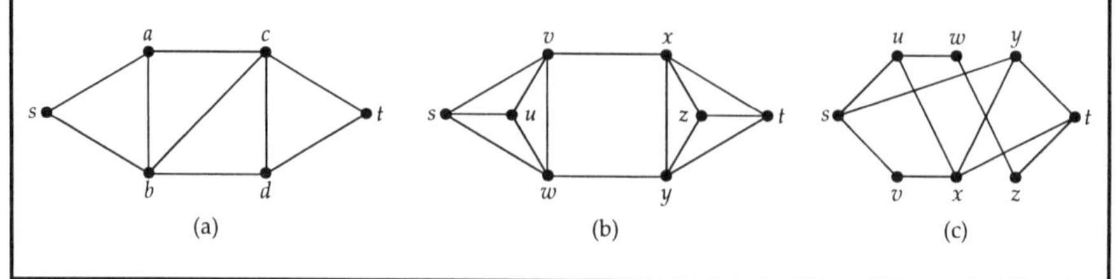

We can use Menger's theorem to obtain a result about edge connectivity. Recall that the edge connectivity $\lambda(G)$ of a connected graph G is the smallest number of edges whose removal disconnects G. So, by Menger's theorem, there are at least $\lambda(G)$ edge-disjoint paths between any given pair of vertices.

We restate this result as follows.

Corollary of Menger's Theorem for Graphs (Edge Form)

A connected graph G has edge connectivity l if and only if there are l or more edge-disjoint paths between each pair of vertices in G, and there are exactly l edge-disjoint paths between at least one pair of vertices in G.

Notice that, in Examples 10.1, 10.2 and 10.3, the edge connectivities are 1, 2 and 3, respectively.

10.3 Some Analogues of Menger's Theorem

We now present some analogues of Menger's theorem, starting with Menger's theorem for digraphs (arc form), and continuing with the vertex forms for both graphs and digraphs. We present all these results without proof.

Menger's Theorem for Digraphs (Arc Form)

Many of the concepts introduced earlier for graphs have analogues for digraphs. For example, the following definitions are almost identical to those given for graphs.

> **Definitions**
>
> Let D be a connected digraph, and let s and t be vertices of D.
> A path from s to t is an ***st*-path**. Two or more st-paths are **arc-disjoint** if they have no arcs in common, and **vertex-disjoint** if they have no vertices in common, apart from s and t.

For example, in the following digraph,

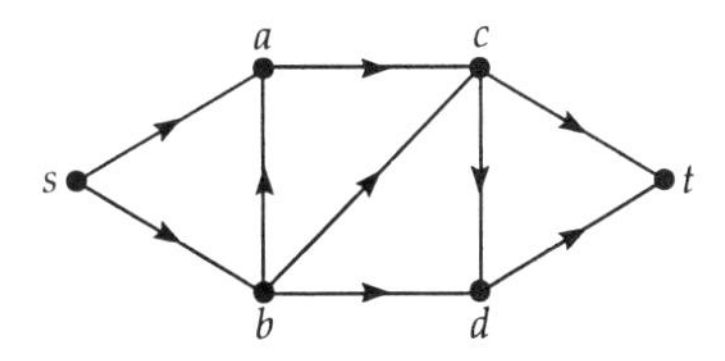

the paths *sact* and *sbdt* are both arc-disjoint and vertex-disjoint st-paths;
the paths *sact* and *sbct* are neither arc-disjoint nor vertex-disjoint;
the paths *sact* and *sbcdt* are arc-disjoint, but not vertex-disjoint.

Remark For convenience, we sometimes abuse the terminology and say 'there is at most *one* vertex-disjoint/arc-disjoint st-path' when there do not exist two vertex-disjoint/arc-disjoint st-paths (see Example 10.4).

> **Definitions**
>
> Let D be a connected digraph, and let s and t be vertices of D.
> Certain *arcs* **separate *s* from *t*** if the removal of these arcs destroys all paths from s to t.
> Certain *vertices* **separate *s* from *t*** if the removal of these vertices destroys all paths from s to t.

For example, in the above digraph,

the arcs *ac*, *bc*, *bd* separate s from t, as do the arcs *sa*, *ac*, *bc*, *bd*, *dt*;
the vertices b and c separate s from t, as do the vertices a, b, d.

Using this terminology, we state the arc form of Menger's theorem for digraphs.

Theorem 10.3: Menger's Theorem for Digraphs (Arc Form)

Let D be a connected digraph, and let s and t be vertices of D.
Then the maximum number of arc-disjoint st-paths is equal to the minimum number of arcs separating s from t.

Remark It follows that, if we can find k arc-disjoint st-paths and k arcs separating s from t (for the same value of k), then k is the *maximum* number of arc-disjoint st-paths and the *minimum* number of arcs separating s from t.

Problem 10.8

By finding k edge-disjoint st-paths, and k arcs separating s from t (for the same value of k), and using the arc-form of Menger's theorem, find the maximum number of arc-disjoint st-paths for each of the following digraphs:

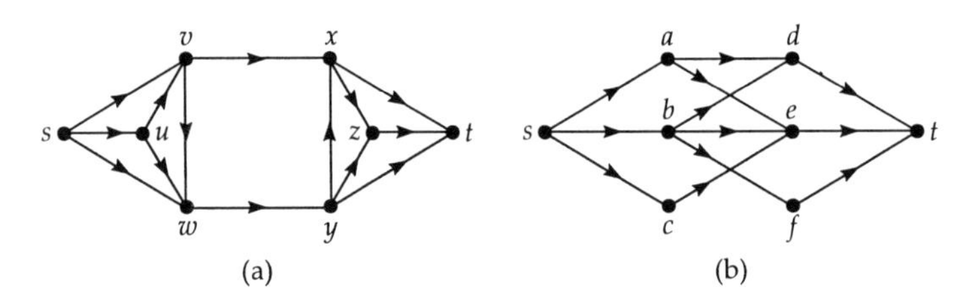

(a) (b)

Menger's Theorem for Graphs (Vertex Form)

We have seen how Menger's theorem (edge form) relates the number of edge-disjoint st-paths in a graph to the smallest number of edges separating s from t, and how this result relates to edge connectivity. We now state an analogous theorem for vertex-disjoint st-paths.

As before, we motivate our discussion with examples.

Example 10.4

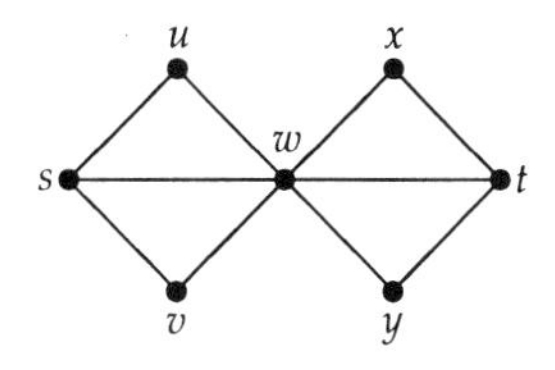

In this graph, the vertex w separates s from t. It follows that *there cannot be two vertex-disjoint st-paths*, since all st-paths must include the vertex w, so there is at most *one* vertex-disjoint st-path.) ❐

Example 10.5

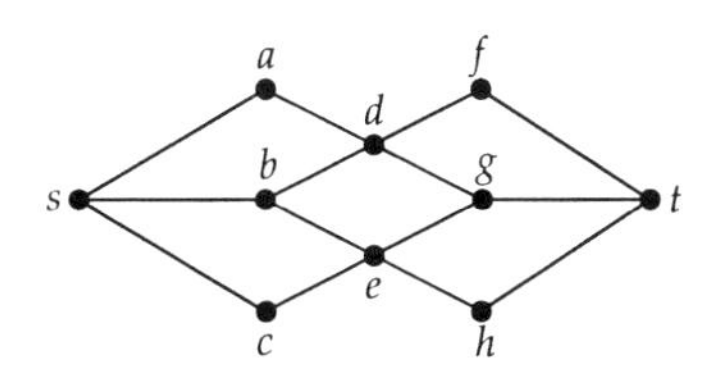

In this graph, the vertices d and e separate s from t. It follows that *there are at most two vertex-disjoint st-paths*, since all st-paths must include one of these vertices. ❐

More generally, consider a set of vertices (excluding s and t) separating non-adjacent vertices s and t in an arbitrary connected graph.

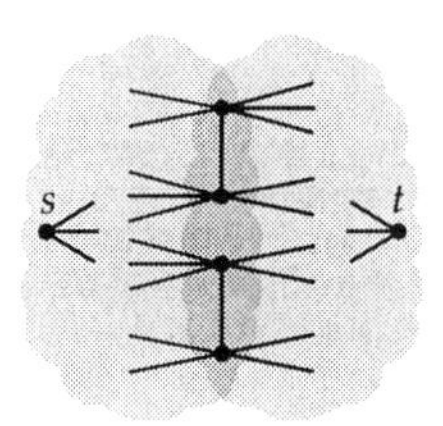

Since the removal of these vertices destroys all paths between s and t, every st-path includes at least one of them. It follows that *the maximum number of vertex-disjoint st-paths cannot exceed the number of vertices in this set.*

As with the edge form of Menger's theorem, these numbers are, in fact, equal. This is the *vertex form* of Menger's theorem, which we state formally as follows.

Theorem 10.4: Menger's Theorem for Graphs (Vertex Form)

Let G be a connected graph, and let s and t be non-adjacent vertices of G.
Then the maximum number of vertex-disjoint st-paths is equal to the minimum number of vertices separating s from t.

Remark It follows that, if we can find k vertex-disjoint st-paths and k vertices separating s from t (for the same value of k), then k is the *maximum* number of vertex-disjoint st-paths and the *minimum* number of vertices separating s from t. These k vertices separating s from t necessarily form a vertex cutset. It follows that, when looking for them, we need consider only vertex cutsets whose removal disconnects G into two or more components, one containing s and another containing t.

Problem 10.9

By finding k vertex-disjoint st-paths, and k vertices separating s from t (for the same value of k), and using the vertex form of Menger's theorem, find the maximum number of vertex-disjoint st-paths for the following graph:

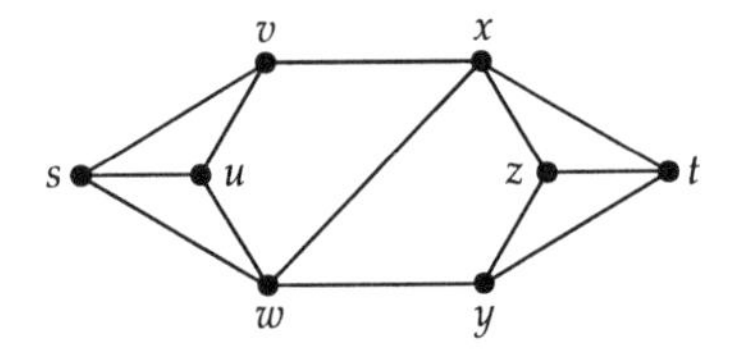

We can use Menger's theorem to obtain a result about vertex connectivity. Recall that the connectivity $\kappa(G)$ of a graph G (other than a complete graph) is the smallest number of vertices whose removal disconnects G. So, by the vertex form of Menger's theorem, there are at least $\kappa(G)$ vertex-disjoint paths between any given pair of vertices.

We restate this result as follows.

Corollary of Menger's Theorem for Graphs (Vertex Form)

A connected graph G (other than a complete graph) has vertex connectivity k if and only if every non-adjacent pair of vertices in G is joined by k or more vertex-disjoint paths, and at least one non-adjacent pair of vertices is joined by exactly k vertex-disjoint paths.

Notice that, in Examples 10.4 and 10.5, the vertex connectivities are 1 and 2, respectively.

Historical Note

The vertex form for graphs is the version of Menger's theorem actually proved by K. Menger in 1927. The corollary was proved five years later by H. Whitney. The edge form and arc form of Menger's theorem were proved in 1955 by L.R. Ford and D.R. Fulkerson.

Menger's Theorem for Digraphs (Vertex Form)

Finally, for completeness, we present the vertex form of Menger's theorem for digraphs. This is almost identical to the vertex form for graphs.

Theorem 10.5: Menger's Theorem for Digraphs (Vertex Form)

Let D be a connected digraph, and let s and t be non-adjacent vertices of D. Then the maximum number of vertex-disjoint st-paths is equal to the minimum number of vertices separating s from t.

Problem 10.10

By finding k vertex-disjoint st-paths, and k vertices separating s from t (for the same value of k), and by using the vertex form of Menger's theorem, find the maximum number of vertex-disjoint st-paths for the following digraph:

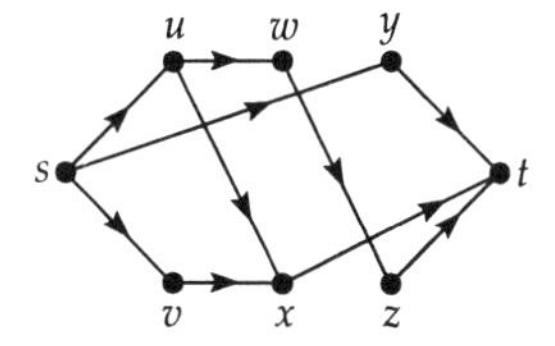

10.4 Case Study

Reliable Telecommunication Networks

In this section we look at the use of graph theory to represent telecommunication networks and show how it can be of value in the design of efficient networks. We find that the notion of connectivity is important in this context.

A graph of a telecommunication network may contain a very large number of vertices and edges. The vertices may represent telephone exchanges and subscribers; the edges represent links between them. It is important that such a network should be reliable. One aspect of reliability is that calls between subscribers should be possible even if a few exchanges or links fail. Consider the following graph, which represents a possible interconnection of exchanges:

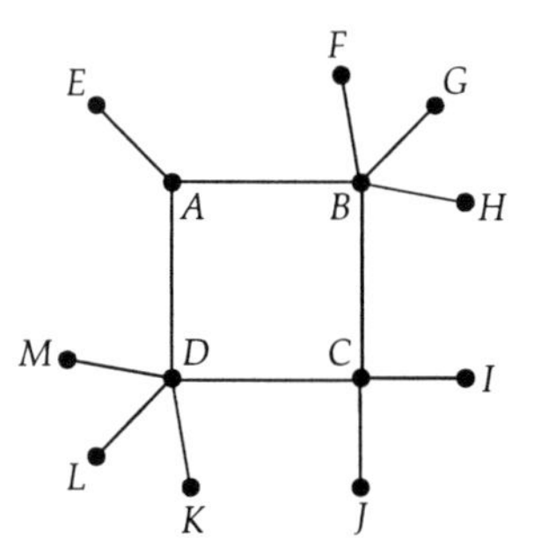

If the link AB fails, then communication between exchanges A and B is still possible, but if the link EA fails, then exchange E cannot communicate with any other exchange. The minimum number of links whose failure prevent the system from functioning fully is equal to the *edge connectivity* of the corresponding graph. Similarly, the *vertex connectivity* tells us how many exchanges must fail before there is a breakdown in communications between the remaining exchanges.

So it is necessary to provide alternative paths between exchanges so that communication between the exchanges is still possible if one path fails. Furthermore, a particular link or exchange in the path between two exchanges may also form part of the path between another pair of exchanges. The link or the exchange may therefore be already used to capacity by other calls when a new call is attempted, thus preventing the new call from being made. In such a case we say that the new call is *blocked*.

For any two particular exchanges, the maximum number of alternative paths, no two of which pass through the same intermediate exchange, is (in the terminology of graph theory) the maximum number of vertex-disjoint paths between the corresponding vertices of the graph. By the corollary to the vertex form of Menger's theorem, the smallest number of such paths between any two exchanges is the *vertex connectivity*. Similarly, the maximum number of alternative paths, no two of which use the same link between exchanges, is the maximum number of edge-disjoint paths between the corresponding vertices of the graph. By the corollary to the edge form of Menger's theorem, the smallest number of such paths between any two exchanges is the *edge connectivity*.

If reliability were the only consideration, a telecommunication system would have as many alternative paths as possible between exchanges – it would have the largest possible vertex connectivity and the largest possible edge connectivity. To achieve this, each exchange would need to be connected to every other exchange, and the corresponding graph would be a complete graph. This is usually impracticable. Designers try to achieve the largest possible values of the vertex connectivity $\kappa(G)$ and the edge connectivity $\lambda(G)$ for a graph G with a given number of vertices and edges.

We know from Theorem 10.1 that, for any connected graph G,

$$\kappa(G) \leq \lambda(G) \leq \delta(G),$$

where $\delta(G)$ is the minimum vertex degree in G. Suppose that G has n vertices and m edges. Then, by the handshaking lemma, the sum of all the vertex degrees is $2m$. It follows that the *average* of the vertex degrees is $2m/n$, so the *minimum* degree of the vertices cannot be greater than this. Combining these results, we obtain the following inequalities:

$$\kappa(G) \leq \lambda(G) \leq \delta(G) \leq 2m/n.$$

A graph G for which these inequalities are all equalities has the maximum vertex connectivity and the maximum edge connectivity possible for any graph with n vertices and m edges. Such a graph is said to have **optimal connectivity**. To show that a graph has optimal connectivity, it is sufficient to show that $\kappa(G) = 2m/n$, as then the above inequalities guarantee that $\kappa(G) = \lambda(G) = \delta(G)$.

All graphs with optimal connectivity are regular graphs (since the smallest vertex degree equals the average of the vertex degrees), but not every regular graph has optimal connectivity. For example, the following graph G is regular of degree 3, but $\kappa(G) = \lambda(G) = 2$, so G does not have optimal connectivity.

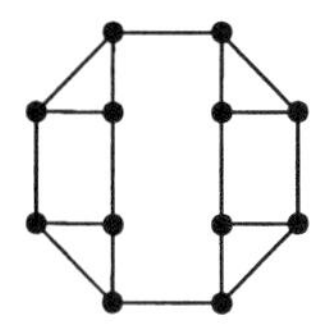

Problem 10.11

Show that the following regular graphs all have optimal connectivity:

(a) the cycle graph C_n ($n \geq 3$);
(b) the complete graph K_n ($n \geq 3$);
(c) the complete bipartite graph $K_{r,r}$ ($r \geq 2$).

Problem 10.12

(a) There are two non-isomorphic simple graphs with 6 vertices and 9 edges that have optimal connectivity. Draw them.
(b) Draw a regular graph with 7 vertices and 14 edges that does not have optimal connectivity.

We have seen that the values of $\kappa(G)$ and $\lambda(G)$ for a graph G give us information about the reliability of the corresponding telecommunication system. However, these values do not tell us the whole story: two graphs with the same number of vertices, the same number of edges, and the same values of $\kappa(G)$ and $\lambda(G)$, may not correspond to equally reliable systems.

Consider, for example, the following two graphs, each of which has 10 vertices and 12 edges, and satisfies $\kappa(G) = \lambda(G) = 2$.

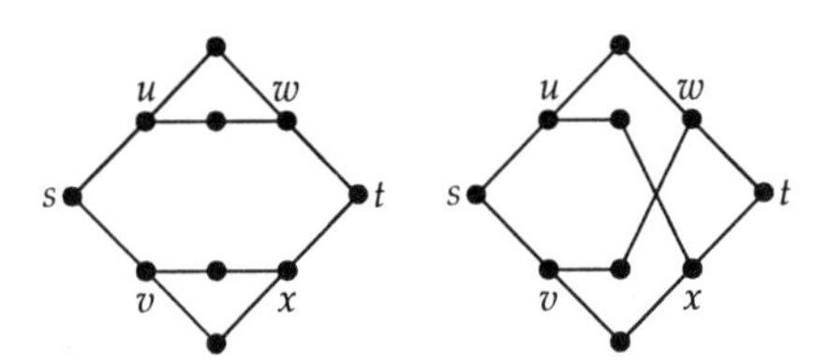

In the first graph, all paths from the vertex s to the vertex t can be destroyed by removing the edges of any one of the four cutsets with two edges separating s from t – $\{su, sv\}$, $\{wt, xt\}$, $\{su, xt\}$ and $\{sv, wt\}$. However, the second graph has only two cutsets with two edges separating s from t – $\{su, sv\}$ and $\{wt, xt\}$. So if all the edges contained in these cutsets have equal likelihood of being blocked or damaged, we would expect the second graph to correspond to a more reliable system than the first, and this is indeed the case.

In order to have full information about the reliability of a network represented by a graph, we need to know not only the vertex connectivity and edge connectivity, but also all the cutsets of the graph. Although this account is necessarily simplified, the ideas presented here have proved to be of great importance in the design and analysis of telecommunication systems.

Exercises 10

Connectivity of Graphs

10.1 Write down the values of $\kappa(G)$ and $\lambda(G)$ for each of the following graphs G:

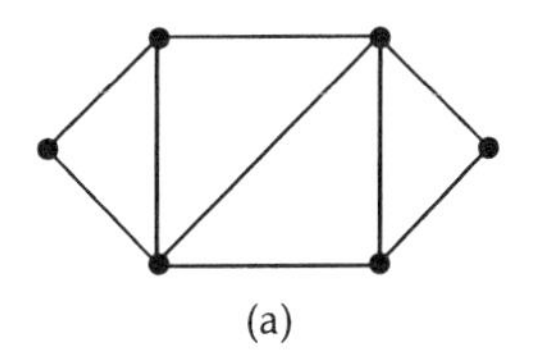
(a)

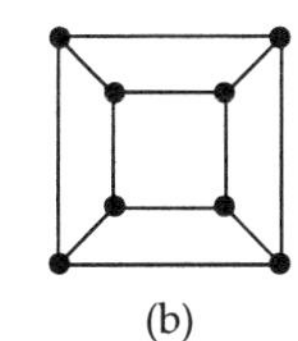
(b)

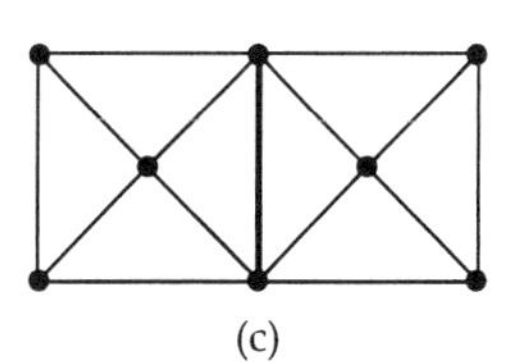
(c)

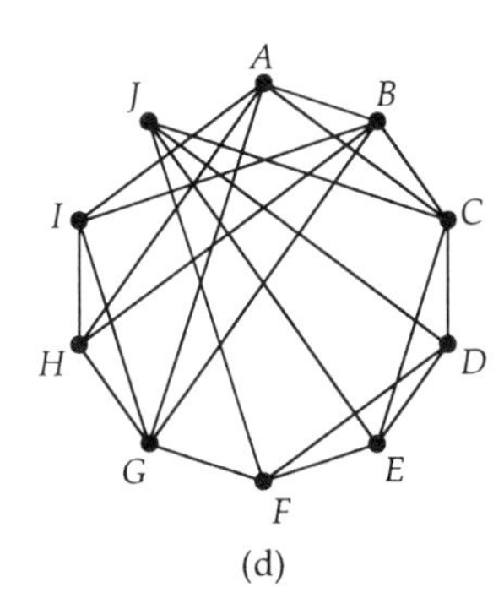

(d)

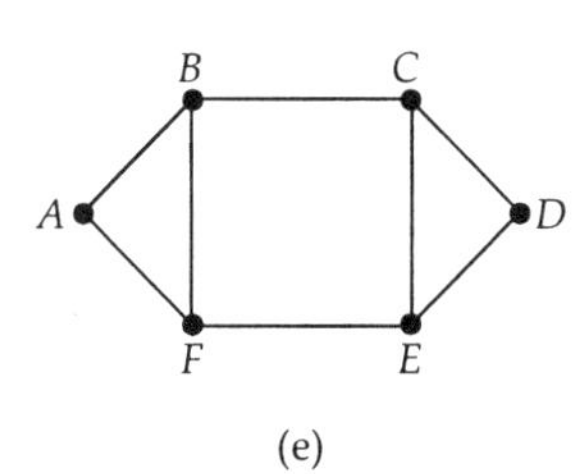

(e)

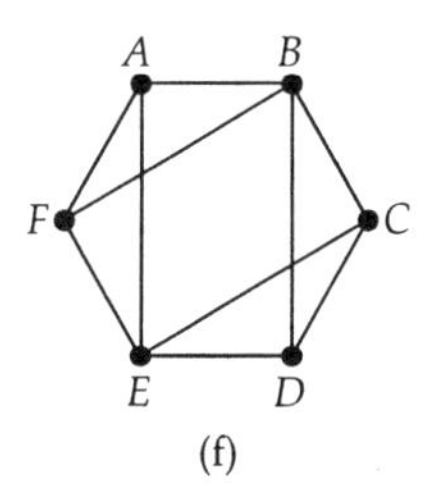

(f)

10.2 Give an example (where possible) of a graph G for which:

(a) $\kappa(G) = 2$, $\lambda(G) = 3$, $\delta(G) = 4$;
(b) $\kappa(G) = 3$, $\lambda(G) = 2$, $\delta(G) = 4$;
(c) $\kappa(G) = 2$, $\lambda(G) = 2$, $\delta(G) = 4$.

10.3 In the Petersen graph, find:

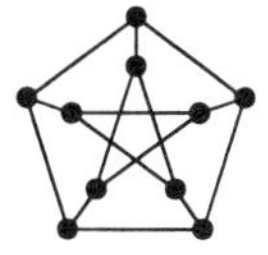

(a) a cutset with 3 edges;
(b) a cutset with 4 edges;
(c) a cutset with 5 edges;
(d) a cutset with 6 edges.

Menger's Theorem and its Analogues

10.4 In the complete bipartite graph $K_{10,13}$, let v be any vertex in the set with 10 vertices, and w be any vertex in the other set.

(a) Find the maximum number of edge-disjoint vw-paths.

(b) Find the maximum number of vertex-disjoint vw-paths.

Hint First remove the edge vw, apply the vertex form of Menger's theorem, then restore the edge vw.

10.5 (a) By finding k arc-disjoint st-paths and k arcs separating s from t (for the same value of k), find the maximum number of arc-disjoint st-paths in the following digraph:

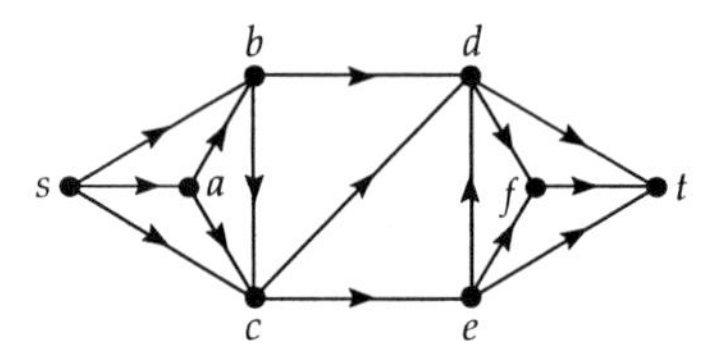

(b) Similarly, find the maximum number of vertex-disjoint st-paths in the above digraph.

Case Study

Telecommunication Networks

10.6 The following diagrams represent telecommunication networks:

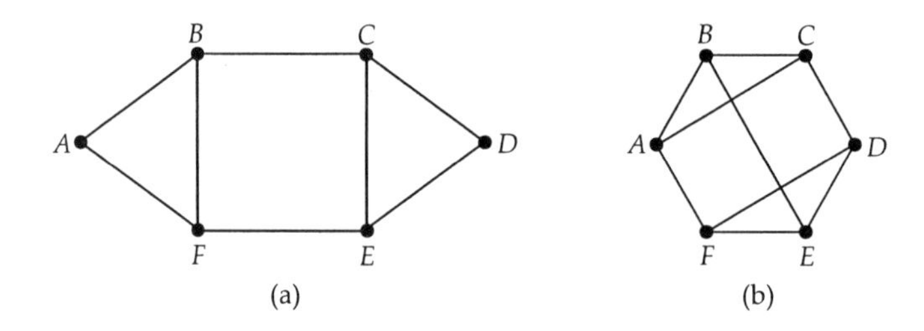

In each case, find:

(a) the smallest number of *links* whose failure would separate the network into two parts, and the corresponding links;

(b) the smallest number of *exchanges* whose failure would separate the network into two parts, and the corresponding exchanges.

10.7 Which of the graphs in Exercise 10.1 have optimal connectivity?

10.8 Which of the Platonic graphs have optimal connectivity?

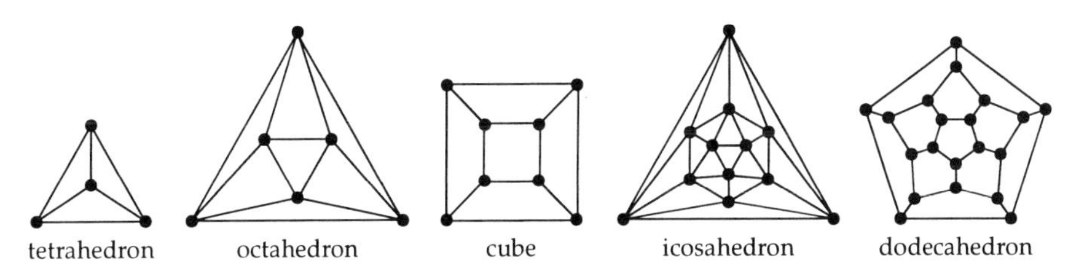

10.9 Determine whether the complete bipartite graph $K_{4,6}$ has optimal connectivity.

Chapter 11
Planarity

After studying this chapter, you should be able to:

- explain the terms *planar graph, non-planar graph, plane drawing, face, infinite face, degree of a face, subdivision of a graph* and *contraction of a graph*;
- state and use the *handshaking lemma* for planar graphs;
- state and use Euler's formula and the corollaries to Theorem 11.2;
- understand the statement of Kuratowski's theorem;
- explain the term *dual graph* and describe its properties;
- recognize the five *regular polyhedra*, and the duality relationships between them;
- state and use Euler's polyhedron formula;
- state and use the handshaking lemma for polyhedra.

In Chapter 1, you met a problem related to the theme of this chapter – the *utilities problem*. In this problem, three neighbours wish to be connected to the three utilities *gas, water* and *electricity*, in such a way that the connections do not cross.

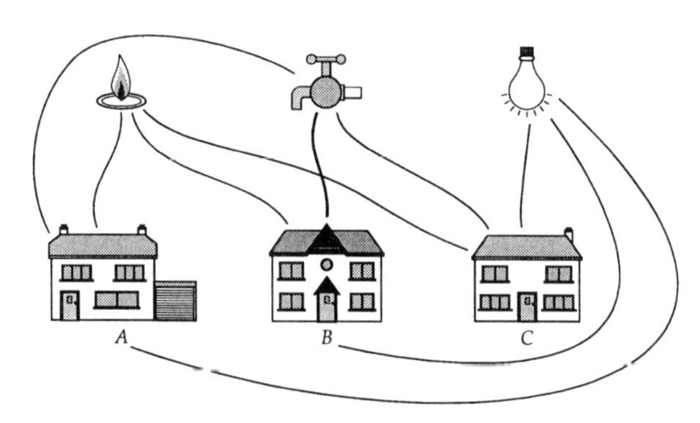

A more practical version of this type of problem arises in the design of printed circuit boards, on which electronic components are connected by means of conducting strips printed directly onto the flat board of insulating material. Such printed connections may not cross, since this would lead to undesirable electrical contact at crossing points.

In this chapter we investigate the properties of graphs that can be drawn in the plane without any of their edges crossing; such graphs are called *planar graphs*. In particular, we determine whether the complete bipartite graph $K_{3,3}$ is planar, thereby solving the utilities problem. We discuss Euler's formula and Kuratowski's theorem; the latter is an important theoretical result which gives a necessary and sufficient condition for a graph to be planar. We also describe a heuristic algorithm that can be used to determine whether a given graph is planar.

11.1 Planar Graphs

Suppose that you wish to design a nine-hole golf course. It is advisable to do so in such a way that no two of the fairways intersect, as this would cause inconvenience and possible danger to the golfers. For example, the first diagram below would be unsuitable, whereas the second would be more appropriate.

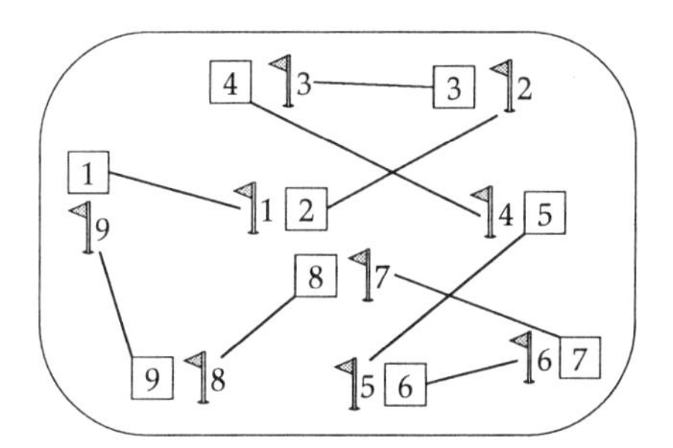

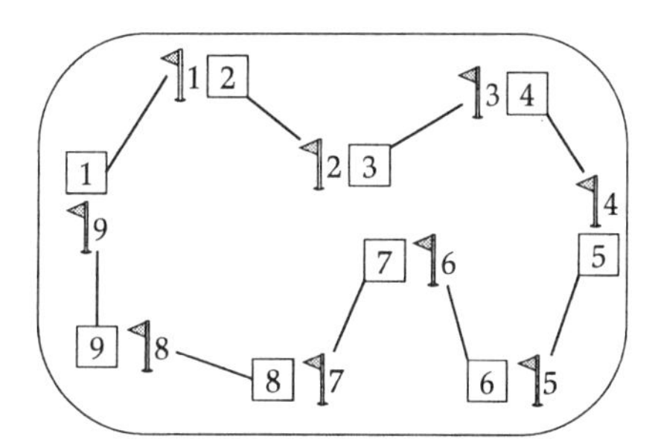

We can represent each of these layouts by the cycle graph C_9; the nine vertices correspond to the tees (or greens), and the edges correspond to the fairways. In the first drawing, some of the edges cross, whereas in the second drawing there are no crossings.

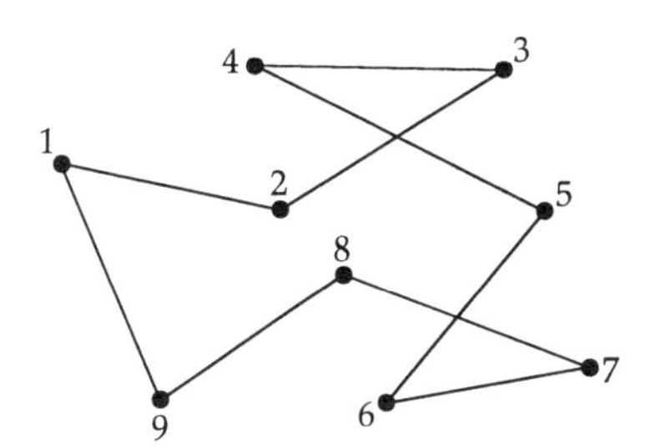

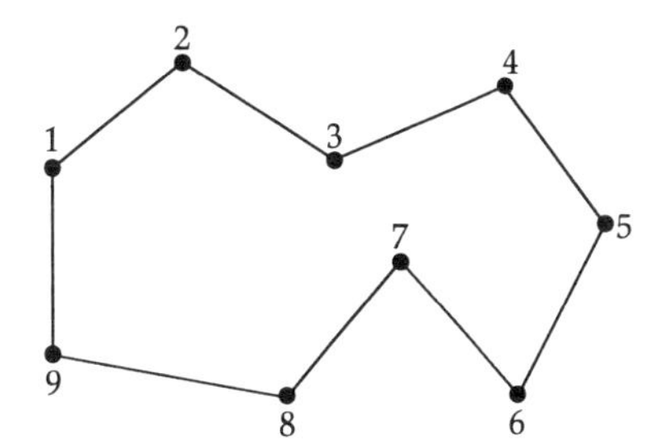

You have seen several instances of graphs drawn in more than one way. For example, the complete graph K_4 and the complete bipartite graph $K_{3,3}$ can be drawn as follows.

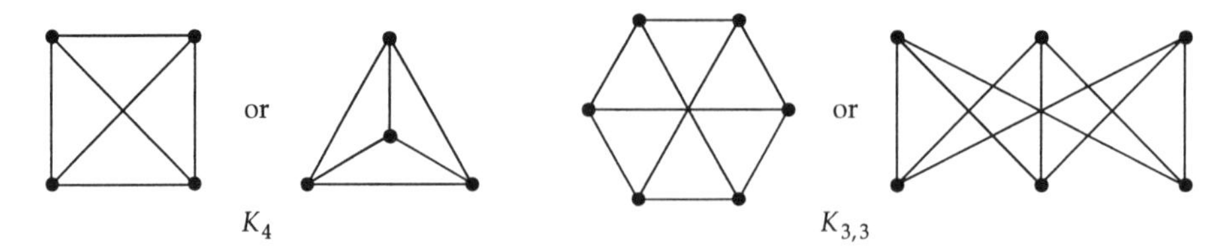

The particular drawing we choose often depends on the use to which the graph is to be put. As we have seen, it is sometimes useful to know whether we can draw a graph in such a way that no two edges cross. For some graphs, such as K_4, it is possible to find a drawing that involves no crossings, whereas for others, such as $K_{3,3}$, there are no such drawings, as you will see. This leads to the following definitions.

Definitions

A graph *G* is **planar** if it can be drawn in the plane in such a way that no two edges meet except at a vertex with which they are both incident. Any such drawing is a **plane drawing** of *G*.

A graph *G* is **non-planar** if no plane drawing of *G* exists.

For example, the graph K_4 is planar, since it can be drawn in the plane without edges crossing. The following diagram shows three plane drawings of K_4.

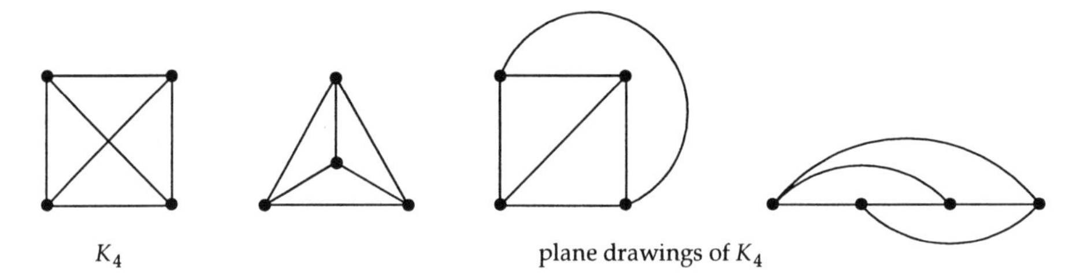

The graphs of the cube and dodecahedron are planar, since they can be drawn as follows.

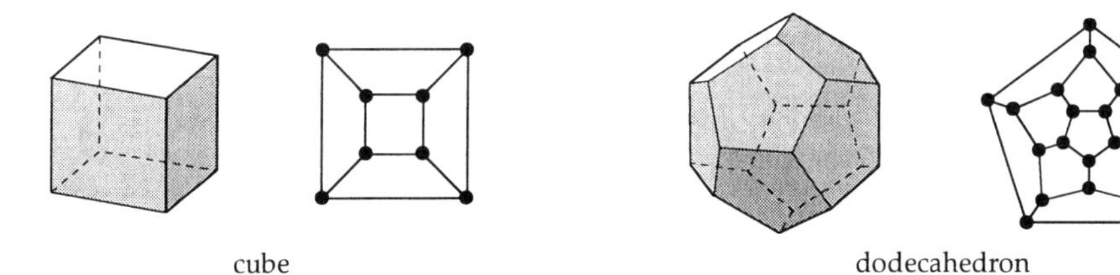

Similarly, the following graph is planar, since it can be 'unravelled' as shown.

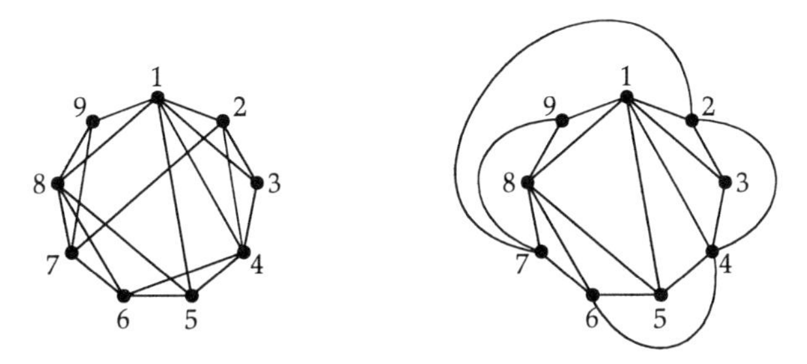

Problem 11.1

Show that the following graphs are planar, by finding a plane drawing of each.

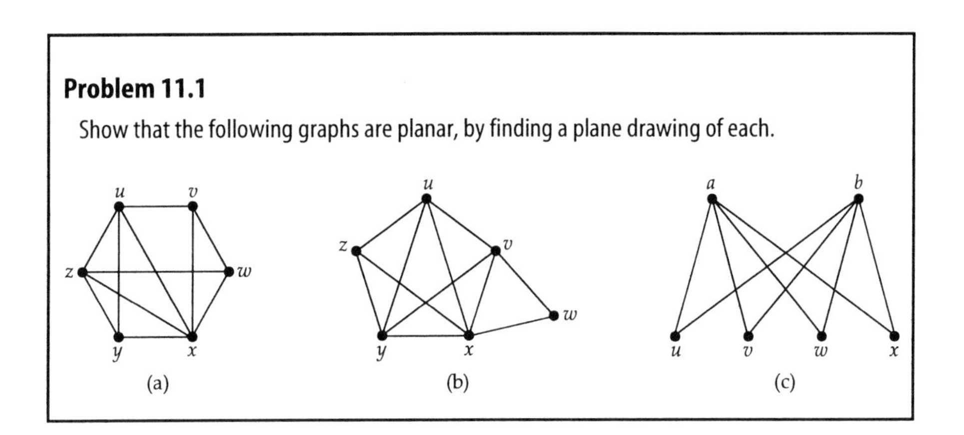

(a) (b) (c)

On the other hand, the complete bipartite graph $K_{3,3}$ is non-planar, since every drawing of it must contain at least one crossing. To see why this is, note that $K_{3,3}$ has a cycle of length 6 – the cycle *uavbwcu*. In any plane drawing, this cycle must appear as a hexagon (not necessarily regular).

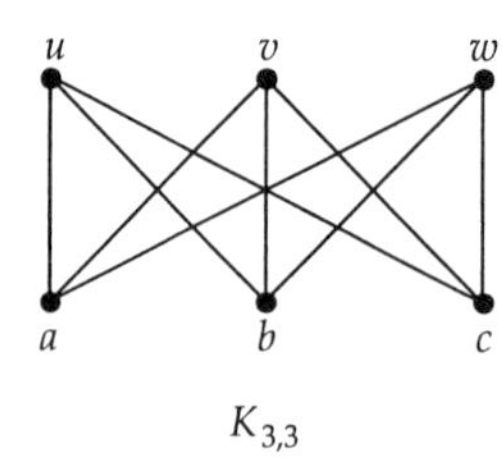

$K_{3,3}$

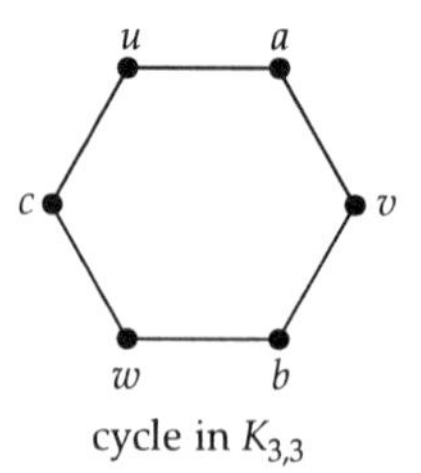

cycle in $K_{3,3}$

We must now insert the edges *ub*, *vc*, *wa*. Only one of them can be drawn *inside* the hexagon, since two would cross. Similarly, only one of them can be drawn *outside* the hexagon, since two would cross.

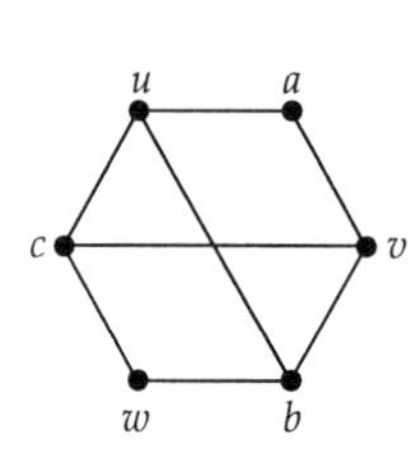

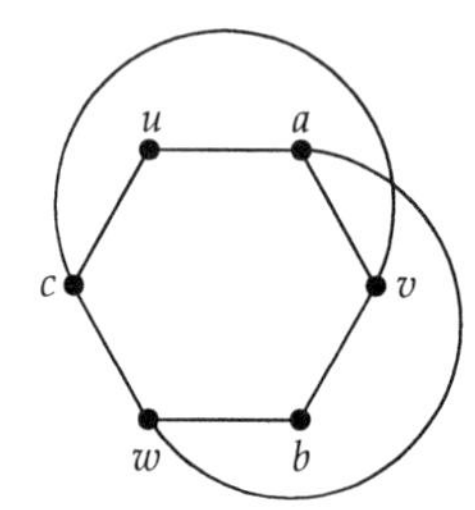

It is therefore impossible to insert all three of these edges without creating a crossing. This demonstrates that $K_{3,3}$ is non-planar.

Problem 11.2

Explain why the utilities problem has no solution – that is, why it is not possible to connect each of the three houses to the three utilities so that no connections cross..

Problem 11.3

Give an explanation, similar to that given for $K_{3,3}$, to demonstrate that the complete graph K_5 is non-planar.

Problem 11.4

There was once a king with five sons. In his will he stated that after his death each son should build a castle, and that the five castles should be connected in pairs by non-intersecting roads. Can the terms of the will be satisfied?

When studying planar graphs, we may restrict our attention to simple graphs whenever it is convenient to do so. If a planar graph has multiple edges or loops, we replace the multiple edges by a single edge and remove the loops. After drawing the resulting simple graph without crossings, we can then insert the loops and multiple edges, as follows.

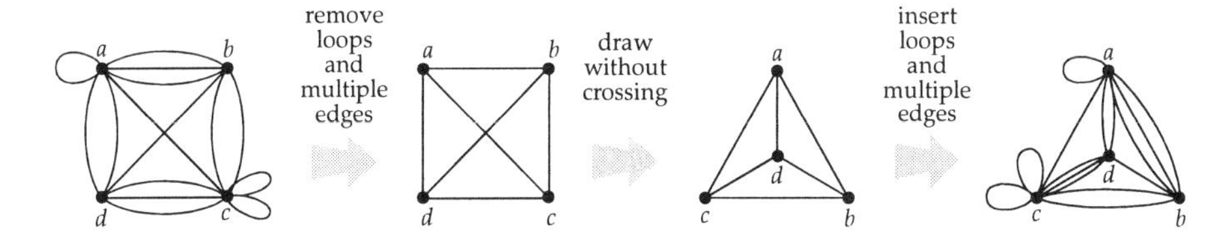

Problem 11.5

Decide whether each of the following statements is true or false, and give a reason or counter-example as appropriate.

(a) Every subgraph of a planar graph is planar.
(b) Every subgraph of a non-planar graph is non-planar.
(c) Every graph that contains a planar subgraph is planar.
(d) Every graph that contains a non-planar subgraph is non-planar.

Problem 11.6

(a) Which trees are planar?
(b) For which values of n is the cycle graph C_n planar?
(c) For which values of n is the complete graph K_n planar?
(d) For which values of s are the complete bipartite graphs $K_{1,s}$ and $K_{2,s}$ planar?
(e) For which values of r and s ($r \leq s$) is the complete bipartite graph $K_{r,s}$ planar?

11.2 Euler's Formula

In this section we introduce *Euler's formula*, which relates the numbers of vertices, edges and faces of a plane drawing of a planar graph. First, we introduce the idea of a *face* of such a drawing.

Every plane drawing of a planar graph divides the plane into a number of regions. For example, any plane drawing of K_4 divides the plane into four regions – three triangles (3-cycles) and one 'infinite region':

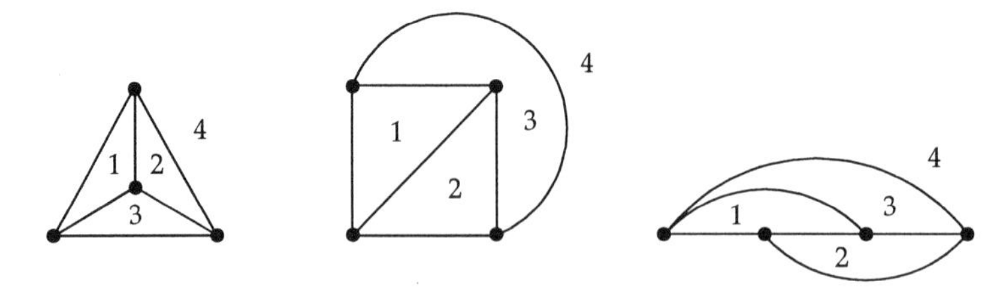

Similarly, any plane drawing of $K_{2,5}$ divides the plane into five regions – four quadrilaterals and one 'infinite region':

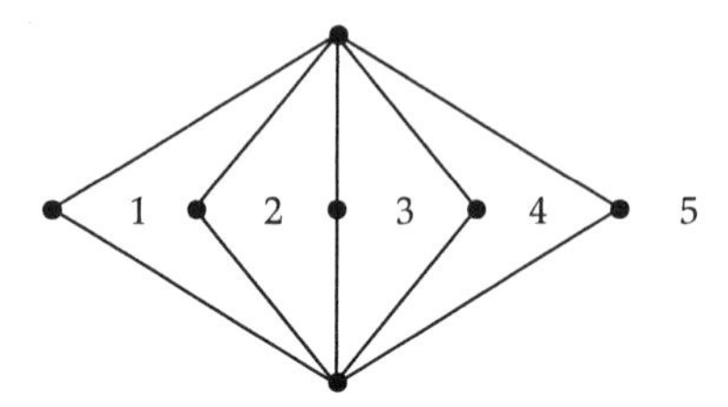

We make these ideas precise as follows.

Definitions

Let G be a planar graph. Then any plane drawing of G divides the set of points of the plane not lying on G into regions, called **faces**; one face is of infinite extent and is the **infinite face**.

Remark The regions do not include the vertices and edges forming their boundaries.

For example, the graph in diagram (a) below has four faces, f_1, f_2, f_3 and f_4 where f_4 is the infinite face. An alternative drawing of G, in which the faces have the same boundaries but f_3 is the infinite face, is given in diagram (b).

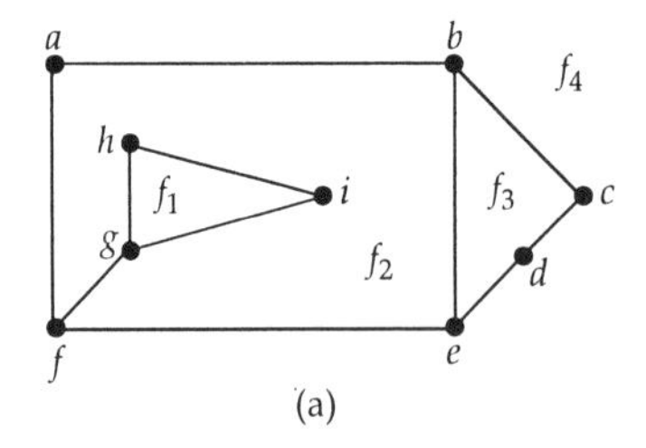

(a)

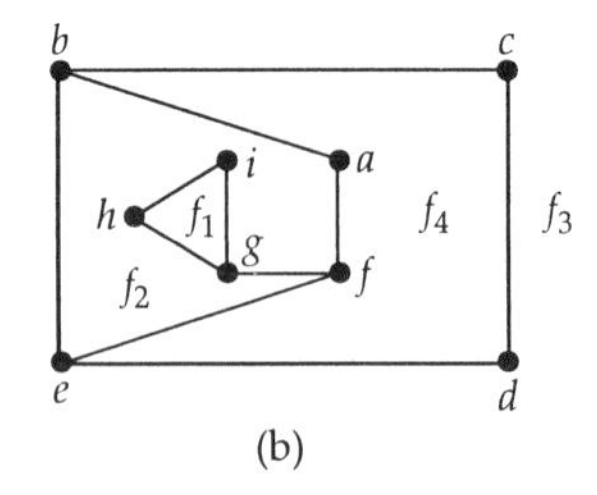

(b)

Problem 11.7

Find plane drawings of the above graph in which:

(a) f_2 is the infinite face;
(b) f_1 is the infinite face.

We define the *degree* of each face of a plane drawing of a connected planar graph as follows.

Definitions

Let G be a connected planar graph, and let f be any face of a plane drawing of G. Then the **degree of *f***, denoted by **deg *f***, is the number of edges encountered in a walk around the boundary of the face f.
If all faces have the same degree g, then G is **face-regular of degree *g***.

For example, in each drawing of the graph G in diagrams (a) and (b) above,

$$\deg f_1 = 3 \quad \text{and} \quad \deg f_3 = 4.$$

Note that both sides of the edge gf lie on the boundary of the face f_2, so must be counted *twice* as we walk around the boundary of the face; thus, $\deg f_2 = 9$.

If we find the sum of all the face degrees, we obtain $3 + 4 + 9 + 6 = 22$, which is exactly twice the number of edges of G. This makes us suspect that the handshaking lemma for graphs has a 'face analogue' for the faces in a plane drawing of a planar graph. This is indeed the case, and we refer to it as the *handshaking lemma for planar graphs*.

Theorem 11.1: Handshaking Lemma for Planar Graphs

In any plane drawing of a planar graph, the sum of all the face degrees is equal to twice the number of edges.

Proof In any plane drawing of a planar graph, each edge has two sides (which may lie on the boundary of a single face or on the boundaries of two different faces), so it contributes exactly 2 to the sum of the face degrees. The result follows immediately. ■

Problem 11.8

Verify the above version of the handshaking lemma for each of the following plane drawings of planar graphs.

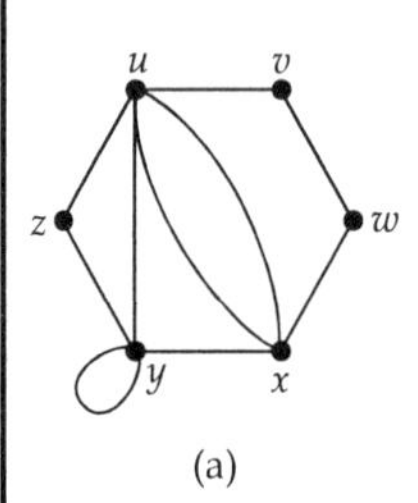

(a)

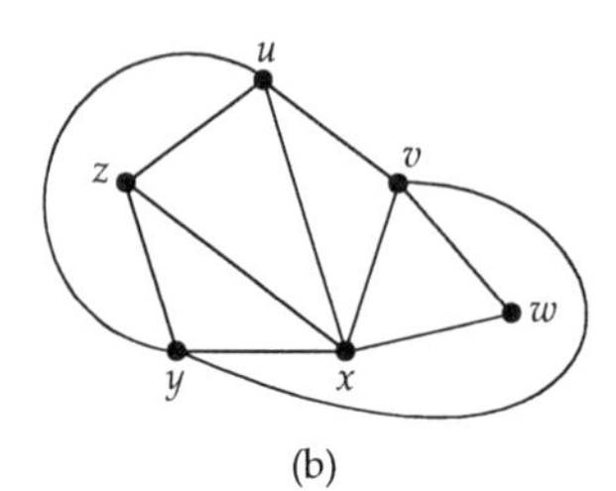

(b)

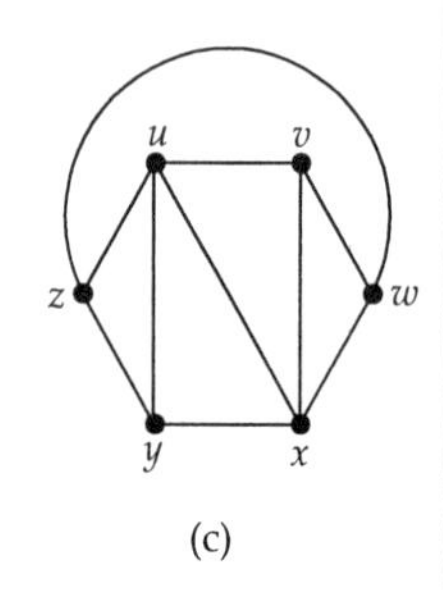

(c)

Problem 11.9

For each of the plane drawings in Problem 11.8, count the numbers of vertices, edges and faces, and find the value of

(number of vertices) – (number of edges) + (number of faces).

In the solution to Problem 11.9, you saw that, for each of the plane drawings under consideration,

(number of vertices) – (number of edges) + (number of faces) = 2.

This equation holds for any plane drawing of a connected planar graph, and is known as *Euler's formula*. It tells us that each plane drawing of a given connected planar graph with n vertices and m edges must have the same number of faces – namely, $2 - n + m$.

Theorem 11.2: Euler's Formula for Planar Graphs

Let G be a connected planar graph, and let n, m and f denote, respectively, the numbers of vertices, edges and faces in a plane drawing of G. Then

$$n - m + f = 2.$$

Proof A plane drawing of any connected planar graph G can be constructed by taking a spanning tree of G and adding edges to it, one at a time, until a plane drawing of G is obtained.

We prove Euler's formula by showing that:

(a) for any spanning tree, $n - m + f = 2$;
(b) adding an edge does not change the value of $n - m + f$.

First, we prove statement (a). Let T be any spanning tree of G; then we may draw T in the plane without crossings.

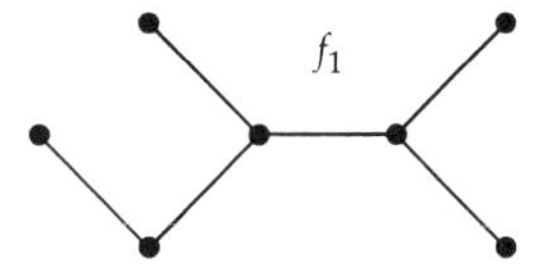

Since T has n vertices and $n - 1$ edges, and there is only 1 face (the infinite face), we have

$$n - m + f = n - (n - 1) + 1 = 2,$$

as required.

Now we prove statement (b) by adding in the other edges one at a time until the graph G is obtained. At each stage the added edge either joins two different vertices:

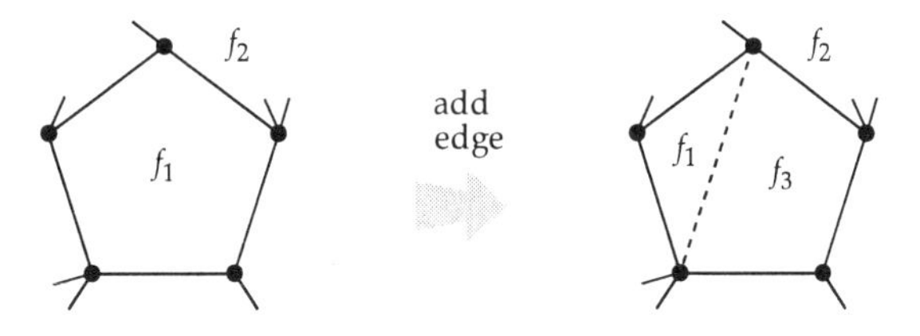

or joins a vertex to itself (if it is a loop):

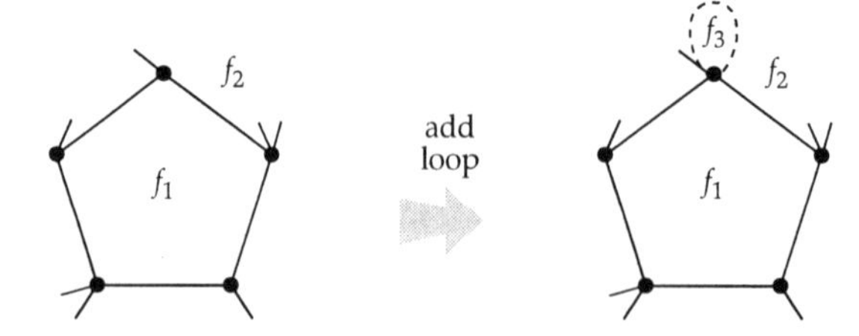

In each case, since we have a plane drawing of G, the added edge cuts an existing face in two, as illustrated above. This leaves n unchanged, increases m by 1, and increases f by 1, and so leaves $n - m + f$ unchanged. Since $n - m + f = 2$ throughout the process, the result follows. ■

Problem 11.10

Verify Euler's formula for each of the following graphs:

(a) the octahedron graph;
(b) the wheel with k spokes;
(c) the complete bipartite graph $K_{2,k}$;
(d) the graph formed from the vertices and edges of a $k \times k$ square lattice.

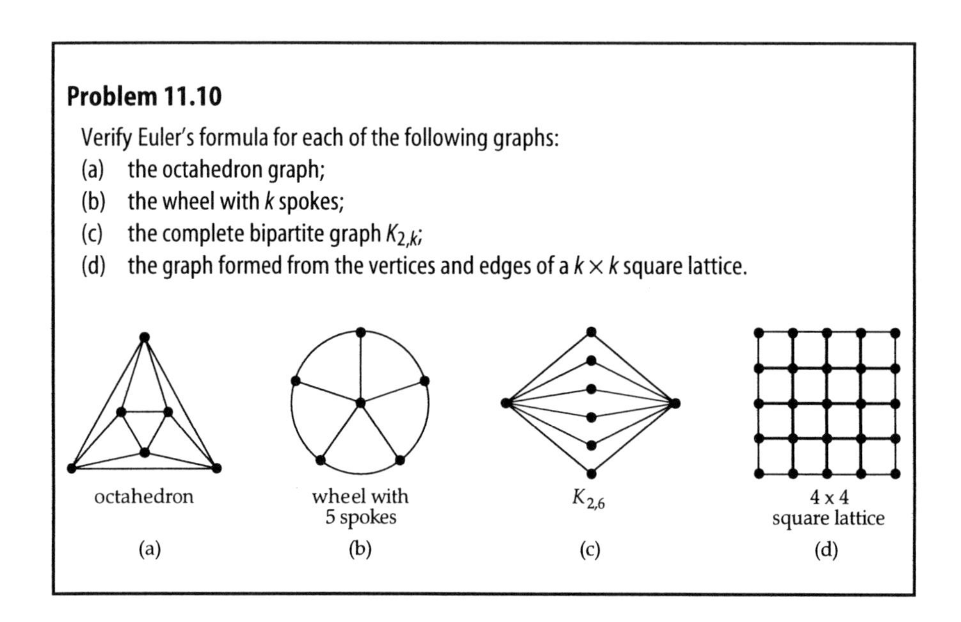

We now show how Euler's formula can be used to prove that certain graphs are *non-planar*. We first derive two corollaries of Theorem 11.2 that give upper bounds for the number of edges of a planar graph.

Corollary 11.1

Let G be a simple connected planar graph with n (≥ 3) vertices and m edges. Then $m \leq 3n - 6$.

Proof Consider a plane drawing of a simple connected planar graph G with f faces. Since a simple graph has no loops or multiple edges, the degree of each face is at least 3. It follows from the handshaking lemma for planar graphs that

$$3f \leq 2m.$$

Substituting for f from Euler's formula $f = m - n + 2$, we obtain

$$3m - 3n + 6 \leq 2m,$$

and hence

$$m \leq 3n - 6,$$

as required. ■

Using Corollary 11.1, we can prove that the complete graph K_5 is non-planar.

Example 11.1: K_5 is Non-Planar

The proof is by contradiction.

Suppose that K_5 is planar. Since K_5 is a simple connected graph with 5 vertices and 10 edges, it follows from Corollary 11.1 that

$$10 \leq (3 \times 5) - 6 = 9,$$

which is false. This contradiction shows that K_5 is non-planar. ❐

We cannot use Corollary 11.1 to prove that the complete bipartite graph $K_{3,3}$ is non-planar, since $K_{3,3}$ has 6 vertices and 9 edges, and the inequality

$$9 \leq (3 \times 6) - 6 = 12$$

is true. However, we can prove that $K_{3,3}$ is non-planar by using the following corollary for graphs with no triangles.

Corollary 11.2

Let G be a simple connected planar graph with n (≥ 3) vertices, m edges and no triangles. Then $m \leq 2n - 4$.

Proof Consider a plane drawing of a simple connected planar graph G with f faces and no triangles. The degree of each face of such a graph is at least 4. It follows from the handshaking lemma for planar graphs that

$$4f \leq 2m,$$

so

$$2f \leq m.$$

Substituting for f from Euler's formula $f = m - n + 2$, we obtain

$$2m - 2n + 4 \leq m,$$

and hence

$$m \leq 2n - 4,$$

as required. ■

Using Corollary 11.2, we can prove that the complete bipartite graph $K_{3,3}$ is non-planar.

Example 11.2: $K_{3,3}$ is Non-Planar

The proof is by contradiction.

Suppose that $K_{3,3}$ is planar. Since $K_{3,3}$ is a simple connected graph with 6 vertices, 9 edges and no triangles, it follows from Corollary 11.2 that

$$9 \leq (2 \times 6) - 4 = 8,$$

which is false. This contradiction shows that $K_{3,3}$ is non-planar. ❐

Problem 11.11

Under what conditions do Corollaries 11.1 and 11.2 give equalities

$$m = 3n - 6 \quad \text{and} \quad m = 2n - 4$$

rather than inequalities?

Problem 11.12

(a) Let G be a simple connected planar graph with n (≥ 5) vertices, m edges and shortest cycle length 5. Prove that

$$m \leq \tfrac{5}{3}(n - 2).$$

Hint Use the method of proof of Corollaries 11.1 and 11.2.

(b) Hence show that the Petersen graph is non-planar.

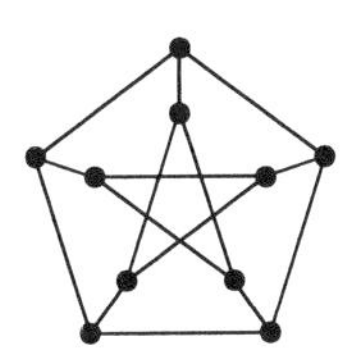

We can prove the following result similarly.

Corollary 11.3

Let G be a simple connected planar graph. Then G contains a vertex of degree 5 or less.

Problem 11.13

Prove Corollary 11.3.

Hint Use Corollary 11.1 and give a proof by contradiction.

Problem 11.14

Give an example of each of the following:

(a) a simple connected planar graph in which each vertex has degree 5;
(b) a non-simple connected planar graph in which each vertex has degree 6.

The restrictions on the number of edges of a planar graph given in Corollaries 11.1 and 11.2 are useful for showing that certain graphs are *non*-planar. For example, we used them to show that K_5 and $K_{3,3}$ are non-planar. Unfortunately, the method does not work the other way round – there are graphs (such as the Petersen graph) that satisfy these inequalities but are non-planar. Because of this, we now turn our attention to other ways of determining whether a given graph is planar.

11.3 Cycle Method for Planarity Testing

In many practical applications it is important to test whether a given graph is planar. There are several methods for this purpose in current use, and we present one of these, the *cycle method*, informally here. It is a heuristic algorithm that can be applied to any small graph containing a Hamiltonian cycle. There exist algorithms that are faster and work in all cases, but they are too complicated to be described here.

Given a graph G that we wish to test for planarity, we look for a Hamiltonian cycle, draw this cycle as a regular polygon, and then try to draw the remaining edges so that no edges cross.

Having chosen a Hamiltonian cycle C, we list the remaining edges of G, and try to divide them into two disjoint sets A and B, as follows:

A is a set of edges that can be drawn *inside* C without crossing;
B is a set of edges that can be drawn *outside* C without crossing.

If this is possible, the graph G is planar, and we can use the sets A and B to obtain a plane drawing of G. If this is not possible, the graph G is non-planar.

You met an example of this procedure earlier when we tested the complete bipartite graph $K_{3,3}$ for planarity. We started by noting that $K_{3,3}$ has a Hamiltonian cycle C of length 6, which we drew in the plane as a regular hexagon *uavbwc*. We then tried to draw in the three remaining edges *ub*, *vc* and *wa*; but only one of these edges can lie inside C and only one can lie outside C, since otherwise two of them cross. Thus, if we put *ub* in the set A and *vc* in the set B, then we cannot allocate *wa* to either set; it follows that we cannot draw in all three edges without crossings, so the graph $K_{3,3}$ is non-planar.

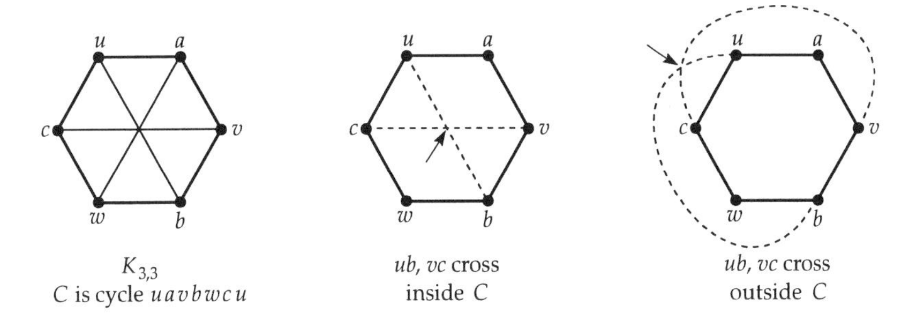

$K_{3,3}$
C is cycle *uavbwcu*

ub, vc cross inside C

ub, vc cross outside C

We say that the edges *ub* and *vc* are *incompatible*, since they cannot both be drawn inside C, or both be drawn outside C, without crossing. Similarly, the edges *ub* and *wa* are incompatible, as are the edges *vc* and *wa*. Edges that can be drawn both inside C or both outside C without crossing are *compatible*.

The following example shows how this idea of incompatible edges can be used to test the planarity of more complicated graphs.

Example 11.3

We determine whether the following graph G is planar.

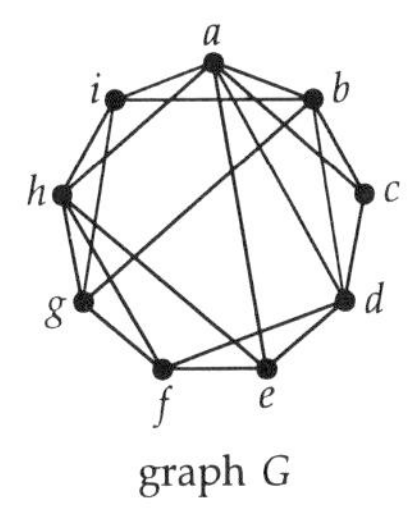

graph G

We choose a suitable cycle C in G; it is natural to choose the Hamiltonian cycle *abcdefghia*.

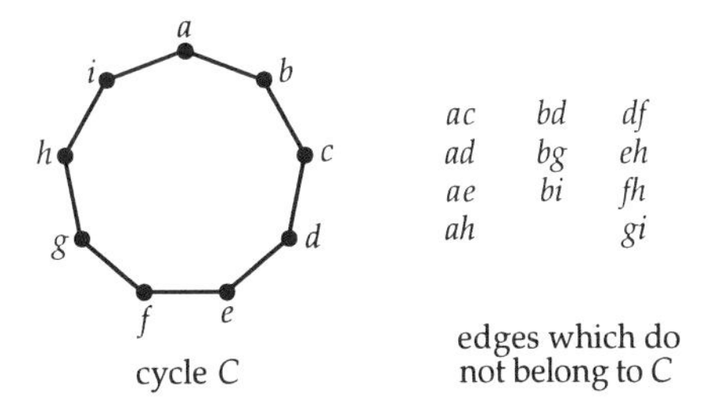

cycle C

edges which do not belong to C

We list the edges that do not belong to C, as shown above.

We put the first edge in the list, *ac*, in a set *A*, and delete this edge from the list:

list: *ad, ae, ah, bd, bg, bi, df, eh, fh, gi.*

The edge *ac* is incompatible with *bd*, *bg* and *bi*, so we put the edges *bd*, *bg* and *bi* in a set *B*. All the edges in *B* are compatible with each other, so we delete the edges *bd*, *bg* and *bi* from the list:

list: *ad*, *ae*, *ah*, *df*, *eh*, *fh*, *gi*.

We now have the following situation:

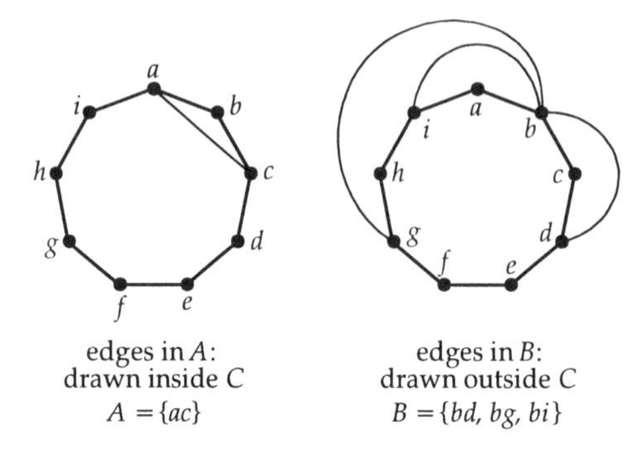

edges in *A*: drawn inside C
$A = \{ac\}$

edges in *B*: drawn outside C
$B = \{bd, bg, bi\}$

We now consider each edge of *B* in turn.
The edge *bd* is compatible with each edge in the list.
The edge *bg* is incompatible with *ad*, *ae*, *eh* and *fh*, so we put the edges *ad*, *ae*, *eh* and *fh* into *A*. All the edges in *A* are compatible with each other, so we delete the edges *ad*, *ae*, *eh* and *fh* from the list:

list: *ah*, *df*, *gi*.

The edge *bi* is incompatible with *ah*, so we put the edge *ah* into *A*. All the edges in *A* are compatible with each other, so we delete the edge *ah* from the list:

list: *df*, *gi*.

The situation is now as follows:

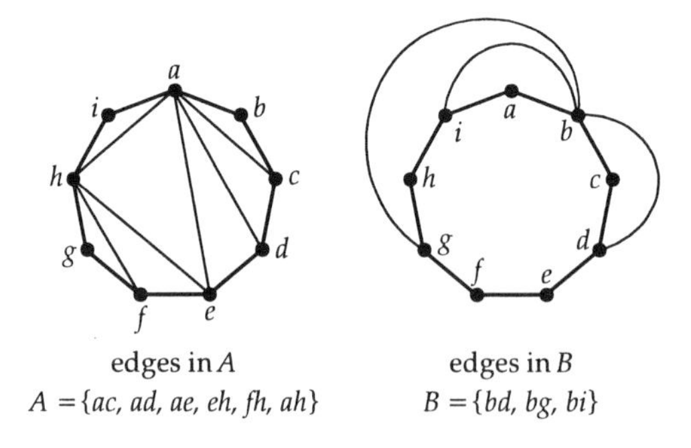

edges in *A*
$A = \{ac, ad, ae, eh, fh, ah\}$

edges in *B*
$B = \{bd, bg, bi\}$

We now consider each edge of A in turn.
The edge ad is compatible with each edge in the list.
The edge ae is incompatible with df, so we put the edge df into B. All the edges in B are compatible with each other, so we delete the edge df from the list:

list: gi.

The edge eh is incompatible with gi, so we put the edge gi into B. All the edges in B are compatible with each other, so we delete the edge gi from the list.

The list is now empty, and we have:

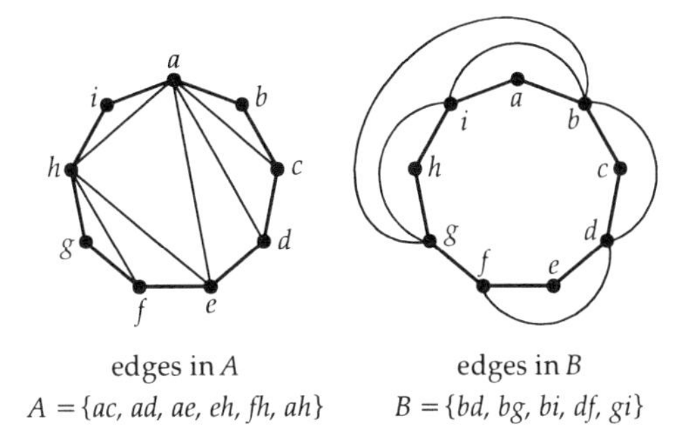

edges in A $\quad$ edges in B

$A = \{ac, ad, ae, eh, fh, ah\}$ $\quad$ $B = \{bd, bg, bi, df, gi\}$

All the edges in A are compatible and all the edges in B are compatible, so G is planar. To obtain a plane drawing of G, we combine the above two figures, as follows.

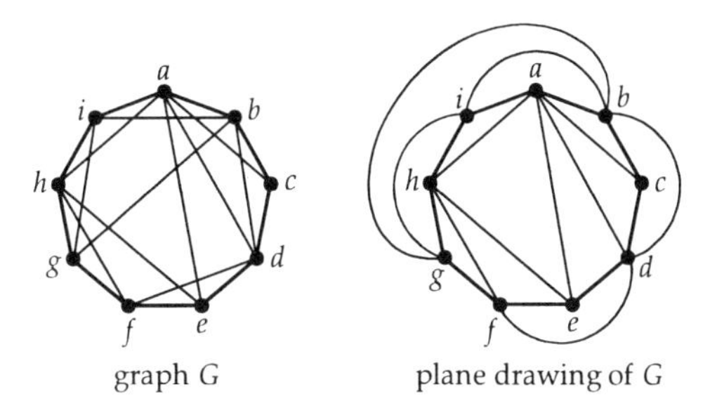

graph G $\quad$ plane drawing of G

❐

Problem 11.15

Use the cycle method to determine whether each of the following graphs is planar. If it is, give a plane drawing.

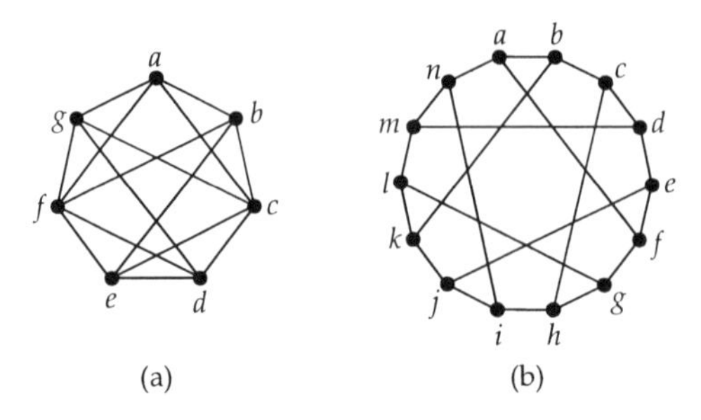

11.4 Kuratowski's Theorem

We now describe two *theoretical* methods for determining planarity.

The first method involves the insertion of vertices of degree 2 into the edges of a graph G, as shown in the following diagram.

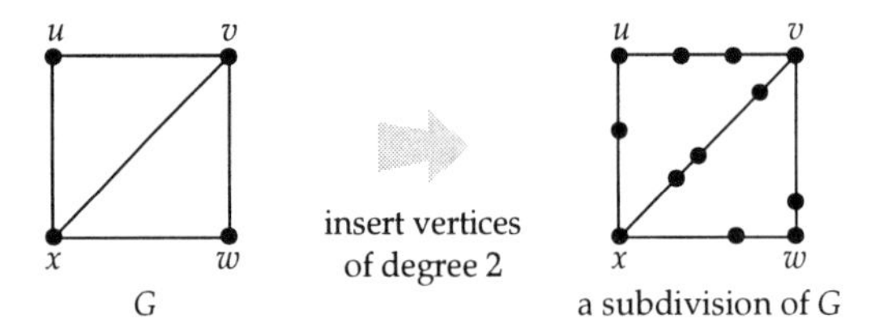

Any graph formed from G in this way is called a **subdivision** of G. Since the insertion of a vertex of degree 2 does not affect the planarity or non-planarity of a graph, we deduce the folowing result.

- If G is a planar graph, then every subdivision of G is planar.

This is often stated in the following alternative form.

- If G is a subdivision of a non-planar graph, then G is non-planar.

For example, the following graphs are non-planar, since the first is a subdivision of K_5 and the second is a subdivision of $K_{3,3}$.

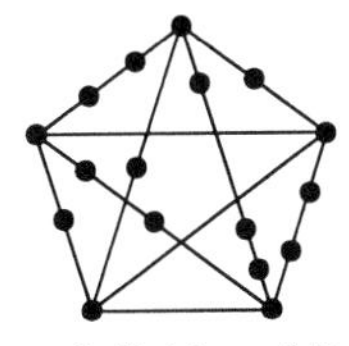

a subdivision of K_5

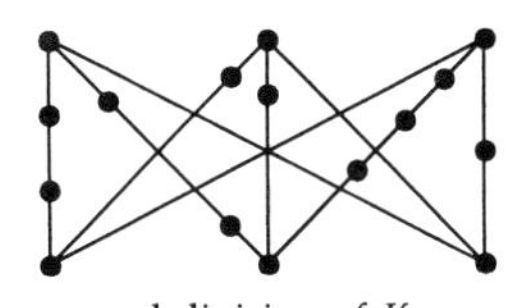

a subdivision of $K_{3,3}$

It follows from these two observations that

- If G is a graph that contains a subdivision of K_5 or $K_{3,3}$, then G is non-planar.

For example, the following graphs are non-planar, since the first contains a subdivision of K_5 and the second contains a subdivision of $K_{3,3}$.

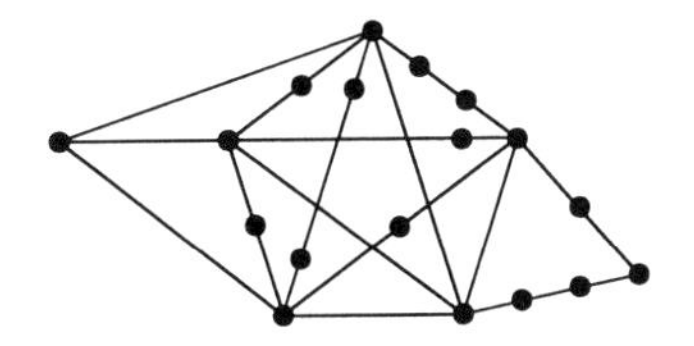

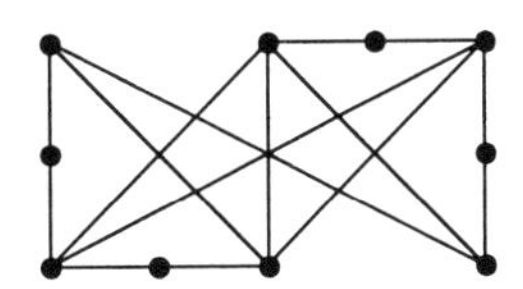

You may be wondering why we are so concerned with K_5 and $K_{3,3}$ and their subdivisions. The reason is that *every* non-planar graph is obtained in the way we have just described – namely, by adding vertices and edges to a subdivision of K_5 or $K_{3,3}$.

- If G is a non-planar graph, then G contains a subdivision of K_5 or $K_{3,3}$.

This result appeared in 1930, and is due to the Polish mathematician Kazimierz Kuratowski. We state it formally but omit the proof, which is rather long and complicated.

Theorem 11.3: Kuratowski's Theorem

A graph is planar if and only if it contains no subdivision of K_5 or $K_{3,3}$.

Problem 11.16

Use Kuratowski's theorem to prove that each of the following graphs is non-planar.

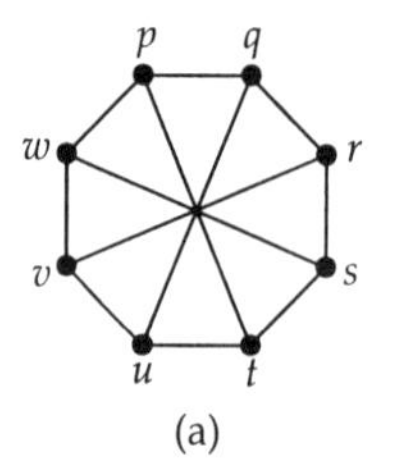

(a)

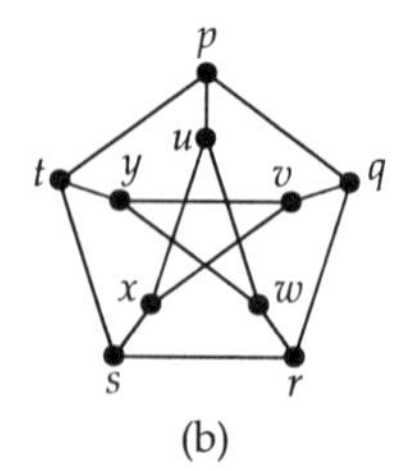

(b)

Hint For graph (b), consider the subgraph obtained by deleting the two 'horizontal' edges.

Another characterization of planar graphs involves the notion of 'contracting' an edge. This is done by bringing one vertex closer and closer to the other vertex until they coincide, and then coalescing any resulting multiple edges into a single edge. In the following diagrams, we contract the edge vw.

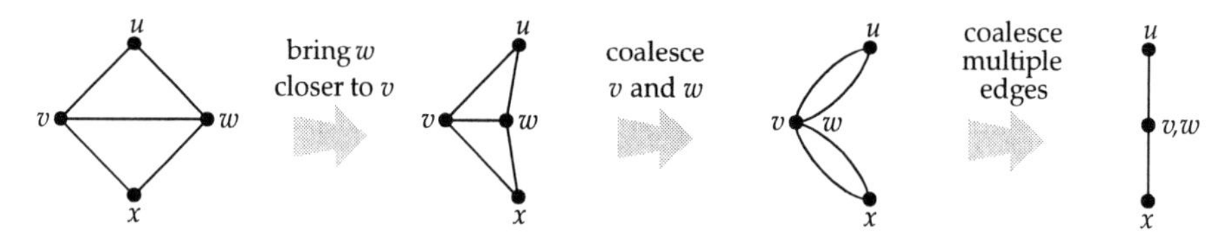

A **contraction** of a graph is the result of a sequence of such edge contractions. For example, K_5 is a contraction of the Petersen graph, since it is the result of contracting each of the five 'spokes' (the edges joining the inner and outer 5-cycles).

We now state the following analogue of Kuratowski's theorem. We omit the proof.

Theorem 11.4

A graph is planar if and only if it contains no subgraph that has K_5 or $K_{3,3}$ as a contraction.

The importance of Theorems 11.3 and 11.4 is that they give necessary and sufficient conditions for a graph to be planar in graph-theoretic terms (subgraph, subdivision, contraction of a graph), rather than in geometric terms (crossing, drawing in the plane). They also provide a convincing demonstration that a given graph is non-planar, if we happen to spot a subgraph that is a subdivision of K_5 or $K_{3,3}$ or a subgraph that has K_5 or $K_{3,3}$ as a contraction.

However, Theorems 11.3 and 11.4 do not provide an easy way of showing that a given graph is planar, since this would involve looking at a large number of subgraphs and verifying that none of them is a subdivision of K_5 or $K_{3,3}$ or contains K_5 or $K_{3,3}$ as a contraction. For this reason, no currently used algorithm for testing the planarity of a graph is based on either of these theorems.

Finally, we make a few observations that simplify the task of determining whether a given graph is planar.

- A disconnected graph is planar if and only if each of its components is planar; for example, the following graph is non-planar, because one of its components is a subdivision of K_5.

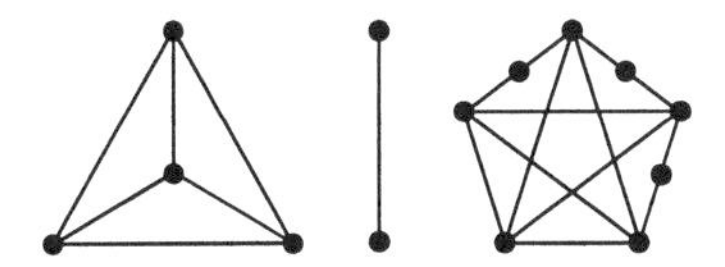

- A graph that has a cut vertex (a vertex whose removal disconnects the graph) is planar if and only if each of the subgraphs obtained when the graph is disconnected at the cut vertex is planar; for example, the following graph is non-planar, because one of these subgraphs is $K_{3,3}$.

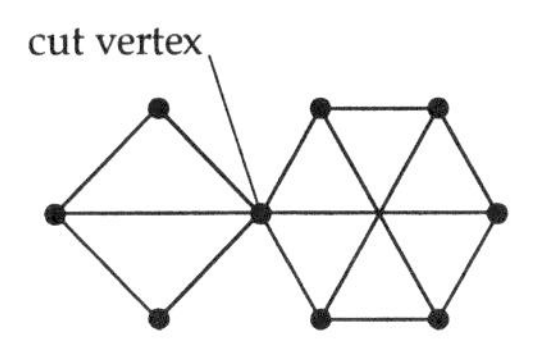

- A graph that has loops or multiple edges is planar if and only if the graph obtained by removing the loops and coalescing the multiple edges is planar; for example, the following graph is non-planar, because the resulting graph is the Petersen graph.

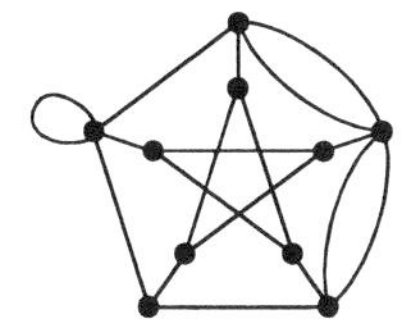

Using these observations, we can sometimes reduce a given graph to a number of smaller graphs that we can deal with more easily.

11.5 Duality

We next introduce the idea of *duality* for plane drawings of planar graphs.

To illustrate this idea, we consider the graph of the cube. If we place a new vertex within each face (including the infinite face) and join the pairs of new vertices in adjacent faces, we obtain the graph of the octahedron, and *vice versa*, as follows.

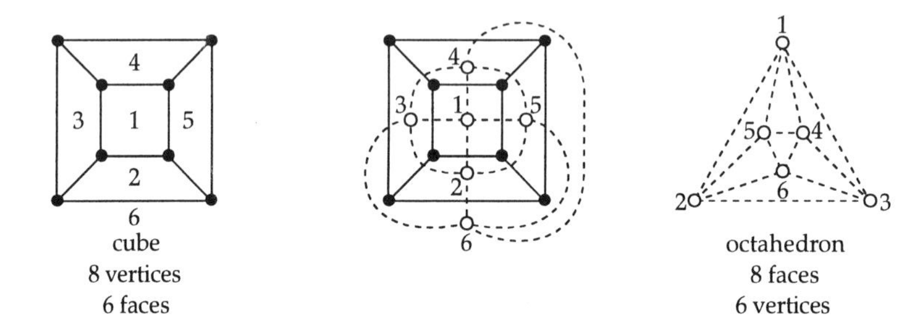

(The new vertices are represented by small circles, and the lines joining them are indicated by dashed lines.)

More generally, for any connected planar graph G, we define the corresponding *dual graph* G^* as follows.

> **Definition**
>
> Let G be a connected planar graph. Then a **dual graph** G^* is constructed from a plane drawing of G, as follows.
>
> Draw one new vertex in each face of the plane drawing: these are the vertices of G^*.
>
> For each edge e of the plane drawing, draw a line joining the vertices of G^* in the faces on either side of e: these lines are the edges of G^*.

Remark We always assume that we have been presented with a plane drawing of G.

The procedure is illustrated below.

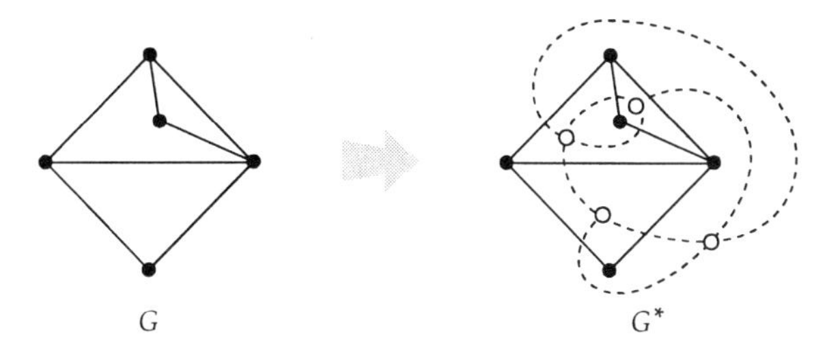

Problem 11.17

Draw the dual of each of the following plane drawings of planar graphs.

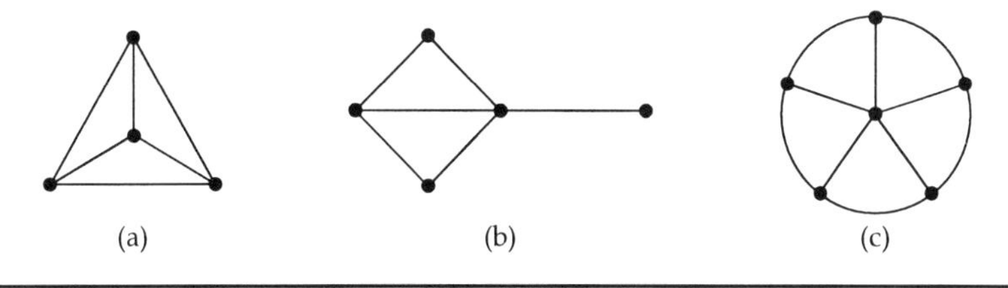

Problem 11.18

The following diagrams show two different plane drawings of a planar graph. Show that their duals are not isomorphic.

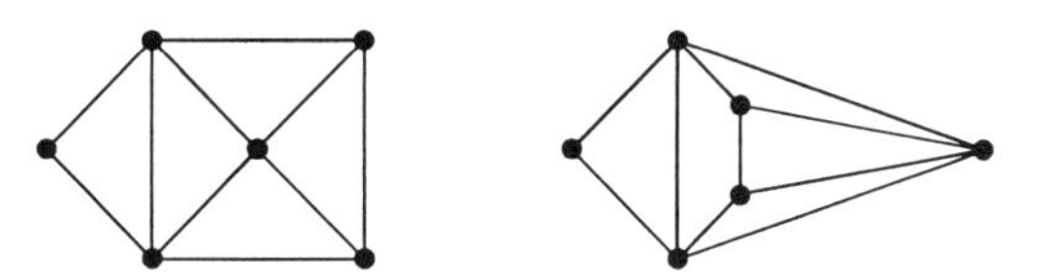

Different plane drawings of a planar graph G may give rise to non-isomorphic dual graphs G^*, as we saw in the above problem.

Also, if G is a plane drawing of a connected planar graph, then so is its dual G^*, and we can thus construct $(G^*)^*$, the dual of G^*.

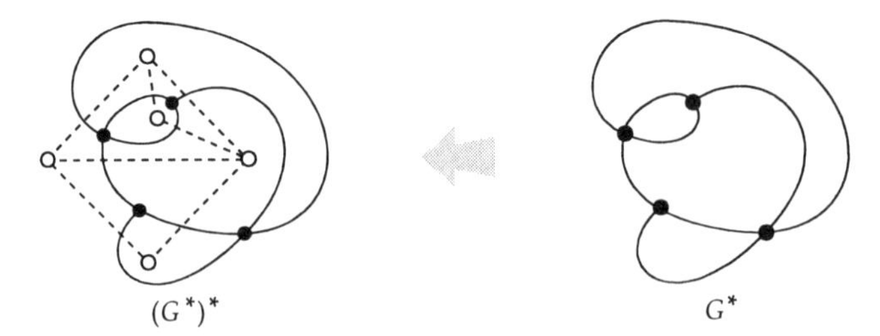

The above diagrams demonstrate that the construction that gives rise to G^* from G can be reversed to give G from G^*; for example, the dual of the octahedron graph is the cube graph. It follows that $(G^*)^*$ is isomorphic to G.

There is a simple relationship between the numbers of vertices, edges and faces of a graph and its dual. In the above example, G has 5 vertices, 7 edges and 4 faces (including the infinite face), and G^* has 4 vertices, 7 edges and 5 faces. In general, we have the following result.

Theorem 11.5

Let G be a plane drawing of a connected planar graph with n vertices, m edges and f faces. Then G^* has f vertices, m edges and n faces.

Proof It follows directly from the construction of G^* that G^* has f vertices and m edges. If G^* has f^* faces, then, by applying Euler's formula to both G and G^*, we obtain

$$\text{for } G\text{: } n - m + f = 2; \qquad \text{for } G^*\text{: } f - m + f^* = 2.$$

Comparing these, we obtain $f^* = n$, as required. ■

In fact, a vertex of degree k in G corresponds to a face of degree k in G^*, and *vice versa*. The following diagram illustrates this correspondence for $k = 5$.

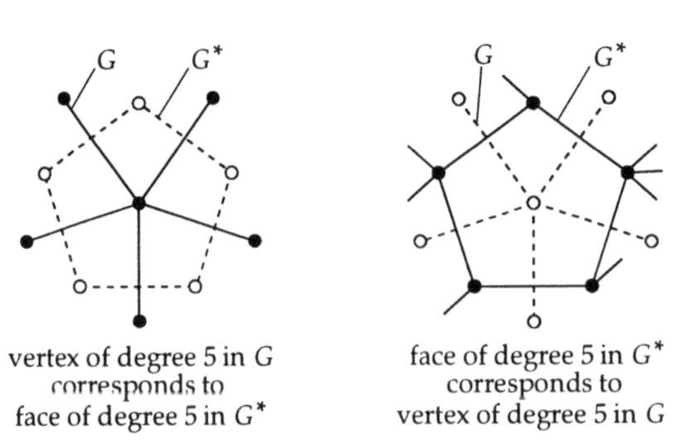

vertex of degree 5 in G corresponds to face of degree 5 in G^*

face of degree 5 in G^* corresponds to vertex of degree 5 in G

Further, a cycle of length k in G corresponds to a cutset with k edges in G^*, and *vice versa*. Again, we illustrate this correspondence for $k = 5$.

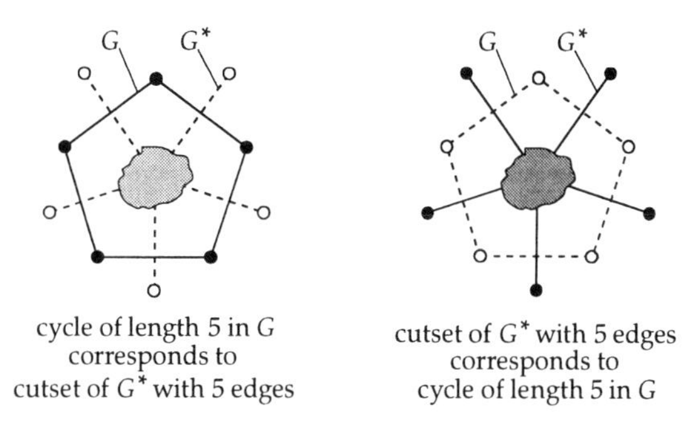

cycle of length 5 in G corresponds to cutset of G^* with 5 edges

cutset of G^* with 5 edges corresponds to cycle of length 5 in G

To obtain the first of the above correspondences, we take a cycle in G (with solid edges); the corresponding edges of G^* (the dashed edges) form a cutset whose removal separates the set of vertices inside the cycle from those outside. To obtain the second correspondence, we interchange the roles of G and G^*.

We summarize these correspondences as follows.

plane drawing G		dual graph G^*
an edge of G	corresponds to	an edge of G^*
a vertex of degree k in G	corresponds to	a face of degree k in G^*
a face of degree k in G	corresponds to	a vertex of degree k in G^*
a cycle of length k in G	corresponds to	a cutset of G^* with k edges
a cutset of G with k edges	corresponds to	a cycle of length k in G^*

We can use these correspondences to obtain new results from old ones. For example, we can reword Corollary 11.1 as follows.

Let G be a connected planar graph with n (≥ 3) vertices and m edges, and with no loops or multiple edges. Then $m \leq 3n - 6$.

Now, loops (cycles of length 1) and pairs of multiple edges (cycles of length 2) in G correspond to cutsets with 1 and 2 edges in G^*.

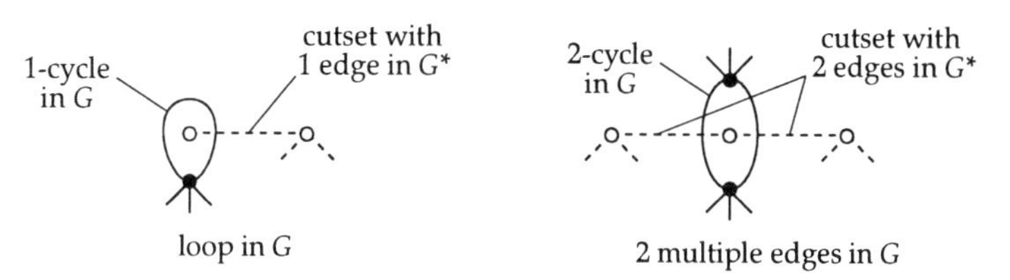

The above correspondence therefore gives the following 'dual' theorem.

Theorem 11.6

Let G^* be a connected planar graph with f faces and m edges, and with no cutsets with 1 or 2 edges. Then $m \leq 3f - 6$.

Conversely, we can *dualize* Theorem 11.6 to obtain Corollary 11.1.
Similarly, we can reword Corollary 11.3 as follows.

Let G be a connected planar graph with no loops or multiple edges. Then G has a vertex of degree 5 or less.

Dualizing this result, we deduce the following theorem.

Theorem 11.7

Let G^* be a connected planar graph with no cutsets with 1 or 2 edges. Then G^* has a face of degree 5 or less.

Problem 11.19

Dualize Corollary 11.2.

11.6 Convex Polyhedra

Several of the results obtained above for planar graphs have analogues for convex polyhedra.

Consider the cube shown below.

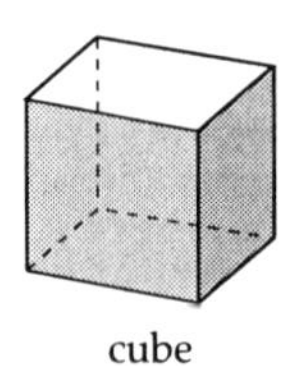

cube

The faces are all congruent regular polyhedra (squares), and the same number of them (three) meet at each vertex. Also, the cube is *convex*: it has no dents or spikes. These are the defining features of a *regular polyhedron*.

Definition

A **regular polyhedron** is a convex polyhedron in which all the polygonal faces are congruent regular polygons, and each vertex has exactly the same arrangement of polygons around it.

Remark At least three polygons must meet at each vertex, for if only two polygons, P and Q, were to meet at a vertex v, then the two edges of P incident with v would coincide with the two edges of Q incident with v, and so P and Q would lie in the same plane and could not enclose a volume.

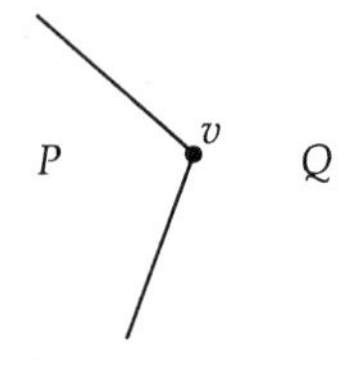

The following table gives the interior angles θ of the n-sided regular polygons, for $n = 3, 4, 5, 6$. Using this information, we can determine which groupings of congruent regular polygons can fit in edge-to-edge contact around a vertex in a regular polyhedron, so that there are no gaps or overlaps.

n	3	4	5	6
θ	$\pi/3$	$\pi/2$	$3\pi/5$	$2\pi/3$

In fact, there are only five types of vertex that can occur in a regular polyhedron:

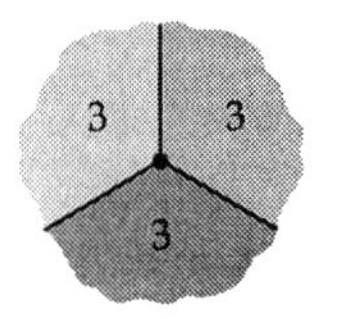

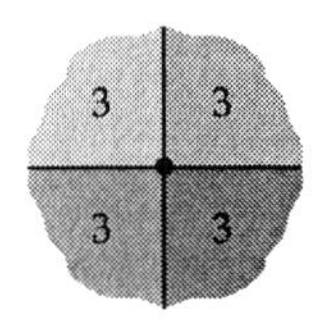

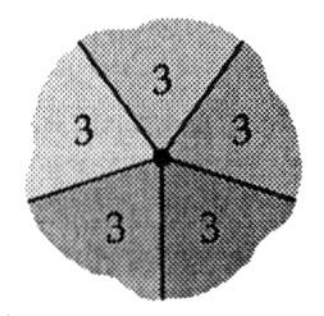

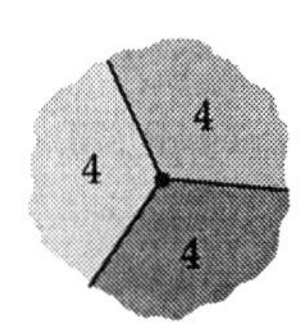

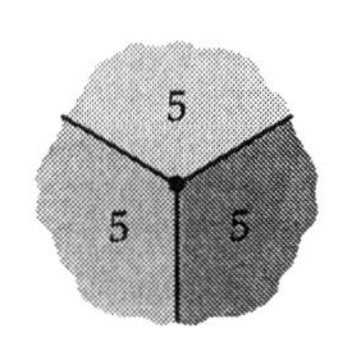

What regular polyhedra can be formed from them?

From the definition, every vertex must be of the same type, so there can be at most five regular polyhedra, corresponding to the five vertex types. It is perhaps surprising that all five do exist; they are the Platonic solids, introduced in Chapter 2:

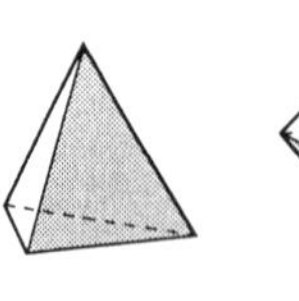
tetrahedron

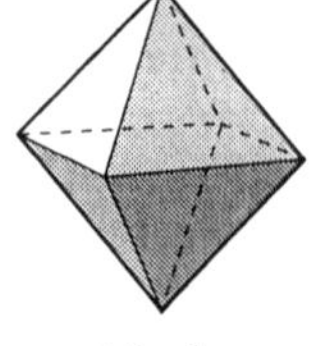
octahedron

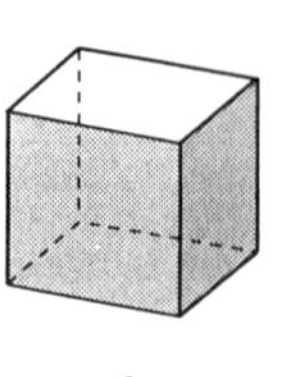
cube

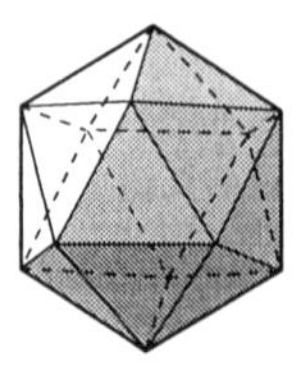
icosahedron

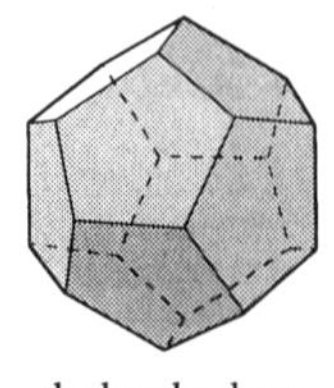
dodecahedron

The numbers of vertices, edges and faces of the Platonic solids are listed below.

polyhedron	vertices	edges	faces
tetrahedron	4	6	4
octahedron	6	12	8
cube	8	12	6
icosahedron	12	30	20
dodecahedron	20	30	12

Euler's Polyhedron Formula

There is a simple relationship between the numbers of faces, vertices and edges of any convex polyhedron. It is the analogue for polyhedra of Theorem 11.2.

Theorem 11.8: Euler's Polyhedron Formula

Let v, e and f denote, respectively, the numbers of vertices, edges and faces of a convex polyhedron. Then

$$v - e + f = 2.$$

Historical Note

The polyhedron formula first appeared in this form in a letter from Leonhard Euler to the number theorist Christian Goldbach in November 1750. At that time, Euler was unable to prove the result, but he presented a proof two years later. Unfortunately, Euler's proof was deficient, but a correct proof was obtained by Adrien Marie Legendre in 1794.

The connection between Euler's formula for planar graphs and Euler's formula for polyhedra is immediate, because we can represent any convex polyhedron as a planar graph by projecting it down onto a plane; this method of projection, called *stereographic projection*, does not alter the value of $n - m + f$.

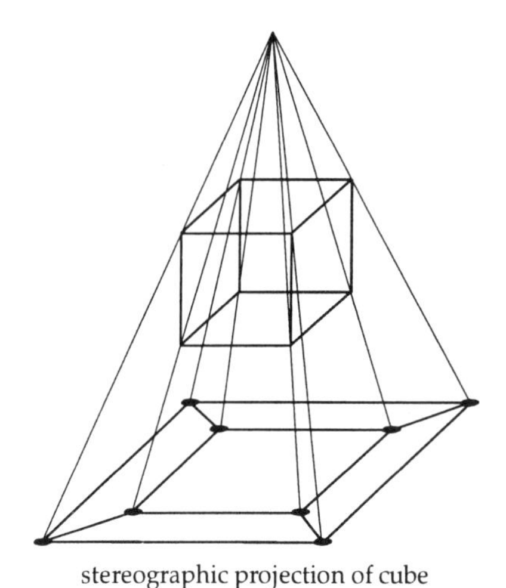

stereographic projection of cube

We can use Euler's polyhedron formula to prove that there are only five regular convex polyhedra. To do this, we also need the following result, which is an analogue of the handshaking lemma for graphs. In order to state it, we define the **degree** of a face of a polyhedron to be the number of edges around it: a triangular face has degree 3, a square face has degree 4, and so on.

Theorem 11.9: Handshaking Lemma for Polyhedra

In any polyhedron, the sum of all the face degrees is equal to twice the number of edges.

Proof In any polyhedron, each edge has two sides, so it contributes exactly 2 to the sum of the face degrees. The result follows immediately. ■

We can now prove the following theorem.

Theorem 11.10

There are only five regular polyhedra:

- three with triangular faces – the tetrahedron, the octahedron and the icosahedron;
- one with square faces – the cube;
- one with pentagonal faces – the dodecahedron.

We prove the first part of the theorem and leave you to prove the rest (Exercise 11.15).

Proof Let v, e and f denote, respectively, the numbers of vertices, edges and faces of a regular polyhedron with triangular faces. It follows from Theorem 11.9 that

$$3f = 2e.$$

If exactly d edges meet at each vertex, then it follows from the handshaking lemma for graphs that

$$dv = 2e.$$

These two results can be rewritten as

$$f = 2e/3 \quad \text{and} \quad v = 2e/d.$$

Substituting these two results into Euler's polyhedron formula, $v - e + f = 2$, we obtain

$$2e/d - e + 2e/3 = 2,$$

which, after division by $2e$, can be rewritten as

$$1/d - 1/6 = 1/e.$$

Since $1/e > 0$, it follows that

$$1/d > 1/6,$$

so $d < 6$. This means that the only possible values of d are 3, 4 and 5.

We consider each case in turn.

Case 1: $d = 3$: then $1/e = 1/3 - 1/6 = 1/6$, so $e = 6$;
it follows that $f = 4$ and $v = 4$ – this gives the tetrahedron.

Case 2: $d = 4$: then $1/e = 1/4 - 1/6 = 1/12$, so $e = 12$;
it follows that $f = 8$ and $v = 6$ – this gives the octahedron.

Case 3: $d = 5$: then $1/e = 1/5 - 1/6 = 1/30$, so $e = 30$;
it follows that $f = 20$ and $v = 12$ – this gives the icosahedron. ■

Dual Polyhedra

There is a duality construction for convex polyhedra, similar to that for graphs. The **dual** of a convex polyhedron can be constructed by placing a vertex at the centre of each face of the original polyhedron, and joining a pair of vertices with a line-segment whenever the corresponding faces of the original polyhedron are adjacent along an edge. For example, the dual of a cube is an octahedron, as shown below.

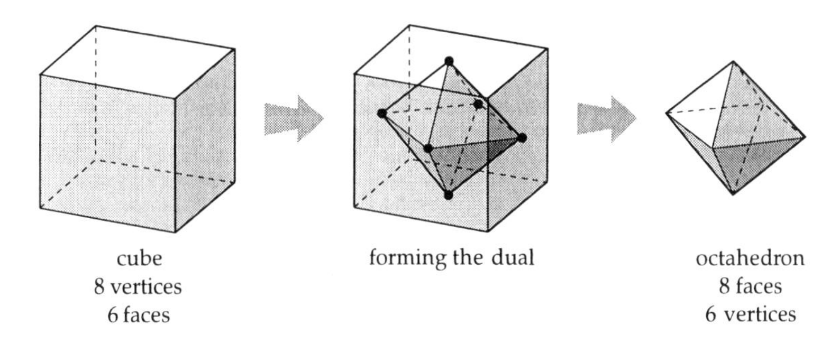

The duals of all the Platonic solids are themselves Platonic solids. The roles of the vertices and faces are exchanged, so we can read off the duality relations from the table given earlier.

Problem 11.20

Identify the dual of each of the Platonic solids.

Exercises 11

Planar Graphs

11.1 Decide which of the following graphs are planar.

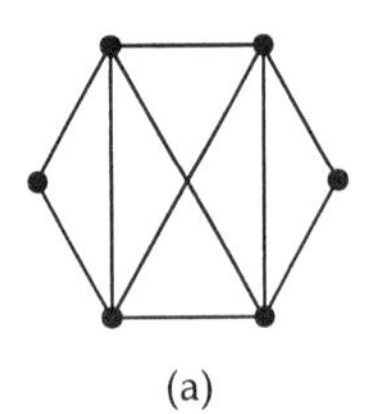
(a)

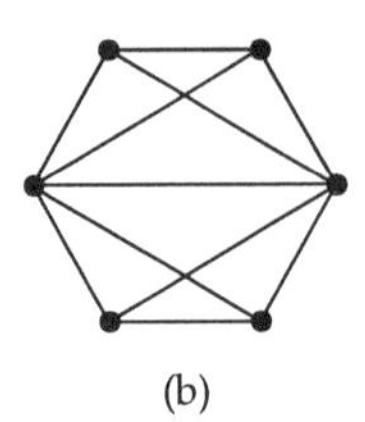
(b)

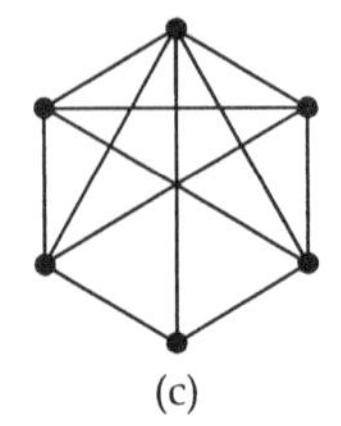
(c)

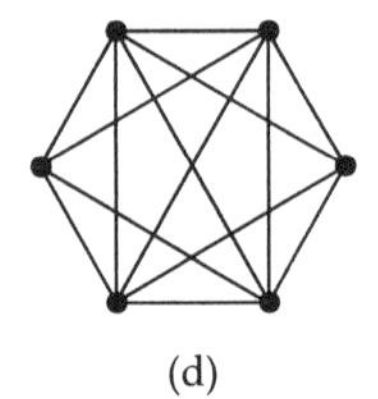
(d)

For each planar graph, give a plane drawing.

11.2 By finding a plane drawing, show that the following graph is planar.

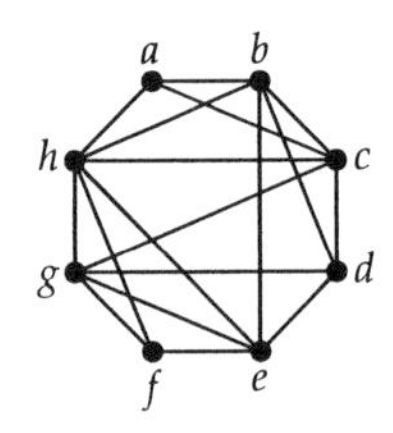

Euler's Formula

11.3 Let G be a simple connected planar graph with n (≥ 3) vertices and m edges, and let g be the length of the shortest cycle in G. Prove that

$$m \leq g(n-2)/(g-2).$$

Hint Imitate the proof of Corollary 11.1.

11.4 Use the result of Exercise 11.3 to prove that K_5, $K_{3,3}$ and the Petersen graph are non-planar.

11.5 Let G be a simple connected graph with n (≥ 4) vertices and m edges and no cycles of length 5 or less. Prove that

$$m \leq (3n/2) - 3.$$

Deduce that G has at least one vertex of degree 1 or 2.

11.6 Prove the following statement.

Let G be a simple connected planar graph with n vertices. If $n \leq 11$, then G contains at least one vertex of degree 4 or less.

Hint Use a proof by contradiction.

11.7 Let G be a planar graph with k components, and let n, m and f denote, respectively, the numbers of vertices, edges and faces in a plane drawing of G.

(a) Show that if each component has at least three vertices, then Euler's formula has the form

$$n - m + f = k + 1.$$

(b) Deduce that if G is simple and each vertex has degree at least 2, then

$$m \leq 3n - 3(k + 1).$$

11.8 Let G be a connected planar graph with 11 vertices. Use the result of Exercise 11.7(b) to prove that the complement of G is non-planar. (Note that the complement of a connected graph need not be connected.)

11.9 Give an example of a connected planar graph G with 7 vertices such that its complement is also planar.

Cycle Method

11.10 Use the cycle method to determine whether each of the following graphs is planar. If it is, give a plane drawing.

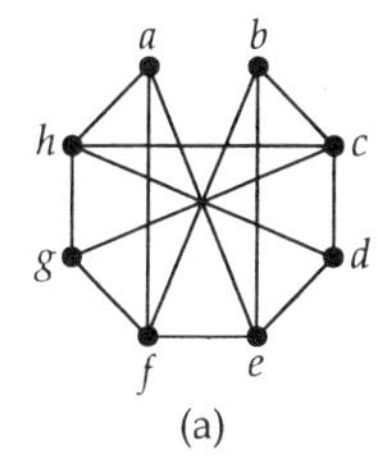

(a)

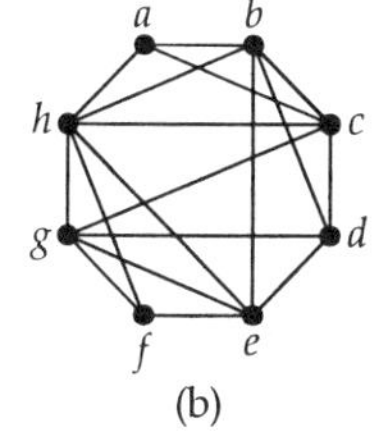

(b)

Kuratowski's Theorem

11.11 For each non-planar graph in Exercise 11.1, verify Kuratowski's theorem by finding a subgraph that is a subdivision of K_5 or $K_{3,3}$.

Duality

11.12 Draw the dual of each of the following plane drawings of planar graphs.

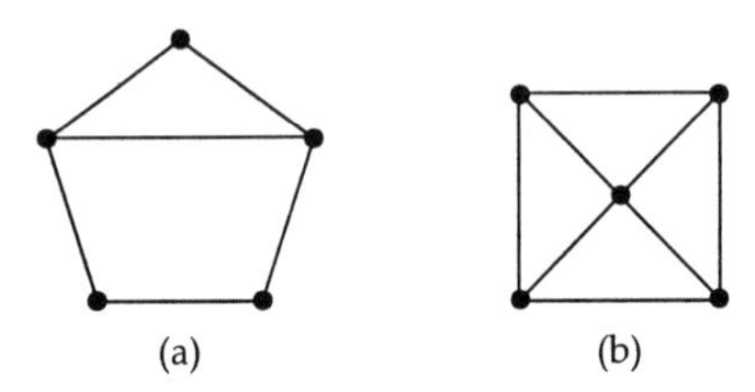

11.13 Dualize the statement given in Exercise 11.6.

11.14 Let G be a plane drawing of a connected planar graph. Prove that G is bipartite if and only if its dual G^* is Eulerian.

Convex Polyhedra

11.15 By imitating the proof of the first part of Theorem 11.10, show that

(a) the cube is the only regular polyhedron with square faces;
(b) the dodecahedron is the only regular polyhedron with pentagonal faces.

11.16 Prove that if a convex polyhedron has only square and hexagonal faces, and if exactly three faces meet at each vertex, then it has exactly 6 square faces.

11.17 Prove that if a convex polyhedron has only pentagonal and hexagonal faces, and if exactly three faces meet at each vertex, then it has exactly 12 pentagonal faces.
What does this have to do with the game of soccer?

Chapter 12

Vertex Colourings and Decompositions

After studying this chapter, you should be able to:

- explain the terms *vertex colouring*, *k-colouring* and *chromatic number*;
- use Brooks' theorem;
- use the greedy algorithm to colour the vertices of a graph;
- explain what are meant by *colouring problems*, the *map colouring problem* and *domination problems*, and how they can be represented as *vertex decomposition problems*.

In this chapter, we consider problems involving the colouring of the *vertices* of a graph, and we introduce an algorithm for vertex colouring. We then consider problems that involve splitting the set of vertices of a graph into disjoint subsets with particular properties. Such problems include colouring problems and domination problems.

12.1 Vertex Colourings

Example 12.1: Storing Chemicals

A chemical manufacturer wishes to store chemicals in a warehouse. Some chemicals react violently when in contact with each other, and the manufacturer decides to divide the warehouse into a number of areas so as to separate dangerous pairs of chemicals. In the following table, an asterisk indicates those pairs of chemicals that must be kept apart. What is the smallest number of areas needed to store these chemicals safely?

	a	b	c	d	e	f	g
a	–	*	*	*	–	–	*
b	*	–	*	*	*	–	*
c	*	*	–	*	–	*	–
d	*	*	*	–	–	*	–
e	–	*	–	–	–	–	–
f	–	–	*	*	–	–	*
g	*	*	–	–	–	*	–

We note first that chemicals a, b, c, d must all be in separate areas, so at least four areas are necessary. In fact, four areas are sufficient, as the following graph shows; the vertices correspond to the seven chemicals, and two vertices are joined by an edge whenever the corresponding chemicals must be kept separate. If we colour the vertices with the minimum number of colours so that adjacent vertices are coloured differently, we find that 4 colours are needed, as indicated by the numbers next to the vertices in the following graph; the four colours 1, 2, 3, 4 correspond to the four areas.

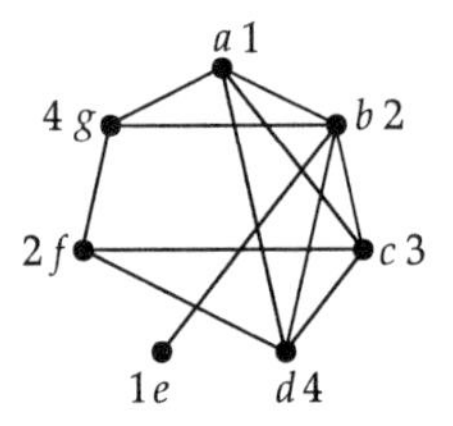

Thus we can split the set of chemicals into four disjoint subsets corresponding to the four areas:

$\{a, e\}, \{b, f\}, \{c\}, \{d, g\}.$

(Other solutions are possible.) □

Chromatic Number

The assignment of colours to chemicals in Example 12.1 illustrates the following definitions.

Definitions

Let G be a simple graph. A ***k*-colouring** of G is an assignment of at most k colours to the vertices of G in such a way that adjacent vertices are assigned different colours. If G has a k-colouring, then G is ***k*-colourable**.

The **chromatic number** of G, denoted by $\chi(G)$, is the smallest number k for which G is k-colourable.

In the above chemical storage example, the graph has chromatic number 4.

Remark The above definitions are given only for *simple* graphs. Loops must be excluded since, in any k-colouring, the vertices at the ends of each edge must be assigned different colours, so the vertex at both ends of a loop would have to be assigned a different colour from itself. We also exclude multiple edges, since the presence of one edge between two vertices forces them to be coloured differently, and the addition of further edges between them is then irrelevant to the colouring. *We therefore restrict our attention to simple graphs.*

We usually show a k-colouring by writing the numbers 1, 2, ..., k next to the appropriate vertices. For example, diagrams (a) and (b) below illustrate a 4-colouring and a 3-colouring of a graph G with five vertices; note that diagram (c) is *not* a 3-colouring of G, since the two vertices coloured 2 are adjacent.

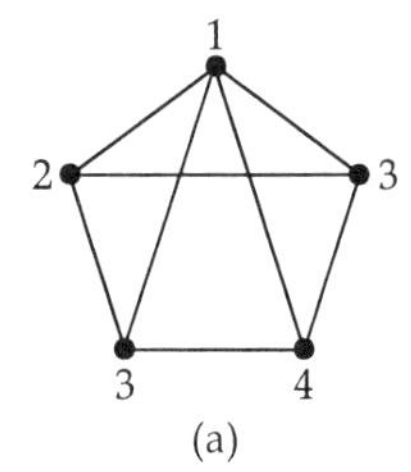

(a)

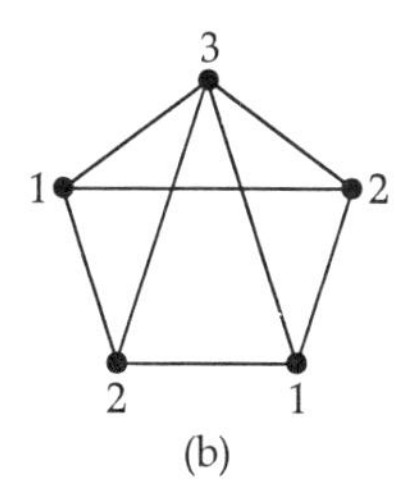

(b)

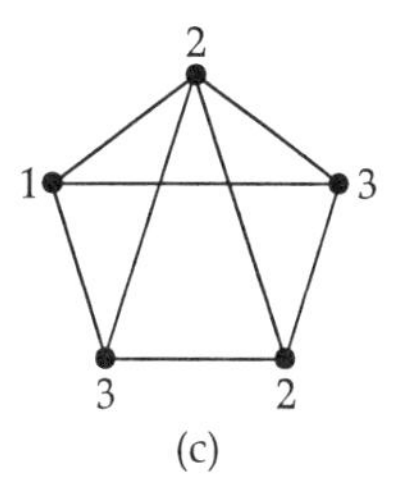

(c)

Since G has a 3-colouring, $\chi(G) \leq 3$; thus 3 is an *upper bound* for $\chi(G)$. Also, G contains three mutually adjacent vertices (forming a triangle) that must be assigned different colours, so $\chi(G) \geq 3$; thus 3 is a *lower bound* for $\chi(G)$. Combining these inequalities, we obtain $\chi(G) = 3$.

Problem 12.1

Determine $\chi(G)$ for each of the following graphs G.

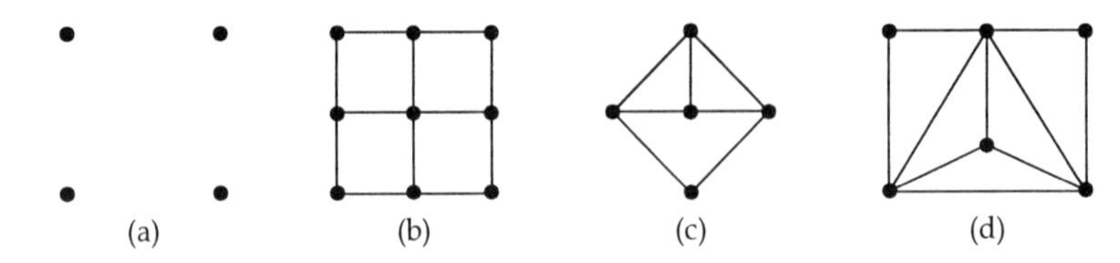

Hint For each graph, devise a suitable colouring and explain why there is no colouring with fewer colours.

Problem 12.2

What can you say about the graphs G for which

(a) $\chi(G) = 1$? (b) $\chi(G) = 2$?

Problem 12.3

Write down the chromatic number of each of the following graphs:

(a) the complete graph K_n;
(b) the complete bipartite graph $K_{r,s}$;
(c) the cycle graph C_n $(n \geq 3)$;
(d) a tree.

Problem 12.4

Decide whether each of the following statements about a graph G is true or false, and give a proof or counter-example, as appropriate.

(a) If G contains the complete graph K_r as a subgraph, then $\chi(G) \geq r$.
(b) If $\chi(G) \geq r$, then G contains the complete graph K_r as a subgraph.

Given a particular graph G, how can we determine its chromatic number? We have seen that an upper bound for $\chi(G)$ may be obtained by construction:

to obtain an upper bound for $\chi(G)$, construct an explicit colouring for the vertices of G.

A lower bound for $\chi(G)$ may be obtained using the result of Problem 12.4(a):

to obtain a lower bound for $\chi(G)$, find the number of vertices in the largest complete subgraph of G.

For example, if G contains a triangle (K_3), then $\chi(G) \geq 3$.

If we can find an upper bound and a lower bound that are the same, then $\chi(G)$ is equal to this common value. For example, the vertices of the graph G below can be coloured with four colours, as shown, so $\chi(G) \leq 4$. But G cannot be coloured with fewer than four colours, since G contains the complete graph K_4, so $\chi(G) \geq 4$. Combining these two inequalities, we obtain $\chi(G) = 4$.

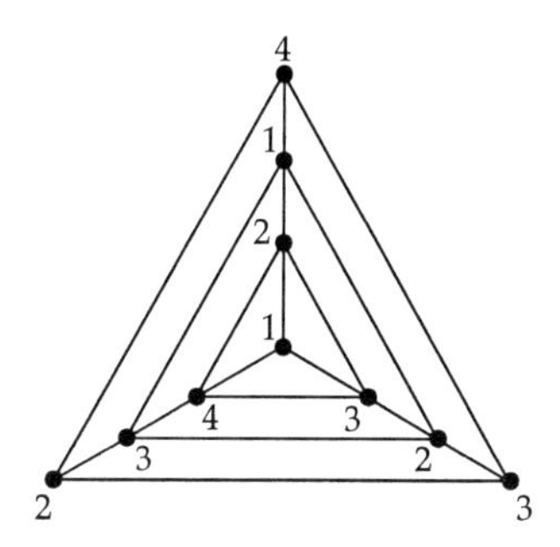

Note that if a graph G has n vertices, then $\chi(G) \leq n$. However, this upper bound is usually poor, except when G has many edges. This inequality becomes an equality ($\chi(G) = n$) only when G is the complete graph K_n.

We can improve on this upper bound considerably if we know the largest vertex degree in G, as our next theorem shows.

Theorem 12.1

Let G be a simple graph whose maximum vertex degree is d. Then

$$\chi(G) \leq d + 1.$$

Proof The proof is by mathematical induction on n, the number of vertices of G.

Step 1 The statement is true for K_1, the simple graph with one vertex, since $\chi(K_1) = 1$ and $d = 0$.
Step 2 We assume that $\chi(H) \leq d + 1$ for all simple graphs H with fewer than n vertices. We wish to show that $\chi(G) \leq d + 1$ for all simple graphs G with n vertices.

Let G be a simple graph with n vertices and maximum vertex degree d, and let H be any graph obtained from G by removing a vertex v and the edges incident with it.

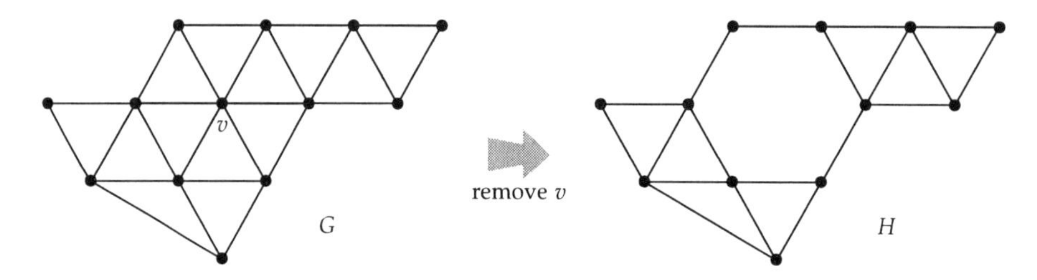

The graph H has fewer than n vertices and maximum vertex degree d (or less) so, by our assumption, $\chi(H) \leq d + 1$ – that is, the graph H is $(d + 1)$-colourable. We can now obtain a $(d + 1)$-colouring of G by colouring v with any colour not assigned to the (at most d) vertices adjacent to v, since these vertices can be coloured with at most d colours. It follows that $\chi(G) \leq d + 1$. Thus if the statement is true for all simple graphs with fewer than n vertices, then it is true for all simple graphs with n vertices. This completes Step 2.

Therefore, by the principle of mathematical induction, the statement is true for all simple graphs with n vertices, for each positive integer n. ∎

(An alternative method of proof, using a greedy algorithm, is outlined in Section 12.2.)

With a lot more effort, we can prove the following slightly stronger theorem, proved by L. Brooks in 1941. We omit the proof.

Theorem 12.2: Brooks' Theorem

Let G be a connected simple graph whose maximum vertex degree is d. If G is neither a cycle graph with an odd number of vertices, nor a complete graph, then $\chi(G) \leq d$.

To illustrate the use of Brooks' theorem, we consider again the graph G below. We have already observed that $\chi(G) \geq 4$, since G contains the complete graph K_4. On the other hand, G satisfies the conditions of Brooks' theorem with $d = 4$, so $\chi(G) \leq 4$. It follows that $\chi(G) = 4$.

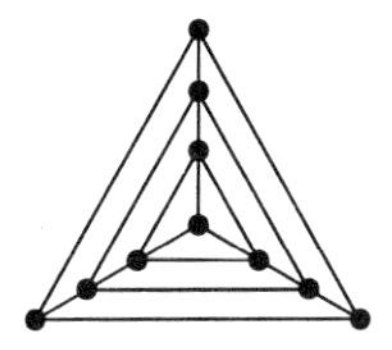

Unfortunately, the situation is not always as satisfactory as this. In particular, if G contains a few vertices of high degree, then the bound given by Brooks' theorem may be very poor. For example, if G is the bipartite graph $K_{1,12}$, then Brooks' theorem gives the upper bound $\chi(G) \leq 12$, whereas the actual value of $\chi(G)$ is 2.

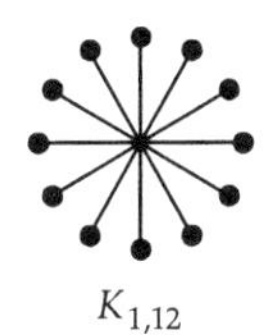

$K_{1,12}$

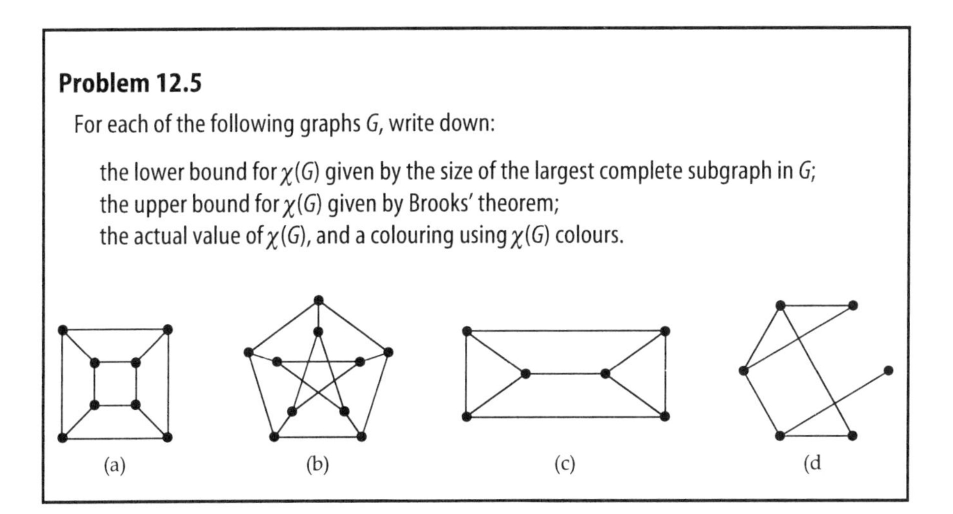

Problem 12.5

For each of the following graphs G, write down:

the lower bound for $\chi(G)$ given by the size of the largest complete subgraph in G;
the upper bound for $\chi(G)$ given by Brooks' theorem;
the actual value of $\chi(G)$, and a colouring using $\chi(G)$ colours.

We summarize the above results as follows.

To find the chromatic number $\chi(G)$ of a simple graph G

Try to find an upper bound and a lower bound that are the same; then $\chi(G)$ is equal to this common value.

Possible upper bounds for $\chi(G)$

- the number of colours in an explicit vertex colouring of G;
- the number n of vertices in G;
- $d + 1$, where d is the maximum vertex degree in G (Theorem 12.1);
- d, where d is the maximum vertex degree in G, provided that G is not C_n (for odd n) or K_n (Brooks' theorem).

Possible lower bound for $\chi(G)$

- the number of vertices in the largest complete subgraph in G.

Colouring Planar Graphs

It seems natural to conjecture that the more complicated a graph, the more colours are needed to colour its vertices. In this subsection we show that this conjecture is false for *planar* graphs – the chromatic number of any planar graph is 'small'.

Our first result of this type shows that every planar graph is 6-colourable.

Theorem 12.3: Six Colour Theorem for Planar Graphs

The vertices of any simple connected planar graph G can be coloured with six (or fewer) colours in such a way that adjacent vertices are coloured differently.

Proof The proof is by mathematical induction on n, the number of vertices of G.

Step 1 The statement is trivially true when $n = 1$. (In fact, the statement is obviously true for all graphs with up to six vertices, since a different colour can be used for each vertex.)

Step 2 We assume that the vertices of all simple connected planar graphs with fewer than n vertices can be coloured with six (or fewer) colours. We wish to show that the vertices of all simple connected planar graphs with n vertices can be coloured with six (or fewer) colours.

Let G be a simple connected planar graph with n vertices. It follows from Corollary 11.3 that G contains a vertex v of degree 5 or less. We remove v and its incident edges; then the resulting planar graph H has fewer than n vertices.

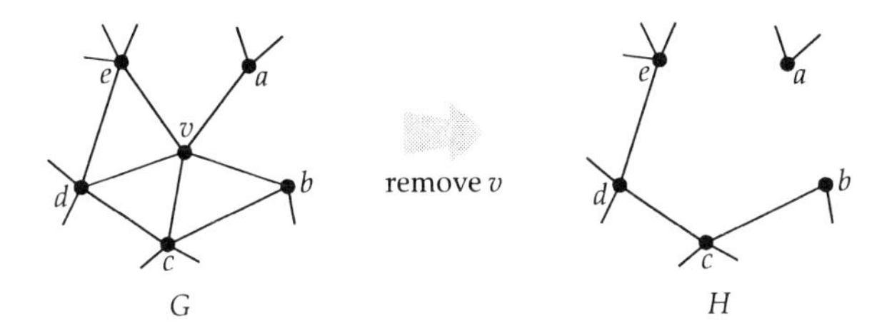

By our assumption, the vertices of H (or of each component of H, if H is disconnected) can be coloured with six colours in such a way that adjacent vertices are coloured differently. We now reinstate the vertex v.

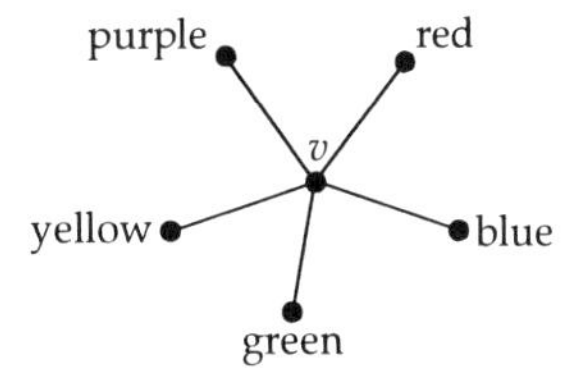

Since v has at most five neighbours, and six colours are available, there is a spare colour that can be used to colour v. This gives a 6-colouring of the vertices of G, as required. Thus if the statement is true for all simple connected planar graphs with fewer than n vertices, then it is true for all simple connected planar graphs with n vertices. This completes Step 2.

Therefore, by the principle of mathematical induction, the statement is true for all simple connected planar graphs with n vertices, for each positive integer n. ■

With a little more effort, we can prove the following stronger theorem.

Theorem 12.4: Five Colour Theorem for Planar Graphs

The vertices of any simple connected planar graph G can be coloured with five (or fewer) colours in such a way that adjacent vertices are coloured differently.

Proof The proof is by mathematical induction on n, the number of vertices of G.

Step 1 The statement is trivially true when $n = 1$. (In fact, the statement is obviously true for all graphs with up to five vertices, since a different colour can be used for each vertex.)

Step 2 We assume that the vertices of all simple connected planar graphs with fewer than n vertices can be coloured with five (or fewer) colours. We wish to show that the vertices of all simple connected planar graphs with n vertices can be coloured with five (or fewer) colours.

Let G be a simple connected planar graph with n vertices. It follows from Corollary 11.3 that G contains a vertex v of degree 5 or less. We remove v and its incident edges; then the resulting planar graph H has fewer than n vertices.

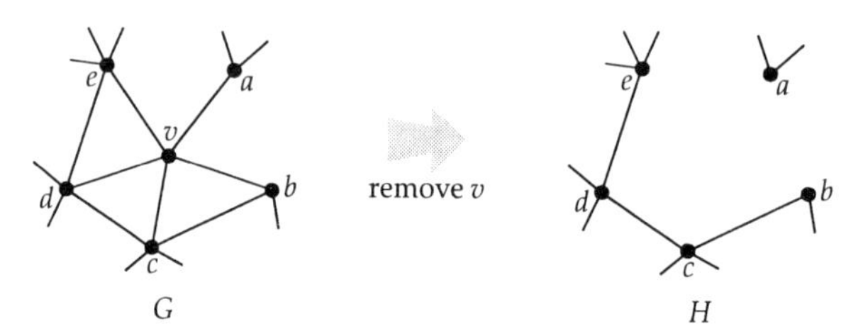

By our assumption, the vertices of H (or of each component of H, if H is disconnected) can be coloured with five colours in such a way that adjacent vertices are coloured differently. We now reinstate the vertex v.

Since there are at most five vertices adjacent to v, and five colours are available, there is a spare colour that can be used to colour v, *unless v is surrounded by five vertices of different colours*; in this case, there is no spare colour that can be used to colour v.

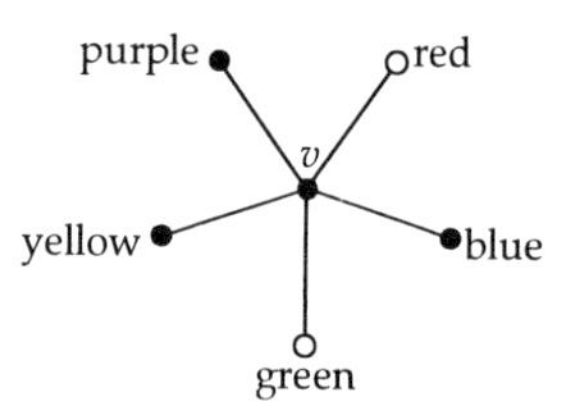

To overcome this difficulty, we consider just the red and green vertices adjacent to v, and investigate whether there is a path of red and green vertices between the adjacent red vertex and the adjacent green vertex. There are two cases that can arise.

In case (a), all the red and green vertices reachable from the adjacent red vertex are different from those reachable from the adjacent green vertex, so there is no such red–green path. In this case, we interchange the colours in the red–green part at the top, say, as shown below.

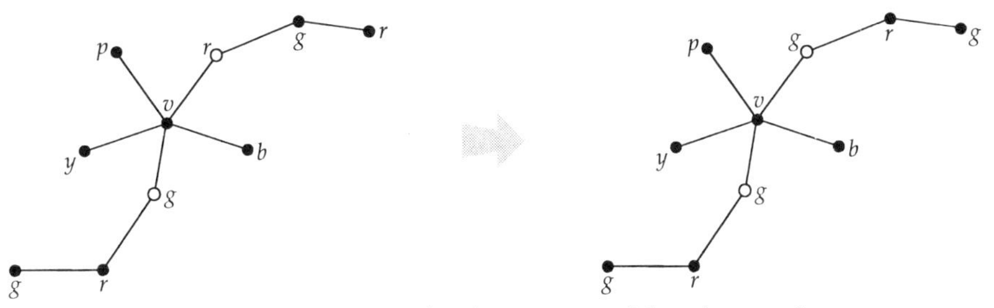

case (a): interchange red and green on top right; colour v red

This replaces the red vertex adjacent to v by a green one, so that v can now be coloured red. This completes the 5-colouring of the vertices of G in this case.

In case (b), the two red–green parts link up, so there is a red–green path, but interchanging the colours does not help us, as the vertex v is still adjacent to a red vertex and a green vertex. However, there can be no path of blue and yellow vertices between the blue and yellow vertices adjacent to v, because the red–green path 'gets in the way'. We can therefore interchange the colours in the blue–yellow part on the right-hand side, say, as shown below.

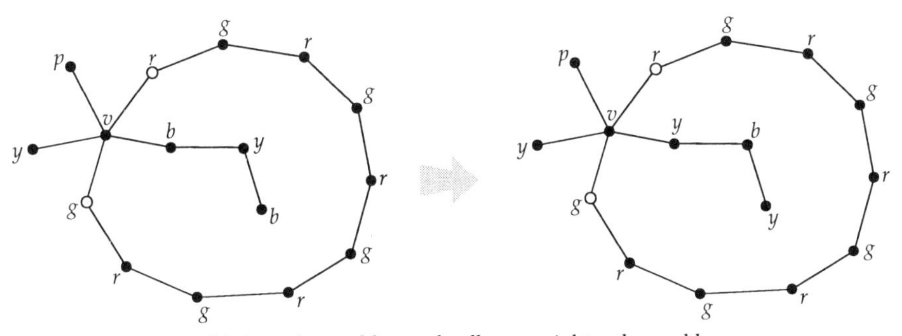

case (b): interchange blue and yellow on right; colour v blue

This replaces the blue vertex adjacent to v by a yellow vertex, so that v can now be coloured blue. This completes the 5-colouring of the vertices of G in this case.

It follows that the statement is true for all simple connected planar graphs with n vertices. This completes Step 2.

Therefore, by the principle of mathematical induction, the statement is true for all simple connected planar graphs with n vertices, for each positive integer n. ∎

We conclude this section by stating without proof the four colour theorem for planar graphs. It is related to the four colour theorem for maps, described later, and is considerably more difficult to prove than the five and six colour theorems for planar graphs.

> **Theorem 12.5: Four Colour Theorem for Planar Graphs**
> The vertices of any simple connected planar graph can be coloured with four (or fewer) colours in such a way that adjacent vertices are coloured differently.

12.2 Algorithm for Vertex Colouring

It is natural to ask whether there are efficient algorithms for colouring the vertices of a graph. Unfortunately, no such efficient algorithms are known. We must therefore seek either inefficient algorithms that give the correct value for the number of colours needed, or heuristic algorithms that are efficient but give only an approximation to the correct value. In this subsection we present such a heuristic algorithm – a straightforward colouring algorithm that usually gives good answers. The method we describe is a greedy algorithm, and may be stated as follows.

> **Greedy Algorithm for Vertex Colouring**
> START with a graph *G* and a list of colours 1, 2, 3, ...
> Step 1 Label the vertices a, b, c, ... in any manner.
> Step 2 Identify the uncoloured vertex labelled with the earliest letter in the alphabet; colour it with the first colour in the list not used for any adjacent coloured vertex.
>
> Repeat Step 2 until all the vertices are coloured, then STOP.
> A vertex colouring of *G* has been obtained. The number of colours used depends on the labelling chosen for the vertices in Step 1.

We present two examples using the same graph with different labellings.

Illustration A

Find a vertex colouring of the following graph *G*.

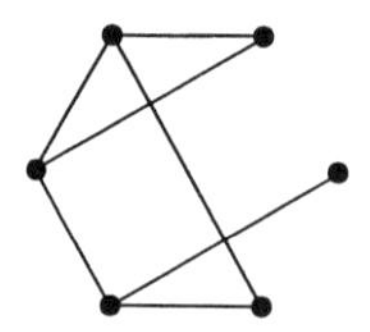

Step 1 We label the vertices $a, ..., f$ as follows.

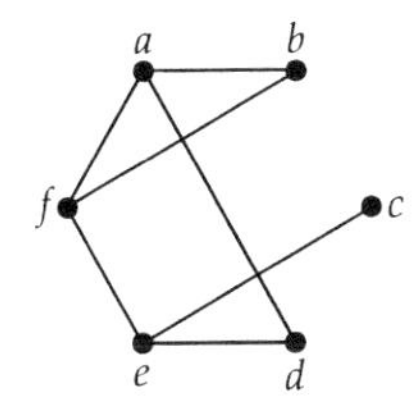

Step 2 We successively colour

vertex a with colour 1,
vertex b with colour 2,
vertex c with colour 1,
vertex d with colour 2,
vertex e with colour 3,
vertex f with colour 4.

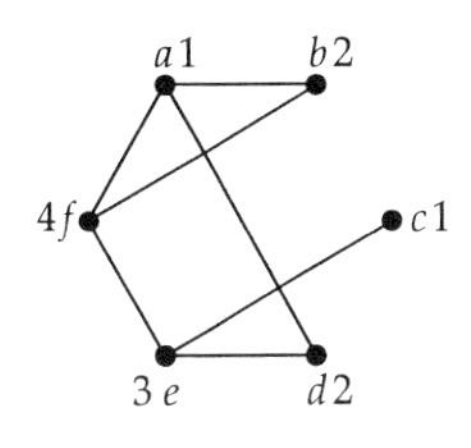

All the vertices are now coloured, so we STOP.

We thus obtain the 4-colouring of G shown above. ❐

Illustration B

Find a vertex colouring of the following graph G.

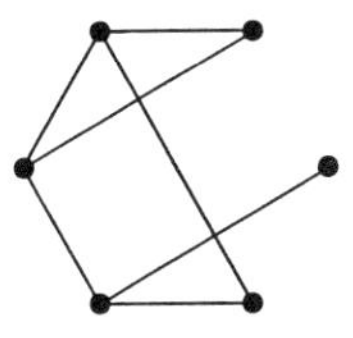

Step 1 We label the vertices $a, ..., f$ as follows.

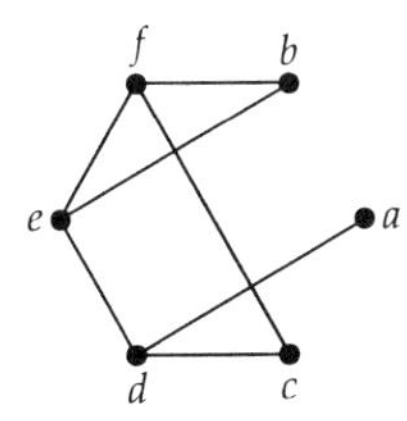

Step 2 We successively colour

vertex a with colour 1,
vertex b with colour 1,
vertex c with colour 1,
vertex d with colour 2,
vertex e with colour 3,
vertex f with colour 2.

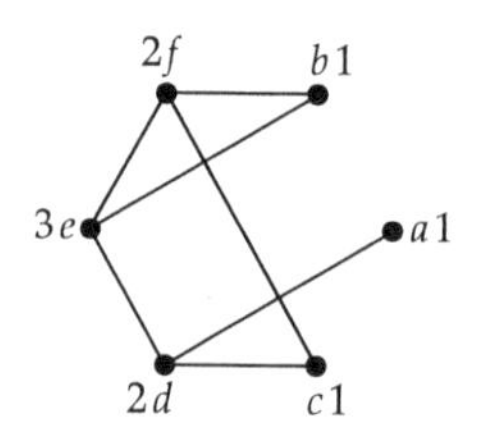

All the vertices are now coloured, so we STOP.

We thus obtain the 3-colouring of G shown above. ❐

Notice that, in the above examples, $\chi(G) = 3$, and in Illustration B we found a vertex colouring of G that uses 3 colours.

Problem 12.6

Use the greedy algorithm to colour the vertices of the following graph G, using each of the given labellings.

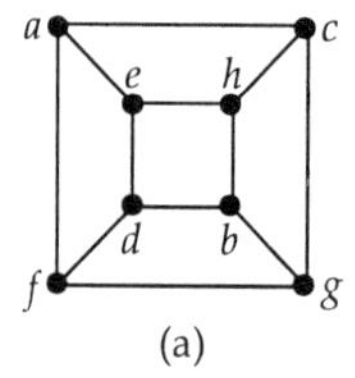

(a)

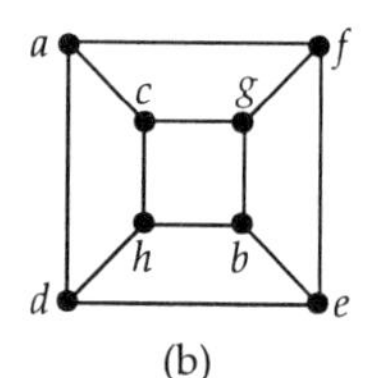

(b)

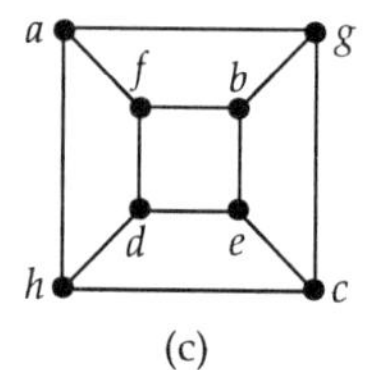

(c)

What is the actual value of $\chi(G)$?

More generally, we have the following theorem.

Theorem 12.6

For any graph G, there is a labelling of the vertices for which the greedy algorithm yields a vertex colouring with $\chi(G)$ colours.

Outline of Proof

Take any vertex colouring of G with $\chi(G)$ colours, denoted by 1, 2, 3, ..., and sequentially label with a, b, c, ... the vertices coloured 1, then the vertices coloured 2, then the vertices coloured 3, and so on. For this labelling, the greedy algorithm assigns the colours 1, 2, 3, ... in that order, so only $\chi(G)$ colours are needed. ■

Problem 12.7

Find a labelling of the vertices of the following graph, for which the greedy algorithm yields a vertex colouring of G with $\chi(G)$ colours.

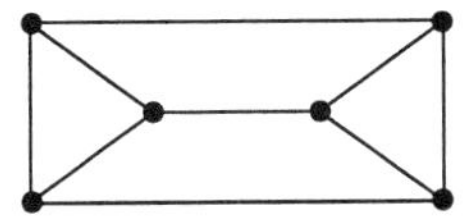

We conclude this section by returning to a result that we proved earlier using the method of mathematical induction. We outline an alternative proof using the greedy algorithm.

Theorem 12.1

Let G be a simple graph whose maximum vertex degree is d. Then

$$\chi(G) \leq d + 1.$$

Outline of Proof

Let a, b, c, ... be any labelling of the vertices of G. Colour these vertices in turn, using the lowest numbered colour available. At each stage the vertex to be coloured has at most d adjacent vertices, and there are $d + 1$ colours, so there is always a colour available. The colouring with $d + 1$ colours can therefore be completed. ■

Brooks' theorem (Theorem 12.2) may be proved similarly; the proof requires a more complicated version of the greedy algorithm.

12.3 Vertex Decompositions

Some of the most interesting problems in graph theory involve the decomposition of a graph G into subgraphs of a particular type. In several of these problems, we split the set of *vertices* of G into disjoint subsets; this is called a **vertex decomposition** of G.

For example, consider the following disconnected graph G.

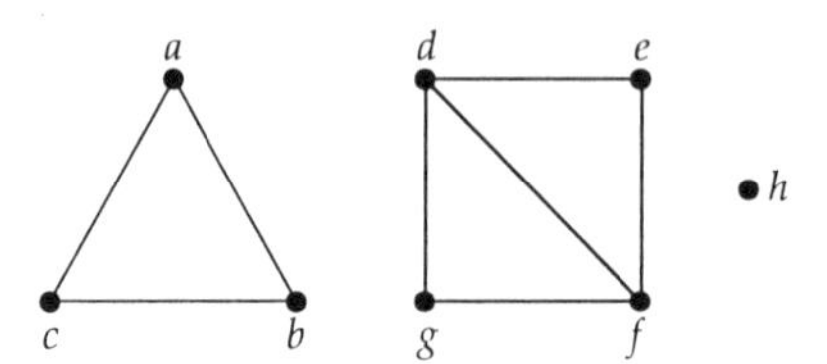

A natural vertex decomposition is to split the set of vertices into the disjoint subsets that correspond to the components of G:

$$\{a, b, c\}, \{d, e, f, g\}, \{h\}.$$

In this section, we adopt a similar approach to several other problems. Each problem can be formulated in graph-theoretic terms, and involves splitting the set of vertices of a graph into disjoint subsets with particular properties. By doing this, we observe similarities between seemingly different problems and can begin to classify them, thereby gaining insight into the nature of the different types of problem.

Colouring Problems

Example 12.1: Storing Chemicals

In Section 12.1 we considered the problem of a chemical manufacturer who wishes to store chemicals $a, b, ..., g$ in a warehouse. Some chemicals react violently when in contact, and the manufacturer divides the warehouse into a number of areas so as to separate certain pairs of chemicals.

In order to determine the smallest number of areas needed to store these chemicals safely, we drew the graph shown below. The vertices correspond to the chemicals, and two vertices are joined by an edge whenever the corresponding chemicals must be kept separate.

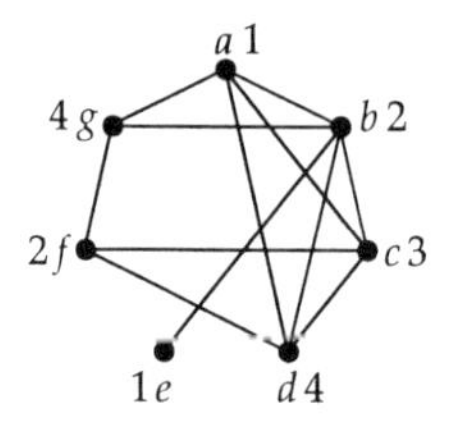

We saw that the assignment of chemicals to areas is a vertex colouring problem in which the colours correspond to the areas. Such a colouring gives rise to a vertex decomposition of the graph in which *no two vertices in the same subset are adjacent*. The vertex decomposition arising from this example is

$$\{a, e\}, \{b, f\}, \{c\}, \{d, g\};$$

the four subsets correspond to the chemicals in the four areas. In such a problem, the minimum number of subsets needed is the *chromatic number* of the corresponding graph. ❐

Example 12.2: Map Colouring

In the *map colouring problem*, we wish to determine the smallest number of colours required to colour the countries of a map in such a way that any two countries with a common boundary are coloured differently. This enables us to distinguish between the various countries, and to locate the boundaries.

Consider the following map of the USA (excluding Alaska and Hawaii):

How many colours are needed to colour the entire map?

This map cannot be coloured with three colours, because three colours are needed for the ring of five states surrounding Nevada, so at least four colours are needed.

However, it can be coloured with just four colours, as follows.

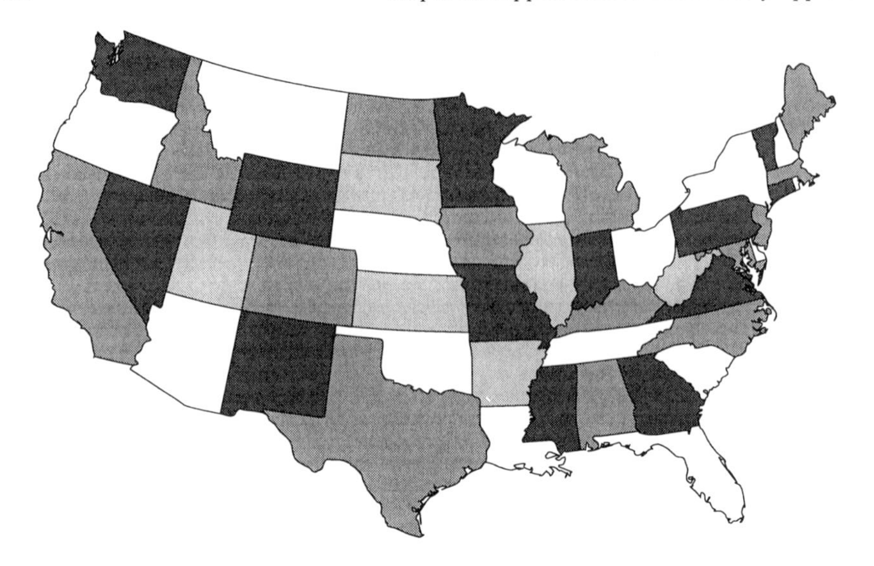

We can represent this situation as a vertex colouring problem by considering the dual problem in which each state is represented by a vertex, and two vertices are joined whenever the corresponding states share a common boundary line. This gives the following graph, in which each vertex has been assigned a symbol to represent the colour of the corresponding state. Since any two neighbouring states in the original map were coloured differently, any two adjacent vertices in this graph must also be assigned different colours.

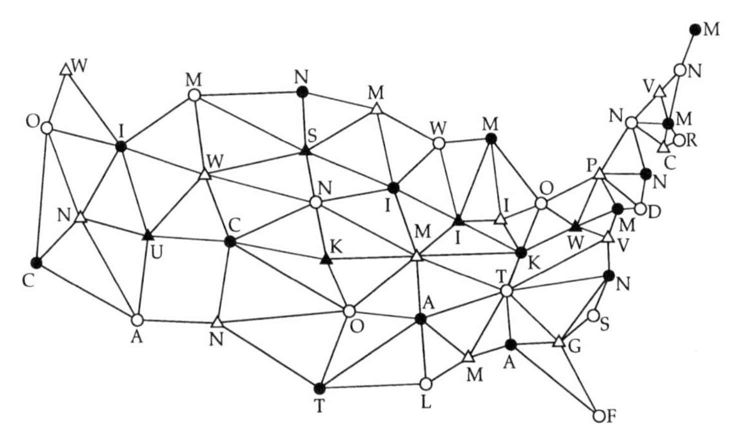

△ Washington, Nevada, Wyoming, New Mexico, Minnesota, Mississippi, Missouri, Indiana, Georgia, Virginia, Pennsylvania, Connecticut, Vermont;

○ Oregon, Montana, Arizona, Nebraska, Oklahoma, Louisiana, Wisconsin, Tennessee, Ohio, Florida, South Carolina, Delaware, New York, Rhode Island, New Hampshire

● California, Idaho, Colorado, North Dakota, Texas, Iowa, Michigan, Alabama, Kentucky, North Carolina, Maryland, New Jersey, Massachusetts, Maine, Arkansas;

▲ Utah, South Dakota, Kansas, Illinois, West Virginia.

Such a colouring of the vertices of the graph splits the set of vertices into four disjoint subsets, corresponding to the four colours. ❐

Problem 12.8

Consider the following map.

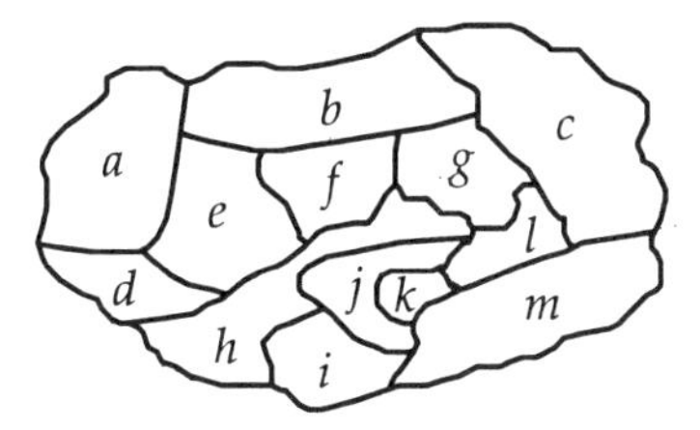

(a) Find a 4-colouring of this map by trial and error.

(b) Draw the corresponding graph, and show how the 4-colouring in part (a) leads to a vertex decomposition of this graph in which no two vertices in the same subset are adjacent.

It is simple to prove that any map can be coloured with six colours so that neighbouring countries are coloured differently; the proof is essentially the 'dual' of the proof of Theorem 12.3. With a little more effort, it can be shown that any map can be coloured with five colours so that neighbouring countries are coloured differently; the proof is essentially the 'dual' of the proof of Theorem 12.4. The celebrated *four colour problem* is related to Theorem 12.5.

Historical Note

In 1852 Francis Guthrie posed the famous *four colour problem*: can all maps be coloured with four colours in such a way that neighbouring countries are coloured differently? This problem was studied by a number of mathematicians, including Augustus De Morgan, Arthur Cayley and Alfred Kempe, but it was not until 1976 that a proof was finally obtained, by Kenneth Appel and Wolfgang Haken. Their proof involved the consideration of nearly 2000 configurations of countries, and made extensive use of a computer. To this day, no 'simple' proof has been discovered.

Vertex decomposition problems also arise in situations that involve planning a tour, such as refuse collection.

Example 12.3: Refuse Collection

A weekly route schedule for refuse collection lorries is to be organized. The daily routes must be different for Monday to Saturday, and some sites need to be visited several times a week. No route is to be too long or too short, every lorry must be used on every working day, and every site must be visited the required number of times. How can a suitable schedule be designed?

In its full complexity, this problem is too hard to be considered here, so we look at just one aspect of it. We investigate whether it is possible to arrange a schedule in such a way that two different lorries do not visit the same site on the same day. To this end, we construct a *tour graph* in which each vertex represents a route, and two vertices are joined by an edge whenever the corresponding routes have a site in common. If the vertices of this tour graph can be coloured with six colours (corresponding to the days Monday to Saturday) so that adjacent vertices are coloured differently, then any such vertex colouring gives rise to a suitable schedule. So the problem reduces to that of a vertex colouring problem. It is therefore again a vertex decomposition problem in which *no two vertices in the same subset are adjacent*. ❐

Problem 12.9

Draw the tour graph for the following routes for refuse vehicles collecting from industrial sites *A*, ..., *R*, and use it to find the minimum number of days needed to ensure that no place is visited more than once on the same day. What is the corresponding vertex decomposition?

route 1	sites *A*, *B*, *C* and *D*	route 2	sites *B*, *E*, *F*, *G*
route 3	sites *B*, *H*, *I*	route 4	sites *B*, *F*
route 5	sites *E*, *G*, *J*	route 6	sites *G*, *K*, *L*
route 7	sites *L*, *M*, *N*	route 8	sites *K*, *N*
route 9	sites *A*, *H*, *O*, *I*	route 10	sites *O*, *P*
route 11	sites *C*, *P*, *Q*	route 12	sites *C*, *Q*, *R*

Domination Problems

Communication Links

Suppose that communication links are to be set up between a number of cities, and transmitting stations are to be built in some of these cities so that each city can receive messages from at least one transmitting station. For reasons of economy, we require the number of transmitting stations to be as small as possible. How can this be done?

We can represent this situation by a graph whose vertices correspond to the cities, and whose edges correspond to pairs of cities that can communicate directly with each other. Since each city must either contain a transmitting station or communicate with a city containing a transmitting station, we wish to find a set of vertices that (between them) are adjacent to all other vertices of the graph.

Example 12.4: Location of Transmitting Stations

Suppose that the following graph represents the communication links between six cities, A, ..., F.

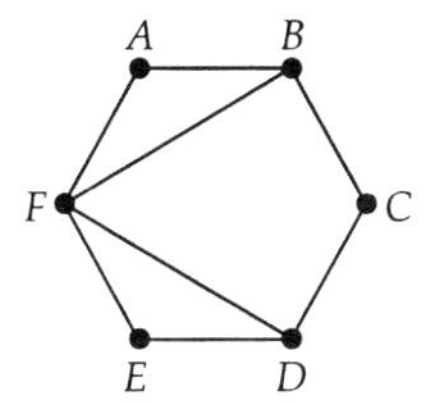

We can locate the transmitting stations at A, C, E, since each of the other vertices (B, D, F) is adjacent to at least one of these vertices; we say that the vertices A, C and E form a *dominating set*. We thus obtain a vertex decomposition into subsets of cities served by the same transmitting station:

$$\{A, B, F\}, \{C, D\}, \{E\}.$$

We obtain a more economical solution by taking just two transmitting stations and locating them at A and D. As before, each of the other vertices (B, C, E, F) is adjacent to at least one of these vertices. Thus the vertices A and D form a *dominating set* that is smaller than the one given above. A corresponding vertex decomposition is

$$\{A, B, F\}, \{D, C, E\}.$$

There is no dominating set comprising just one vertex, so we say that the vertices A and D form a *minimum dominating set*. The number of vertices in a minimum dominating set is the *dominating number* – in this case, 2.

Notice that, in each of the above vertex decompositions, *each subset contains a vertex adjacent to all the other vertices in that subset.* ❐

Problems that reduce to that of finding a minimum dominating set in a given graph occur in many guises. For example, suppose that a number of locations in a nuclear power plant are fitted with warning lights, and that sensors are to be stationed in various places to keep watch on these lights. We can minimize the number of sensors by finding a minimum dominating set in the corresponding graph and positioning the sensors accordingly. Any light that comes on can then be seen by at least one sensor, and appropriate action can be taken.

Problem 12.10

Find a minimum dominating set in each of the following graphs.

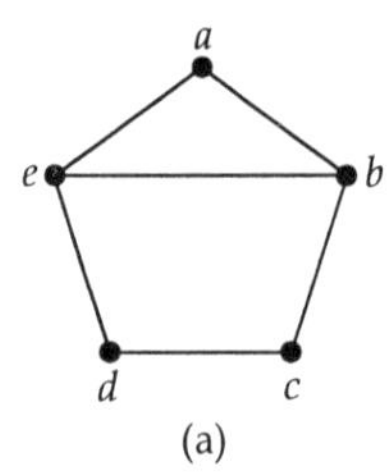

(a)

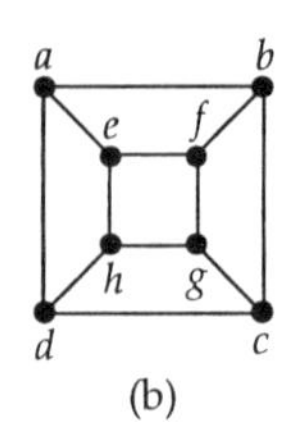

(b)

In each case, write down a vertex decomposition in which each subset contains a vertex adjacent to all the other vertices in that subset.

Note that the type of vertex decomposition described for domination problems is different from that described for colouring problems. For colouring problems, *in each subset, no two vertices are adjacent*. For domination problems, *each subset contains a vertex adjacent to all the other vertices in that subset.*

Exercises 12

Vertex Colourings

12.1 A zoo-keeper wishes to place eight animals $A, B, ..., H$ into enclosures. For safety reasons, some of the animals cannot be placed in the same enclosure. In the following table, crosses indicate pairs of animals that must be placed in different enclosures.

	A	B	C	D	E	F	G	H
A	–	×	–	–	×	×	–	×
B	×	–	×	–	–	×	–	×
C	–	×	–	×	–	×	×	×
D	–	–	×	–	×	×	×	–
E	×	–	–	×	–	×	×	–
F	×	×	×	×	×	–	–	–
G	–	–	×	×	×	–	–	×
H	×	×	×	–	–	–	×	–

By drawing a suitable graph, determine the least number of enclosures that are needed to house all the animals, and an appropriate placing of the animals.

12.2 Determine $\chi(G)$ for each of the following graphs G.

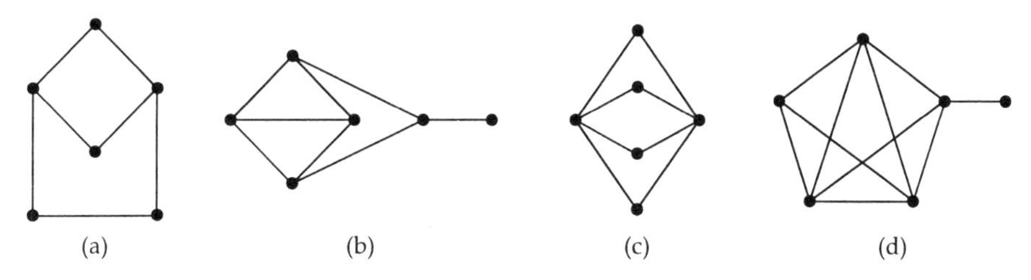

12.3 Consider the following graph G.

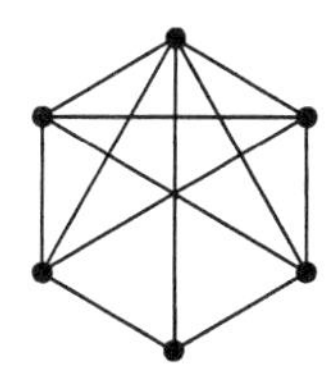

(a) Use results from Section 12.1 to obtain lower and upper bounds for $\chi(G)$.

(b) What is the actual value of $\chi(G)$?

12.4 Draw two non-isomorphic simple connected graphs G with five vertices and maximum vertex degree d for which $\chi(G) = d + 1$.

12.5 Let G be the graph obtained by removing an edge from the complete graph K_n. By Brooks' theorem, $\chi(G) \leq n - 1$. Give a method for $(n - 1)$-colouring G, and test your method by 6-colouring K_7 with one edge removed.

12.6 Prove that if G is an r-regular graph with n vertices, then

$$\chi(G) \geq n/(n - r).$$

Algorithm for Vertex Colouring

12.7 Use the greedy colouring algorithm to colour the vertices of each of the following labelled graphs.

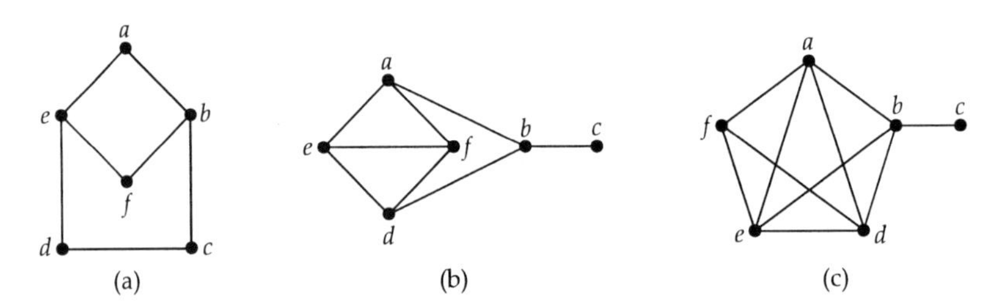

Comment on your results.

Vertex Decompositions

12.8 A youth club organizer wishes to arrange outings to the Zoo for nine children: Andrew, Bill, Catherine, Deirdre, Edward, Fiona, Gina, Harry and Iris. Unfortunately, Catherine refuses to go on an outing with any of the boys, Andrew will not go if there are any girls (except Deirdre), Edward and Harry must not be allowed to go together since they will cause havoc, Fiona cannot stand Bill or Gina, and Bill and Edward both dislike Iris. Express this information in terms of a suitable graph, find the minimum number of outings needed, and write down the corresponding vertex decomposition.

12.9 Each of ten students $A, B, \ldots, J$, must attend three (out of eight) lectures, as indicated by the crosses in the following table:

	1	2	3	4	5	6	7	8
A	×	×	–	–	×	–	–	–
B	×	–	–	–	×	×	–	–
C	×	–	–	–	–	×	×	–
D	–	–	–	–	–	×	×	×
E	–	–	×	–	–	–	×	×
F	–	–	×	×	–	–	–	×
G	–	×	×	×	–	–	–	–
H	–	×	–	×	×	–	–	–
I	–	–	–	×	×	–	–	×
J	–	–	–	–	×	×	–	×

Each student can attend only one lecture per day. By drawing a suitable graph, find the minimum number of days needed to timetable all the lectures and write down a suitable timetable.

12.10 The following map shows sixteen countries, numbered from 0 to 15. Each country other than country 0 has been allocated one of the colours red (r), blue (b), yellow (y), green (g).

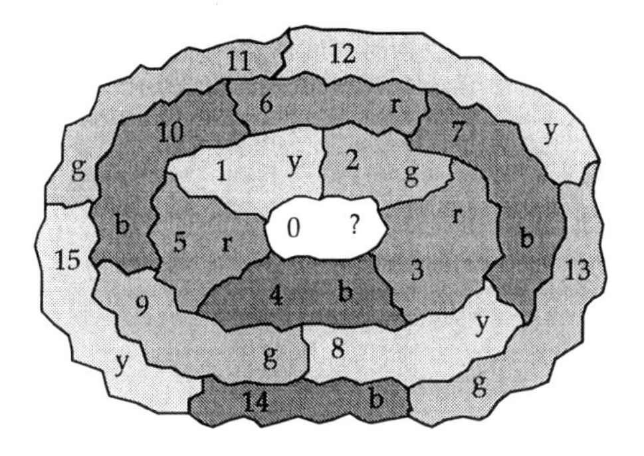

(a) Write down the numbers of the countries in the blue–green part of the map adjacent to country 0, and the blue–yellow part adjacent to country 0.

Explain why an interchange of colours on just one of these two-coloured parts does not help to find a 4-colouring of the map.

(b) By interchanging the colours of countries 1 and 2 and recolouring countries 3 and 5, find a 4-colouring of the map.

(c) Write down a vertex decomposition of the coresponding graph with the property that no two vertices in the same subset are adjacent.

12.11

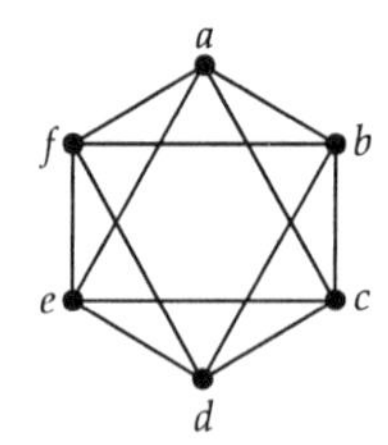

For the octahedron graph shown above, find, if possible:

(a) a vertex decomposition in which no two vertices in the same subset are adjacent;

(b) a vertex decomposition in which each subset contains a vertex adjacent to each of the other vertices in the subset.

12.12 Find a minimum dominating set and the dominating number for each of the graphs in Exercise 12.7.

12.13 Find a minimum dominating set and the dominating number for the octahedron graph in Exercise 12.11.

12.14 Find the dominating number of:

(a) the Petersen graph; (b) the 4-cube graph Q_4.

Chapter 13

Edge Colourings and Decompositions

After studying this chapter, you should be able to:

- explain the terms *edge colouring, k-edge colouring* and *chromatic index;*
- use Vizing's theorem (both versions), Shannon's theorem and König's theorem;
- use the greedy algorithm to colour the edges of a graph;
- explain what are meant by the *printed circuits problem, matching problems,* and various *bus route problems,* and how they can be represented as *edge decomposition problems.*

In this chapter, we consider problems involving the colouring of the *edges* of a graph, and we introduce an algorithm for edge colouring. We then consider problems that involve splitting the set of edges of a graph into disjoint subsets with particular properties. Our discussion involves problems relating to printed circuits, matchings, and the scheduling of examinations.

13.1 Edge Colourings

Example 13.1: Wire Colouring

An engineer wishes to make a display panel on which electrical components a, b, ... are to be mounted and then interconnected. The connecting wires are formed into a cable, with the wires to be connected to a emerging through one hole in the panel, those connected to b emerging through another hole, and so on. In order to distinguish the wires that emerge from the same hole, they are coloured differently. What is the minimum number of colours necessary for the whole system? (This problem was posed by C. E. Shannon in 1949, in a paper on electrical networks.)

In order to investigate this problem, we represent the connection points by the vertices of a graph and the wires by edges. For example, the following graph represents a panel with six components, $a, ..., f$.

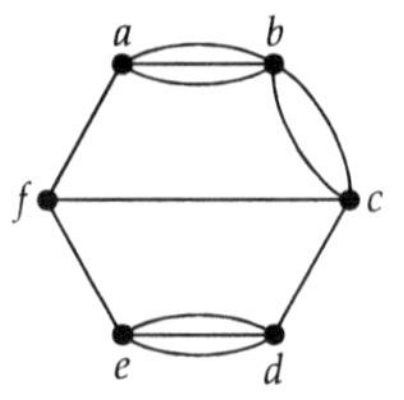

Since vertex b has five edges incident with it, and since these edges must all be coloured differently, at least five colours are necessary. In fact, five colours are sufficient, as the following diagram shows; the numbers on the edges correspond to the five colours.

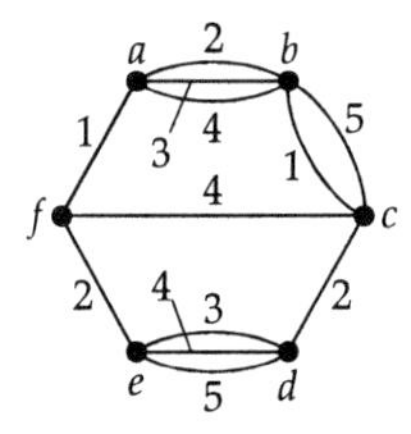

❐

Chromatic Index

The assignment of colours to wires in Example 13.1 illustrates the following definitions.

> **Definitions**
>
> Let G be a graph without loops. A ***k*-edge colouring** of G is an assignment of at most k colours to the edges of G in such a way that any two edges meeting at a vertex are assigned different colours. If G has a k-edge colouring, then G is ***k*-edge colourable**.
> The **chromatic index** of G, denoted by $\chi'(G)$, is the smallest number k for which G is k-edge colourable.

In the above wire colouring example, the graph has chromatic index 5.

Remark The above definitions are given only for graphs *without loops*. Loops must be excluded since, in any k-edge colouring, the edges meeting at a vertex must be assigned different colours. However, we sometimes wish to consider graphs with multiple edges, since the introduction of multiple edges may alter the chromatic index, as in the wire colouring problem.

We usually show a k-edge colouring by writing the numbers $1, 2, \ldots, k$ next to the appropriate edges. For example, diagrams (a) and (b) below illustrate a 5-edge colouring and a 4-edge colouring of a graph G with eight edges; note that diagram (c) is *not* a 5-edge colouring of G, since two of the edges coloured 2 meet at a vertex.

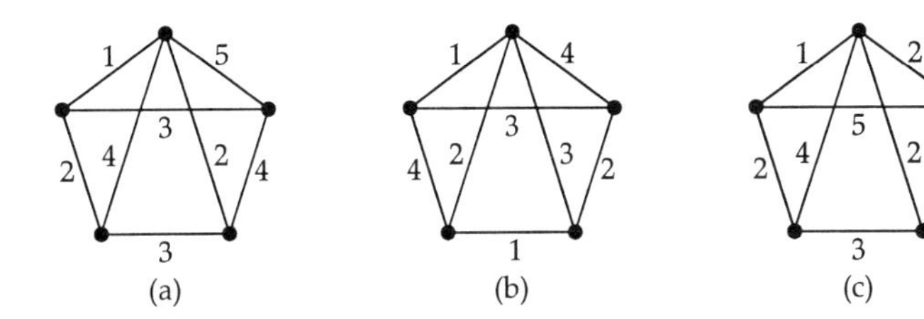

(a) (b) (c)

Since G has a 4-edge colouring, $\chi'(G) \leq 4$; thus 4 is an *upper bound* for $\chi'(G)$. Also, G contains four edges meeting at a common vertex (a vertex of degree 4) that must be assigned different colours, so $\chi'(G) \geq 4$; thus 4 is a *lower bound* for $\chi'(G)$. Combining these inequalities, we obtain $\chi'(G) = 4$.

Problem 13.1

Determine $\chi'(G)$ for each of the following graphs G.

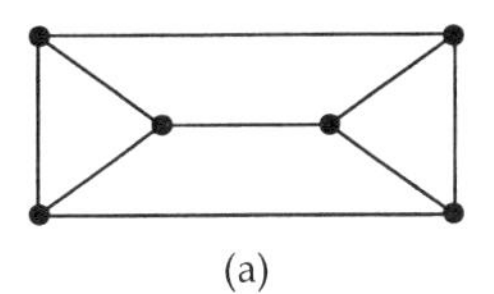

(a)

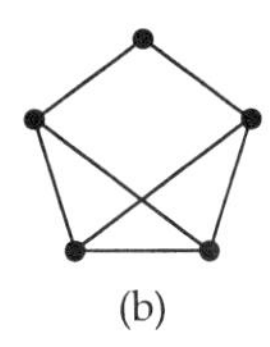

(b)

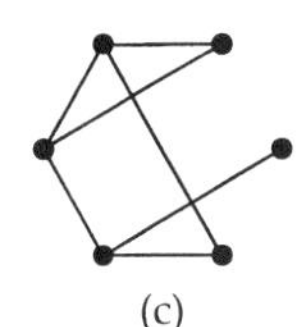

(c)

Hint For each graph, devise a suitable edge colouring and explain why there is no edge colouring with fewer colours.

Problem 13.2

What can you say about the graphs G for which

(a) $\chi'(G) = 1$? (b) $\chi'(G) = 2$?

Problem 13.3

Write down the chromatic index of each of the following graphs:

(a) the complete graph K_4;
(b) the complete bipartite graph $K_{2,3}$;
(c) the cycle graph C_6.

Problem 13.4

Decide whether each of the following statements about a graph G is true or false, and give a proof or counter-example, as appropriate.

(a) If G contains a vertex of degree r, then $\chi'(G) \geq r$.
(b) If $\chi'(G) \geq r$, then G contains a vertex of degree r.

Given a particular graph G, how can we determine its chromatic index? We have seen that an upper bound for $\chi'(G)$ may be obtained by construction:

to obtain an upper bound for $\chi'(G)$, construct an explicit colouring for the edges of G.

A lower bound for $\chi'(G)$ may be obtained using the result of Problem 13.4(a):

to obtain a lower bound for $\chi'(G)$, find the largest vertex degree in G.

For example, if G contains a vertex of degree 3, then $\chi'(G) \geq 3$.

If we can find an upper bound and a lower bound that are the same, then $\chi'(G)$ is equal to this common value. For example, the edges of the graph G below can be coloured with five colours, as shown, so $\chi'(G) \leq 5$. But G cannot be coloured with fewer than 5 colours, since G contains a vertex of degree 5, so $\chi'(G) \geq 5$. Combining these two inequalities, we obtain $\chi'(G) = 5$.

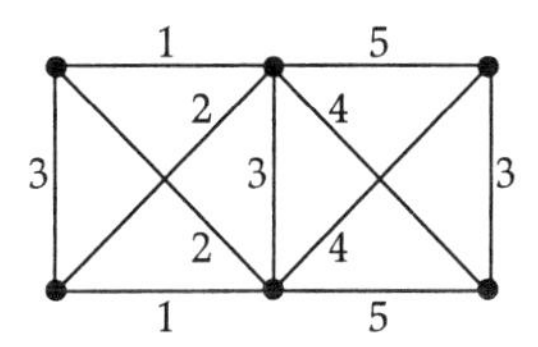

Note that if a graph G has m edges, then $\chi'(G) \leq m$. However, this upper bound is usually poor. This inequality becomes an equality ($\chi'(G) = m$) only when G is a complete bipartite graph of the form $K_{1,m}$.

Much better upper bounds have been established by V. G. Vizing and by C. E. Shannon. For simple graphs, Vizing proved the following result in 1963, which we state without proof.

Theorem 13.1: Vizing's Theorem

Let G be a simple graph whose maximum vertex degree is d. Then

$$d \leq \chi'(G) \leq d + 1.$$

This remarkable result tells us that, if G is any simple graph with maximum vertex degree d, then the chromatic index of G is either d or $d + 1$. This classifies simple graphs into two classes: those for which $\chi'(G) = d$, and those for which $\chi'(G) = d + 1$. The graphs in Problem 13.1 show that both possibilities occur, but it is not known in general which graphs belong to which class.

Problem 13.5

For each of the following simple graphs G, write down:

the lower and upper bounds for $\chi'(G)$ given by Vizing's theorem;
the actual value of $\chi'(G)$, and an edge colouring using $\chi'(G)$ colours:

(a) the cycle graph C_7;
(b) the complete bipartite graph $K_{2,4}$;
(c) the complete graph K_6.

Before investigating the problem of classifying simple graphs into those with $\chi'(G) = d$ and those with $\chi'(G) = d + 1$, we state (without proof) two results that give upper bounds for the chromatic index of a graph with multiple edges. The first of these is an extension of Vizing's theorem; it reduces to the earlier version of Vizing's theorem when G is a simple graph.

Theorem 13.2: Vizing's Theorem (Extended Version)

Let G be a graph whose maximum vertex degree is d, and let h be the maximum number of edges joining a pair of vertices. Then

$$d \leq \chi'(G) \leq d + h.$$

For example, for the graph G shown below, $d = 3$ and $h = 2$, since there are two edges joining a pair of vertices, so the lower bound is 3 and the upper bound is 5; in fact, $\chi'(G) = 4$ for this particular graph.

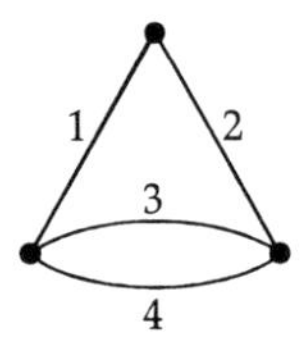

Another upper bound for the chromatic index of a graph was obtained by Shannon in his paper on the wire colouring problem.

Theorem 13.3: Shannon's Theorem

Let G be a graph with maximum vertex degree d. Then

$$d \leq \chi'(G) \leq 3d/2, \quad \text{if } d \text{ is even;}$$
$$d \leq \chi'(G) \leq (3d-1)/2, \quad \text{if } d \text{ is odd.}$$

For example, for the graph G above, $d = 3$, and $(3d - 1)/2 = 4$. So the lower bound is 3 and the upper bound is 4; in fact, $\chi'(G) = 4$.

Problem 13.6

For each of the following graphs G, write down:

the lower and upper bounds for $\chi'(G)$ given by Vizing's theorem (extended version);
the lower and upper bounds for $\chi'(G)$ given by Shannon's theorem;
the actual value of $\chi'(G)$, and a colouring using $\chi'(G)$ colours.

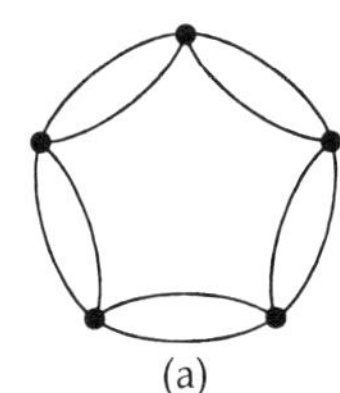
(a)

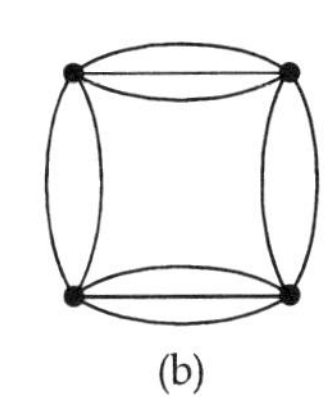
(b)

We summarize the above results as follows.

To find the chromatic index $\chi'(G)$ of a graph G without loops

Try to find an upper bound and a lower bound that are the same; then $\chi'(G)$ is equal to this common value.

Possible upper bounds for $\chi'(G)$

- the number of colours in an explicit edge colouring of G;
- the number m of edges in G;
- $d+1$, where d is the maximum vertex degree in G, provided that G has no multiple edges (Vizing's theorem);
- $d+h$, where d is the maximum vertex degree in G and h is the maximum number of edges joining a pair of vertices (Vizing's theorem, extended version);
- $3d/2$, where d is the maximum vertex degree and d is even (Shannon's theorem);
- $(3d-1)/2$, where d is the maximum vertex degree and d is odd (Shannon's theorem).

Possible lower bound for $\chi'(G)$

- d, the maximum vertex degree in G.

Classifying Some Simple Graphs

We now return to the problem of classifying simple graphs into two classes: those with $\chi'(G) = d$ and those with $\chi'(G) = d + 1$. For some types of graph, this is straightforward; for example, it is easy to show that, for the cycle graphs C_n ($n \geq 3$),

$$\chi'(C_n) = \begin{cases} 2 & \text{if } n \text{ is even;} \\ 3 & \text{if } n \text{ is odd.} \end{cases}$$

For example, for C_5 and C_6, we have the following edge colourings:

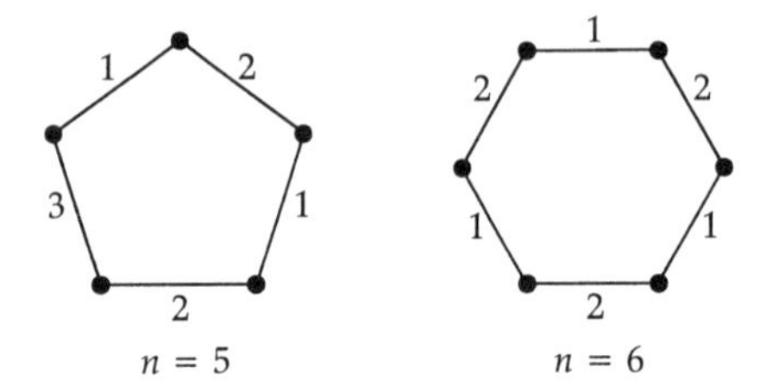

For the complete graphs K_5 and K_6, we have the following edge colourings:

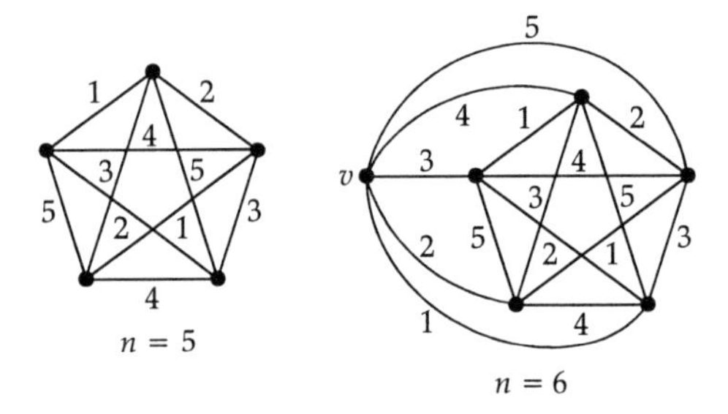

More generally, we have the following theorem.

Theorem 13.4

For the complete graph K_n,

$$\chi'(K_n) = \begin{cases} n-1 & \text{if } n \text{ is even;} \\ n & \text{if } n \text{ is odd.} \end{cases}$$

Proof Since each vertex has degree $n - 1$, it follows from Vizing's theorem that $\chi'(K_n)$ is either $n - 1$ or n.

If n is odd, then the maximum number of edges that can be assigned the same colour is $(n-1)/2$, since otherwise two of these edges meet at a common vertex. But K_n has exactly $n(n-1)/2$ edges, so the number of colours must be at least n. Hence $\chi'(K_n) \geq n$.

We can obtain an explicit n-edge colouring of K_n by drawing the vertices in the form of a regular n-gon and colouring the edges of the boundary using a different colour for each edge. Each of the remaining edges is then assigned the same colour as the boundary edge parallel to it. It follows that $\chi'(K_n) \leq n$. Combining the above inequalities, we deduce that $\chi'(K_n) = n$, if n is odd.

If n is even, we prove that $\chi'(K_n) = n - 1$, by explicitly constructing an $(n-1)$-edge colouring of K_n. If $n = 2$, this is trivial. If $n > 2$, we choose any vertex v and remove it, together with its incident edges. This leaves a complete graph K_{n-1} with an *odd* number of vertices, whose edges can be coloured with $n-1$ colours, using the above construction. At each vertex there is exactly one colour missing, and these missing colours are all different. The edges of K_n incident to v can therefore be coloured using these missing colours. It follows that $\chi'(K_n) = n-1$, if n is even. ■

Problem 13.7

(a) Suppose that 31 teams take part in a competition in which each team must play exactly one match against each of the other 30 teams. If no team can play more than one match a day, how many days are needed?

(b) What is the corresponding answer if there are 32 teams, each of which must play exactly one match against each of the other 31 teams?

We conclude this section with a theorem of Dénes König, a Hungarian mathematician who wrote the first comprehensive treatise on graph theory, *Theorie der Endlichen und Unendlichen Graphen* (Theory of Finite and Infinite Graphs) in 1936. His theorem tells us that the edges of any *bipartite* graph (not necessarily simple) with maximum vertex degree d can be coloured with just d colours.

Theorem 13.5: König's Theorem

Let G be a bipartite graph whose maximum vertex degree is d. Then

$$\chi'(G) = d.$$

Proof The proof is by mathematical induction on m, the number of edges of G.

Step 1 The statement is true for $m = 1$ since, for the bipartite graph G with one edge, $\chi'(G) = 1$ and $d = 1$.

Step 2 We assume that $\chi'(G) = d$ for all bipartite graphs with fewer than m edges. We wish to show that $\chi'(G) = d$ for all bipartite graphs with m edges.

Let G be a bipartite graph with m edges and maximum vertex degree d, and let H be the graph obtained from G by removing an edge e adjacent to the vertices v and w:

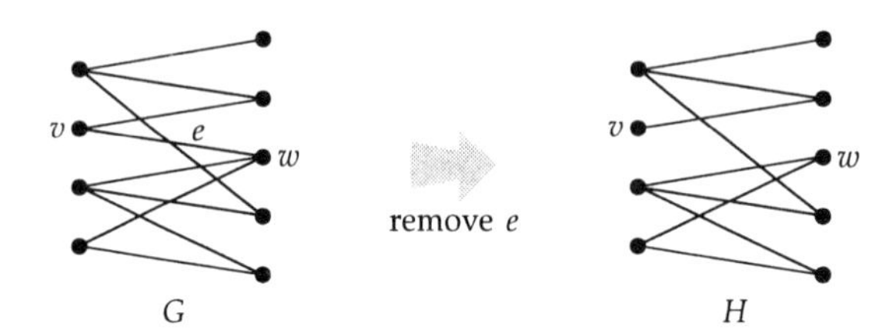

Since H has fewer than m edges and maximum vertex degree d (or less), it follows from our assumption that $\chi'(H) \leq d$; that is, H is d-edge colourable. We now colour the edges of H with d colours, and replace the edge e. If we can colour e with one of the d colours, then we obtain a d-edge colouring of G, as required.

To show that the edge e can always be coloured in this way, we argue as follows. Since H is obtained from G by removing the edge e, there must be at least one colour missing at v, and at least one colour missing at w.

If there is some colour missing at *both* v and w, then we can assign this colour to the edge e, thereby completing the d-edge colouring of G.

If there is no colour missing at *both* v and w, suppose that the colour blue is missing at v, and the colour red is missing at w, and consider the path starting at v and consisting entirely of red and blue edges. The edges in such a path must alternate in colour, and must alternate between the vertices on the left and those on the right of the bipartite graph. Since there are no blue edges at v, the colour red must appear there. It follows that w *cannot* be reached from v by such a red-blue path, since w would have to be reached by a red edge.

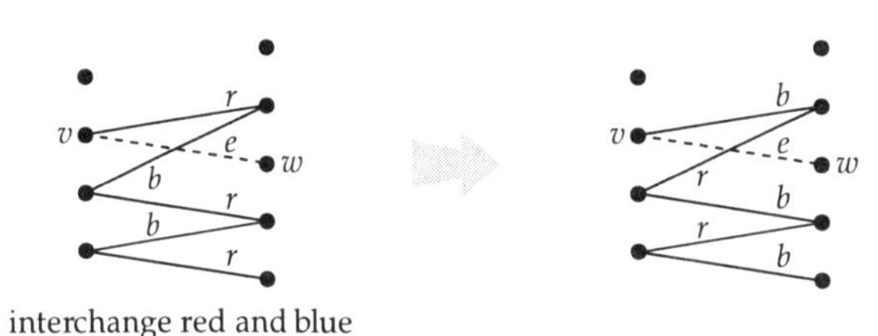

We now interchange the colours on this path, so that the blue edges become red, and the red edges become blue. Then the colours appearing at w are unchanged, and the colour red is now missing at both v and w. We can therefore assign to the edge e the colour red, thereby completing the colouring of the edges of G.

It follows that the statement is true for all bipartite graphs with m edges. This completes Step 2.

Therefore, by the principle of mathematical induction, the statement is true for all bipartite graphs with m edges, for each positive integer m. ■

Problem 13.8

Use König's theorem to write down the chromatic index of each of the following graphs:

(a) the complete bipartite graph $K_{r,s}$ $(r \leq s)$;
(b) the graph of the cube;
(c) the k-cube Q_k.

13.2 Algorithm for Edge Colouring

In Section 12.2 we presented a greedy algorithm for vertex colouring. We now present a corresponding greedy algorithm for edge colouring.

Greedy algorithm for edge colouring

START with a graph G and a list of colours 1, 2, 3, ...

Step 1 Label the edges $a, b, c, \ldots$ in any manner.
Step 2 Identify the uncoloured edge labelled with the earliest letter in the alphabet; colour it with the first colour in the list not used for any coloured edge that meets it at a vertex.

Repeat Step 2 until all the edges are coloured, then STOP.
An edge colouring of G has been obtained. The number of colours used depends on the labelling chosen for the edges in Step 1.

We present two examples using the same graph with different labellings.

Illustration A

Find an edge colouring of the following graph G.

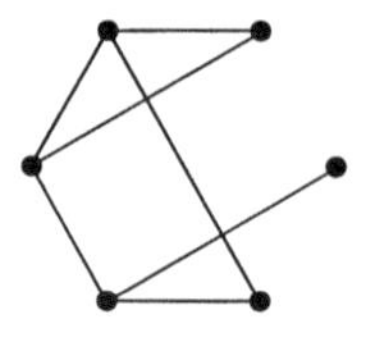

Step 1 We label the edges $a, ..., g$ as follows.

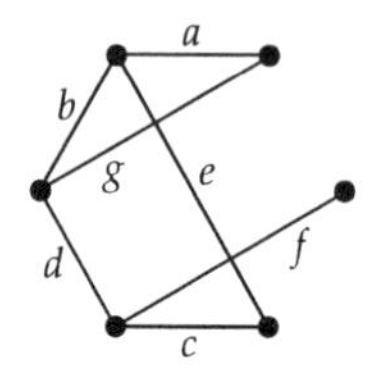

Step 2 We successively colour

edge a with colour 1,
edge b with colour 2,
edge c with colour 1,
edge d with colour 3,
edge e with colour 3,
edge f with colour 2,
edge g with colour 4.

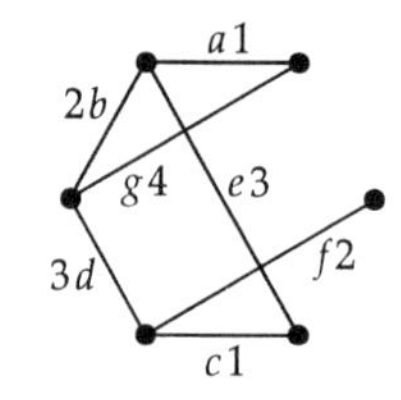

All the edges are now coloured, so we STOP.

We thus obtain the 4-edge colouring of G shown above. ❐

Illustration B

Find an edge colouring of the following graph G.

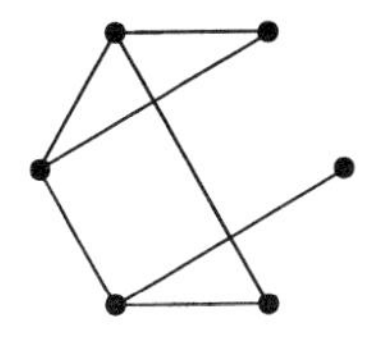

Step 1 We label the edges $a, \ldots, g$ as follows.

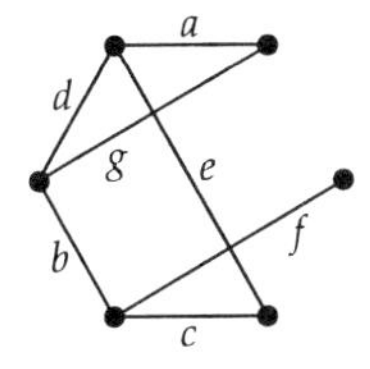

Step 2 We successively colour

edge a with colour 1,
edge b with colour 1,
edge c with colour 2,
edge d with colour 2,
edge e with colour 3,
edge f with colour 3,
edge g with colour 3.

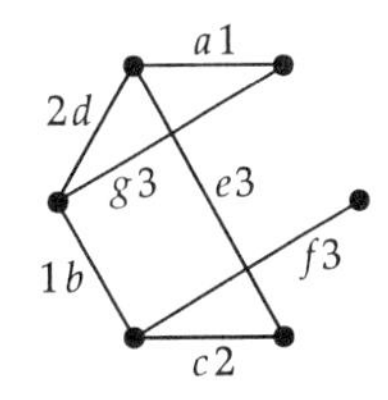

All the edges are now coloured, so we STOP.

We thus obtain the 3-edge colouring of G shown above. ❐

Notice that, in the above examples, $\chi'(G) = 3$, and in Illustration B we found an edge colouring of G that uses 3 colours.

Problem 13.9

Use the greedy algorithm to colour the edges of the following graph G, using each of the given labellings.

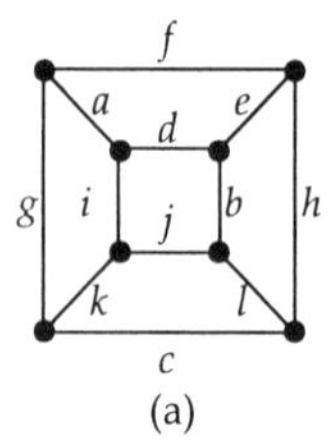

(a)

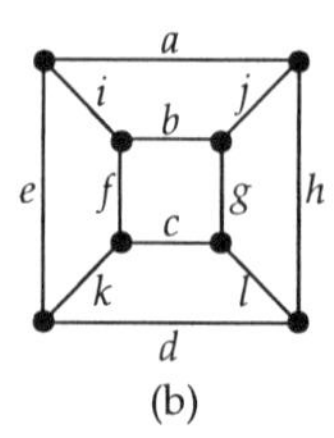

(b)

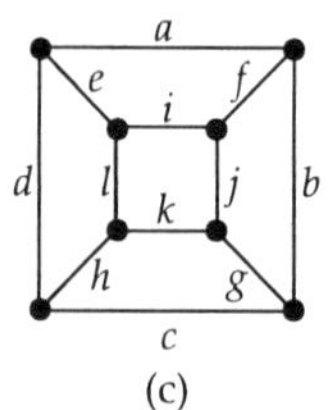

(c)

What is the actual value of $\chi'(G)$?

More generally, we have the following theorem.

Theorem 13.6

For any graph G, there is a labelling of the edges for which the greedy algorithm yields an edge colouring with $\chi'(G)$ colours.

Outline of Proof

Take any edge colouring of G with $\chi'(G)$ colours, denoted by 1, 2, 3, ..., and sequentially label with a, b, c, ... the edges coloured 1, then the edges coloured 2, then the edges coloured 3, and so on. For this labelling, the greedy algorithm assigns the colours 1, 2, 3, ... in that order, so only $\chi'(G)$ colours are needed. ■

Problem 13.10

Find a labelling of the edges of the following graph, for which the greedy algorithm yields an edge colouring of G with $\chi'(G)$ colours.

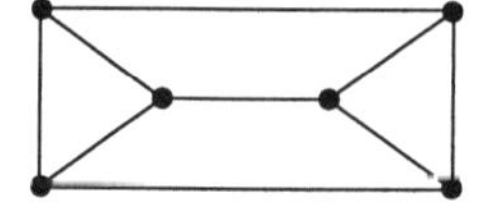

13.3 Edge Decompositions

Some of the most interesting problems in graph theory involve the decomposition of a graph G into subgraphs of a particular type. In several of these problems, we split the set of *edges* of G into disjoint subsets; this is called an **edge decomposition** of G.

For example, consider the following disconnected graph G.

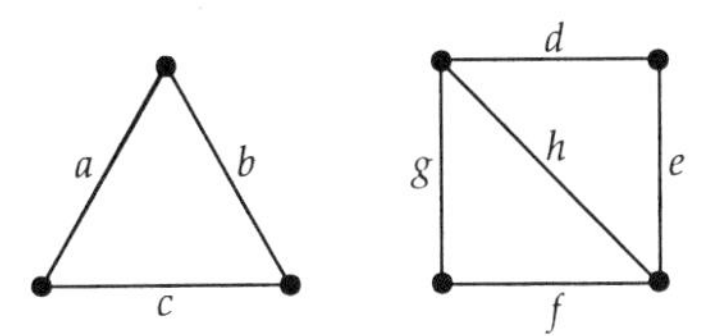

A natural edge decomposition is to split the set of edges into disjoint subsets that correspond to the components of G:

$\{a, b, c\}, \{d, e, f, g, h\}$.

Another natural edge decomposition arises from the idea of an Eulerian graph. In Chapter 3 we investigated conditions under which a given connected graph is Eulerian, and showed that every Eulerian graph can be split into disjoint cycles – this means that we can split the set of edges of G into disjoint subsets.

For example, for the Eulerian graph G shown below, there are five edge decompositions of G into disjoint cycles:

$\{a, b, c, d, e, f\}, \{g, h, i\}$;

$\{a, f, i\}, \{b, c, g\}, \{d, e, h\}$;

$\{a, f, h, g\}, \{b, c, d, e, i\}$;

$\{b, c, h, i\}, \{a, f, e, d, g\}$;

$\{d, e, i, g\}, \{a, b, c, h, f\}$. ❐

In this section, we adopt a similar approach to several other problems. Each problem can be formulated in graph-theoretic terms, and involves splitting the set of edges of a graph into disjoint subsets with particular properties. By doing this, we observe similarities between seemingly different problems and can begin to classify them, thereby gaining insight into the nature of the different types of problem.

Decomposition Into Matchings

The following diagram shows the cube graph and three sets of edges indicated by thick lines.

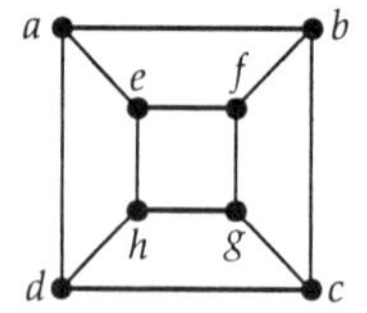

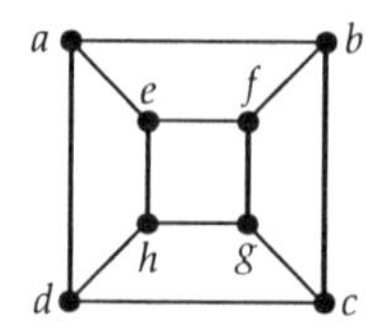

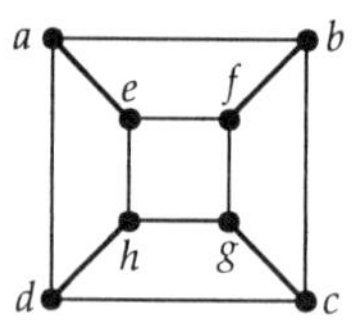

These three sets have the property that each edge of the graph appears in just one of them, and this leads to the following edge decomposition:

$$\{ab, cd, ef, gh\}, \{ad, bc, eh, fg\}, \{ac, bf, cg, dh\}.$$

Each of the above sets consists of edges that have no vertex in common. Such a set of edges is called a *matching*.

> **Definition**
> A **matching** in a graph G is a set of edges of G, no two of which have a vertex in common.

Every graph can be decomposed into matchings, since if there are m edges, then we can simply take m matchings, each consisting of a single edge. However, the problem of determining the *minimum* number of matchings needed to decompose a given graph can be much more difficult, and is unsolved in general. This question is of more than academic interest, and has arisen in several contexts, two of which we consider below.

Notice that the problem of decomposing a graph into the minimum number of matchings is an edge colouring problem in which the edges of each matching are assigned the same colour.

Example 13.1: Wire Colouring

In Section 13.1 we considered a display panel on which six electrical components $a, \ldots, f$ are mounted and then interconnected.

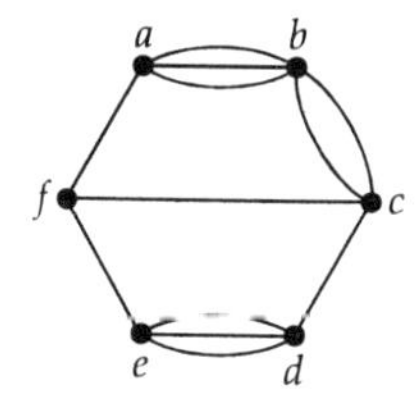

The connecting wires are first formed into a cable, with the wires to be connected to a emerging through one hole in the panel, those connected to b emerging through another hole, and so on. In order to distinguish the wires that emerge from the same hole, they are coloured differently.

In order to determine the minimum number of colours necessary for the whole system, we represented the connection points by the vertices of a graph and the wires by edges. We found that five colours are necessary to colour the wires in the system. The following diagram shows the edges of each colour.

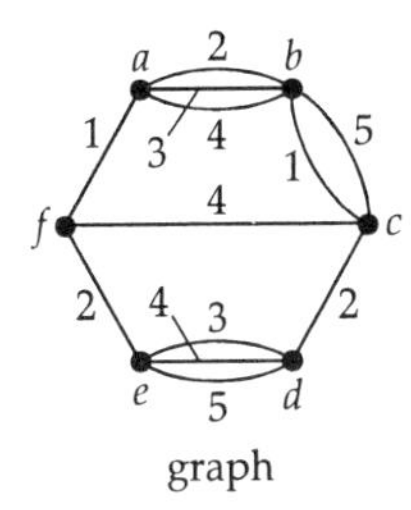

graph

The edge decomposition corresponding to the above edge colouring consists of the five subsets of edges coloured with each of the five colours:

$$\{af, bc\}, \{ab, cd, ef\}, \{ab, cd\}, \{ab, cf, de\}, \{bc, de\}.$$ ❐

In a wire colouring problem, the edges of each colour form a matching, so the problem of finding the smallest number of colours needed to colour the wires is the same as that of determining the minimum number of matchings needed to decompose the graph. In other words, it is an edge decomposition of the graph in which *the edges in each subset form a matching.*

Since the graphs considered in wire colouring problems usually have multiple edges, the best we can say is that the number of matchings is limited by the bounds for the chromatic index given by the extended version of Vizing's theorem (Theorem 13.2) and Shannon's theorem (Theorem 13.3):

$$d \le \chi'(G) \le d + h \quad \text{and} \quad d \le \chi'(G) \le \tfrac{3}{2}\, d,$$

where d is the maximum vertex degree in the graph G and h is the maximum number of edges joining a pair of vertices.

It is possible to find graphs attaining any of these bounds, so we cannot obtain better results than this in general.

Example 13.2: Scheduling Examinations

At the end of an academic year, all students have to take an hour-long examination with each of their tutors. How many examination periods are required?

To see what is involved, consider a simple example with four students a, b, c, d and three tutors A, B, C. We represent the students and tutors by the vertices of a bipartite graph, and join a student vertex to a tutor vertex whenever the student needs to be examined by the tutor. An example of such a graph is:

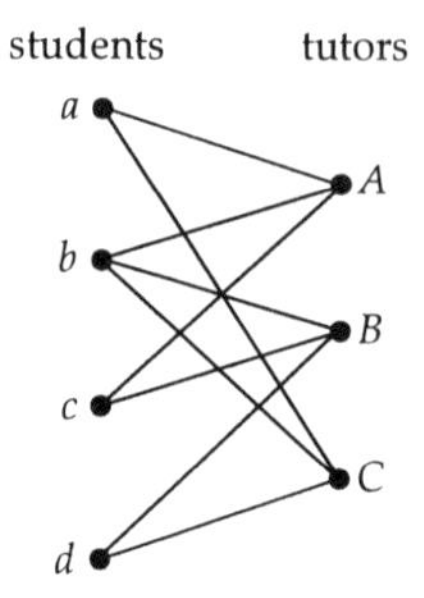

If two edges meet at a common vertex, then the corresponding examinations cannot take place simultaneously. So the problem is an edge decomposition problem in which we must split the graph into subgraphs in which no two edges meet – that is, into matchings. In this particular case, the minimum number of matchings is 3, and a suitable timetable is as follows.

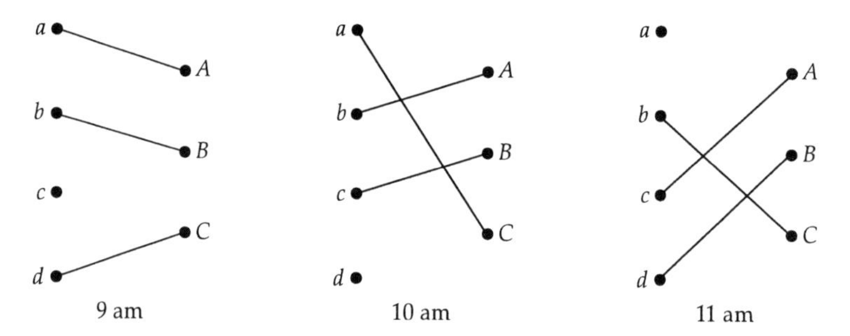

The corresponding edge decomposition is

$$\{aA, bB, dC\}, \{aC, bA, cB\}, \{bC, cA, dB\}.$$

This problem can also be thought of as an edge colouring problem. If we colour the 9 am edges red, the 10 am edges yellow, and the 11 am edges blue, then the colours appearing at each vertex (student or tutor) are different. All edges of the same colour form a matching. ❐

In a scheduling problem of the above type, the graphs under consideration are bipartite graphs. The problem therefore reduces to that of finding the chromatic index of a bipartite graph, and this problem is answered by König's theorem (Theorem 13.5) – the smallest number of matchings needed is equal to the largest vertex degree in the bipartite graph.

> **Problem 13.11**
>
> Five students $a, \ldots, e$, are to be examined by five tutors $A, \ldots, E$:
>
> tutor A must examine students b and d;
> tutor B must examine students a, b and e;
> tutor C must examine students b, c and e;
> tutor D must examine students a and c;
> tutor E must examine students b, d and e.
>
> If each examination takes the same amount of time, find the minimum number of examination periods needed, and devise a suitable schedule.

Decomposition Into Planar Subgraphs

Printed Circuits Problem

Recall that in printed circuits, electronic components are connected by conducting strips printed directly onto a flat board of insulating material. Such printed connectors may not cross, since this would lead to undesirable electrical contact at crossing points.

Circuits in which many crossings are unavoidable may be printed on several boards that are then sandwiched together. Each board consists of a printed circuit without crossings. What is the smallest number of such layers needed for a given circuit?

We illustrate this problem with a particular example.

Example 13.3: Printed Circuits

Consider a printed circuit that has 36 interconnections and is represented by the complete graph K_9. It is impossible to arrange all these interconnections in one layer, or even two; three layers are needed, and a solution is given below. Note that each edge of K_9 is included on just one of the layers – for example, the edge 28 appears on layer 2, and the edge 69 appears on layer 3.

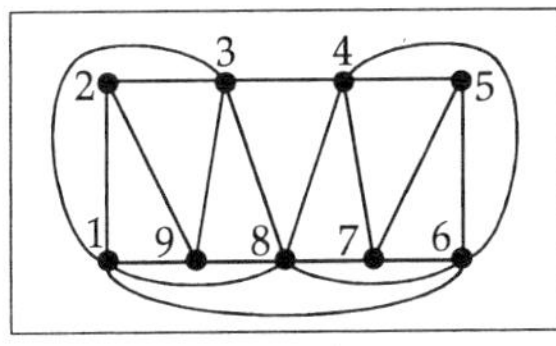

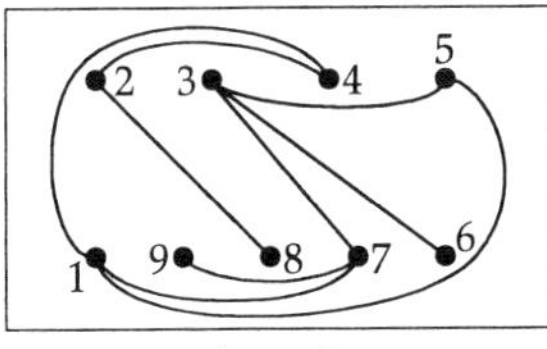

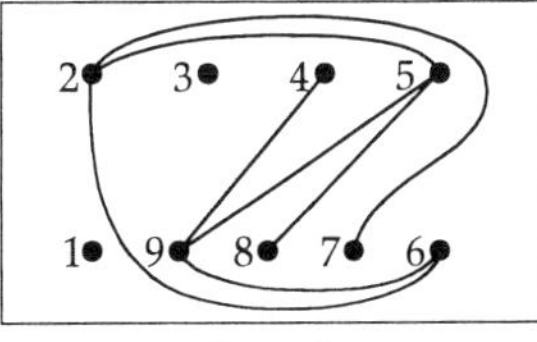

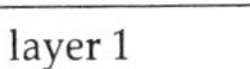
layer 1 layer 2 layer 3

Each of these three graphs is a planar graph. So the printed circuits problem reduces to that of decomposing the graph into smaller graphs, each of which is planar. In other words, it is an edge decomposition problem in which *the edges in each subset form a planar graph*. In the case of K_9, we get the following edge decomposition corresponding to the three layers shown above.

{12, 13, 16, 18, 19, 23, 29, 34, 38, 39, 45, 46, 47, 48, 56, 57, 67, 68, 78, 89},
{14, 15, 17, 24, 28, 35, 36, 37, 79},
{25, 26, 27, 49, 58, 59, 69}. ❐

Problem 13.12

Show that K_6 can be 'printed' in two layers, and write down a corresponding edge decomposition.

The above idea of splitting a graph into planar graphs leads us to define the **thickness** of a graph G, denoted by $t(G)$, to be the minimum number of planar graphs that can be superimposed to form G. For example, the thickness of any planar graph is 1, and the thickness of the complete graph K_9 is 3.

Problem 13.13

Determine the thickness of each of the following graphs:

(a) the complete graph K_5;
(b) the complete bipartite graph $K_{3,3}$;
(c) the Petersen graph.

In general, there is no known formula that gives the thickness of a graph G. However, we can easily obtain a *lower bound* for $t(G)$ that often coincides with the correct value. We restrict our attention to simple graphs, since we can collapse multiple edges to a single edge and remove loops, as we did in Section 11.1. We adopt the following notation.

Let a be any positive number. Then $\lfloor a \rfloor$ is the integer obtained by 'rounding a down', and $\lceil a \rceil$ is the integer obtained by 'rounding a up'; for example,

$$\lfloor \pi \rfloor = 3,\ \lfloor 6.2 \rfloor = 6,\ \lfloor 4 \rfloor = 4 \quad \text{and} \quad \lceil \pi \rceil = 4,\ \lceil 6.2 \rceil = 7,\ \lceil 4 \rceil = 4.$$

The connection between these functions is given by

$$\lceil a/b \rceil = \lfloor (a/b) + (b-1)/b \rfloor;$$

for example,

$$\lceil 7/5 \rceil = \lfloor 7/5 + 4/5 \rfloor = \lfloor 11/5 \rfloor = 2.$$

Note that, if a is an integer, then $\lfloor a \rfloor = \lceil a \rceil = a$.

We can now prove the following result.

Theorem 13.7

Let G be a simple connected graph with n (≥ 3) vertices and m edges. Then

(a) $t(G) \geq \lceil m/(3n-6) \rceil$;

(b) $t(G) \geq \lceil m/(2n-4) \rceil$, if G has no triangles.

Proof

(a) By Corollary 11.1, the number of edges in a simple connected planar graph with n (≥ 3) vertices and m edges is at most $3n - 6$; thus the number of edges on each 'layer' of G is at most $3n - 6$. Since there are m edges altogether, the number of planar graphs must be at least $m/(3n - 6)$. However, the number of planar graphs is an integer, so $t(G) \geq \lceil m/(3n - 6) \rceil$.

(b) By Corollary 11.2, the number of edges in a simple connected planar graph with n (≥ 3) vertices, m edges and no triangles is at most $2n - 4$. Since there are m edges altogether, the number of planar graphs must be at least $m/(2n - 4)$. However, the number of planar graphs is an integer, so $t(G) \geq \lceil m/(2n - 4) \rceil$. ∎

We can now deduce lower bounds for the thickness of K_n and that of $K_{r,s}$.

Theorem 13.8

(a) $t(K_n) \geq \lfloor (n + 7)/6 \rfloor$;

(b) $t(K_{r,s}) \geq \lceil rs/(2r + 2s - 4) \rceil$.

Proof

(a) If $G = K_n$, then $m = \frac{1}{2}n(n-1)$. It follows from part (a) of Theorem 13.7 that

$$t(K_n) \geq \lceil \tfrac{1}{2} n(n - 1)/(3n - 6) \rceil$$

Using $\lceil a/b \rceil = \lfloor (a/b) + (b-1)/b \rfloor$, we can rewrite the expression on the right as follows:

$$\begin{aligned}
&\lceil \tfrac{1}{2} n(n-1)/(3n-6) \rceil \\
&= \lfloor \tfrac{1}{2} n(n-1)/(3n-6) + (3n-7)/(3n-6) \rfloor \\
&= \lfloor (\tfrac{1}{2}(n^2 - n) + (3n-7))/(3n-6) \rfloor \\
&= \lfloor ((n^2 - n) + (6n-14))/2(3n-6) \rfloor \\
&= \lfloor (n^2 + 5n - 14)/2(3n-6) \rfloor \\
&= \lfloor (n+7)(n-2)/6(n-2) \rfloor \\
&= \lfloor (n+7)/6 \rfloor
\end{aligned}$$

Thus

$$t(K_n) \geq \lfloor (n+7)/6 \rfloor.$$

(b) If $G = K_{r,s}$, then $m = rs$ and G has no triangles.
It follows from part (b) of Theorem 13.7 that

$$t(K_{r,s}) \geq \lceil m/(2n-4) \rceil = \lceil rs/(2(r+s)-4) \rceil.$$

Thus

$$t(K_{r,s}) \geq \lceil rs/(2r+2s-4) \rceil.$$

■

It can be shown that $t(K_n) = \lfloor (n+7)/6 \rfloor$ for *all* n, except for $n = 9$ and $n = 10$, when $t(K_n) = 3$.

It is not known whether the inequality in part (b) is always an equality, but it certainly is for all complete bipartite graphs with fewer than 48 vertices.

So, to sum up, although we cannot solve the printed circuits problem in general, we have obtained a lower bound for the solution, and this bound coincides with the correct value surprisingly often.

Decomposition Into Spanning Subgraphs

Bus Route Problems

In a certain county there are a number of rival bus companies. Each company wishes to run a service that includes every town in the county, in such a way that passengers using that company can get from any town to any other town. However, the County Council will not allow different companies to operate along the same stretch of road. How many companies can be accommodated?

We solve this problem by drawing a graph whose vertices correspond to the towns and whose edges correspond to the roads joining them.

Example 13.4: Bus Routes

The following graph represents a county containing 11 towns joined by 22 roads.

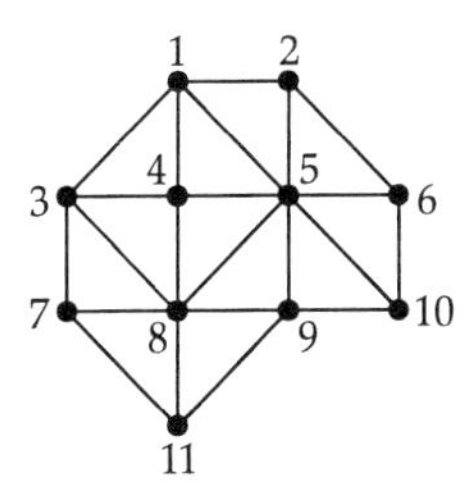

Each bus company needs a network that connects all 11 towns, so each company must be assigned at least 10 of the interconnecting roads. Since there are only 22 roads, the maximum number of companies that can be accommodated is 2. The following diagram shows an appropriate allocation of roads to the two companies.

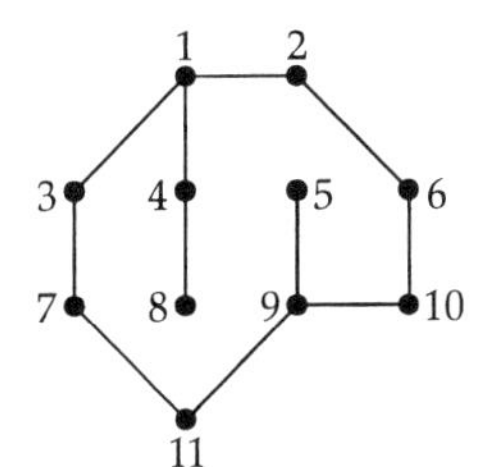

Red Devil bus company

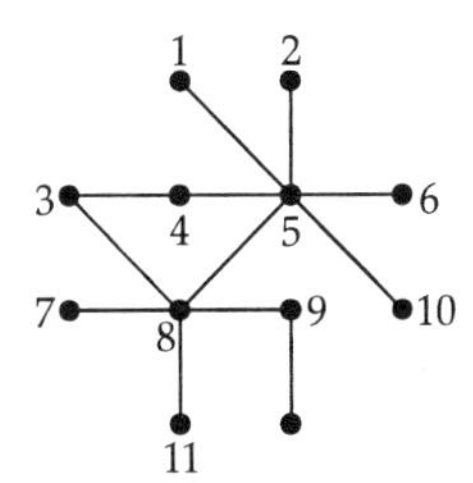

Purple Peril bus company

❐

Such an allocation of roads to companies produces an edge decomposition of the original graph. Each subgraph in this decomposition must include edges incident with all the vertices and must be connected, so that a passenger can travel from any town to any other by the buses of each company. So the problem reduces to that of decomposing the graph into the maximum number of connected subgraphs, each of which includes every vertex of the graph; such subgraphs are called *spanning subgraphs*.

We denote the number of spanning subgraphs of a graph G by $\boldsymbol{s(G)}$. An expression for the number $s(G)$ was obtained by W. T. Tutte, who proved the following result in 1961.

Theorem 13.9

Let G be a connected graph with n vertices. Then $s(G)$ is the largest integer for which the following statement is true:

for each positive integer $k \leq n$, at least $(k - 1) \times s(G)$ edges must be removed in order to disconnect G into k components.

To illustrate this result, we consider the following graph G, for which $s(G) = 2$, as we saw in Example 13.4.

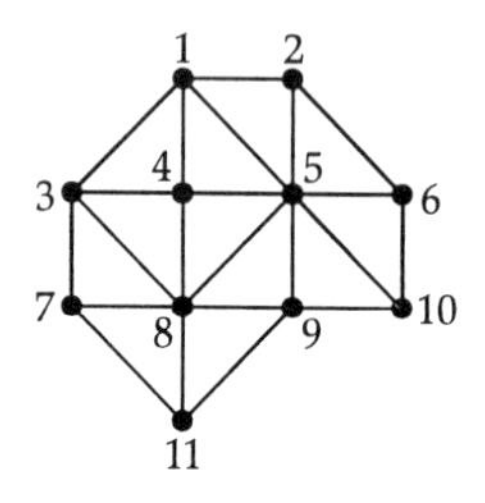

To disconnect G into

2 components, we must remove at least 3 edges, so

$$s(G) \leq 3/(2-1) = 3;$$

3 components, we must remove at least 5 edges, so

$$s(G) \leq 5/(3-1) = 5/2;$$

4 components, we must remove at least 7 edges, so

$$s(G) \leq 7/(4-1) = 7/3;$$

$$\vdots$$

11 components, we must remove all 22 edges, so

$$s(G) \leq 22/(11-1) = 22/10.$$

The largest integer $s(G)$ that satisfies all these inequalities is 2.

The formal proof of the above result is too complicated to include here, but the following remark indicates why the condition is necessary.

Assume that the graph G has been disconnected into k components by the removal of r edges. In order to have a connected system, each bus company must have at least $k - 1$ linking edges between the various components. Thus, if there are $s(G)$ bus companies, then

$$r \geq (k - 1) \times s(G).$$

Problem 13.14

Find the value of $s(G)$ for the following road network G.

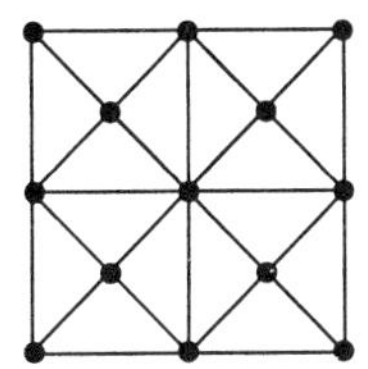

Decomposition into Spanning Trees

Several variations of the above problem lead to interesting mathematical results. For example, suppose that each bus company operates from a depot in one of the towns and chooses each of its routes to be a path out to another vertex, returning the same way. Then *each of the connected subgraphs must be a tree* – in other words, the graph can be decomposed into spanning trees. Such a decomposition is possible only when the number of edges in the graph is a multiple of the number of edges in a spanning tree; if the graph has n vertices and m edges, then m must be a multiple of $n - 1$.

Example 13.5: Bus Routes – A Variation

In the above example, where $n = 11$ and $m = 22$, the graph can be decomposed into spanning trees only if two of the roads are not used by either company. For example, if the roads 3–8 and 5–6 are removed from the graph, the resulting graph can be decomposed into the following spanning trees.

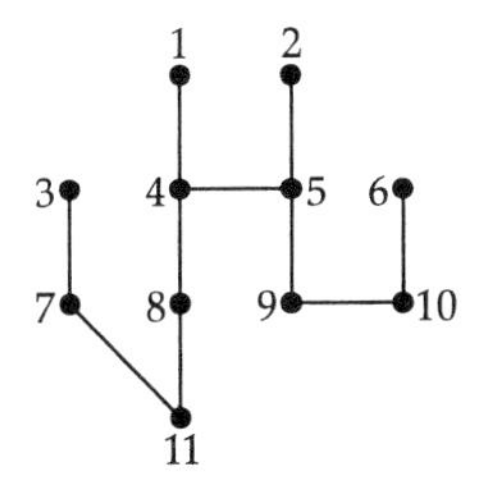

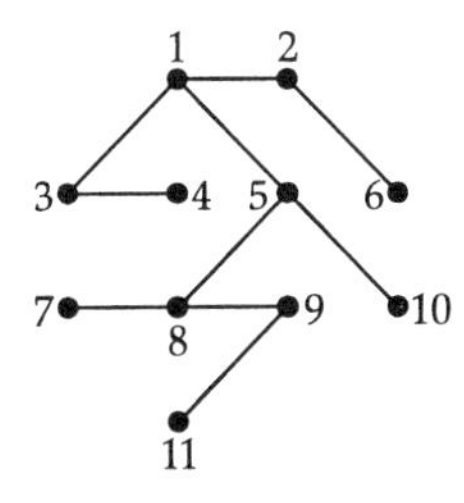

❒

Problem 13.15

Decompose the following graph into disjoint spanning trees.

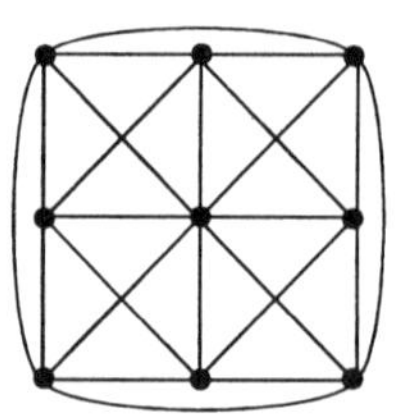

The following theorem gives a necessary and sufficient condition for the existence of a decomposition into spanning trees.

Theorem 13.10

Let G be a connected graph with n vertices and $s(n-1)$ edges. Then G can be decomposed into s spanning trees if and only if

for each positive integer $k \leq n$, at least $(k-1) \times s$ edges must be removed in order to disconnect G into k components.

Proof By Theorem 13.9, this theorem asserts that G can be decomposed into s spanning trees if and only if $s = s(G)$. However, if G can be decomposed into s connected subgraphs each of which includes every vertex of the graph, then each such subgraph must have $n-1$ edges, and must therefore be a spanning tree, since there are no edges left to form any cycles. ■

We have now found an expression for the maximum number of bus companies that can be accommodated in the first type of problem, and we have obtained a necessary and sufficient condition for the existence of a solution to the second type of problem.

Exercises 13

Edge Colourings

13.1 Determine $\chi'(G)$ for the following graph G.

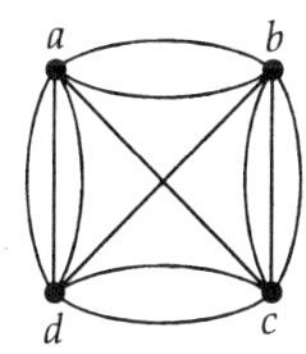

13.2 Determine $\chi'(G)$ for each of the following graphs G.

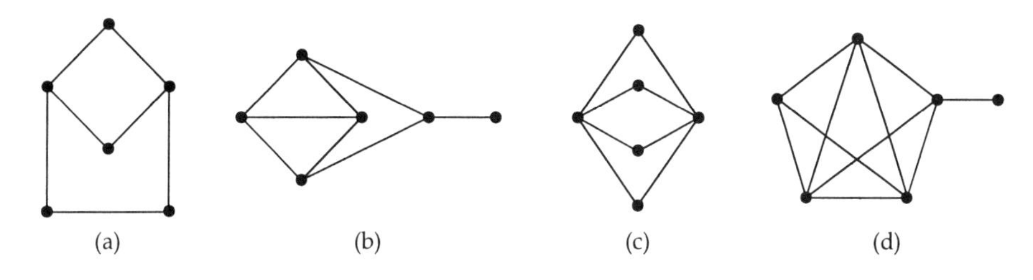

13.3 Consider the following graph G.

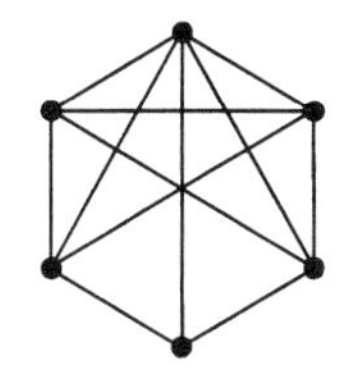

(a) Use Vizing's theorem to obtain lower and upper bounds for $\chi'(G)$.
(b) What is the actual value of $\chi'(G)$?

13.4 Consider the following graph G.

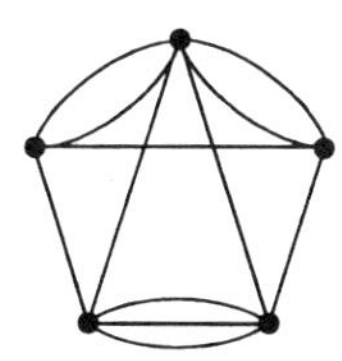

(a) Use the theorems of Vizing and Shannon to obtain lower and upper bounds for $\chi'(G)$.
(b) What is the actual value of $\chi'(G)$?

13.5 Prove that the Petersen graph has chromatic index 4.

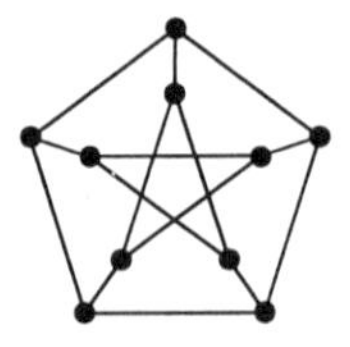

Hint Apply Vizing's theorem. Then assume that the chromatic index is 3, and note that there is essentially only one way to 3-edge colour the outside pentagon.

13.6 Let G be a 3-regular Hamiltonian graph. Show that $\chi'(G) = 3$.

Algorithm for Edge Colouring

13.7 Use the greedy colouring algorithm to colour the edges of each of the following labelled graphs.

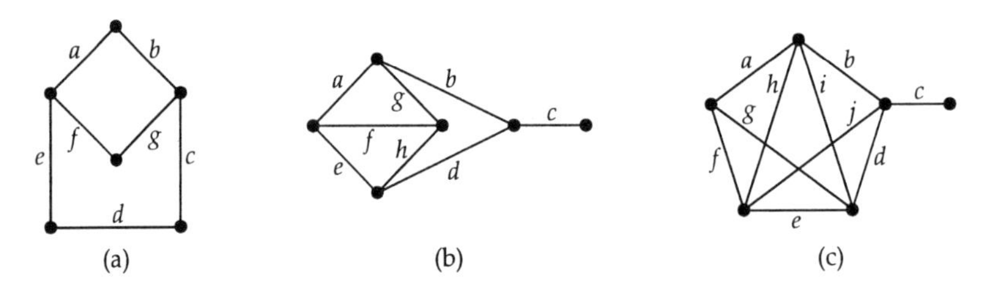

Comment on your results.

Edge Decompositions

13.8 Five students are to be examined by four tutors:

tutor A must examine students a, b and e;
tutor B must examine students a, c and d;
tutor C must examine students b, c and e;
tutor D must examine students b, c and d.

If each examination takes the same amount of time, find the minimum number of examination periods needed, and devise a suitable schedule.

13.9 For the octahedron graph shown below, find an edge decomposition into each of the following, where possible:

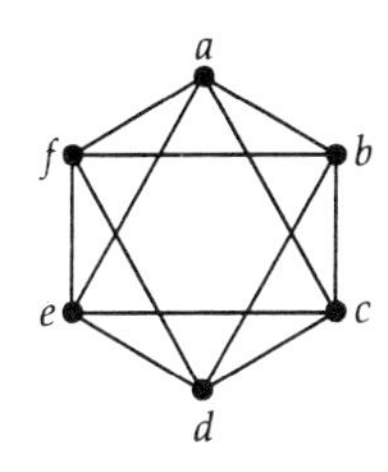

(a) disjoint cycles;
(b) planar subgraphs;
(c) subsets in which no two edges in any subset meet;
(d) connected subgraphs that include every vertex;
(e) spanning trees.

13.10 Show how the complete graph K_7 can be 'printed' in two layers, and write down a corresponding edge decomposition.

13.11 Use Theorem 13.8 to determine the thickness of each of the following graphs:

(a) K_{20} ; (b) $K_{20,20}$.

13.12 Determine the thickness of the complete bipartite graph $K_{10,40}$.

Hint To obtain an upper bound, split $K_{10,40}$ into a number of copies of the planar graph $K_{2,40}$.

13.13 Decompose the following graph into disjoint spanning trees.

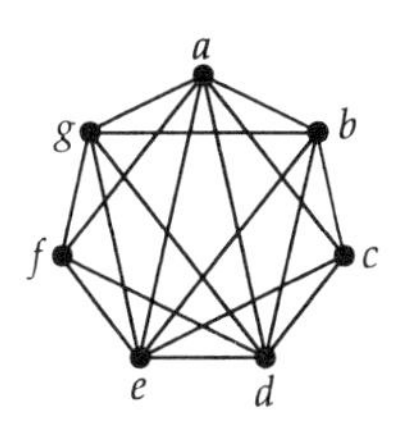

13.14 Verify that Theorem 13.10 holds for the following graph.

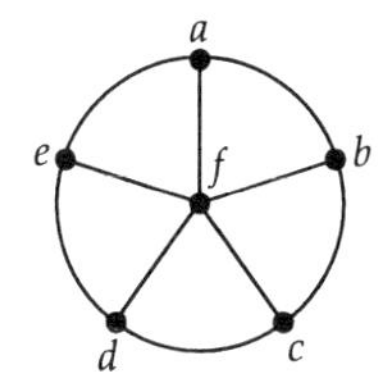

Chapter 14
Conclusion

After working through this chapter, you should be able to:

- explain what is meant by the *efficiency* of an algorithm;
- understand what are meant by a *P-problem*, an *NP-problem* and an *NP-complete problem*.

We conclude by presenting two classifications of some of the problems introduced in this book. First, we use the classification introduced in Chapter 1 and describe the problems as *existence, construction, enumeration* and *optimization* problems; we discuss briefly the ways of solving each class of problem. Then we introduce a new classification that indicates whether it is 'easy' or 'hard' to solve particular problems.

14.1 Classification of Problems

In Chapter 1 we classified problems as follows.

Existence problems	Does there exist ...? Is it possible to ...?
Construction problems	If ... exists, how can we construct it?
Enumeration problems	How many ... are there? Can we list them all?
Optimization problems	If there are several ..., which one is the best?

We now review some problems of each type.

Existence Problems

From a historical point of view, some of the problems that we now regard as part of graph theory arose as recreational puzzles. For instance:

Königsberg bridges problem (Chapters 1 and 3)
Does there exist a closed trail that crosses each of the seven bridges exactly once?

Knight's tour problem (Chapter 3)
Does there exist a sequence of knight's moves that visits each square of an 8×8 chessboard exactly once and returns to the starting point?

Utilities problem (Chapters 1 and 11)
Does there exist a way of connecting three neighbours to three utilities so that no two connections cross?

Four colour problem (Chapter 12)
Does there exist a map that requires more than four colours to colour it so that neighbouring countries are coloured differently?

The methods used to answer such questions vary considerably, even for different instances of the same problem. For example, if the answer is yes, as in the knight's tour problem for an 8×8 chessboard, then it is sufficient to produce a single example. This may not be easy to do in practice – for example, it may take a lot of trial and error to find a knight's tour – but once a single solution has been found, the problem is answered.

If the answer is no, then a different approach is required. It may be sufficient to make a simple observation, or we may need to produce a formal proof. For the Königsberg bridges problem, it is sufficient to observe that when we enter one part of the city we must be able to leave it again, so each vertex of the corresponding graph must have even degree; but in the corresponding graph each vertex has odd degree, so a solution cannot exist. For the utilities problem, we need to show that the complete bipartite graph is non-planar, and this can be done either directly or by using Euler's formula. Finally, in the case of the four colour problem, it was a major task lasting many years to prove that no map needing more than four colours exists.

It is instructive to generalize such specific problems. For example, we can generalize the Königsberg bridges problem by asking whether *any* given graph has an Eulerian trail. We answered this question in Theorem 3.1 which gives a simple test that can be used to determine whether a given connected graph is Eulerian.

We can pose the knight's tour problem for 'chessboards' of other sizes. There exist tours for some chessboards, but not for others; to establish whether or not a tour exists in a particular instance, we may have to guess the answer and then attempt to show that our guess is correct. Even more generally, we may ask whether *any* given graph has a Hamiltonian cycle. Unlike the Eulerian problem, no useful test is known for determining whether a given graph is Hamiltonian, although there are some sufficient conditions, such as those given by Dirac's or Ore's theorem, that work well in particular cases.

The problem of deciding whether a given graph is planar generalizes the utililities problem. We have a result that answers the question in principle – Kuratowski's theorem (Theorem 11.3): a graph is planar if and only if it contains no subdivision of K_3 or $K_{3,3}$. Unfortunately, it is usually very difficult to recognize subdivisions of K_3 or $K_{3,3}$ in a given graph, so this theorem is almost useless in practice. We therefore have to resort to other means, such as using Euler's formula to show that a particular graph is non-planar. Alternatively, there are a number of planarity algorithms that can be used.

The problem of determining the chromatic number of a given graph generalizes the four colour problem. There is *no* simple method for determining the chromatic number of *any* given graph.

In general, we cannot even determine when two given graphs are essentially the same. There is no simple method for determining whether there exists an isomorphism between two given graphs, although we can answer this question in particular instances – for example, no isomorphism exists if the graphs have different degree sequences.

We note, finally, that for *every* property that a graph G may possess (planar, Eulerian, Hamiltonian. k-colourable, etc.), there is a corresponding existence problem; for example:

planar	does there exist a plane drawing of G?
Eulerian	does there exist an Eulerian trail in G?
Hamiltonian	does there exist a Hamiltonian cycle in G?
k-colourable	does there exist a k-colouring of G?

Construction Problems

For each type of problem, we may be able to construct the required solution for small instances of the problem by trial and error. For example, if you are given a graph with six vertices, then to determine whether it is Eulerian, planar or 3-colourable, it is probably easiest to use inspection, rather than a systematic method. However, for large graphs, we need to use an algorithm. In order to introduce the idea of an *efficient* algorithm, we need some way of specifying the number of operations that may be involved.

We say that a graph algorithm involves $\boldsymbol{O(n)}$ operations when the number of operations a computer uses in applying the algorithm to a given graph is at most cn, where c is some fixed constant (depending on the particular algorithm) and n is a parameter related to the input required (usually the number of vertices or edges of the given graph). Similarly, a graph algorithm involves

$O(n^k)$ operations when the number of operations a computer uses when applying it is at most cn^k, for some fixed constant c.

The following algorithms are examples of algorithms used for problems where a solution is known to exist and we wish to find one.

Spanning Tree Algorithms

In Chapter 6 we described two methods for finding a spanning tree in a connected graph. In the *building-up* method, we start with no edges and add edges one at a time in such a way that no cycles are created; in the *cutting-down* method, we start with the graph and remove one edge at a time in such a way that the resulting graph is never disconnected. These algorithms are easy to apply by hand, or to adapt for computer use, and are efficient algorithms involving $O(n^2)$ operations, where n is the number of vertices in the graph.

Fleury's Algorithm

This algorithm, presented in Chapter 9, is used to find an Eulerian trail in an Eulerian graph; it involves $O(m)$ operations, where m is the number of edges in the graph.

Enumeration Problems

The subject of graphical enumeration is a major one, although it has not featured prominently in this book. However, we have looked at a few important problems, which we now summarize.

Labelled Graphs

A simple graphical enumeration problem is that of determining the number of simple labelled graphs with n vertices. Since each of the $\frac{1}{2}n(n-1)$ possible edges is either present or absent, and the number of distinct subsets of k objects is 2^k, there are $2^{n(n-1)/2}$ such graphs. The number of simple labelled graphs with n vertices and m edges is the binomial coefficient

$$\binom{\frac{1}{2}n(n-1)}{m},$$

since each choice of m of the $\frac{1}{2}n(n-1)$ possible edges determines a different labelled graph with exactly m edges.

Labelled Digraphs

Similarly, there are $2^{n(n-1)}$ labelled digraphs with n vertices, and the number of labelled digraphs with n vertices and m arcs is the binomial coefficient

$$\binom{n(n-1)}{m}.$$

Labelled Trees

In Chapter 7 we proved that the number of labelled trees with n vertices is n^{n-2} (Cayley's theorem).

Unlabelled Graphs

The corresponding enumeration problems for unlabelled graphs are far more difficult. The numbers of simple graphs of various types are given in the following table.

Number of vertices	1	2	3	4	5	6	7	8
labelled graphs	1	2	8	64	1024	32768	2097152	268435456
unlabelled graphs	1	2	4	11	34	156	1044	12346
unlabelled connected graphs	1	1	2	6	21	112	853	11117
unlabelled regular graphs	1	2	2	4	3	8	6	20
unlabelled Eulerian graphs	1	0	1	1	4	8	37	184
unlabelled Hamiltonian graphs	1	0	1	3	8	48	383	6020
labelled trees	1	1	3	16	125	1296	16807	262144
unlabelled trees	1	1	1	2	3	6	11	23
labelled digraphs	1	4	64	4096	2^{20}	2^{30}	2^{42}	2^{56}
unlabelled digraphs	1	3	16	218	9608	1540944	$\sim 9\times 10^{9}$	$\sim 2\times 10^{12}$

In practice, we are interested not just in counting various types of graph, but also the number of solutions to various problems. For example, in Chapter 7 we considered the problems of counting binary trees and counting alkanes. Also, it follows from our discussion in Chapter 6 that the number of distinct minimum bracings of a rectangular framework is the number of non-isomorphic spanning trees in the corresponding bipartite graph. We may also be interested in other attributes of a given graph; for example, the number of minimum connectors, or the number of shortest paths between two given vertices.

Optimization Problems

We now review some problems for which solutions are known to exist, and we wish to find the 'best' solution.

Minimum Connector Problem

We introduced this problem in Chapter 8, and presented Prim's algorithm for its solution. This algorithm is efficient and involves $O(n^2)$ operations, where n is the number of vertices of the graph.

Travelling Salesman Problem

There is no efficient algorithm known for the travelling saleman problem. As we saw in Chapter 8, lower and upper bounds for the solution can be derived fairly easily.

Shortest Path Problem

In Chapter 9 we presented an algorithm for finding the shortest path(s) between two vertices of a weighted digraph. This is an efficient algorithm that involves $O(n^2)$ operations, where n is the number of vertices of the digraph.

Longest Path Problem

We can modify the shortest path algorithm to obtain an algorithm for finding the *longest* path(s) between two vertices of a weighted digraph that does not contain a cycle. (The concept of a *longest* path is needed in certain scheduling problems.) However, the procedure does not terminate for a digraph containing a cycle; such a digraph has no longest path, because we can traverse the cycle indefinitely.

Chinese Postman Problem

In Chapter 9 we showed how to solve this problem for simple instances. An algorithm employing computations on shortest paths and weighted matchings was developed by J. Edmonds; it involves $O(n^3)$ operations, where n is the number of vertices of the graph.

Dominating Set Problem

In Chapter 12 we introduced the idea of a *dominating set* of vertices of a graph G – a subset of the set of *vertices* of G such that each *vertex* of G is adjacent to at least one of the vertices in the subset; we then looked for *minimum* dominating sets. There is no known algorithm for solving this problem in general.

Edge-dominating Set Problem

Similarly, we can define an *edge-dominating set* of a graph G to be a subset of the set of *edges* of G such that each *edge* of G has at least one endpoint in common with some edge in the subset; we can then look for *minimum* edge-dominating sets. There is no known algorithm for solving this problem in general.

14.2 Efficiency of Algorithms

Next, we consider the nature and efficiency of some of the algorithms presented in this book.

Once an algorithm has been constructed for solving a particular problem, a number of questions arise as to the efficient implementation of the algorithm:

- can moderately large instances of the problem be solved in a reasonable time?
- can we construct another algorithm that solves the problem more quickly?

What does it mean to say that an algorithm is *efficient*, and why is this important?

In this book you have met many algorithms, some more efficient than others. For some problems, we can find an algorithm where the time taken is proportional to n^2, or to n^3, or more generally to n^k, for some fixed number k; here, n is some parameter associated with the amount of input data required for the problem, such as the number of cities, or the number of vertices of a graph.

Such algorithms are called **polynomial-time algorithms**, because the time taken is bounded by a polynomial in n. Examples of polynomial-time algorithms are Prim's algorithm for the minimum connector problem, and the algorithm given for the shortest path problem.

Algorithms with a time proportional to a *power* of n (such as 2^n, 3^n) or to $n!$ are known as **exponential-time algorithms**.

The times taken by exponential-time algorithms for problem instances of moderate size may be so long that problems for which no polynomial-time algorithm exists are generally considered to be intractable. So we make the distinction between *polynomial-time algorithms*, which are normally efficient, and *exponential-time algorithms*, which are usually so inefficient as to be of little practical use except for problem instances of small size.

Some caution must be exercised in applying this general conclusion to particular problems. For example, it is possible for the time taken by a polynomial-time algorithm to involve a large coefficient of proportionality (such as $10^{100}n$) or a very large exponent (such as n^{100}), so that the algorithm is not efficient in practice. However, such cases are rare, and most polynomial-time algorithms met in practice have a reasonably small coefficient and an exponent not greater than 3. Further, some exponential-time algorithms are sometimes efficient in practice.

The following table compares the approximate times for two polynomial-time algorithms (with times n and n^3) and two exponential-time-algorithms (with times 2^n and 3^n), using a computer performing a thousand operations per second:

	$n = 10$	$n = 50$
n	0.01 seconds	0.05 seconds
n^3	1 second	2 minutes
2^n	1 second	35 years
3^n	1 minute	2.3×10^{13} years

14.3 Another Classification of Problems

NP-Problems

We remarked earlier that nobody has been able to find a polynomial-time algorithm for the travelling salesman problem. Suppose that, instead of the usual travelling salesman problem:

find a sequence of cities that forms a Hamiltonian cycle of minimum length,

we consider the related problem:

is there a Hamiltonian cycle of total length less than k?

where k is some given number. This is called the corresponding **decision problem**.

Suppose that we have a particular instance of this problem, and that we are *given* a cycle whose total length is less than k. If we wish to *check* that the given cycle really does have total length less than k, then this is straightforward: the checking can certainly be done in polynomial time. (Consider the effort involved in *originating* the solution to a problem as compared to *checking* someone else's solution.) Any problem like this whose solution *when given* can be *checked* in polynomial time (even if it took exponential time to find that solution originally) is called a **non-deterministic polynomial-time problem**, abbreviated to **NP-problem**. The class of all such problems is denoted by **NP**.

In particular, any problem that can be solved in polynomial time is an NP-problem, because we can certainly *check* the given solution in polynomial time if it took only polynomial time to find it in the first place! A problem whose solution can be found using a polynomial-time algorithm is called a **polynomial-time problem** or **P-problem**. The class of all such problems is denoted by **P**.

To compare the performance of algorithms for problems in the class NP with that of algorithms for problems in the class P, logicians have introduced the concept of a *non-deterministic computer*. This is a model of computation that is deliberately unrealistic and does not correspond to any existing physical computing device. An ordinary computer is a *deterministic* machine – one in which the state at a particular time is determined in a predictable manner by the state and input to the machine at a previous time. A non-deterministic computer is a hypothetical device that has the remarkable ability to 'guess' the answer to a problem.

For the decision-problem form of the travelling salesman problem, the 'guess' takes the form of a sequence of cities. The computer can then verify that the 'guess' is a solution – that is, that the sequence forms a cycle with total length less than k. This *checking stage* can be carried out in polynomial time by a deterministic process. Thus the non-deterministic computer operates in two stages: a 'guessing' stage and a checking stage, in which the computer verifies that the 'guess' is in fact a solution.

The non-deterministic computer always 'guesses' correctly. The reason for having a checking stage in the non-deterministic computer is so that the performance of an algorithm for a non-deterministic computer can be compared with the performance of an algorithm for an ordinary deterministic machine.

It is assumed that the 'guessing' stage takes no appreciable time, so in this case, since the checking stage can be carried out in polynomial time, the non-deterministic computer can answer the decision form of the travelling salesman problem in polynomial time. Thus an NP-problem (non-deterministic polynomial-time problem) is one that can be solved in polynomial time on a non-deterministic computer.

We can use the non-deterministic computer to solve a decision problem – if the answer is yes, the computer produces a 'guess' and verifies it in polynomial time, while if the answer is no, the computer either produces this answer or does not stop running. For example, if we want to find the chromatic number of a graph G, then we can repeatedly bisect the range of the value of $\chi(G)$, asking: is $\chi(G)$ less than 20? is $\chi(G)$ between 11 and 20? and so on.

Another example of a problem that cannot be solved in polynomial time is the graph isomorphism decision problem:

given two graphs G and H, are they isomorphic?

To date, the best known algorithm for solving this problem is an exponential-time algorithm. But suppose that, for a particular instance of this problem, an isomorphism between G and H is *given*. Then it takes only polynomial time to verify that this one–one correspondence really is an isomorphism. A non-deterministic computer can therefore solve the graph isomorphism decision problem by producing a one–one correspondence in the 'guessing' stage, and by verifying that this is an isomorphism in the checking stage. Hence this problem is an NP-problem.

Some Important Results

We now state and discuss three important results concerning these classes of problems.

The class P is contained in NP

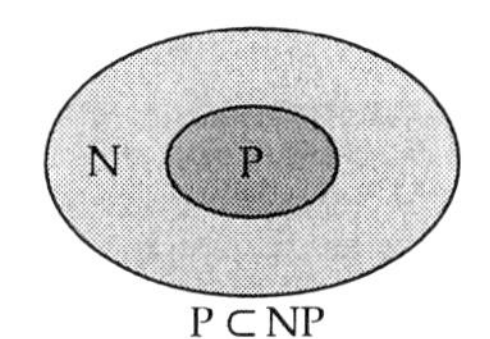

P ⊂ NP

Is P a proper subclass of NP?

Every decision problem in P is also in NP. This is obvious, since if a problem is solvable in polynomial time on a deterministic computer, we can solve it using a non-deterministic computer by ignoring the guessing stage and using the deterministic algorithm instead of the checking stage. The question arises as to whether there are any problems in NP that are not in P. Algorithms for a non-deterministic computer seem to be very much more powerful than algorithms for a deterministic computer, so we expect NP to contain more problems than P. However, nobody has yet been able to prove this, so it remains a conjecture, although one that is generally accepted to be true.

A polynomial-time algorithm for a non-deterministic computer can be converted to an exponential-time algorithm for a deterministic computer

To convert an algorithm for a non-deterministic computer to one for a deterministic machine, we must replace the guessing stage by a deterministic process. The only obvious way of doing this is to try all possible guesses. Unfortunately, the number of possible guesses is usually an exponential function of the problem size. It can be shown that a problem of size n in NP can be solved by an algorithm for a deterministic computer that has time-complexity $2^{p(n)}$, where $p(n)$ is a polynomial function. It seems likely, therefore, that there are problems in NP for which the only possible algorithms for a deterministic machine have exponential time complexity.

A decision problem is no harder than the corresponding optimization problem
We can associate a decision problem with any optimization problem. For example, if an optimization problem requires a solution which has minimum cost or minimum length, we can associate with it a decision problem which asks whether there is a solution whose cost or length is not more than a given bound k. Decision problems can be associated in a similar way with maximization problems by replacing 'not more than' by 'not less than'. Provided that the cost or length of a solution is easy to evaluate, a decision problem can be no harder than the corresponding optimization problem. For example, if we have solved an instance of the travelling salesman problem, all we have to do to answer the corresponding decision problem is to compare the length of a minimum cycle with the bound k in the decision problem. So, although the theory of NP-complete problems applies to decision problems, we can extend many of the results about the difficulty of a problem to the corresponding optimization problems.

Polynomial-Time Reducibility

We have seen that polynomial-time algorithms are generally considered to be efficient, and that exponential-time algorithms are generally considered to be inefficient. In comparing the difficulty of two problems, a useful technique is to try to reduce one problem to the other – that is, to find a transformation that converts any instance of one problem to an instance of the other. If such a transformation can be carried out by a polynomial-time algorithm, and if the first problem can be solved in polynomial time, then so can the second. Polynomial-time reducibility plays an important part in the theory of NP-complete problems, as we shall see shortly.

NP-Complete Problems

The basis of the theory of NP-completeness was provided in 1971 by Stephen Cook; he proved that one particular problem in NP, called the *satisfiability problem*, has the property that every other problem in NP can be polynomially reduced to it. The implication of this is that if the satisfiability problem can be solved in polynomial time (on a deterministic computer), then so can every other problem in NP, so NP = P. Also, if any problem in NP is intractable, in the sense that it can be solved only in exponential time, then the satisfiability problem must also be intractable, so NP is strictly larger than P.

Subsequently, a large number of problems (including the decision form of the travelling salesman problem) have been shown to share this property of the satisfiability problem. This class of problems is called the class of **NP-complete problems**.

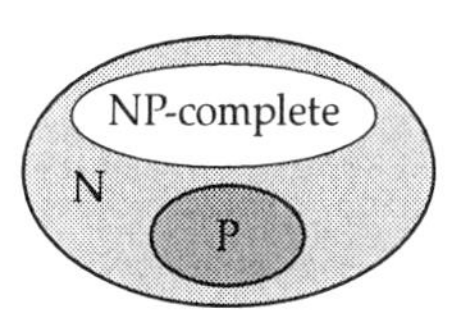

It follows that if a polynomial-time algorithm can be found for *any one* of these NP-complete problems, then *every one* of them must be solvable in polynomial time; conversely, if *any one* of them can be proved to be intractable, then they must *all* be intractable. The question of whether NP-complete problems are intractable is generally considered to be one of the principal unsolved problems of the theory of algorithms. However, as the number of problems shown to be NP-complete grows, and no polynomial-time algorithm is found for any of them, it seems more and more likely that NP-complete problems are intractable.

There are many problems that are known to be NP-complete. Those related to problems described in this book are listed at the end of this chapter.

But suppose that we are presented with a *new* problem. How do we attempt to classify it as a P-problem, NP-problem or NP-complete problem? It is prudent to try both the practical and theoretical approaches described in Chapter 1. One the one hand, we try to construct a polynomial-time algorithm to solve it: on the other, we try to prove that the problem is NP-complete.

Grant me
The serenity to accept the problems that I cannot solve
The persistence to solve the problems that I can
And the wisdom to tell the difference.

NP-Complete Problems

Travelling Salesman Problem (Chapter 1)

given a set C of cities, the distances between each pair of cities, and a positive integer k

problem is there a tour of C with total length not exceeding k?

Subgraph Isomorphism (related to Chapter 2)

given two graphs G_1 and G_2

problem does G_1 contain a subgraph isomorphic to G_2?

comment can be solved in polynomial time if G_1 is a forest and G_2 is a tree

Generalized 4 Cubes Problem (related to Chapter 2)

given a finite set C of k colours, and a set Q of k cubes, with each side of each cube in Q assigned a colour in C

problem can the cubes in Q be stacked in a vertical column such that each of the colours in C appears exactly once on each of the four sides of the column?

Hamiltonian Cycle (Chapter 3)

given a graph G

problem does G contain a Hamiltonian cycle?

Isomorphic Spanning Tree (related to Chapter 6)

given a graph G and a tree T

problem does G contain a spanning tree isomorphic to T?

comment remains NP-complete even if T is a path

Longest Path (related to Chapter 9)

given a digraph G, the length of each edge, a positive integer k, and two specified vertices s and t

problem is there an st-path in G of length at least k?

comment remains NP-complete when the length of each edge is 1, as does the corresponding problem for paths in graphs

Planar Subgraph (Chapter 11)

given a graph G with m edges and a positive integer $k < m$

problem is there a planar subgraph of G with the same vertices as G and more than k edges?

Graph k-Colourability (Chapter 12)

given a graph G with n vertices and a positive integer $k < n$

problem is G k-colourable?

comment solvable in polynomial time for $k = 2$

Dominating Set (Chapter 12)

given a graph G with n vertices and a positive integer $k < n$

problem is there a dominating set of size k or less for G? – that is, is there a subset S of V of size k or less such that each vertex in V is either in S or adjacent to a vertex in S?

comment the corresponding problem for trees is solvable in polynomial time

Suggestions for Further Reading

There are many books on graphs and digraphs and their applications. Two books at an elementary level are:

G Chartrand, *Introductory Graph Theory*, Dover, New York, 1985.
O Ore, *Graphs and their Uses*, revised ed., New Mathematical Library 10, Mathematical Association of America, Washington DC, 1990.

Standard texts in graph theory include:

J Clark and DA Holton, *A First Look at Graph Theory*, World Scientific Publishing, Singapore, 1991.
RJ Wilson, *Introduction to Graph Theory*, 4th ed., Addison-Wesley Longman, Harlow, Essex, 1996.
C Berge, *Graphs*, North-Holland, Amsterdam-New York, 1985.
JA Bondy and USR Murty, *Graph Theory with Applications*, American Elsevier, New York, 1979.
G Chartrand and L Lesniak, *Graphs & Digraphs*, 3rd ed., Wadsworth & Brooks/Cole, Monterey, California, 1996.
F Harary, *Graph Theory*, Addison-Wesley, Reading, Massachusetts, 1969.
D West, *Introduction to Graph Theory*, Prentice Hall, Upper Saddle River, NJ.
VK Balakrishnan, *Introductory Discrete Mathematics*, Prentice Hall International Inc., 1991.

A historical approach to graph theory can be found in:

NL Biggs, EK Lloyd and RJ Wilson, *Graph Theory 1736-1936*, paperback ed., Clarendon Press, Oxford, 1998.

Applications of graph theory and the use of algorithms are discussed in:

A Dolan and J Aldous, *Networks : An Introductory Approach*, John Wiley & Sons, New York, 1990.
G Chartrand and OR Oellermann, *Applied and Algorithmic Graph Theory*, McGraw-Hill, 1993.

SB Maurer and A Ralston, *Discrete Algorithmic Mathematics*, Addison-Wesley, 1991.

S Even, *Graph Algorithms*, Computer Science Press, Potomac, Maryland, 1979.

MR Garey and DS Johnson, *Computers and Intractability. A Guide to the Theory of NP-Completeness*, WH Freeman, San Francisco, 1979.

A Gibbons, *Algorithmic Graph Theory*, Cambridge University Press, Cambridge, 1985.

EL Lawler, JK Lenstra, AHG Rinnooy Kan and DB Shmoys (eds.), *The Travelling Salesman Problem*, John Wiley & Sons, New York, 1985.

FS Roberts, *Discrete Mathematical Models, with Applications to Social, Biological and Environmental Problems*, Prentice Hall, Englewood Cliffs, New Jersey, 1976.

A Tucker, *Applied Combinatorics*, 2nd ed., John Wiley & Sons, New York, 1984.

MN Swamy and K Thulasiraman, *Graphs, Networks and Algorithms*, John Wiley & Sons, New York, 1981.

RJ Wilson and LW Beineke (eds.), *Applications of Graph Theory*, Academic Press, London, 1979.

F Harary, RZ Norman and D Cartwright, *Structural Models: An Introduction to the Theory of Directed Graphs*, John Wiley & Sons, New York, 1965.

Specialist texts on some of the topics in this book include:

LW Beineke and RJ Wilson (eds.), *Selected Topics in Graph Theory*, Academic Press, London, Vol. 1, 1978, Vol. 2, 1983, Vol. 3, 1988.

TR Jensen and B Toft, *Graph Coloring Problems*, John Wiley & Sons, New York, 1995.

F Harary and EM Palmer, *Graphical Enumeration*, Academic Press, 1973.

RC Read and RJ Wilson, *An Atlas of Graphs*, Clarendon Press, Oxford, 1998.

The following book relates graph theory to other subjects:

LW Beineke and RJ Wilson (eds.), *Graph Connections*, Clarendon Press, Oxford, 1997.

Appendix

Methods of Proof

Mathematical Statements

The basic ingredients of mathematical reasoning are sentences called *statements*. A mathematical statement is a sentence that is either TRUE or FALSE.

To prove that a given statement is FALSE, it is enough to produce a single example for which the statement fails to be true; such an example is called a *counter-example*. For example, consider the statement:

Every bipartite graph is a tree.

To prove that this statement is FALSE, it is sufficent to produce just one counter-example, such as $K_{3,3}$.

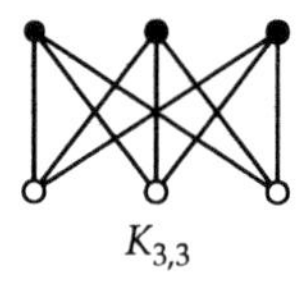

$K_{3,3}$

To prove that a given statement is TRUE, we need to provide a convincing argument, or *proof*, that covers all possibilities. For example, consider the statement

Every tree is a bipartite graph.

This statement is TRUE, and to establish this we must give a proof applicable to *all* trees.

Our aim here is to explain what such a proof entails, and to describe some methods of proof in the context of graph theory.

Methods of Proof

The four types of proof that appear in this book are *direct proofs*, *indirect proofs* (proofs by contradiction), *proofs by mathematical induction* and *if and only if* proofs. We look at each of these in turn.

Direct Proofs

In a *direct* proof (the most common type of proof), we start with the information we are given and proceed by logical steps, using known facts and theorems, to the result required. A simple example of such a proof is our proof of Theorem 2.2. (Most direct proofs are longer!)

Theorem A.1

Let G be an r-regular graph with n vertices. Then G has $nr/2$ edges.

Proof Let G be a graph with n vertices, each of degree r; then the sum of the degrees of all the vertices is nr. By the handshaking lemma, the number of edges is one-half of this sum, which is $nr/2$. ■

Indirect Proofs

Indirect proofs are often called *proofs by contradiction*, or *proofs by the method of 'reductio ad absurdum'*. We begin by assuming that the given statement is FALSE and follow a logical argument until we obtain a contradiction or absurdity; we conclude that the original statement must be TRUE. An example of such a proof is given in the solution to Problem 2.11(a).

Theorem A.2

In any graph, the number of vertices of odd degree is even.

Proof Let G be a graph in which the number of vertices of odd degree is odd; then the sum of all the vertex degrees is also odd. But we know from the handshaking lemma that the sum of the vertex degrees is twice the number of edges, and is therefore even. We have obtained a contradiction, so no such graph G exists. Thus, in any graph, the number of vertices of odd degree is even. ■

Proofs by Mathematical Induction

Many mathematical statements include an integer variable n. Suppose that we wish to prove a statement concerning a particular type of graph with n vertices – for example,

the complete graph K_n has $n(n-1)/2$ edges

or

a tree with n vertices has $n-1$ edges.

We wish to prove that such a statement is true for all allowable values of the integer n.

One approach to proving results of this kind is to use the *principle of mathematical induction*. (The method described here uses a version of this called the *principle of strong induction*.) Such a proof is in two parts:

Step 1 Show that the statement is true for the appropriate graph with one vertex;
Step 2 Show that, for each integer $n > 1$, if the statement is true for the appropriate graphs with fewer than n vertices, then it must also be true for the appropriate graphs with exactly n vertices.

We can thus deduce successively that:

since the statement is true for the appropriate graphs with fewer than two vertices (Step 1), it must be true for the appropriate graphs with two vertices (Step 2);
since the statement is true for the appropriate graphs with fewer than three vertices (shown above), it must be true for the appropriate graphs with three vertices (Step 2);
since the statement is true for the appropriate graphs with fewer than four vertices (shown above), it must be true for the appropriate graphs with four vertices (Step 2);
and so on.

We thus deduce, by the principle of mathematical induction, that the statement is true for the appropriate graphs with any given number of vertices n.

We illustrate the method by proving the following statement.

Theorem A.3

A tree with n vertices has $n - 1$ edges.

Proof

Step 1 The statement is true when $n = 1$, since the only tree with one vertex is K_1, which has no edges.

Step 2 We assume that the statement is true for trees with fewer than n vertices – that is, that every tree with k vertices has $k - 1$ edges whenever $k < n$. We wish to deduce that every tree T with n vertices has $n - 1$ edges.

To do this, we consider a tree T with n vertices and remove any edge e of T. Since T has no cycles, this disconnects T and gives two trees, with k_1 and k_2 vertices, say, where $k_1 + k_2 = n$. These trees have fewer than n vertices so, by our assumption, they have $k_1 - 1$ and $k_2 - 1$ edges, respectively. Reinstating the edge e, we restore T, with a total of

$$(k_1 - 1) + (k_2 - 1) + 1 = k_1 + k_2 - 1 = n - 1 \text{ edges.}$$

Thus if the statement is true for trees with fewer than n vertices, then it is true for trees with n vertices. This completes Step 2.

Therefore, by the principle of mathematical induction, the statement is true for all positive integers n. ■

A similar approach can be used when we wish to prove a result concerning particular types of graph with a general number of *edges*, rather than a general number of vertices. In such proofs, we usually replace the words 'with one vertex' in Step 1 by 'with no edges'. For example, we can adapt the proof of Theorem A.3 to show that every tree with m edges has $m + 1$ vertices.

Theorem A.4

A tree with m edges has $m + 1$ vertices.

Proof

Step 1 The statement is true when $m = 0$, since the only tree with no edges is K_1, which has one vertex.

Step 2 We assume that the statement is true for trees with fewer than m edges – that is, that every tree with k edges has $k + 1$ vertices whenever $k < m$. We wish to deduce that every tree T with m edges has $m + 1$ vertices.

To do this, we consider a tree T with m vertices and remove any edge e of T. Since T has no cycles, this disconnects T and gives two trees, with k_1 and k_2 edges say, where $k_1 + k_2 = m - 1$. These trees have fewer than m edges so, by our assumption, they have $k_1 + 1$ and $k_2 + 1$ vertices, respectively. Reinstating the edge e, we restore T, with a total of

$$(k_1 + 1) + (k_2 + 1) = k_1 + k_2 + 2 = m + 1 \text{ vertices.}$$

Thus if the statement is true for trees with fewer than m edges, then it is true for trees with m edges. This completes Step 2.

Therefore, by the principle of mathematical induction, the statement is true for all non-negative integers m. ■

Proofs Involving 'If and Only If'

Consider the statement

A graph G is bipartite IF and ONLY IF every cycle of G has even length.

This statement is equivalent to the following two statements, so to prove it we have to establish two things:

(a) IF every cycle of a graph G has even length, THEN G is bipartite.
(b) A graph G is bipartite ONLY IF every cycle of G has even length.

Statement (b) is the *converse* of statement (a), and we usually write it as follows.

if G is bipartite, then each cycle of G has even length.

Statement (a) tells us that 'having every cycle of even length' is a *sufficient* condition for G to be bipartite. (*Sufficient* means *enough*.)

Statement (b) tells us that 'having every cycle of even length' is a *necessary* condition for G to be bipartite. (*Necessary* means *essential*.)

So we sometimes say that

'having every cycle of even length' is a *necessary and sufficient* condition for a graph G to be bipartite.

In general, to prove a result of the form

a is true IF AND ONLY IF b is true

we must prove two separate statements – a statement and its converse:

1. *a is true* IF b is true
   – that is, we must prove that IF *b* is true, THEN *a* is true.
   (Here, *b is true* is SUFFICIENT to ensure that *a is true*.)
2. *a is true* ONLY IF *b is true*
   – that is, we must prove that IF *a is true*, THEN *b is true*.
   (Here, *b is true* is NECESSARY to have *a is true*.)

Some proofs of this type are given in Chapter 3; there the two statements are proved directly, but any of the above types of proof may be used.

Computing Notes

The accompanying software is supplied on a CD-ROM for a PC. It is designed to be used with Microsoft Windows 95, 98, NT or a later version; we assume that you are familiar with the basic facilities of Windows†.

These notes introduce the software and suggest some activities – an introductory activity associated with Chapter 1, followed by activities on selected chapters.

Installing the Software

- Insert the supplied disk into the appropriate disk drive (A:, B:, ...).
- Click on **Start**, **Run**.
- Type A:\Graphs.exe in the dialogue box that appears, where A is the CD drive letter.

Two programs are provided: the ***Graph Editor*** and the ***Graph Database***. Supporting Windows DLLs are also included and will be installed automatically. Please follow the procedure shown on the screen; suggested defaults may be overwritten in your particular case.

A folder with the two programs will be created. It can be accessed by clicking/selecting from the **Start** button – for example,

Start/Programs/Graphs software/... .

Computer Activities

These notes contain details of suggested activities on selected chapters.

†The CD-ROM is compatible with Windows 3.1 and 3.11 but with minor modifications to the printed instructions shown here. These modifications can be supplied on request. Contact the publisher at Springer-Verlag London Ltd, Sweetapple House, Catteshall Road, Godalming, Surrey GU7 3DH, UK, or by email at postmaster@svl.co.uk.

Each activity number corresponds to a chapter.
Please read the relevant chapter before you attempt each activity.
The solutions to these activities are given at the end of the Computing Notes.

Tool Bar Buttons

For reference, we list the most commonly used tool bar buttons.

 Open — allows you to open a stored file.

 Save — allows you to save the contents of the top window as a file.

 Print — allows you to print the top window.

 Undo — allows you to undo the last operation.

 Reset — resets the top window to its initial state.

 Cut — allows you to delete a selected (highlighted) item (and temporarily stores it in the clipboard).

 Copy — allows you to copy an item from the top window to the clipboard.

 Paste — allows you to paste an item from the clipboard into the top window.

 Cascade — allows you to rearrange the open windows in cascade fashion.

 Tile — allows you to tile the screen with the open windows.

 Help — allows you access to the help files.

Computer Activity for Chapter 1

Activity 1

In this activity, you can:

- use the ***Graph Editor*** to create, label and colour a graph or digraph;
- save and print a graph or digraph.

Creating Your Own Graphs

The *Graph Editor* enables you to construct and modify graphs and digraphs. We begin by describing how to draw a graph.

Drawing Vertices and Edges

Note that the right-hand button of the mouse is used *only* for creating new vertices or edges. At all other times, use the left-hand button.

1 Select from the **Start** button by clicking Start/Programs/Graphs Software/Editor.

A title page is displayed as a dialogue box.

Press **OK** to proceed.

You should now see a blank screen.

Select **New** from the **File** menu.

A dialogue box appears asking you to select the kind of graph you wish to construct.

Select **Graph** and click on **OK**.

You now have an empty window for drawing. Draw some vertices in selected positions, as follows.

2 Position the mouse somewhere in the window where you wish to draw a vertex and click the *right*-hand button.

Draw some more vertices in the same way.

[If you make an error, click on the vertex you wish to delete, then click on the **Cut** button in the tool bar or on **Cut** or **Delete** from the **Edit** menu, or press the Delete key. To delete the *last* vertex you drew, click on the **Undo** button or on **Undo** from the **Edit** menu.]

Notice that newly created vertices are highlighted.

You now have several vertices. Join some of them by edges, as follows.

3 Position the mouse over any vertex, depress the *right*-hand button, drag the cursor to another vertex, and release the button.

Draw some more edges in the same way.

[If you make an error, click on the edge you wish to delete, then click on the **Cut** button in the tool bar or on **Cut** or **Delete** from the **Edit** menu, or press the Delete key. To delete the *last* edge you drew, click on the **Undo** button or on **Undo** from the **Edit** menu.]

Notice that newly created edges are highlighted.

Next, add a loop at one of the vertices.

4 Position the mouse over a vertex, depress the *right*-hand button, drag away from the vertex, drag back to the vertex, and release the button.

Labelling Vertices and Edges

Next, label the vertices. For example, label the vertices 1, 2, ... or A, B, ...; you may use longer labels if you wish.

5 Double-click on a vertex (using the *left*-hand button of the mouse) and use the keyboard to type a label for the vertex into the dialogue box that appears. Press the Return key or click on **OK** to apply that label to the vertex.

Repeat for the other vertices by double-clicking on each in turn.

[If you make an error, double-click on the vertex again and relabel it.]

An alternative to double-clicking is to click once on a vertex and then to select **Label** from the **Edit** menu. (The selected vertex is highlighted by a coloured ring round it.) The same dialogue box then appears.

A similar procedure is used to label edges.

6 Double-click on an edge (using the *left*-hand button of the mouse) and use the keyboard to type a label for that edge. Press the Return key or click on **OK** to apply that label to the edge.

Repeat for the other edges by double-clicking on each in turn.

An alternative to double-clicking is to click once on an edge and then to select **Label** from the **Edit** menu. (The selected edge is highlighted by a strip of colour either side of it.)
The labelling dialogue boxes can also be used to *change* the labels on vertices and edges. Simply double-click on a vertex or edge (or click and use **Label** from the **Edit** menu) and type in the new label.

7 Experiment with changing the labels on vertices and edges.

Colouring Vertices and Edges

All the vertices in a graph are, by default, coloured cyan (pale blue); the edges are, by default, coloured black. You can change the colours of vertices or edges, as follows.

8 Click on a vertex, to select it, then colour it by clicking on one of the **Colour** buttons in the tool bar, or one of the items from the **Colour** menu. Then click on another vertex and another colour, and so on, until all the vertices are coloured.

Similarly, click on each edge, to select it, and then on a colour to colour each edge of the graph.

You should now have a fully labelled and coloured graph.

Selecting Vertices and Edges

It is possible to select *all* the vertices and edges of a graph, by clicking on **Select All** from the **Edit** menu. It is also possible to select several of the vertices and/or edges of a graph by clicking on them in turn while holding down the Control key. You can then apply the various editing and colouring facilities to *all the selected* vertices and/or edges at once. You can also use the Control key in this way to add vertices and/or edges to those already selected.

You can also select several items at once by using the Shift key. Clicking on an edge while holding down the Shift key also selects the vertices incident with the edge. Also, if one or more vertices have already been selected, clicking on another vertex while holding down the Shift key also selects any edges joining that vertex to the previously selected vertices.

Deleting and Moving Vertices and Edges

You can delete a vertex (and the edges joined to it) or an edge by clicking on it and then clicking on the **Cut** button or on the **Cut** item or **Delete** item from the **Edit** menu. You can move a vertex (and the edges joined to it) around the window by *dragging* it, using the *left*-hand button of the mouse.

To space the vertices regularly, select **Snap to Grid** from the **Edit** menu, and move the vertices to the grid that appears.

9 Experiment with deleting vertices and edges and with moving vertices. When you have finished experimenting, make sure you have a graph in the window.

Saving and Printing Graphs

The graph you have created is given the name **Untitled** at the top of the window. In order to give your graph a name, so that you can find it again easily, use **Save As** from the **File** menu.

Look at the **File Name** *edit box* at the top left, where you will see, highlighted:

***.grf**

Use the keyboard to type a name to replace the asterisk, before saving your graph. You can use up to eight characters for the name. (Some characters are not allowed in the name; you'll get a warning message in a dialogue box if you try to use a disallowed character.) The three letters after the dot – **grf** in this instance – are known as the *extension* and indicate the type of object you have saved – in this instance, a graph.

10 From the **File** menu, use **Save As** to save this graph as '**my. grf**'.

Another way of saving something for future reference is to print it. You can print the contents of the top window by using **Print** from the **File** menu, or the **Print** button.

11 Select **Print** from the **File** menu or click on the **Print** button.

You should now see the print dialogue box.
[You may find that you need different default printer settings for this work than for other work on your computer.]

After doing any necessary checking and adjusting of the information in the **Print** dialogue box, print **my.grf** by clicking on **OK** or pressing the Return key.

Digraphs and Networks

The *Graph Editor* package can also be used to create and edit weighted graphs, digraphs and networks by selecting the appropriate item from the menu of graph types.

To create a digraph, click on **New** in the **File** menu of the *Graph Editor* package. Then, in the dialogue box that appears, click on the arrow, to display a menu of graph types. Select **Digraph** and then click on **OK**. You can now create a digraph in the top window.

Digraphs are created in a similar manner to graphs. The only difference is that, when adding *arcs*, you must drag the pointer (using the *right*-hand button of the mouse) *in the appropriate direction*. For example, if you want to add an arc from vertex 1 to vertex 2, you must drag from 1 to 2. Dragging from 2 to 1 adds an arc from vertex 2 to vertex 1.

Weighted graphs, digraphs and networks are created similarly. Each arc of a weighted graph or digraph originally appears with zero weight, and each arc of a network originally appears with zero flow and zero capacity; to change a value, click on it and type the desired number(s).

Computer Activities for Chapter 2

Using the Graph Database

The ***Graph Database*** package contains a database of all simple unlabelled graphs with up to seven vertices – it contains 1252 graphs, numbered G1 to G1252. They are ordered according to:

- number of vertices n
- number of edges m
- degree sequence

For example, all the graphs with 4 vertices come before all those with 5 vertices; all the 5-vertex graphs with 6 edges come before all the 5-vertex graphs with 7 edges; and the 5-vertex, 6-edge graphs are ordered according to their degree sequences as follows:

G39	(0, 3, 3, 3, 3)
G40	(1, 2, 2, 3, 4)
G41	(1, 2, 3, 3, 3)
G42	(2, 2, 2, 2, 4)
G43	(2, 2, 2, 3, 3)
G44	(2, 2, 2, 3, 3)

When two degree sequences are the same, the order of the two graphs is arbitrary.

The next two activities use the ***Graph Database*** package. They both start with the assumption that you are already running the *Graph Database* package, and that you have the ***Graph Database*** window on your screen. To obtain this:

Select from the **Start** button by clicking
Start/Programs/Graphs software/Database.

Activity 2A

In this activity, you can:

- use the graph database to find graphs with a given number of vertices, number of edges and/or degree sequence.

The main window for the ***Graph Database*** package contains the numbers and other information on all 1252 graphs in the database. You can find the graph with a given number by scrolling through this window.

1 Scroll through the window until you see graph G39.
The information given for graph G39 is:

G39 $n = 5$ $m = 6$ $(0, 3, 3, 3, 3)$

2 Click on G39.

You should now see graph G39 in a window on the right of your screen, together with the above information.

3 Remove graph G39 from the screen by clicking on the close icon ☒ in its title bar.

You can observe two or more graphs at the same time.

4 Scroll through the main window until you reach G208, and then click on G208 to display that graph. Continue scrolling until you reach G431, and display that too.

You should now see the windows containing graphs G208 and G431 in *tile* form – that is, side by side.

To see them in *cascade* form, click on **Cascade** from the **Window** menu or the **Cascade** button in the tool bar.

5 Remove graphs G208 and G431 from the screen by clicking on the close icon for each of the corresponding windows.

Another way to observe two or more graphs at the same time is to use the **Select** menu or the corresponding buttons in the tool bar. The first three items in this menu or the corresponding buttons enable you to reduce all those graphs *currently* listed in the main window to those with one or more of the following properties:

property	*menu item*	*button label*
a given number of vertices	**No. Vertices (n)**	**n**
a given number of edges	**No. edges (m)**	**m**
a given degree sequence	**Degree sequence**	**DS**

The fourth menu item **All** returns you to the full list of 1252 graphs; the corresponding button in the tool bar is the **Reset** button.

6 Click on **Select** in the menu bar.

Click on **No. Vertices** from the **Select** menu or on the **n** button, then type 5 in the dialogue box that appears, and then click on **OK** (or press the Return key), to enter 5 for the number of vertices.

After a short delay, you should see, in the main window, a list of all 34 graphs with exactly 5 vertices. The window's title bar tells you how many such graphs there are.

7 Click on **No. Edges** from the **Select** menu or on the **m** button, then enter 6 for the number of edges.

You should now see, in the main window, a list of all six graphs with 5 vertices that have 6 edges.

8 Click on **Degree Sequence** from the **Select** menu or on the **DS** button, then type 22233 or (2,2,2,3,3) and click on **OK** (or press the Return key), to enter (2, 2, 2, 3, 3) for the degree sequence.

You should now see, in the main window, a list of the two graphs with 5 vertices and 6 edges that have degree sequence (2, 2, 2, 3, 3).

9 Display these graphs by clicking on each one in the list in turn.

Note that, to obtain all the graphs with 5 vertices, 6 edges and degree sequence (2, 2, 2, 3, 3), you do not need to enter values for the number of vertices, for the number of edges *and* for the degree sequence; the degree sequence alone is sufficient. The degree sequence (2, 2, 2, 3, 3) contains 5 values, so there are 5 vertices; also, since the sum of the vertex degrees is $2 + 2 + 2 + 3 + 3 = 12$, the number of edges is $12/2 = 6$, by the handshaking lemma. So, if the computer searches the database for all graphs with degree sequence (2, 2, 2, 3, 3), it finds only those obtained in Step 8.

10 Confirm this by selecting **All** from the **Select** menu or by clicking on the **Reset** button in the tool bar, to return to the complete list of 1252 graphs, and then repeating Step 8.

11 Use the **Select** menu and/or the corresponding buttons to find:

(a) all the graphs with 1 edge and not more than 7 vertices;

(b) all the graphs with 7 vertices and 15 edges;

(c) all the graphs with degree sequence (2, 3, 3, 3, 4, 4, 5).

Activity 2B

In this activity, you can:

- use the graph database to locate all the following simple graphs with up to six vertices: null graphs, path graphs, cycle graphs, complete graphs, regular graphs and trees.

Use the **Select** menu and/or the corresponding buttons, together with what you know about the number of edges and the degree sequence of null graphs, path graphs, cycle graphs, complete graphs and regular graphs, and with what you know about the number of edges of a tree with n vertices, to help you to fill in the blank spaces in the following table.

Hint Use the following results.

An r-regular graph with n vertices has $nr/2$ edges (Theorem 2.2).
A tree with n vertices has $n - 1$ edges (Problem 2.23(b)).

n	null graphs	path graphs	cycle graphs	complete graphs	trees	regular graphs
1	G1	G1	–	G1	G1	G1
2	G2		–		G3	G2, G3
3			G7		G6	G4, G7
4					G13, G14	G8, G11, G16, G18
5					G29, G30, G31	G19, G38, G52
6						

Activity 2C

In this activity, you can:

- locate the graph in the database that is isomorphic to a given unlabelled graph.

This activity uses both the ***Graph Database*** and the ***Graph Editor*** packages. It starts with the assumption that you are running the ***Graph Database*** package, and have the ***Graph Database*** window on your screen.

Consider the following unlabelled graph:

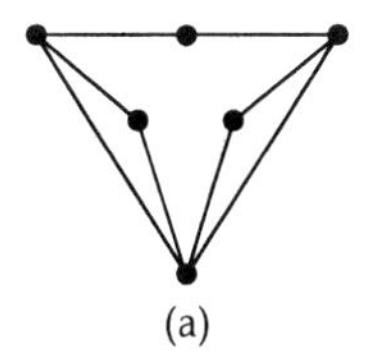

(a)

All the graphs in the database are different, so *in the database* there is only *one* graph isomorphic to a given graph with at most 7 vertices.

When looking for a graph isomorphic to a given graph, we first identify those graphs with the same degree sequence as the given graph.

1 Use the **Select** menu and/or the corresponding tool bar buttons to list all those graphs in the database with the same degree sequence as graph (a), and then display them.

Next, we examine specific features of the graphs, in order to try to find an appropriate one-one correspondence between the vertices. This one-one correspondence is shown by giving the same label to the corresponding vertices in the graphs. For example, if each graph has just one vertex of degree 6, then the vertex of degree 6 should be given the same label in each graph.

2 Use a pencil to label the vertices of graph (a) above.

In order to label the vertices of a graph in the database, we first have to transfer it to the ***Graph Editor***.

3 Choose one of the displayed graphs and make its window the top one by clicking somewhere in it, then copy it to the clipboard by selecting **Copy** from the **Edit** menu or by clicking on the **Copy** button. Minimize the ***Graph Database*** package by clicking on its minimize icon ▪.

A Graph Database icon should appear on the task bar. You will need this to return to the ***Graph Database*** later; it enables you to return to that package with the same selected graphs displayed as when you left it.

4 Run the ***Graph Editor*** package.

5 Select **New** from the **File** menu and click on **OK** in the dialogue box that appears. Select **Paste** from the **Edit** menu or click on the **Paste** button to paste your chosen graph into the new window.

Your chosen graph should now be displayed.

To label a vertex, double-click on the vertex (or click on it and then click on **Label** in the **Edit** menu) and then type your chosen label and click on **OK** (or press the Return key).

[Click on **Cancel** if you change your mind.]

6 Try to label the vertices of the graph in the window so that there is a one-one correspondence between the vertices of that graph and those of graph (a) that is an isomorphism.

The package enables you to check your potential isomorphism by manipulating your chosen graph so that it matches the given graph exactly. You can use the ***Graph Editor*** to do this, by using the mouse to drag the vertices (and their incident edges) around the window.

7 Use the mouse to drag the vertices (and edges) around the window to try to obtain an exact copy of graph (a).

If you failed to find an isomorphism, try again with another of the graphs you selected in Step 1, as follows.

8 Return to the ***Graph Database*** package by clicking on the Graph Database icon on your task bar. Choose another of the displayed graphs and copy it to the clipboard.

9 Return to the ***Graph Editor*** package by clicking on the Graph Editor icon on your task bar.

Repeat Steps 5 to 7 for your newly chosen graph.

10 Repeat Steps 8 and 9 until you think you have found the graph isomorphic to graph (a).

With practice, you should not need to go through all the above steps in order to find the graph isomorphic to a given graph. You may be able to eliminate some graphs immediately by inspection, and to eliminate some others and spot the isomorphism after labelling just a few vertices and/or doing a little manipulation.

11 Repeat some or all of Steps 1 to 10 for each of the following unlabelled graphs, in order to find the graph isomorphic to each.

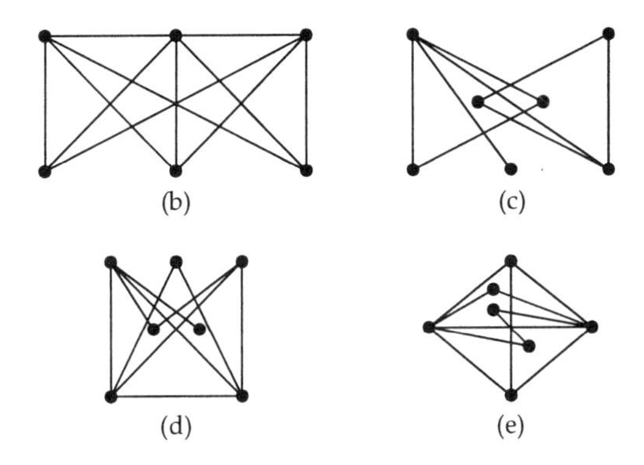

Activity 2D

> In this activity you can:
>
> - determine whether given pairs of labelled graphs are isomorphic.

This activity uses the ***Graph Editor*** package. It starts with the assumption that you are already running that package, and that you have the ***Graph Editor*** window on your screen.

For this activity, the graphs have more than seven vertices, so are not in the graph database. Here your strategy is to draw each pair of graphs in a pair of windows in the ***Graph Editor*** package and then to use the labelling, colouring and manipulation facilities of the graph editor to try to determine an isomorphism between the vertices of the two graphs.

Consider the following two labelled graphs:

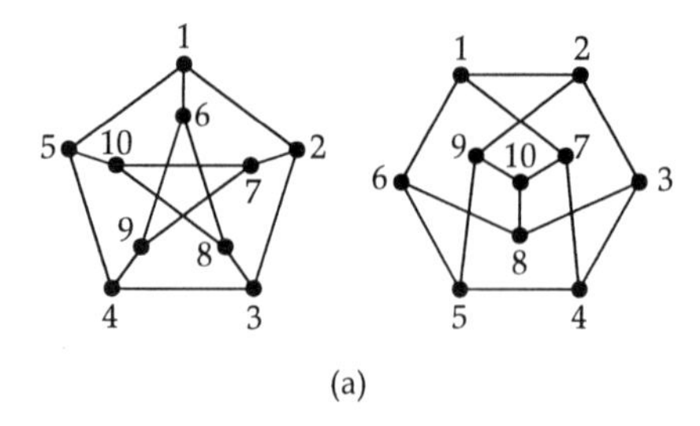

(a)

1 Check that both graphs have the same number of vertices, the same number of edges and the same degree sequence.

If they don't, then you know immediately that the graphs are not isomorphic.

If both graphs have the same degree sequence, the next step is to draw the two graphs on the screen, ready to compare their features. Start with the graph on the left above.

2 Draw and label the left-hand graph, as in Activity 1.

You should now have an exact copy of the left-hand graph in the window.

3 In a new window, draw, *but do not label*, the right-hand graph.

4 Select **Tile** from the **Window** menu or click on the **Tile** button in the tool bar.

You should now see both graphs side by side in their separate windows.

To show that two labelled graphs are isomorphic, we compare the features of the two graphs and then relabel the vertices of one of the graphs to give an appropriate one-one correspondence between the vertices.

To help you identify corresponding features, and hence label corresponding vertices, you may wish to *colour* certain vertices. You can colour all the vertices differently, or you can use different colours for vertices of different degrees, for example.

You may also find it helpful to label edges, by double-clicking on them and typing in the label, and/or to colour edges, by clicking on them and using the colour buttons or the **Colour** menu.

[When you label edges that cross, the labels may overlap. You can usually avoid this overlap by moving some of the vertices.]

You may find it easier to spot corresponding features if you move the vertices of one or both graphs around, to try to change the orientations of the graphs, or to rearrange the edges so that they cross as little as possible.

Another strategy is to colour the two graphs differently, copy both graphs to a single new window and try to manipulate them so that one lies exactly on top of (or nearly on top of) the other. To do this, first select **New** from the **File** menu to create a new window. Then, for each graph in turn, choose **Select All** from the **Edit** menu to select all the vertices and edges of the graph, click on a colour to colour them all, copy the coloured graph to the clipboard, and paste it into the new window. The copying facilities enable you to copy *all* the *selected* vertices and edges.

You should now have both graphs in the new window, differently coloured, and ready to manipulate to try to find an isomorphism.

5 Use the labelling, colouring, moving, copying and pasting facilities to help you determine whether the two labelled graphs are isomorphic.

6 Repeat Steps 1 to 5 for each of the following pairs of labelled graphs.

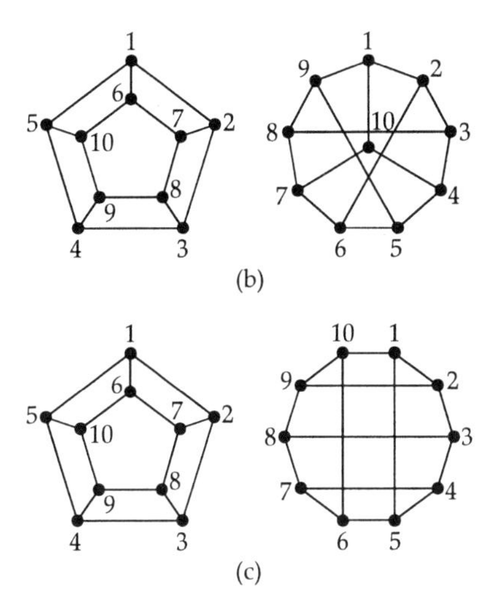

Tracing Paths and Cycles in Graphs

Note that the colouring facilities can be useful in helping you trace trails, paths and cycles in graphs. You may find these facilities particularly helpful when you are asked to find Eulerian trails, Hamiltonian cycles, semi-Eulerian trails and semi-Hamiltonian paths in Chapters 3 and 4.

Computer Activities for Chapter 5

These activities use the ***Graph Editor*** package. This package contains a small database of graphs, but you can also call up any graph from the ***Graph Database*** package or any that you have saved yourself.

Activity 5A

In this activity, you can:

- observe and comment on the form of the adjacency matrices of certain graphs, when the vertices are suitably labelled;
- observe and comment on the effect of relabelling the vertices of a graph on the form of its adjacency matrix;
- conjecture on the form of the adjacency matrices of certain types of graph when the vertices are suitably labelled.

The graphs for this activity can be displayed by use of the **File** menu of the *Graph Editor* package.

1 Select **Open** from the **File** menu of the *Graph Editor* package, or click on the **Open** button, and then select the complete graph K_4 by double-clicking on its file name **k4.grf** in the list box on the left of the dialogue box that appears. Observe its labelling.

2 Click on the **(A)** button in the tool bar, or on **Adjacency Matrix** from the **View** menu, to display its adjacency matrix. Observe its form.
[Clicking again on the **(A)** button or the **Adjacency Matrix** menu item removes the adjacency matrix from the screen.]

The rows and columns of the adjacency matrix are labelled twice: once numerically and once using the vertex labels of the graph.

We now ask you to try different labellings of K_4, place the different versions side by side, and observe the effect on the adjacency matrix.

3 Repeat Steps 1 and 2, to obtain another copy of K_4 and its adjacency matrix, and then select **Tile** from the **Window** menu or click on the **Tile** button to see both side by side.

4 Relabel the vertices of the graph in one of the windows, using the numbers 1, 2, 3, ...

Observe the structure of the new adjacency matrix.

5 Repeat Step 4 for several different labellings.

Does the relabelling affect the form of the adjacency matrix? If so, in what way?

6 Repeat Steps 1 to 5 for the complete graph K_6, and then conjecture on the form of the adjacency matrix of any complete graph, and on whether the labelling of the vertices affects this.

[The filename for K_6 is **k6.grf**.]

7 Repeat Steps 1 to 5 for the cycle graph C_6, and then conjecture on the form of the adjacency matrix of any cycle graph, and on whether the labelling of the vertices affects this form.

[The filename for C_6 is **c6.grf**.]

8 Repeat Steps 1 and 2 for the octahedron graph, and then try to relabel the vertices of the graph so that its adjacency matrix is of a form similar to that of suitably labelled complete graphs and of suitably labelled cycle graphs.

[The filename for the octahedron graph is **octa.grf**.]

9 Repeat Steps 1 to 5 for the complete bipartite graphs $K_{2,4}$ and $K_{3,3}$, and then conjecture on the form of the adjacency matrix of any complete bipartite graph, and on whether the labelling of the vertices affects this form.

[The filenames for $K_{2,4}$ and $K_{3,3}$ are **k24.grf** and **k33.grf**.]

10 Repeat Steps 1 and 2 for the cube graph, and then try to relabel the vertices of the graph so that its adjacency matrix is of a form similar to that of suitably labelled complete bipartite graphs.

If you were successful, why do you think this was so?

[The filename for the cube graph is **cube.grf**.]

11 Conjecture on the form of the adjacency matrix of suitably labelled bipartite graphs, and on what constitutes a suitable labelling.

Activity 5B

In this activity, you can:

- observe and comment on the form of the incidence matrices of certain graphs, when the vertices and edges are suitably labelled;
- observe and comment on the effect of relabelling the vertices and/or edges of a graph on the form of its incidence matrix;
- conjecture on the form of the incidence matrices of certain types of graph when the vertices and edges are suitably labelled.

1 Display the complete graph K_4, and label its *edges* using the numbers 1, 2, 3, ...

2 Click on the **(I)** button in the tool bar, or on **Incidence Matrix** from the **View** menu, to display its incidence matrix. Observe its form.

[Clicking again on the **(I)** button or the **Incidence Matrix** menu item removes the incidence matrix from the screen.]

The rows and columns of the incidence matrix are labelled twice: once numerically and once using the vertex and edge labels of the graph.

3 Relabel the vertices and edges of the graph in several different ways to try to find the most 'structured' form of its incidence matrix.

4 Repeat Steps 1 to 3 for each of the following graphs in turn:

K_6; C_6; octahedron; $K_{2,4}$; $K_{3,3}$; cube.

5 Conjecture on the form of the incidence matrices of:

(a) complete graphs;
(b) cycle graphs;
(c) the octahedron graph;
(d) complete bipartite graphs;
(e) the cube graph;
(f) bipartite graphs;

and on whether the labelling of the vertices and edges affects this form.

Activity 5C

In this activity, given several pairs of isomorphic graphs, you can:

- relabel the vertices of the second graph in each pair so that the labels on corresponding vertices are the same, and observe the effect of the relabelling on the adjacency matrix of the second graph;
- conjecture on the relationship between the adjacency matrices of isomorphic graphs.

Consider the following two isomorphic graphs:

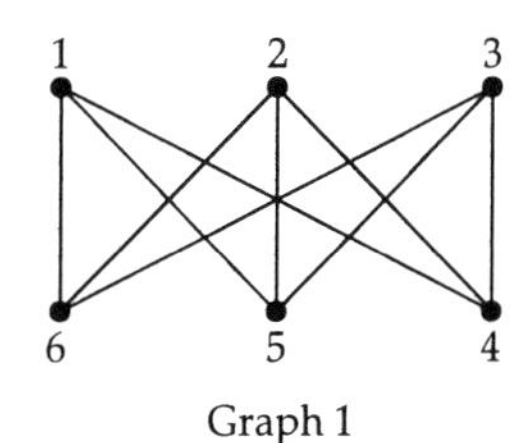

Graph 1

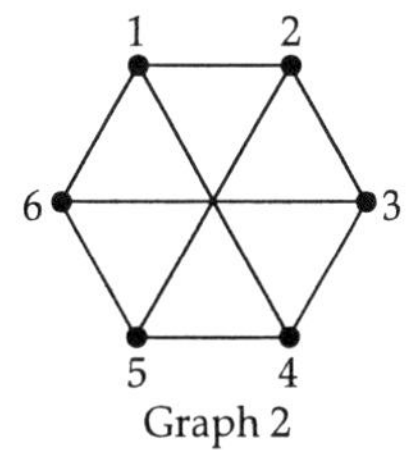

Graph 2

[Their file names are **graph1.grf** and **graph2.grf**.]

1 Use **Open** from the **File** menu of the ***Graph Editor*** package or the **Open** button, and **Tile** from the **Window** menu or the **Tile** button, to display both graphs side by side.

2 Make each window the top window in turn, by clicking on it, and then click on the **(A)** button or **Adjacency Matrix** from the **View** menu to display both adjacency matrices.
Notice that the two matrices are different.

3 Interchange the labels 2 and 5 of Graph 2 and observe the effect on the adjacency matrix.

4 Repeat Steps 1 and 2 for the following pair of isomorphic graphs.

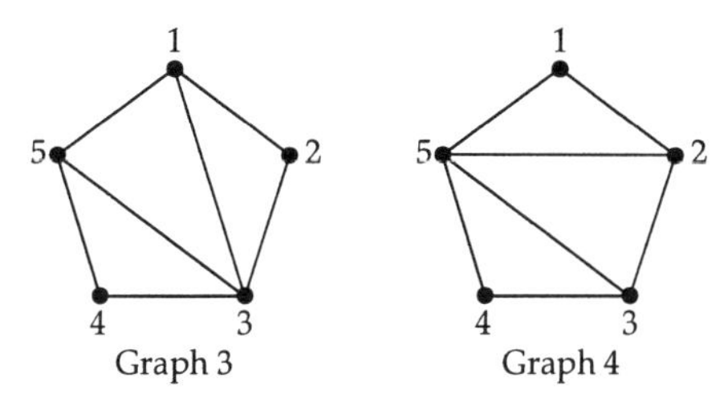

Graph 3 Graph 4

[Their file names are **graph3.grf** and **graph4.grf**.]

5 Relabel the vertices of Graph 4 in stages, so that the labels on only *two* vertices are changed at each stage, until both sets of labels correspond. Observe the effect on the adjacency matrix at each stage.

6 Repeat Steps 4 and 5 for the two pairs of isomorphic graphs available as Graphs 5 and 6 and as Graphs 7 and 8 via **Open** from the **File** menu or the **Open** button.
[Their file names are **graph5.grf**, **graph6.grf**, **graph7.grf**, **graph8.grf**.]

7 Conjecture on the relationship between the adjacency matrices of isomorphic graphs.

From your work on Activity 5C, you may already be able to conjecture on the relationship between the incidence matrices of isomorphic graphs. If not, try the following activity.

Activity 5D

> In this activity, given several pairs of isomorphic graphs, you can:
>
> - relabel the vertices and edges of the second graph in each pair so that the labels on corresponding vertices and edges are the same, and observe the effect of the relabelling on the incidence matrix of the second graph;
> - conjecture on the relationship between the incidence matrices of isomorphic graphs.

1 Use **Open** from the **File** menu of the ***Graph Editor*** package or the **Open** button, and **Tile** from the **Window** menu or the **Tile** button, to display Graphs 1 and 2, the first pair of isomorphic graphs from Activity 5C.
Label the edges of both graphs.

2 Make each window the top window in turn, by clicking on it, and then click on the **(I)** button or **Incidence Matrix** from the **View** menu to display both incidence matrices.
Notice that the two matrices are different.

3 Relabel the vertices and edges of Graph 2 in stages, so that the labels on only *two* vertices or *two* edges are changed at each stage, until both sets of labels correspond. Observe the effect on the incidence matrix at each stage.

4 Repeat Steps 1 to 3 for each of the other three pairs of isomorphic graphs from Activity 5C.

5 Conjecture on the relationship between the incidence matrices of isomorphic graphs.

Computer Activities for Chapter 11

The following activities use the ***Graph Editor*** package.

Activity 11A

This activity is based on the definition of a *planar graph* and on the use of the following results:

a graph is planar if and only if it does not contain a subdivision of K_5 or a subdivision of $K_{3,3}$ (Theorem 11.3);
a graph is planar if and only if it does not have K_5 or $K_{3,3}$ as a contraction (Theorem 11.4).

In this activity, you can determine whether certain graphs are *planar* or *non-planar* by

- redrawing them in planar form

or

- identifying a subgraph that is a subdivision of K_5 or $K_{3,3}$

or

- identifying a subgraph that can be contracted to K_5 or $K_{3,3}$.

To show that a graph you have drawn using the package is *planar*, move its vertices around until no two edges cross.

To show that a graph is non-planar by identifying a subgraph that is a subdivision of K_5 or $K_{3,3}$ can be difficult, as subdivisions of these graphs are not easy to spot. However, if you remove all vertices of degree 2 from the graph, and replace each such vertex and its incident edges by a single edge, then looking for subdivisions of K_5 or $K_{3,3}$ in the graph may become a bit easier.

The package allows you to replace each vertex of degree 2 and its incident edges by an edge, by selecting the vertex and then clicking on **Remove Vertex** from the **Edit** menu.

To show that a graph is non-planar by identifying a subgraph that can be contracted to K_5 or $K_{3,3}$ can also be difficult, as the overall result of a *series of contractions* is not easy to imagine. The package allows you to experiment with different sets of contractions and to see the results. An edge may be contracted by selecting it and then clicking on **Contract Edge** from the **Edit** menu.

(Note that the removal of a vertex of degree 2 is equivalent to a contraction. However, the converse is not true: most contractions are not equivalent to the removal of a vertex of degree 2.)

Consider the following graphs:

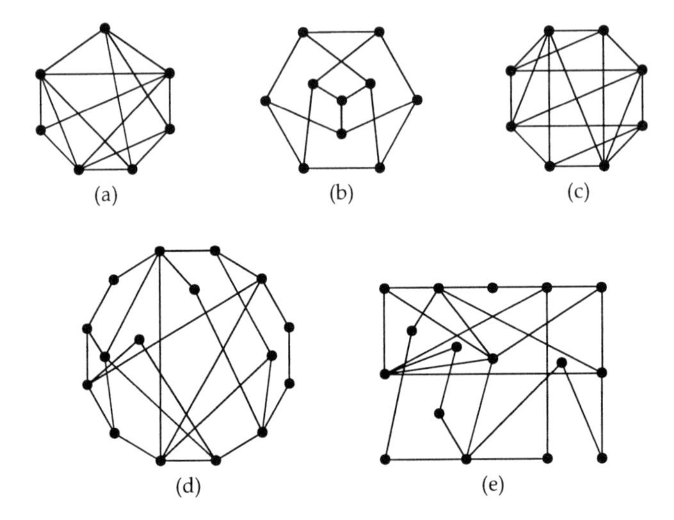

1 Select **New** from the **File** menu and draw graph (a).

[Save each graph once you have drawn it, in case you lose track of your manipulations and need to start again. You will need graphs (a), (c) and (d) again in Activity 11B.]

2 Determine whether graph (a) is *planar* or *non-planar* by:

- moving its vertices until no edges cross;

or

- using **Remove Vertex** and **Contract Edge** to help you to find a subgraph that is a subdivision of, or can be contracted to, K_5 or $K_{3,3}$.

3 Repeat Steps 1 and 2 for each of graphs (b) to (e).

Activity 11B

In this activity, you can:

- use the cycle method to determine whether certain graphs are planar or non-planar.

This activity considers the same five graphs as Activity 11A.

1 Use **Open** from the **File** menu to call up the copy of graph (a) that you saved in Activity 11A.

2 Identify a Hamiltonian cycle in the graph and colour all its edges the same colour.
[The uncoloured edges are those in the 'list' of the cycle method.]

3 If necessary, move the vertices of the graph so that all the uncoloured edges lie inside the cycle.

4 Identify, stage by stage, the edges in the two sets A and B of edges that can be drawn respectively inside and outside the cycle, without crossing, by:

- using a second colour to colour those edges in A;
- using a third colour to colour those edges in B.

[The colours should enable you to spot incompatibilities without having to draw the edges in B outside the cycle.]
Stop when:

EITHER all edges are coloured;
in this case the graph is *planar*

OR an edge not in the cycle cannot be allocated to A or B;
in this case the graph is *non-planar*.

5 If the graph is planar, move its vertices to redraw it in planar form.

6 Repeat Steps 1 to 5 for each of graphs (c) and (d) of Activity 11A. (Graph (b) is the Petersen graph; graphs (b) and (e) are not Hamiltonian.)

Computer Activity for Chapter 12

The following activity uses the *Graph Editor* package.

Activity 12

> In this activity, you can:
>
> - use the greedy algorithm for vertex colouring to help you to determine the chromatic number $\chi(G)$ of various graphs G.

Consider the following graphs:

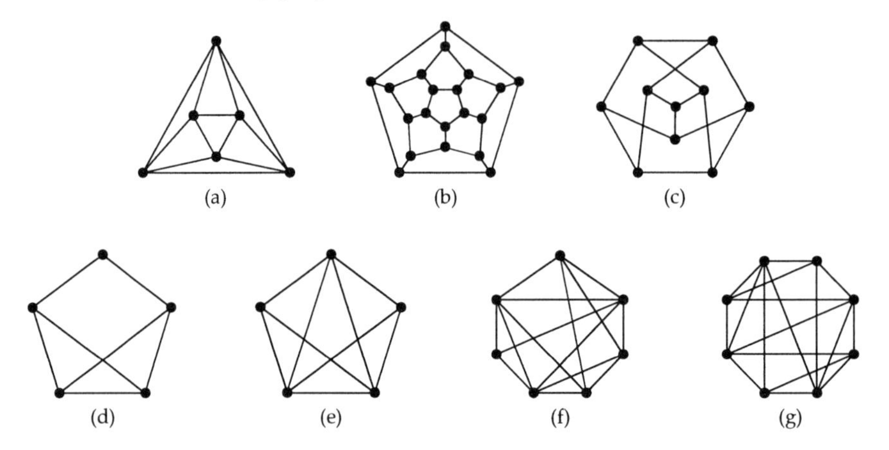

(a) (b) (c)

(d) (e) (f) (g)

1. Select **New** from the **File** menu and draw graph (a).
   [Save each graph once you have drawn it, as you will need each graph again in Activity 13.]
2. Label the vertices with the labels a, b, c, ... in any manner.
3. Colour the vertices in the manner given by the greedy algorithm for colouring vertices, and record an upper bound for the chromatic number.
4. Relabel the vertices of the graph, and make them the same colour. [Use **Select All** to speed up the recolouring process.]
5. Repeat Step 3.
6. Repeat Steps 4 and 5 until you think you know the value of the chromatic number.
7. Repeat Steps 1 to 6 for each of the graphs (b) to (g).

Computer Activity for Chapter 13

The following activity uses the *Graph Editor* package.

Activity 13

In this activity, you can:

- use the greedy algorithm for edge colouring to help you to determine the chromatic index $\chi'(G)$ of various graphs G.

This activity considers the same seven graphs as Activity 12.

1 Use **Open** from the **File** menu to call up your copy of graph (a).

2 Label the edges with the labels $a, b, c, \ldots$ in any manner.

3 Colour the edges in the manner given by the greedy algorithm for colouring edges, and record an upper bound for the chromatic index.

4 Relabel the edges of the graph, and make them the same colour. [Use **Select All** to speed up the recolouring process.]

5 Repeat Step 3.

6 Repeat Steps 4 and 5 until you think you know the value of the chromatic index.

7 Repeat Steps 1 to 6 for each of the graphs (b) to (g).

Solutions to Computer Activities

Activity 2A

(a) 6 (G3, G5, G9, G20, G54, G210);

(b) 41 (G1172–G1212);

(c) 20 (G959–G978).

Activity 2B

n	null graphs	path graphs	cycle graphs	complete graphs	trees	regular graphs
1	G1	G1	–	G1	G1	G1
2	G2	G3	–	G3	G3	G2, G3
3	G4	G6	G7	G7	G6	G4, G7
4	G8	G14	G16	G18	G13, G14	G8, G11, G16, G18
5	G19	G31	G38	G52	G29, G30, G31	G19, G38, G52
6	G53	G83	G105	G208	G77, G78, G79, G80, G81, G83	G53, G61, G105, G106, G174, G175, G204, G208

Activity 2C

(a) G148; (b) G197; (c) G429; (d) G635; (e) G799.

Activity 2D

(a) isomorphic; (b) non-isomorphic; (c) isomorphic.

Activity 5A

General observations on the form of the adjacency matrix of any graph are given in Chapter 5, so here our observations are restricted to those specific to the particular graphs considered in the activity.

Complete Graphs

The adjacency matrix of a complete graph consists of 1s everywhere except along the main diagonal, where there are 0s, irrespective of the labelling of the vertices.

Cycle Graphs

The most 'structured' form of adjacency matrix for any cycle graph corresponds to a labelling of the vertices in order around the cycle. This labelling gives an adjacency matrix with 0s along the main diagonal and 1s along both of the diagonals next to the main one. All other entries are 0s except for 1s in the top right and bottom left corners.

Octahedron Graph

The most 'structured' form of the adjacency matrix and a corresponding labelling are as follows.

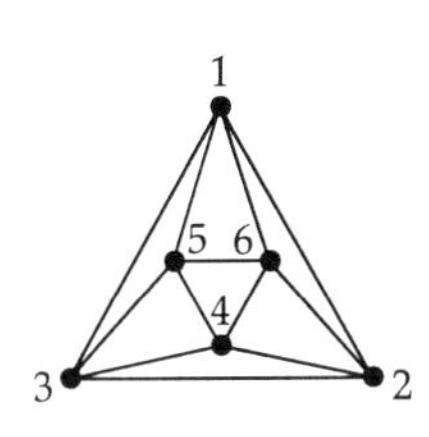

$$\begin{array}{c} \\ 1 \\ 2 \\ 3 \\ 4 \\ 5 \\ 6 \end{array} \begin{array}{c} \begin{array}{cccccc} 1 & 2 & 3 & 4 & 5 & 6 \end{array} \\ \begin{bmatrix} 0 & 1 & 1 & 0 & 1 & 1 \\ 1 & 0 & 1 & 1 & 0 & 1 \\ 1 & 1 & 0 & 1 & 1 & 0 \\ 0 & 1 & 1 & 0 & 1 & 1 \\ 1 & 0 & 1 & 1 & 0 & 1 \\ 1 & 1 & 0 & 1 & 1 & 0 \end{bmatrix} \end{array}$$

Complete Bipartite Graphs

The most 'structured' form of the adjacency matrix of any complete bipartite graph $K_{r,s}$ occurs when the black vertices are labelled $1, \ldots, r$ and the white vertices are labelled $r + 1, \ldots, r + s$, or *vice versa*. The adjacency matrix is partitioned into rectangular blocks of 1s in the top right and bottom left corners (one $r \times s$ and one $s \times r$), and square blocks of 0s in the top left and bottom right corners (one $r \times r$ and one $s \times s$). This reflects the facts that no two black vertices are adjacent and no two white vertices are adjacent.

Cube Graph

The most 'structured' form of the adjacency matrix and a corresponding labelling are as follows.

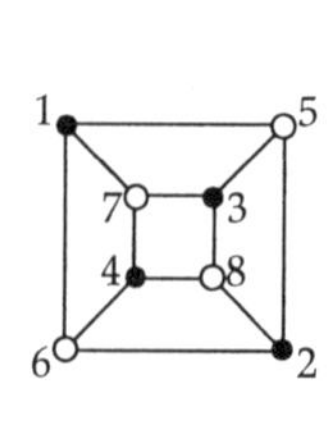

$$
\begin{array}{c} \\ 1 \\ 2 \\ 3 \\ 4 \\ 5 \\ 6 \\ 7 \\ 8 \end{array}
\begin{array}{c}
\begin{array}{cccccccc} 1 & 2 & 3 & 4 & 5 & 6 & 7 & 8 \end{array} \\
\left[\begin{array}{cccc:cccc}
0 & 0 & 0 & 0 & 1 & 1 & 1 & 0 \\
0 & 0 & 0 & 0 & 1 & 1 & 0 & 1 \\
0 & 0 & 0 & 0 & 1 & 0 & 1 & 1 \\
0 & 0 & 0 & 0 & 0 & 1 & 1 & 1 \\
\hdashline
1 & 1 & 1 & 0 & 0 & 0 & 0 & 0 \\
1 & 1 & 0 & 1 & 0 & 0 & 0 & 0 \\
1 & 0 & 1 & 1 & 0 & 0 & 0 & 0 \\
0 & 1 & 1 & 1 & 0 & 0 & 0 & 0
\end{array}\right]
\end{array}
$$

The zeros in the top right and bottom left corners occur because the cube graph is not a complete bipartite graph.

The adjacency matrix can be partitioned as shown above, with zeros in the top left and bottom right corners, because the cube graph is a bipartite graph.

Bipartite Graphs

The adjacency matrix of any bipartite graph with r black vertices labelled $1, \ldots, r$ and s white vertices labelled $r + 1, \ldots, r + s$, or *vice versa*, can be partitioned into square blocks of 0s in the top left and bottom right corners (one $r \times r$ and one $s \times s$).

Activity 5C

The adjacency matrices of two isomorphic graphs can be made the same by interchanging certain rows and the corresponding columns (corresponding to relabelling the vertices) in one of the matrices.

Activity 5D

The incidence matrices of two isomorphic graphs can be made the same by interchanging certain rows (corresponding to relabelling the vertices) and by interchanging certain columns (corresponding to relabelling the edges) in one of the matrices.

Activities 11A and 11B

(a) planar; (b) non-planar; (c) planar;
(d) non-planar; (e) non-planar.

Activity 12

The chromatic numbers are:

(a) 3; (b) 3; (c) 3; (d) 3; (e) 4; (f) 4; (g) 4.

Activity 13

The chromatic indices are:

(a) 4; (b) 3; (c) 4; (d) 4; (e) 5; (f) 5; (g) 5.

Solutions to Problems in the Text

Chapter 1

1.1

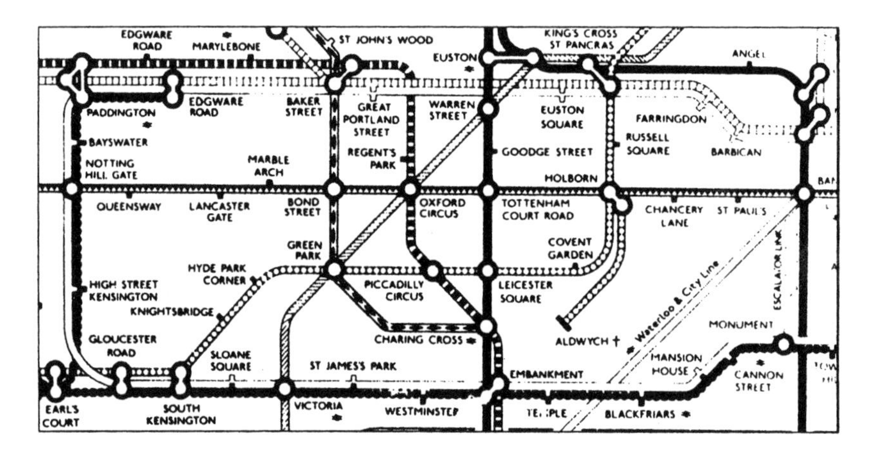

A 'best' route might be one that involves the least number of intermediate stations. One such route is:

Marble Arch → Bond Street → Green Park →

Charing Cross → Embankment → Westminster;

Alternatively, since changing trains takes time, a 'best' route might be one that involves the least number of changes. A route involving only one change is:

Marble Arch → Lancaster Gate → Queensway →

Notting Hill Gate → High Street Kensington →

Gloucester Road → South Kensington → Sloane Square →

Victoria → St James's Park → Westminster.

1.2

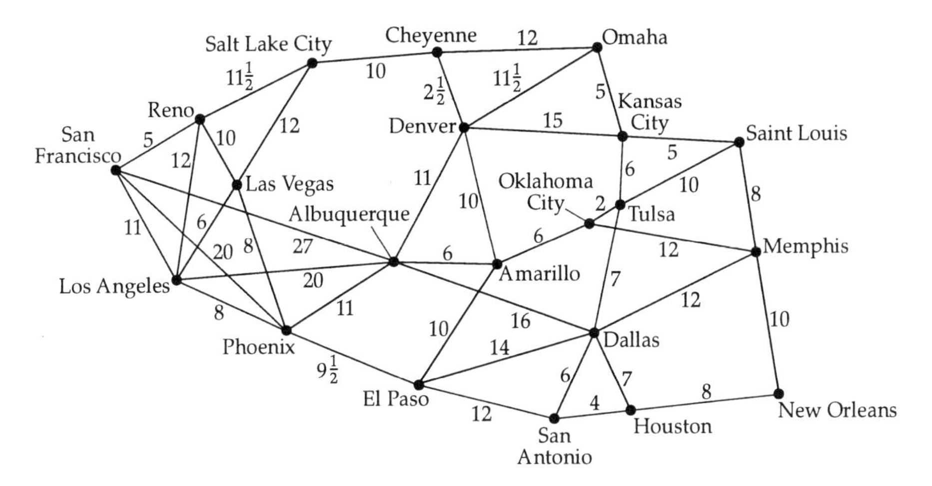

The shortest time to travel from Los Angeles to Amarillo is 25 hours, using the route

Los Angeles → Phoenix → Albuquerue → Amarillo.

The shortest time to travel from San Francisco to Denver is 29 hours, using the route

San Francisco → Reno → Salt Lake City → Cheyenne → Denver.

1.3 For an alkane with formula C_3H_8, the only possible arrangement is as follows:

```
    H   H   H
    |   |   |
H — C — C — C — H
    |   |   |
    H   H   H
```

For an alkane with formula C_4H_{10}, the carbon atoms can be arranged in two different ways, as follows:

```
                                        H
                                        |
                                    H — C — H
    H   H   H   H                   H   |   H
    |   |   |   |                   |   |   |
H — C — C — C — C — H           H — C — C — C — H
    |   |   |   |                   |   |   |
    H   H   H   H                   H   H   H
```

1.4 (a) A suitable repositioning is:

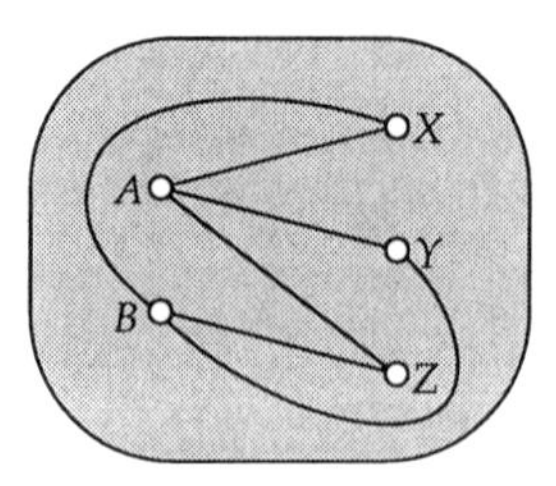

(b) It is not possible to reposition the conducting strips in a way that avoids crossing points.

1.5

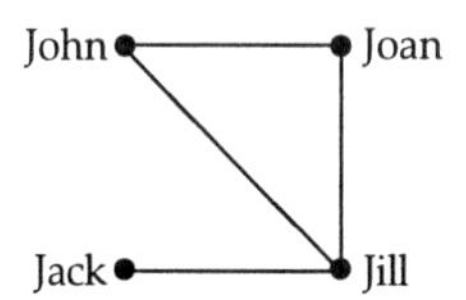

1.6 We take a vertex corresponding to each person, and join two vertices by a thick edge if the corresponding people know each other and by a thin edge otherwise. We must show that there is always *either* a triangle of thick edge *or* a triangle of thin edges.

Let v be any vertex. Then there must be exactly five edges emerging from v, either thick or thin, so at least three of these edges must be of the same type.

Let us assume that there are at least three thick edges, as shown in the following diagram:

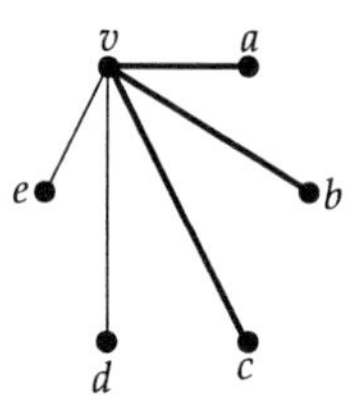

If the people corresponding to vertices a and b know each other, then the edges joining the vertices v, a and b form a triangle of thick edges, as required.

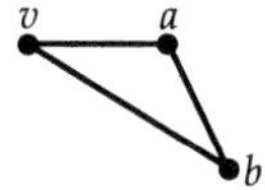

Similarly, if the people corresponding to vertices a and c know each other, then the thick edges joining the vertices v, a and c form a triangle; and if the people corrsponding to vertices b and c know each other, then the thick edges joining the vertices v, b and c form a triangle.

In the remaining case, a and b do not know each other, a and c do not know each other, and b and c do not know each other; but then the thin edges joining the vertices a, b and c form a triangle.

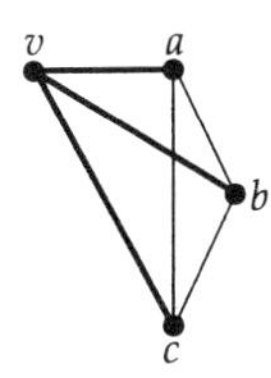

The case of at least three thin edges is analogous.

1.7 Because we do not want routes that involve retracing steps, we can remove the dead-end edges joining B and C, H and I, J and K, and L and M. This gives the following graph:

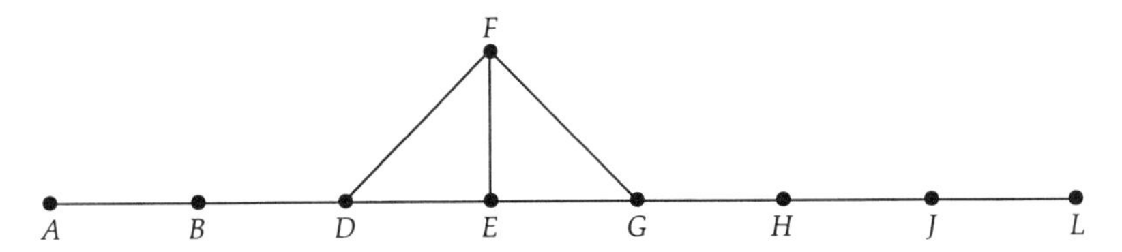

Therefore the only four routes that do not involve retracing steps are:

$A \to B \to D \to E \to G \to H \to J \to L$,
$A \to B \to D \to E \to F \to G \to H \to J \to L$,
$A \to B \to D \to F \to G \to H \to J \to L$,
$A \to B \to D \to F \to E \to G \to H \to J \to L$.

1.8 We leave the solution to this problem open for the time being. A solution is given in Chapter 3.

1.9

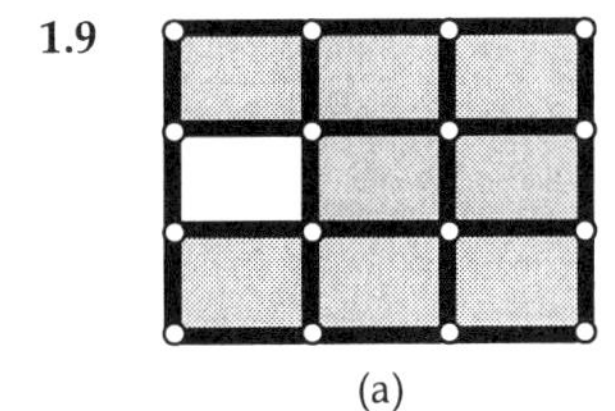

(a)

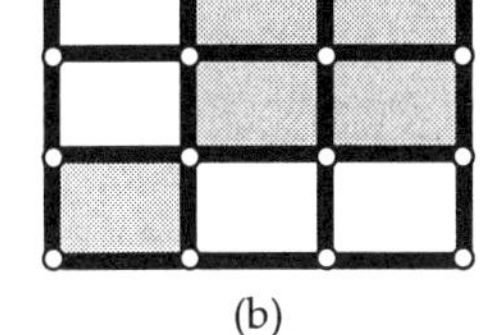

(b)

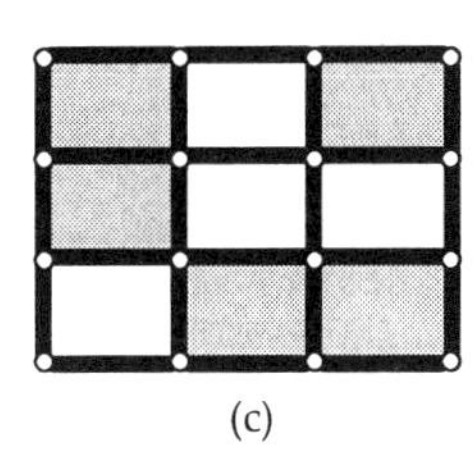

(c)

(a) One possibility is to start by removing the braces from the top-right and bottom-right rectangles. You can then remove any one other brace, except either of the two in the middle row.

(b) One possibility is to add a brace in the top-left corner.

(c) The framework is rigid, but the removal of any one of the five braces destroys the rigidity.

1.10 (a)

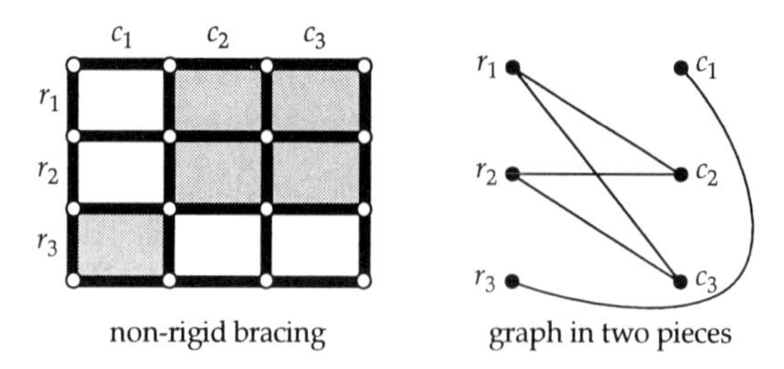

(b) Framework (c) is rigid and the corresponding bipartite graph is 'in one piece'.

Framework (b) is not rigid and the corresponding bipartite graph is 'in two pieces'.

It seems possible that a framework is rigid when the corresponding bipartite graph is 'in one piece'.

1.11

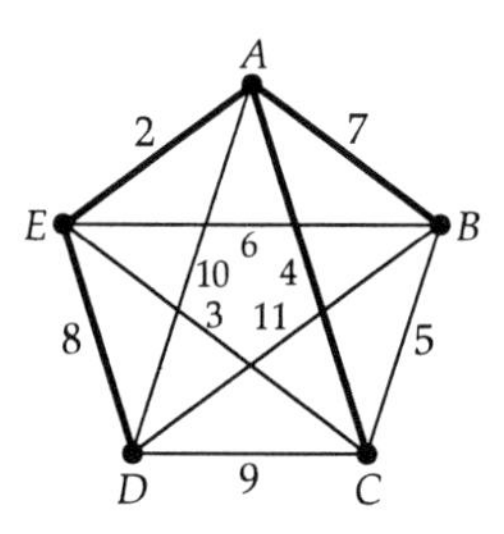

The shortest route is:

$$A \to C \to B \to D \to E \to A,$$

of total length

$$4 + 5 + 11 + 8 + 2 = 30 \text{ metres.}$$

1.12

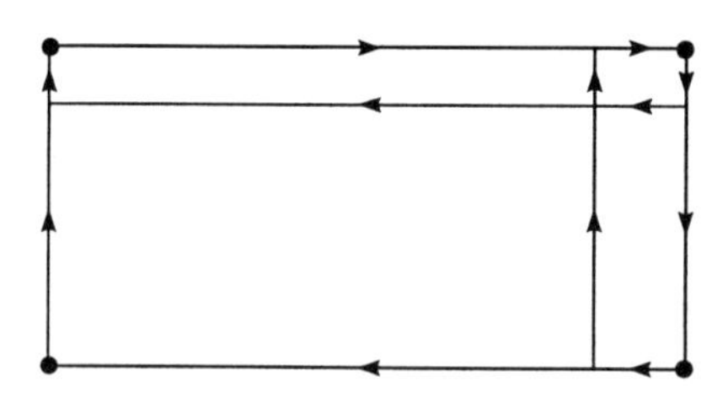

1.13

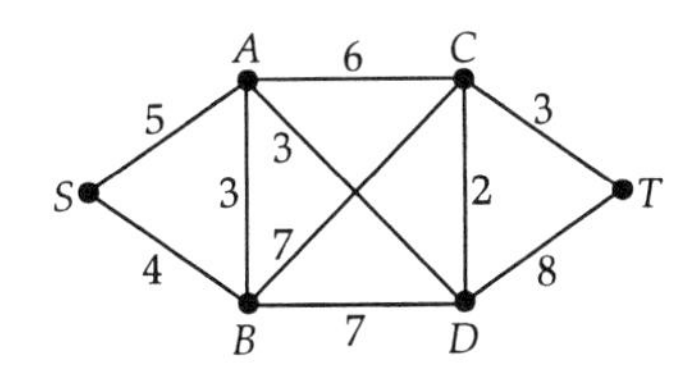

The shortest route from A to T is $ADCT$, with length 8;
the shortest route from B to T is BCT, with length 10;
hence the shortest route from S to T is $SADCT$, with length 13.

Chapter 2

2.1 (a) vertices: $\{a, b, c, d\}$
edges: $\{ab, ad, bc, bd, cc, cd\}$
Graph (a) is not a simple graph, because there is a loop at the vertex c.

(b) vertices: $\{0, 1, 2, 3, 4, 5, 6, 7, 8, 9\}$
edges: $\{01, 04, 05, 12, 16, 23, 27, 34, 38, 49, 57, 58, 68, 69, 79\}$
Graph (b) is a simple graph.

2.2

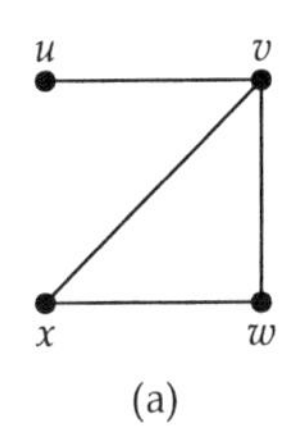

(a)

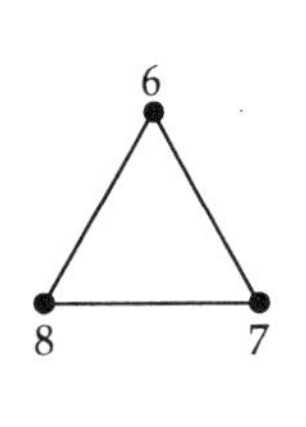

(b)

Graph (a) is a simple graph.
Graph (b) is not a simple graph, because there is a loop at the vertex 2.

2.3 (a) yes; (b) no; (c) yes; (d) no.

2.4 (a) To show that the graphs are isomorphic, we must match up:

the vertices with a loop: 3 and e;

the vertices where four edges meet: 1 and c;

the vertices where three edges meet: 5 and b;

the remaining vertices of the 'triangles': 4 and a;

the other two vertices: 2 and d.

Thus, to show that the two graphs are isomorphic, we use the one-one correspondence:

$1 \leftrightarrow c$
$2 \leftrightarrow d$
$3 \leftrightarrow e$
$4 \leftrightarrow a$
$5 \leftrightarrow b$

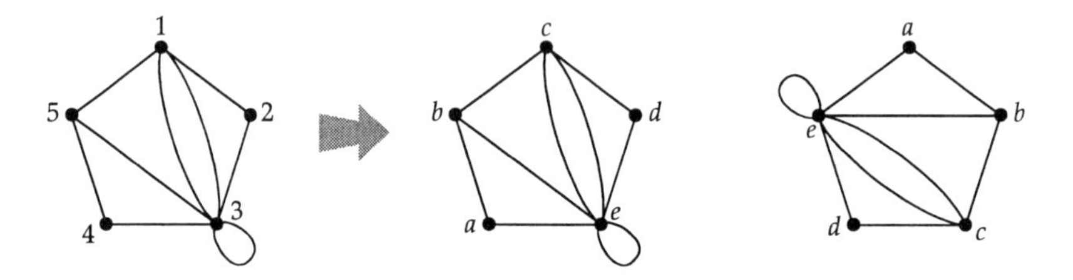

(b) To show that the graphs are isomorphic, we must match up:

the vertices where three edges meet;

the vertices where two edges meet.

Thus

$\{q, s\}$ must correspond to $\{+, -\}$; we can do this in 2 ways.

$\{p, r, t\}$ must correspond to $\{\times, =, \div\}$; we can do this in 6 ways.

There are no other constraints. Thus there are $2 \times 6 = 12$ possible matchings. For example, we can use the one–one correspondence

$p \leftrightarrow \times$
$q \leftrightarrow +$
$r \leftrightarrow =$
$s \leftrightarrow -$
$t \leftrightarrow \div$

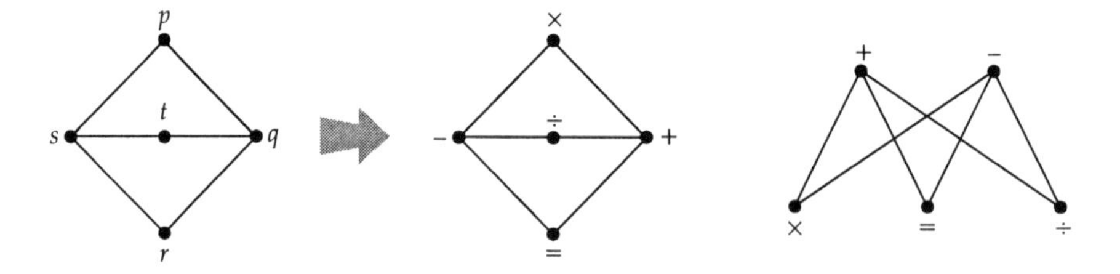

2.5 No, the graphs are not isomorphic. One way of seeing this is to look at the four vertices where just two edges meet – in the first graph they are the vertices 3, 4, 7 and 8, which are adjacent in pairs, whereas in the second graph they are b, d, f and h, and none is adjacent to any other.

2.6 One possible labelling is as follows:

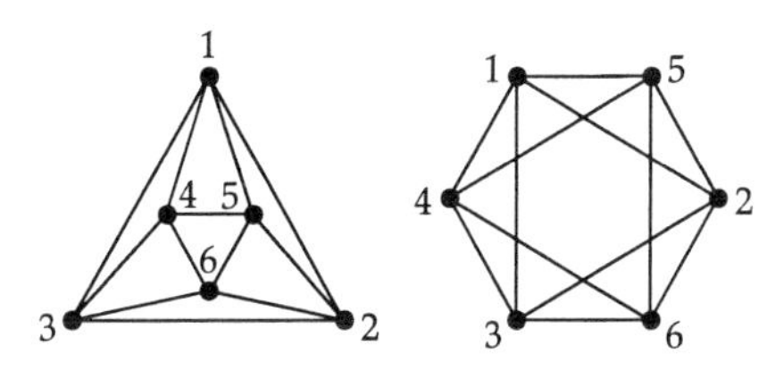

2.7 Graphs (a) and (b) are subgraphs of G; graph (c) is not a subgraph of G, as it contains the edges uw and vx which are not edges of G.

2.8 Graph (c) is a subgraph of H; graphs (a) and (b) are not.

2.9 The degree sequences are:
(a) (1, 1, 1, 1, 1, 1, 2, 4, 4); (b) (4, 4, 4, 4, 4); (c) (0, 1, 3, 4, 4, 5, 5).

2.10

graph	number of edges	sum of the vertex degrees
(a)	8	16
(b)	10	20
(c)	11	22

In each case, the sum of the vertex degrees is exactly twice the number of edges; the reason for this is given in the following text.

2.11 (a) Let G be a graph in which the number of vertices of odd degree is odd; then the sum of all the vertex degrees is also odd. But we know from the handshaking lemma that the sum of the vertex degrees is twice the number of edges, and is therefore even. We have obtained a contradiction, so no such graph G exists. Thus, in any graph, the number of vertices of odd degree is even.

(b) The three graphs (a), (b) and (c) have, respectively, 6, 0 and 4 vertices of odd degree, and these numbers are all even.

2.12 (a) trail, 5, x, y; (b) path, 3, u, z.
Alternative answers are possible; for example, the answer *walk* is appropriate in each case.

2.13 length 3: *stzy*;
length 4: *stzxy* and *svtzy*;
length 5: *stzwxy*, *svtzxy* and *svutzy*;
length 6: *svutzxy* and *svtzwxy*;
length 7: *svutzwxy*.

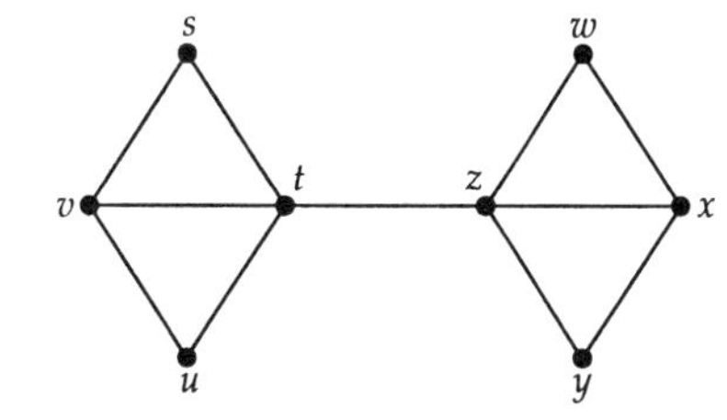

2.14 There are various possibilities – for example:

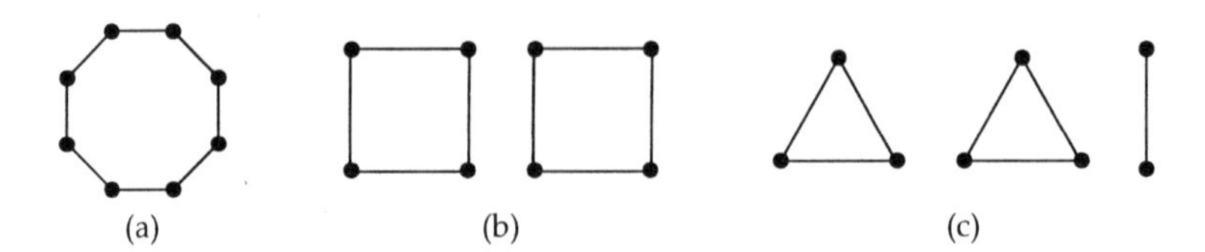

2.15

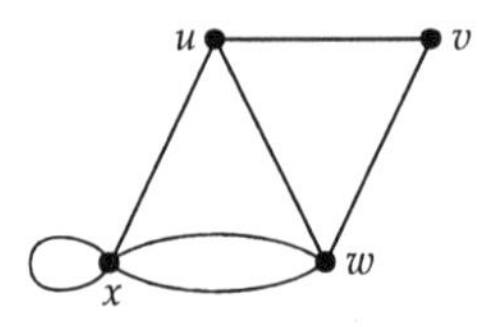

(a) There are various possibilities – for example, *uvu* or *uvwuxwu*.
(b) Again, there are various possibilities – for example, *uwxxu*.
(c) length 1: the loop *xx*;
length 2: the multiple edges *wxw*;
length 3: the triangle *uvwu*, and both of the triangles *uwxu*;
length 4: both of the 'quadrilaterals' *uvwxu*.

2.16 There are various possibilities – for example:

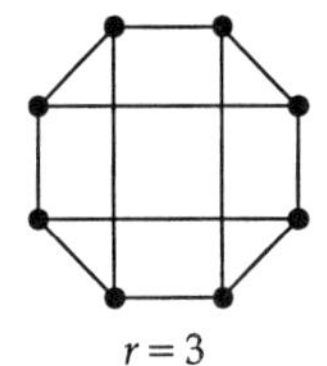
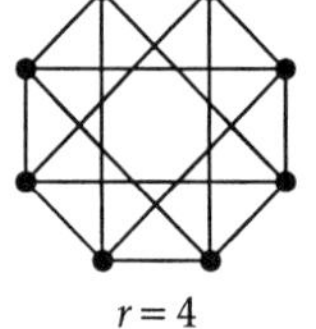
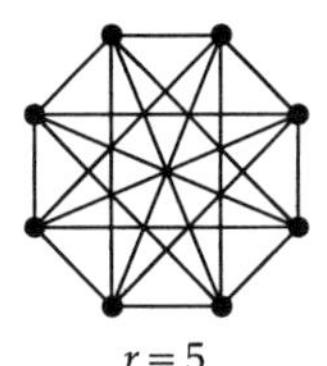

$r=3$ $r=4$ $r=5$

2.17 (a) $n = 5, r = 2$, so the number of edges is $(5 \times 2)/2 = 5$;
(b) $n = 10, r = 3$, so the number of edges is $(10 \times 3)/2 = 15$;
(c) $n = 12, r = 5$, so the number of edges is $(12 \times 5)/2 = 30$.
You can check that these numbers are correct by counting the edges in each case.

2.18 (a) Suppose that such a graph exists. Then it has an odd number of vertices of odd degree, contradicting the result of Problem 2.11(a).
Alternatively, such a graph has $(7 \times 3)/2 = 10\frac{1}{2}$ edges, which is impossible. Thus no such graph exists.
(b) Suppose that such a graph exists. Then it has an odd number of vertices of odd degree, contradicting the result of Problem 2.11(a).
Alternatively, such a graph has $nr/2$ edges; since n and r are odd, so is nr, and hence $nr/2$ is not an integer. Thus no such graph exists.

2.19

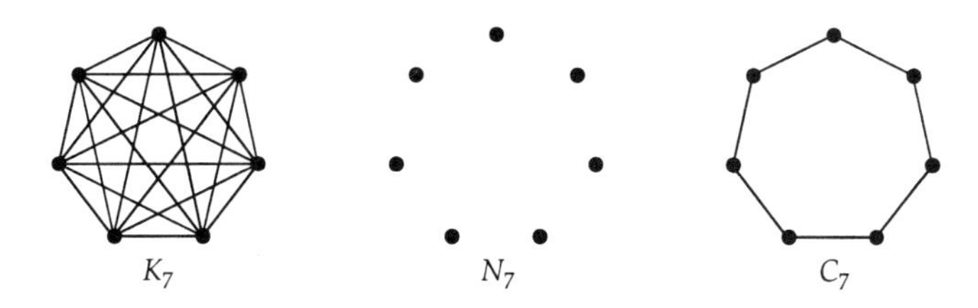

These graphs have, respectively, 21, 0 and 7 edges.

2.20 Let G be a bipartite graph. If we colour the vertices of G black and white, then the vertices in each cycle must alternate between these two colours. This implies that the number of edges in every cycle is even.

2.21 (a) There are various possible drawings – for example:

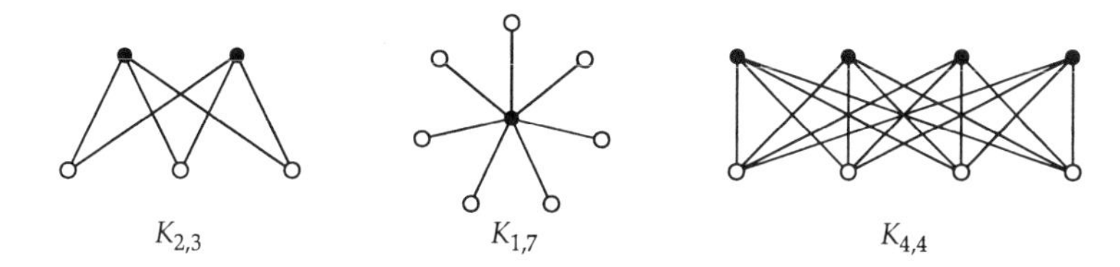

These graphs have, respectively,
5 vertices and 6 edges;
8 vertices and 7 edges;
8 vertices and 16 edges.

(b) $K_{r,s}$ is a regular graph when r and s are equal.

2.22

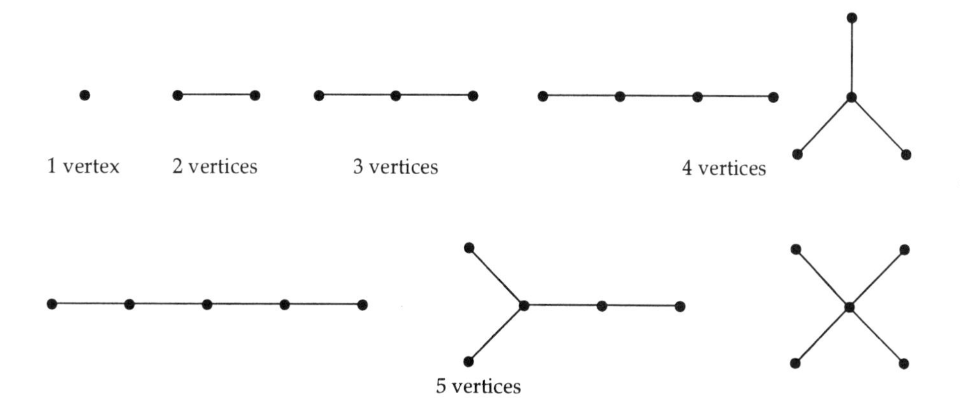

2.23 (a) Choose any vertex v in a tree T and colour it black. Colour all vertices adjacent to v white. Next, colour all vertices adjacent to these black. Continue this process until every vertex has been coloured.

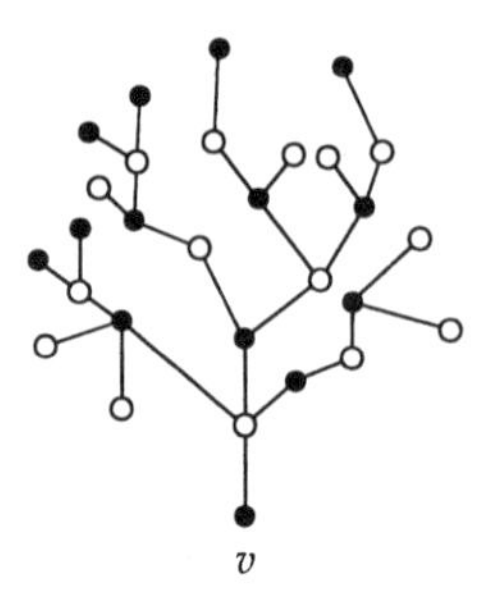

Since T is a tree, there is just one path between any two vertices. Thus, by the way the vertices have been coloured, no two adjacent vertices in T have the same colour. So T is bipartite.

(b) Every tree can be built up from a single vertex by successively adding an edge and a new vertex, as often as necessary. At each stage we increase the number of vertices by 1 and the number of edges by 1. Since we start with 1 vertex and 0 edges, we must end up with n vertices and $n-1$ edges.

2.24 Using the method described in the text, we obtain the following superimposed graph G:

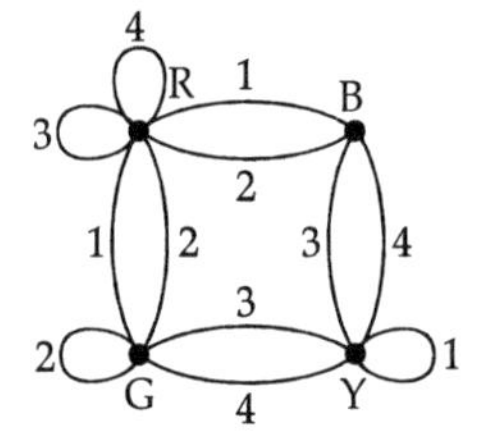

A pair of subgraphs H_1 and H_2 and a corresponding solution are as follows.

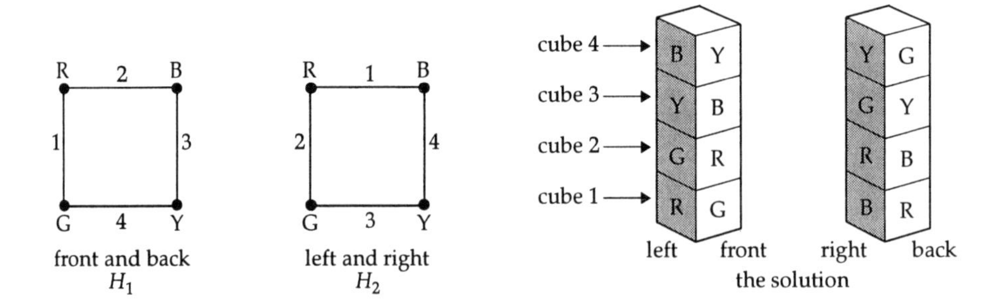

There are several other solutions.

2.25 The signed graphs (a) and (c) are balanced. The corresponding bipartite graphs are

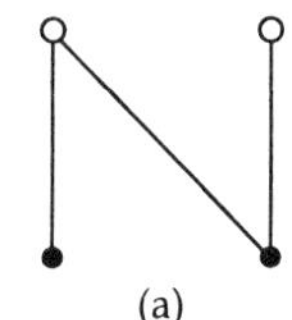
(a)

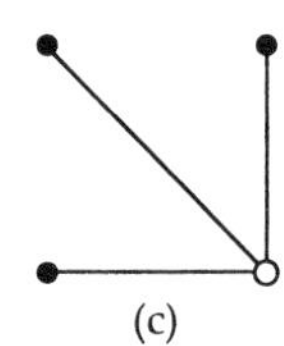
(c)

Chapter 3

3.1

	Eulerian?	Eulerian trail	Hamiltonian?	Hamiltonian cycle
(a)	no	–	yes	*a b c d a*
(b)	yes	*a b c d e a c e b d a*	yes	*a b c d e a*
(c)	no	–	yes	*a b f e h g d a*
(d)	yes	*a b c a d c f b e f d e a*	yes	*a b c d f e a*
(e)	no	–	no	–
(f)	no	–	yes	*a d b e c f a*
(g)	no	–	yes	*a b c d a*

In each case where an Eulerian trail or Hamiltonian cycle exists, there are several different trails/cycles.

3.2 (a) If an Eulerian trail exists in a graph, then whenever you go into a vertex, you must be able to leave it by another edge. It follows that each time you pass through a vertex you contribute 2 to the degree of that vertex. (This is also true of the first and last edges, which contribute 2 to the degree of the starting vertex.) So, in an Eulerian graph, each vertex degree must be a sum of 2s – that is, an even number.

(b) The rule is as follows.

To show that a given connected graph is Eulerian, demonstrate that *all* the vertices have even degree.

To show that a given connected graph is *not* Eulerian, exhibit just *one* vertex of odd degree.

In Problem 3.1,

graph (a) has vertex degrees 3, 3, 3, 3, so is not Eulerian;
graph (b) has vertex degrees 4, 4, 4, 4, 4, so is Eulerian;
graph (c) has vertex degrees 3, 3, 3, 3, 3, 3, 3, 3, so is not Eulerian;
graph (d) has vertex degrees 4, 4, 4, 4, 4, 4, so is Eulerian;
graph (e) has vertex degrees 3, 3, 2, 2, 2, so is not Eulerian;
graph (f) has vertex degrees 3, 3, 3, 3, 3, 3, so is not Eulerian;
graph (g) has vertex degrees 5, 3, 3, 3, so is not Eulerian.

For each of the graphs (a), (c), (e), (f) and (g), it is sufficient to exhibit just *one* vertex of odd degree.

3.3

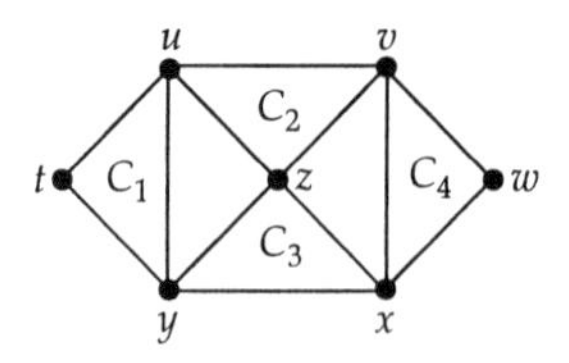

One possibility uses the cycles C_1, C_2, C_3 and C_4.

If we trace around C_1, 'picking up' C_2 and C_3 as we go, we get the closed trail *tuvzuyzxyt*. This trail misses out C_4, which can be inserted, on tracing round this trail, at the vertex v to give the Eulerian trail *tuvwxvzuyzxyt*.

3.4 (a) K_8 is not Eulerian, since it is regular of degree 7;

(b) $K_{8,8}$ is Eulerian, since it is regular of degree 8;

(c) C_8 is Eulerian, since it is regular of degree 2;

(d) the dodecahedron graph is not Eulerian, since it is regular of degree 3;

(e) Q_8 is Eulerian, since it is regular of degree 8.

3.5 Graph (a) is semi-Eulerian, since a and b are the only vertices of odd degree; a suitable open trail is *acbdaeb*, starting at a and ending at b.

Graph (b) is not semi-Eulerian, since it has four vertices of odd degree.

Graph (c) is semi-Eulerian, since the only vertices of odd degree are w and z; a suitable open trail is *wxyzuvwyuwzvxz*.

3.6 There are two such Hamiltonian cycles:

JVTSRWXZQPNMLKFDCBGHJ and *JVTSRWXHGFDCBZQPNMLKJ*.

Note that the letter after R must be W, since otherwise W would have to be omitted.

3.7 There is only one such path – *BCDFGHXZQPNMLKJVWRST*.

3.8 (a) The graph $K_{4,4}$ is Hamiltonian; a suitable Hamiltonian cycle is *ahbgcfdea*.

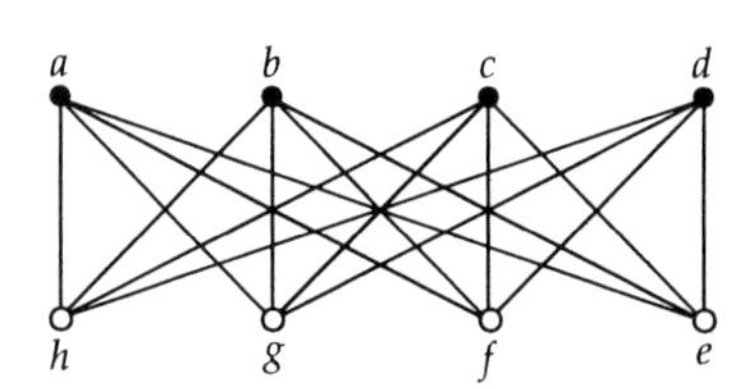

(b) A tree does not contain a cycle, so the only tree which is Hamiltonian is the trivial tree with one vertex and no edges.

3.9 (a) The vertices of any bipartite graph can be split into two sets A and B in such a way that each edge has one end in A and one end in B.

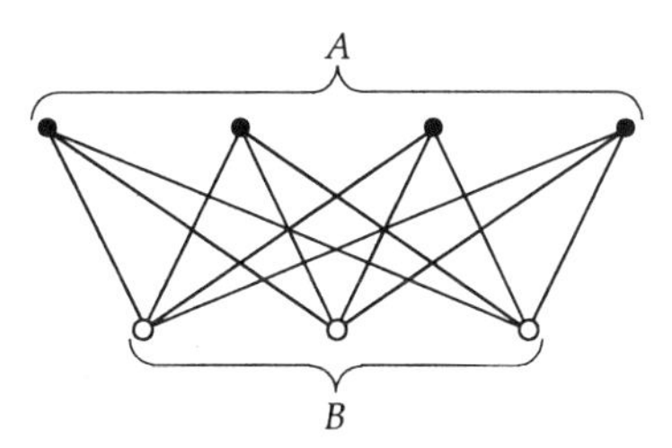

Any Hamiltonian cycle must alternate between these two sets, ending in the same set as it started. It follows that if a bipartite graph is Hamiltonian then the sets A and B must have the same number of vertices. This is impossible if the total number of vertices is odd.

(b)

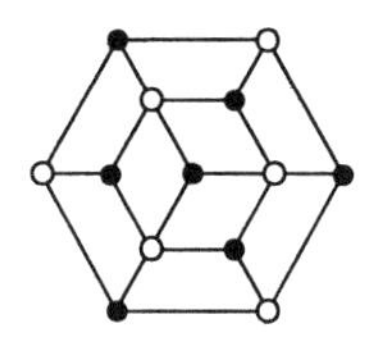

The graph is a bipartite graph with an odd number of vertices, so cannot be Hamiltonian, by part (a).

3.10 (a) If $\deg v \geq n/2$ for each vertex v, then $\deg v + \deg w \geq n$ for each pair of vertices v and w, whether adjacent or not. The result now follows from Ore's theorem.

(b) Any cycle graph C_n, where $n \geq 5$, is Hamiltonian, but does not satisfy the conditions of Ore's theorem, because each vertex has degree 2.

3.11 Graph (a) is semi-Hamiltonian – a suitable path is *cadbe*.
Graph (b) is Hamiltonian – a suitable cycle is *abcda* – so is not semi-Hamiltonian.
Graph (c) is Hamiltonian – a suitable cycle is *vwxyuzv* – so is not semi-Hamiltonian.

3.12 There are many possibilities. For example, if we take the Eulerian trail 01234024130, and add the 'doubles' in a suitable way, we obtain the following ring:

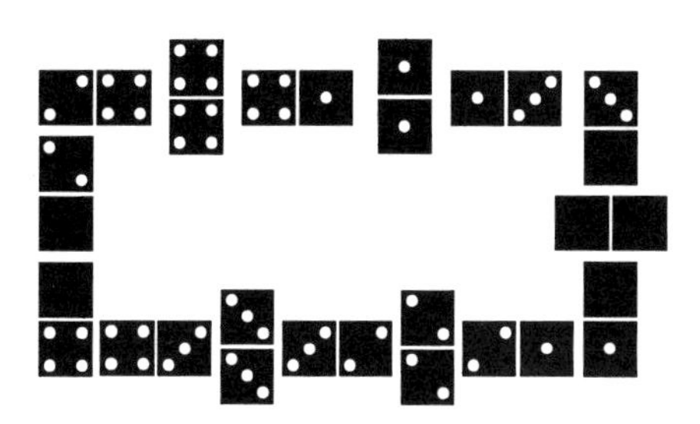

3.13 The diagram has four vertices of odd degree, so at least two continuous pen-strokes are required; in fact, two are sufficient.

3.14 If we add $k/2$ edges to G, joining the k vertices of odd degree in pairs, we obtain a new graph G' in which each vertex has even degree. It follows that G' contains an Eulerian trail. If we now write out this trail, and then omit the added edges, we get the required $k/2$ pen-strokes.

3.15 The graph associated with any chessboard is bipartite, since a knight's move always takes a knight to a square of a different colour. So we can take A to be the set of black squares and B to be the set of white squares. The result now follows immediately from the fact that a bipartite graph with an odd number of vertices is not Hamiltonian. (See Problem 3.9(a).)

(In other words, since a knight always moves from a black square to a white square, or *vice versa*, the number of black squares must equal the number of white squares. But this is impossible for any board with an odd number of squares.)

3.16 We find another Hamiltonian cycle in the 4-cube. One possibility is

$$0000 \to 0100 \to 1100 \to 1110 \to 1111 \to 1011 \to 0011 \to 0001$$
$$\to 1001 \to 1101 \to 0101 \to 0111 \to 0110 \to 0010 \to 1010 \to 1000 \; (\to 0000).$$

Chapter 4

4.1 (a) vertices: $\{a, b, c, d\}$
arcs: $\{ba, bd, cb, da, db, dc\}$
Digraph (a) is a simple digraph.

(b) vertices: $\{0, 1, 2, 3, 4\}$
arcs: $\{10, 12, 32, 40, 43\}$
Digraph (b) is a simple digraph.

4.2

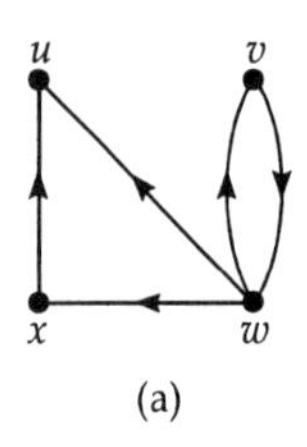

(a)

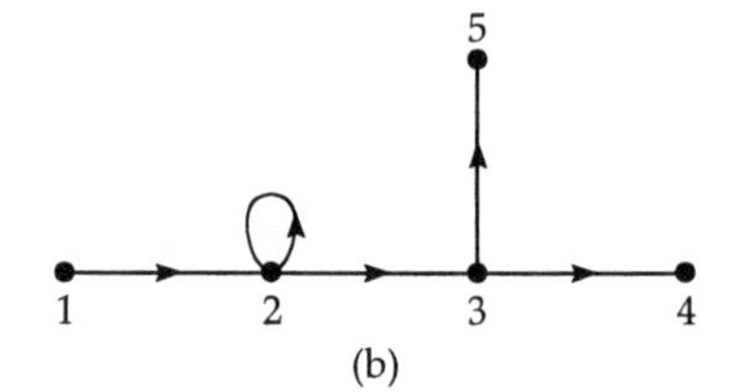

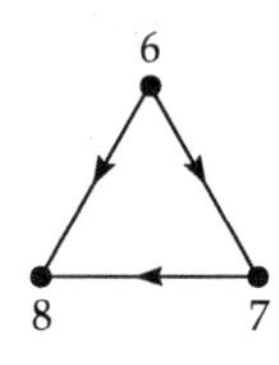

(b)

Digraph (a) is a simple digraph.
Digraph (b) is not a simple digraph, because there is a loop at the vertex 2.

4.3 (a) yes; (b) no; (c) yes; (d) yes.

4.4 To show that the digraphs are isomorphic, we must match up:

the vertices with a loop: c and 1;

the vertices where six arcs meet: d and 3;

the vertices where five arcs meet: a and 2;

the other two vertices: b and 4.

Thus, to show that the two digraphs are isomorphic, we use the one-one correspondence

$a \leftrightarrow 2$

$b \leftrightarrow 4$

$c \leftrightarrow 1$

$d \leftrightarrow 3$

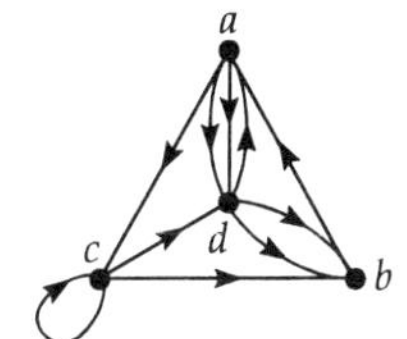

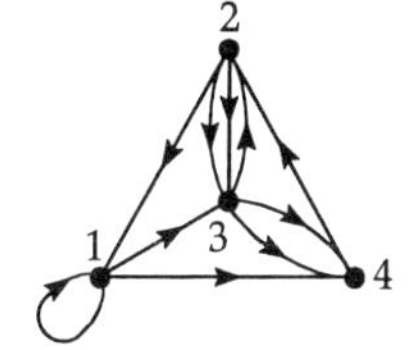

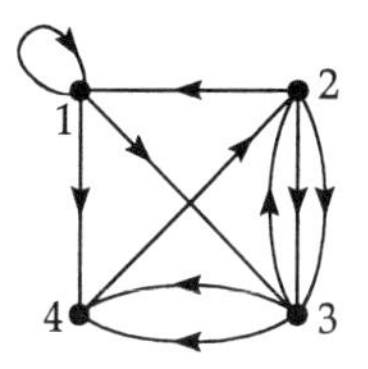

4.5 No, the digraphs are not isomorphic. One way of seeing this is to notice that the second digraph has a vertex with three emerging arcs (vertex 1), whereas the first digraph has no such vertex. Another way is to notice that the first digraph has a 'directed triangle' ($a \rightarrow b \rightarrow c \rightarrow a$), whereas the second digraph has no such triangle.

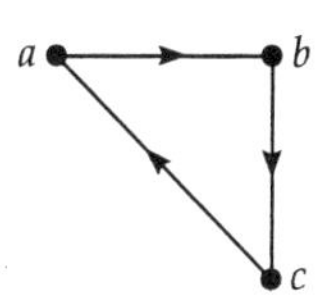

4.6 One possible labelling is as follows:

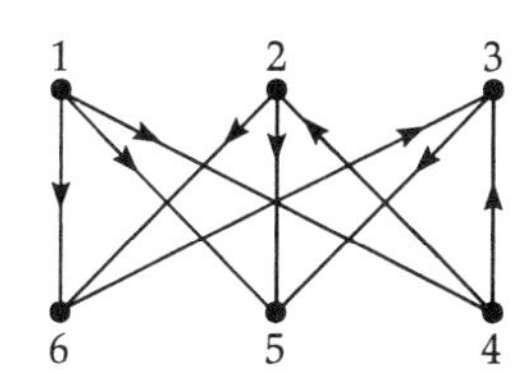

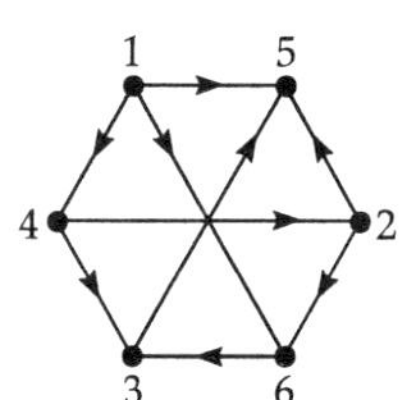

4.7 Digraph (a) is a subdigraph of D; digraph (b) is not a subdigraph of D, as it contains the arc xv which is not an arc of D; digraph (c) is not, as it contains the arc ux which is not an arc of D.

4.8 Digraphs (a) and (b) are subdigraphs of C; digraph (c) is not.

4.9 The out-degree and in-degree sequences are:

(a) out-degree sequence: (0, 1, 1, 1, 1, 1, 1, 1, 1);
in-degree sequence: (0, 0, 0, 0, 0, 0, 1, 3, 4);

(b) out-degree sequence: (1, 2, 2, 2, 3);
in-degree sequence: (1, 2, 2, 2, 3);

(c) out-degree sequence: (1, 1, 2, 2, 2, 3);
in-degree sequence: (0, 0, 2, 3, 3, 3).

4.10

digraph	number of arcs	sum of out-degrees	sum of in-degrees
(a)	8	8	8
(b)	10	10	10
(c)	11	11	11

In each case, the sum of the out-degrees and the sum of the in-degrees are both equal to the number of arcs; the reason for this is given in the following text.

4.11 (a) For any digraph, the handshaking dilemma holds, so the sum of all the out-degrees is equal to the sum of all the in-degrees. But the sum of the out-degrees and the sum of the in-degrees must both be odd. It follows that the number of vertices with odd in-degree must be odd.

(b) For this digraph, the out-degrees and in-degrees are as follows:

vertex	s	t	u	v	w	x	y	z
out-degree	2	1	2	1	2	1	2	2
in-degree	1	2	1	2	1	2	2	2

The number of vertices with odd out-degree is odd (3);
the number of vertices with odd in-degree is odd (3).

4.12 (a) length 5: *tsyzvw*
length 6: *tsxyzvw* and *tsyzuvw*
length 7: *tsxyzuvw*

(b) length 3: *wxyt* and *wzut*
length 5: *wxyzut*;

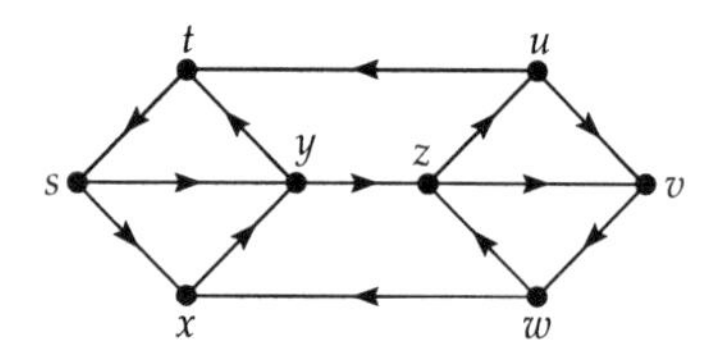

(c) There are two possibilities:
tsyzvwxyt and *tsyzvwzut*.

(d) Any cycle containing both t and w consists of a path from t to w followed by a path from w to t. But all paths from t to w contain both y and z (see part (a)), and all paths from w to t contain y or z (see part (b)), so that either y or z occurs twice. Since this is not allowed, there can be no cycle containing both t and w.

4.13 Digraph (a) is connected, but not strongly connected, since there are no paths from the centre vertex to any other.

Digraph (b) is strongly connected.

Digraph (c) is disconnected.

Digraph (d) is connected, but not strongly connected, since, for example, there are no paths from the top right-hand vertex to any other.

4.14

	Eulerian?	Eulerian trail	Hamiltonian?	Hamiltonian cycle
(a)	no	–	yes	$a\,b\,d\,c\,a$
(b)	yes	$a\,b\,d\,e\,c\,d\,a\,c\,b\,e\,a$	yes	$a\,b\,e\,c\,d\,a$
(c)	no	–	yes	$a\,b\,e\,c\,d\,a$

4.15 (a) A digraph is Eulerian if and only if the out-degree and in-degree of each vertex are equal.

(b) In digraph (a), there is no vertex whose out-degree and in-degree are equal, so digraph (a) is not Eulerian.

In digraph (b), the out-degree and in-degree of each vertex are equal, so digraph (b) is Eulerian.

In digraph (c), the out-degree and in-degree of the vertices a and e are not equal, so digraph (c) is not Eulerian.

4.16

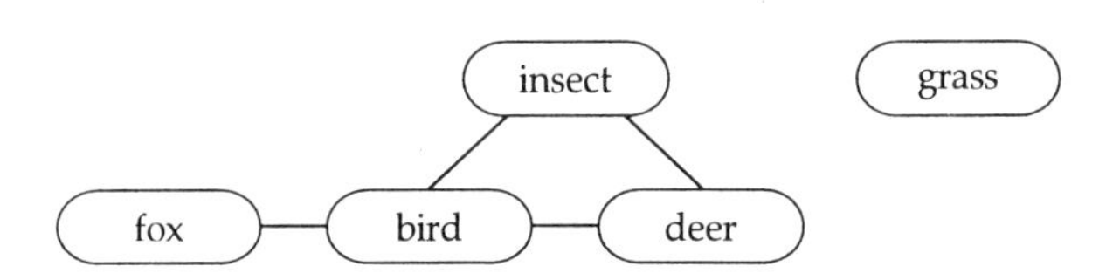

4.17 Positive feedback cycle: *cefhgc*.

Negative feedback cycles: *ghg, ahga, abdfhga, ceifhgc*.

4.18 Another Eulerian trail is

$$101 \to 010 \to 100 \to 001 \to 011 \to 110 \to 100 \to 000 \to$$
$$000 \to 001 \to 010 \to 101 \to 011 \to 111 \to 110 \to 101$$

This can be compressed to give the 16-bit sequence

1 0 1 0 0 1 1 0 0 0 0 1 0 1 1 1.

This leads to the following solution to the rotating drum problem for 16 divisions.

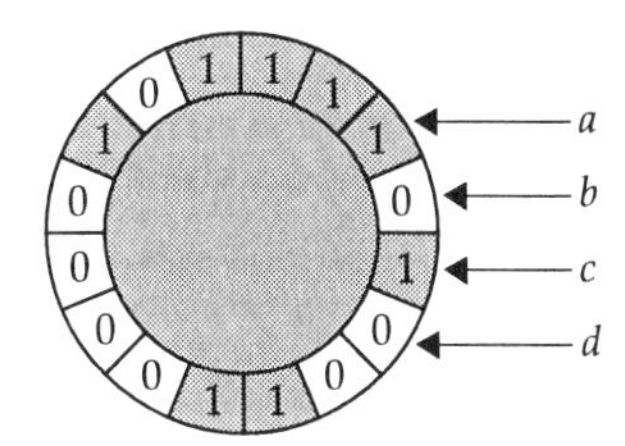

4.19 Five: *abecd, bcaed, becad, cabed, ecabd.*

Chapter 5

5.1

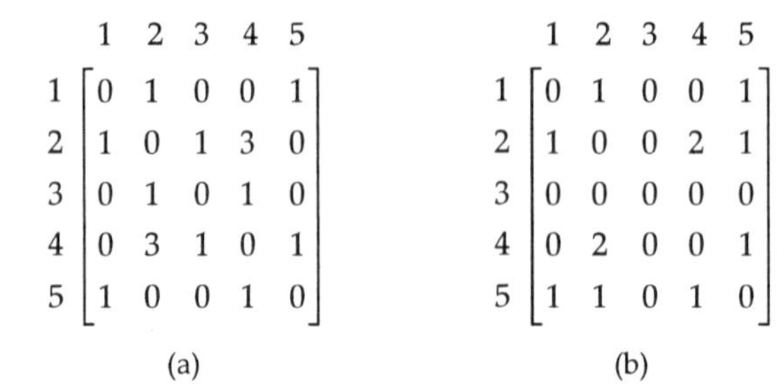

$$\begin{array}{c} \\ 1\\2\\3\\4\\5 \end{array}\begin{array}{c} \begin{array}{ccccc}1&2&3&4&5\end{array}\\ \begin{bmatrix} 0&1&0&0&1\\ 1&0&1&3&0\\ 0&1&0&1&0\\ 0&3&1&0&1\\ 1&0&0&1&0 \end{bmatrix}\end{array} \qquad \begin{array}{c} \\ 1\\2\\3\\4\\5 \end{array}\begin{array}{c} \begin{array}{ccccc}1&2&3&4&5\end{array}\\ \begin{bmatrix} 0&1&0&0&1\\ 1&0&0&2&1\\ 0&0&0&0&0\\ 0&2&0&0&1\\ 1&1&0&1&0 \end{bmatrix}\end{array}$$

(a) (b)

5.2

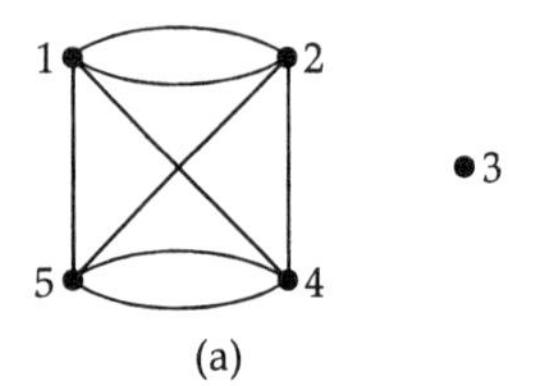

(a)

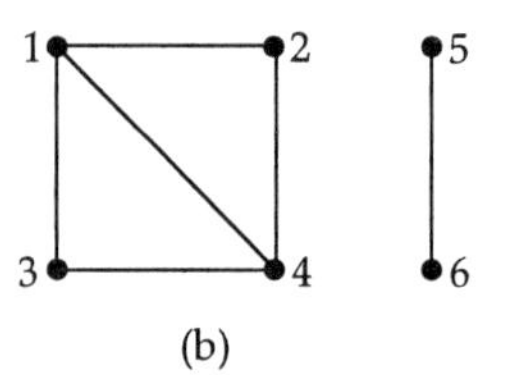

(b)

5.3

$$\begin{array}{c} \\ 1\\2\\3\\4\\5 \end{array}\begin{array}{c} \begin{array}{ccccc}1&2&3&4&5\end{array}\\ \begin{bmatrix} 0&1&0&0&0\\ 0&0&1&1&0\\ 0&0&0&0&0\\ 0&2&1&0&0\\ 1&0&0&1&0 \end{bmatrix}\end{array} \qquad \begin{array}{c} \\ 1\\2\\3\\4\\5 \end{array}\begin{array}{c} \begin{array}{ccccc}1&2&3&4&5\end{array}\\ \begin{bmatrix} 0&0&0&1&0\\ 1&0&0&0&0\\ 1&1&0&1&0\\ 1&0&0&0&1\\ 1&0&0&0&0 \end{bmatrix}\end{array}$$

(a) (b)

5.4

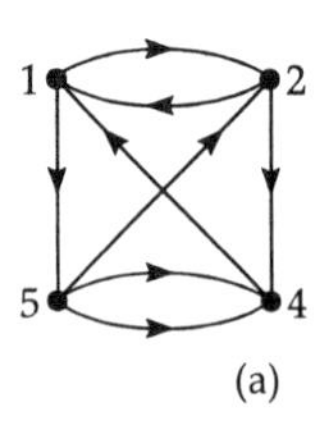

(a)

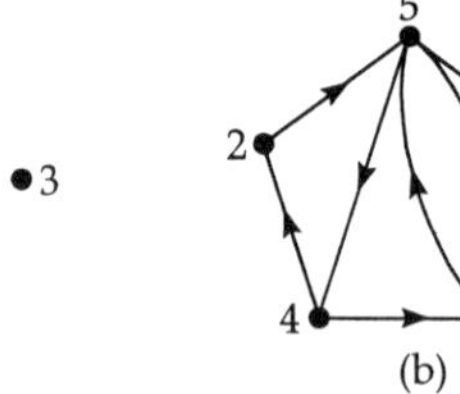

(b)

5.5 (a)

	a	b	c	d
a	0	2	1	0
b	0	0	0	1
c	1	0	0	0
d	2	1	0	0

numbers of walks of length 2

	a	b	c	d
a	2	1	0	0
b	0	2	1	0
c	0	0	0	1
d	1	0	0	2

numbers of walks of length 3

(b) $$\mathbf{A}^2 = \begin{bmatrix} 0 & 2 & 1 & 0 \\ 0 & 0 & 0 & 1 \\ 1 & 0 & 0 & 0 \\ 2 & 1 & 0 & 0 \end{bmatrix}, \quad \mathbf{A}^3 = \begin{bmatrix} 2 & 1 & 0 & 0 \\ 0 & 2 & 1 & 0 \\ 0 & 0 & 0 & 1 \\ 1 & 0 & 0 & 2 \end{bmatrix}$$

(c) The tables in part (a) correspond to the matrix products in part (b). For example, the number of walks of length 2 between vertices d and a is the entry in row 4 (corresponding to d) and column 1 (corresponding to a) in the matrix $\mathbf{A}^2$; a similar result holds for the number of walks of length 3 and the entries in $\mathbf{A}^3$.

5.6

$$\mathrm{A} = \begin{bmatrix} 0 & 1 & 0 & 0 & 1 \\ 1 & 0 & 1 & 0 & 0 \\ 0 & 0 & 0 & 2 & 0 \\ 0 & 0 & 0 & 0 & 1 \\ 0 & 0 & 1 & 1 & 0 \end{bmatrix}, \quad \mathbf{A}^2 = \begin{bmatrix} 1 & 0 & 2 & 1 & 0 \\ 0 & 1 & 0 & 2 & 1 \\ 0 & 0 & 0 & 0 & 2 \\ 0 & 0 & 1 & 1 & 0 \\ 0 & 0 & 0 & 2 & 1 \end{bmatrix},$$

$$\mathbf{A}^3 = \begin{bmatrix} 0 & 1 & 0 & 4 & 2 \\ 1 & 0 & 2 & 1 & 2 \\ 0 & 0 & 2 & 2 & 0 \\ 0 & 0 & 0 & 2 & 1 \\ 0 & 0 & 1 & 1 & 2 \end{bmatrix}, \quad \mathbf{A}^4 = \begin{bmatrix} 1 & 0 & 3 & 2 & 4 \\ 0 & 1 & 2 & 6 & 2 \\ 0 & 0 & 0 & 4 & 2 \\ 0 & 0 & 1 & 1 & 2 \\ 0 & 0 & 2 & 4 & 1 \end{bmatrix}.$$

The numbers of walks from b to d of lengths 1, 2, 3 and 4 are given by the entries in row 2 column 4 of the matrices $\mathbf{A}$, $\mathbf{A}^2$, $\mathbf{A}^3$ and $\mathbf{A}^4$, respectively – namely, 0, 2, 1 and 6.

There is no walk of length 1, 2, 3 or 4 from d to b, because each of the matrices $\mathbf{A}$, $\mathbf{A}^2$, $\mathbf{A}^3$ and $\mathbf{A}^4$ has 0 in row 4 column 2.

5.7

$$\mathbf{B} = \mathbf{A} + \mathbf{A}^2 + \mathbf{A}^3 + \mathbf{A}^4 = \begin{bmatrix} 2 & 2 & 5 & 7 & 7 \\ 2 & 2 & 5 & 9 & 5 \\ 0 & 0 & 2 & 8 & 4 \\ 0 & 0 & 2 & 4 & 4 \\ 0 & 0 & 4 & 8 & 4 \end{bmatrix}.$$

The matrix $\mathbf{B}$ contains some zero entries off the main diagonal, so the digraph is not strongly connected, by Theorem 5.2.

Note that this fact was already clear from the matrix $\mathbf{A}$ in Solution 5.6, so in this case we do not need to go on and calculate $\mathbf{B}$ explicitly.

5.8 The adjacency matrix **A** is a 5×5 matrix, so the digraph has five vertices. We therefore need to find $\mathbf{A}, \mathbf{A}^2, \mathbf{A}^3, \mathbf{A}^4$ and $\mathbf{B} = \mathbf{A} + \mathbf{A}^2 + \mathbf{A}^3 + \mathbf{A}^4$.

$$\mathbf{A} = \begin{bmatrix} 0 & 0 & 0 & 1 & 0 \\ 1 & 0 & 1 & 0 & 0 \\ 0 & 0 & 0 & 1 & 0 \\ 0 & 0 & 0 & 0 & 1 \\ 0 & 1 & 0 & 0 & 0 \end{bmatrix}, \quad \mathbf{A}^2 = \begin{bmatrix} 0 & 0 & 0 & 0 & 1 \\ 0 & 0 & 0 & 2 & 0 \\ 0 & 0 & 0 & 0 & 1 \\ 0 & 1 & 0 & 0 & 0 \\ 1 & 0 & 1 & 0 & 0 \end{bmatrix},$$

$$\mathbf{A}^3 = \begin{bmatrix} 0 & 1 & 0 & 0 & 0 \\ 0 & 0 & 0 & 0 & 2 \\ 0 & 1 & 0 & 0 & 0 \\ 1 & 0 & 1 & 0 & 0 \\ 0 & 0 & 0 & 2 & 0 \end{bmatrix}, \quad \mathbf{A}^4 = \begin{bmatrix} 1 & 0 & 1 & 0 & 0 \\ 0 & 2 & 0 & 0 & 0 \\ 1 & 0 & 1 & 0 & 0 \\ 0 & 0 & 0 & 2 & 0 \\ 0 & 0 & 0 & 0 & 2 \end{bmatrix},$$

$$\mathbf{B} = \begin{bmatrix} 1 & 1 & 1 & 1 & 1 \\ 1 & 2 & 1 & 2 & 2 \\ 1 & 1 & 1 & 1 & 1 \\ 1 & 1 & 1 & 2 & 1 \\ 1 & 1 & 1 & 2 & 2 \end{bmatrix}.$$

All the entries in **B** off the main diagonal are non-zero, so the digraph with adjacency matrix **A** is strongly connected, by Theorem 5.2.

5.9

	1	2	3	4	5	6	7	8
①	1	0	0	0	1	0	0	0
②	1	1	0	0	0	1	1	1
③	0	1	1	0	0	0	0	0
④	0	0	1	1	0	1	1	1
⑤	0	0	0	1	1	0	0	0

(a)

	1	2	3	4	5	6	7
①	1	0	0	0	0	1	1
②	1	1	1	1	0	0	0
③	0	1	1	1	1	0	1
④	0	0	0	0	1	1	0

(b)

5.10

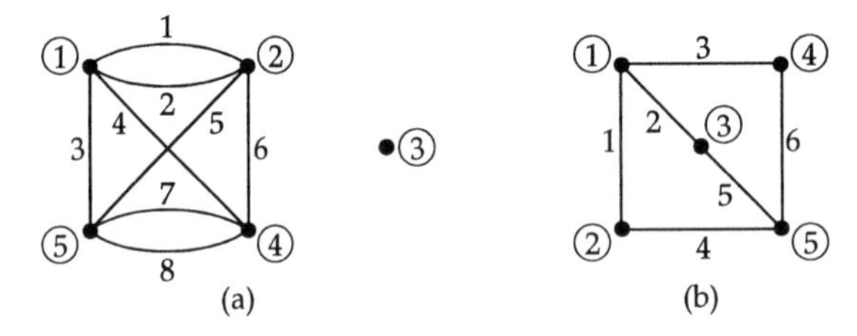

(a) (b)

5.11

	1	2	3	4	5
①	1	0	0	0	–1
②	–1	1	0	1	1
③	0	–1	1	0	0
④	0	0	–1	–1	0

(a)

	1	2	3	4	5	6	7	8
①	1	0	0	0	–1	0	0	0
②	–1	1	0	0	0	1	–1	–1
③	0	–1	–1	0	0	0	0	0
④	0	0	1	–1	0	–1	1	1
⑤	0	0	0	1	1	0	0	0

(b)

5.12

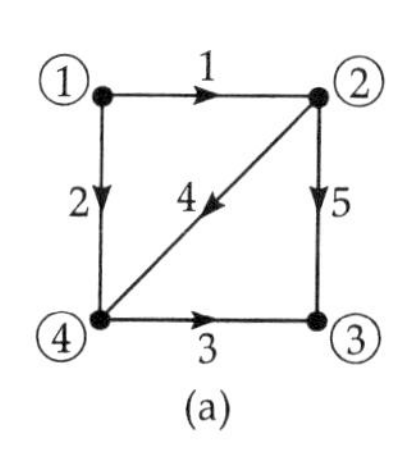

(a)

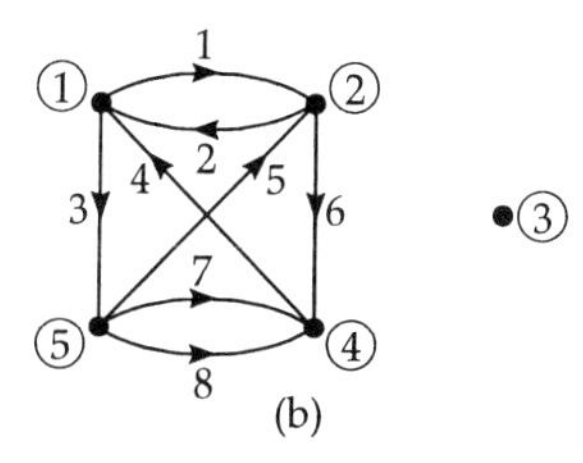

(b)

5.13

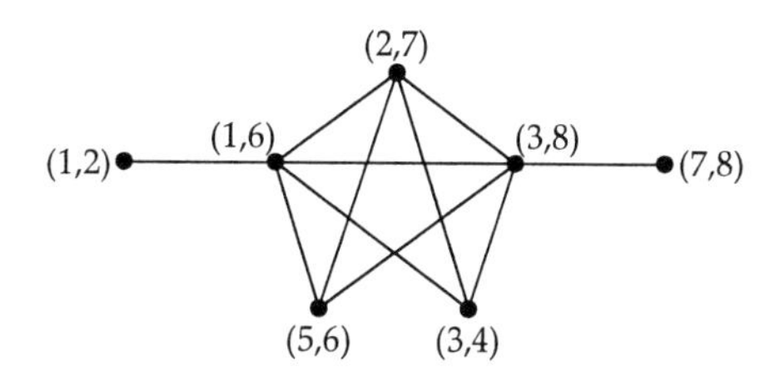

5.14 (a) $$\begin{bmatrix} 1 & 0 & 0 & 0 & 0 & 0 \\ \frac{1}{2} & \frac{1}{6} & \frac{1}{3} & 0 & 0 & 0 \\ 0 & \frac{1}{2} & \frac{1}{6} & \frac{1}{3} & 0 & 0 \\ 0 & 0 & \frac{1}{2} & \frac{1}{6} & \frac{1}{3} & 0 \\ 0 & 0 & 0 & \frac{1}{2} & \frac{1}{6} & \frac{1}{3} \\ 0 & 0 & 0 & 0 & 1 & 0 \end{bmatrix}$$

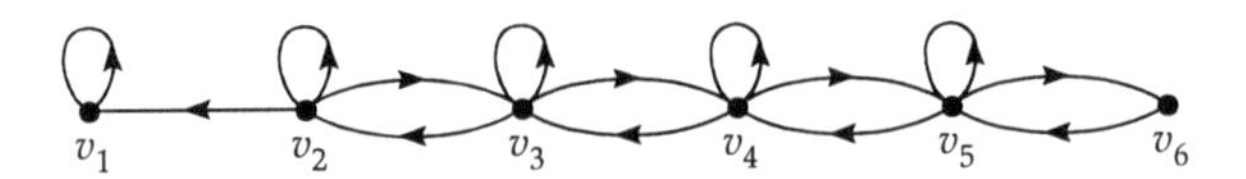

The associated digraph is not strongly connected, as there is no arc out of v_1 to another vertex, so the resulting Markov chain is not irreducible.

(b) $$\begin{bmatrix} 0 & 1 & 0 & 0 & 0 & 0 \\ \frac{1}{2} & \frac{1}{6} & \frac{1}{3} & 0 & 0 & 0 \\ 0 & \frac{1}{2} & \frac{1}{6} & \frac{1}{3} & 0 & 0 \\ 0 & 0 & \frac{1}{2} & \frac{1}{6} & \frac{1}{3} & 0 \\ 0 & 0 & 0 & \frac{1}{2} & \frac{1}{6} & \frac{1}{3} \\ 0 & 0 & 0 & 0 & 1 & 0 \end{bmatrix}$$

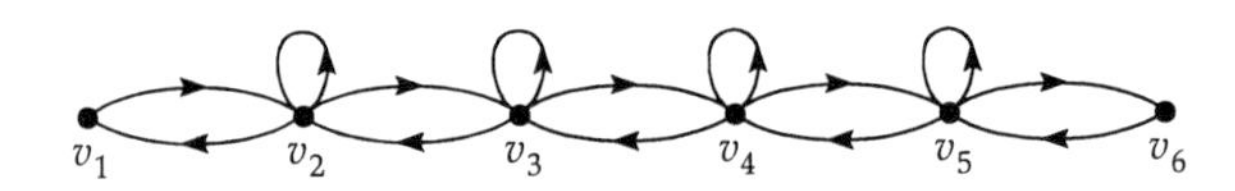

The associated digraph is strongly connected, so the resulting Markov chain is irreducible.

Chapter 6

6.1 The six unlabelled trees with six vertices are as follows.

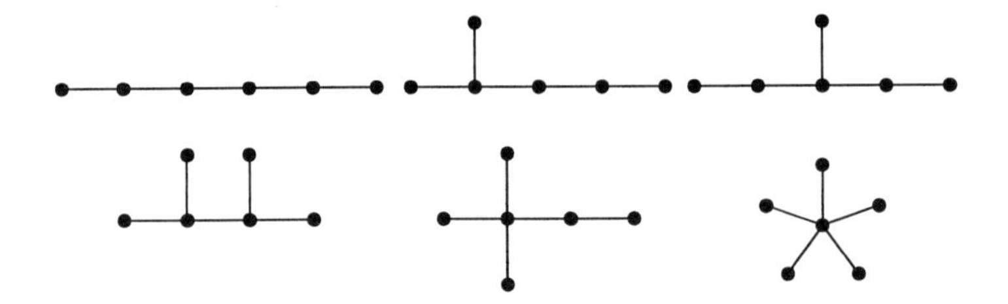

6.2 The eleven unlabelled trees with seven vertices are as follows.

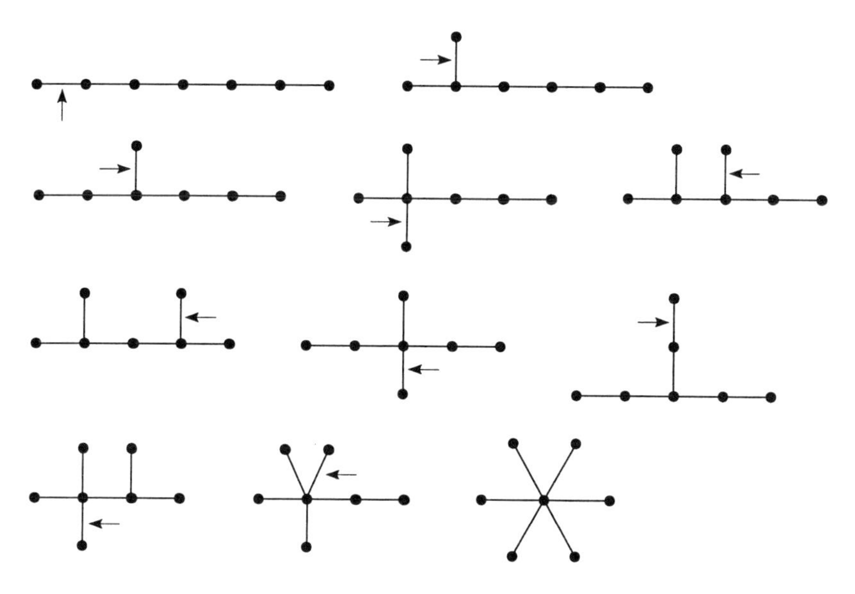

6.3 By definition, a tree is a connected graph that has no cycles.

(a) Suppose that the removal of an edge e disconnects a tree into more than two components:

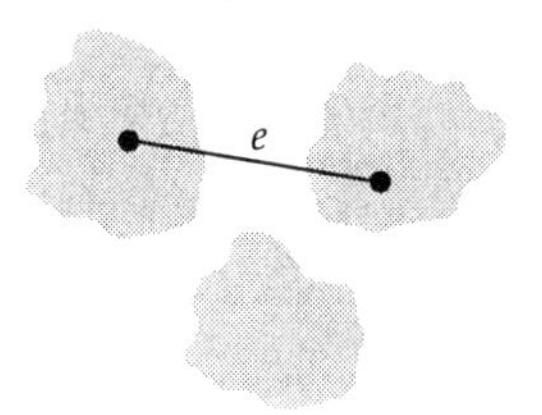

Since e joins only two vertices, it can link at most two of these components, so at least one component remains disconnected from the rest when e is reinstated in the tree. This contradicts the fact that a tree is connected. Thus, the removal of e disconnects the tree into just two components.

(b) Suppose that the addition of an edge e creates two or more cycles:

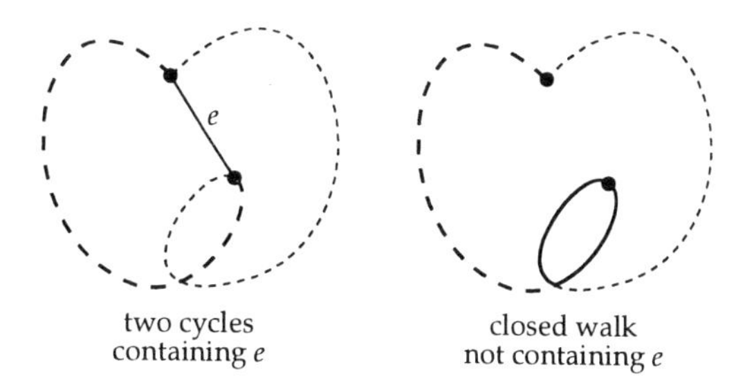

The parts of any two such cycles other than e can be combined into a closed walk that does not contain e, and this closed walk must contain a cycle. This contradicts the fact that a tree contains no cycles. Thus, the addition of a new edge cannot create more than one cycle.

6.4 (a)

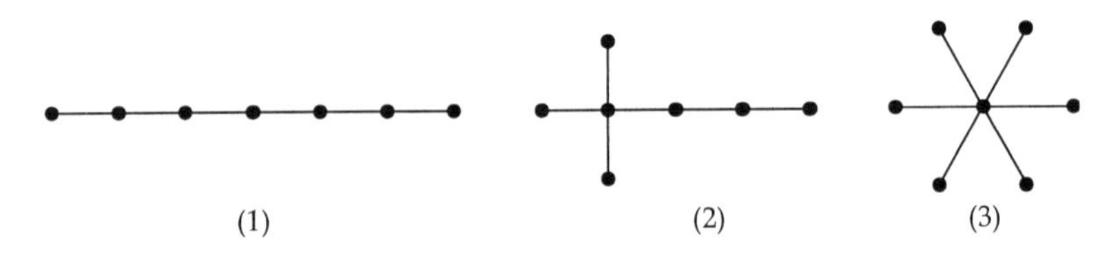

(b) Let T be a tree with n (≥ 2) vertices and at most one vertex of degree 1. Then T has at least $n-1$ vertices of degree 2 or more. It follows that the sum of the vertex degrees is at least $2(n-1)+1$, so, by the handshaking lemma, the number of edges of T is at least $(n-1)+\frac{1}{2}$. This contradicts the fact that T has exactly $n-1$ edges. Thus, T has at least two vertices of degree 1.

6.5

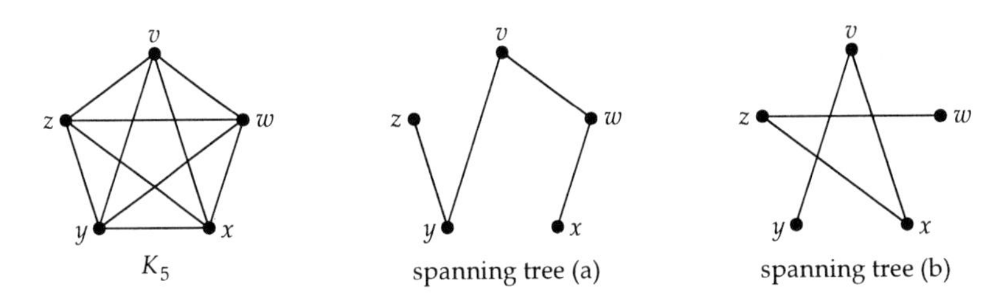

Building-up method: We choose the edges vw, wx, vy and yz; then no cycles are created, and we obtain spanning tree (a) above.

Cutting-down method: We remove the edges

vw (destroying the cycle $vwxv$),
wx (destroying the cycle $wxyw$),
xy (destroying the cycle $xyzx$),
yz (destroying the cycle $vyzv$),
vz (destroying the cycle $vxzv$),
wy (destroying the cycle $vxzwyv$);

then no cycles remain, and we obtain the spanning tree with edges vx, xz, wz, vy, that is, spanning tree (b) above.

6.6 The eighteen spanning trees (other than those depicted in the text) are the following; for clarity, the labels are omitted from these diagrams.

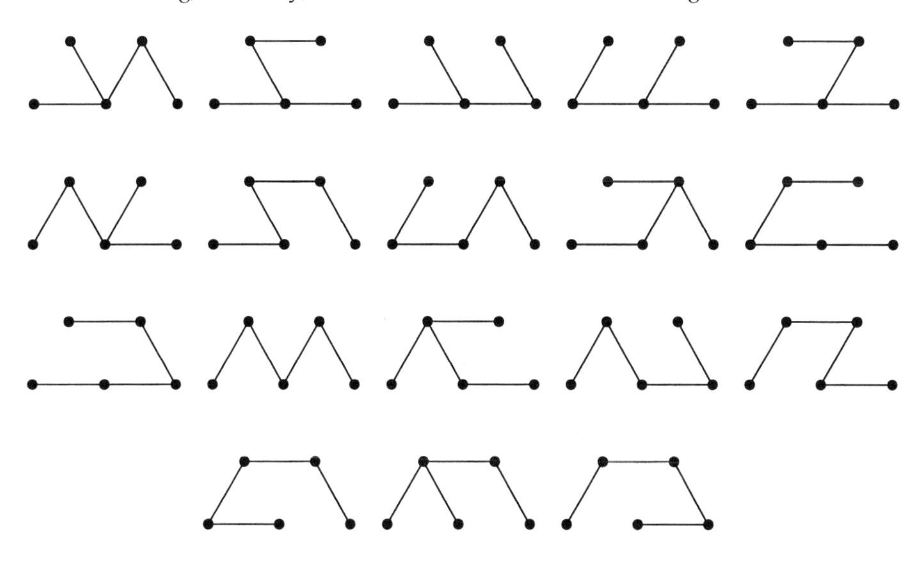

6.7 Three spanning trees of the Petersen graph are:

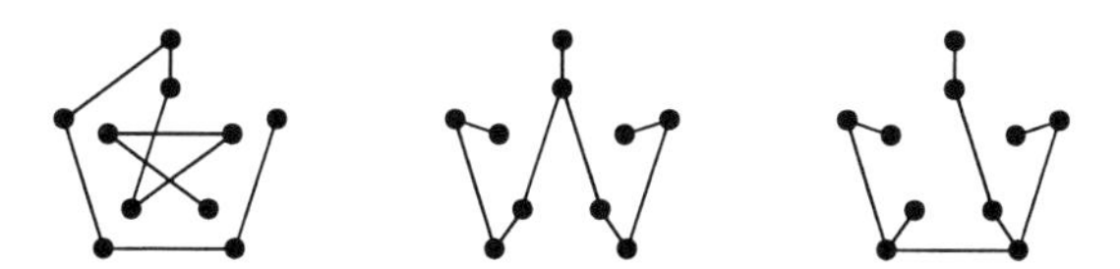

There are 1997 other possibilities, so we cannot show them all! The Petersen graph has 10 vertices, so a spanning tree is a connected graph with 9 edges linking all the vertices. Check that your trees have this property.

6.8 In the branching tree representing the outcomes of two throws of a six-sided die, there are three 'levels' (including the root) with six downward edges from each vertex:

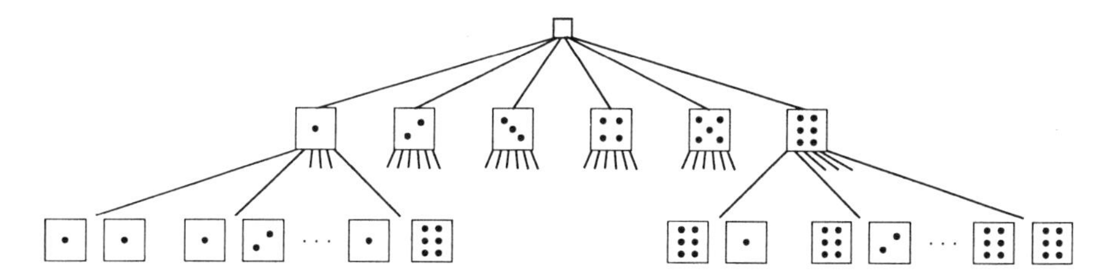

6.9 The subsets of a set corresponding to this tree can be drawn as follows.

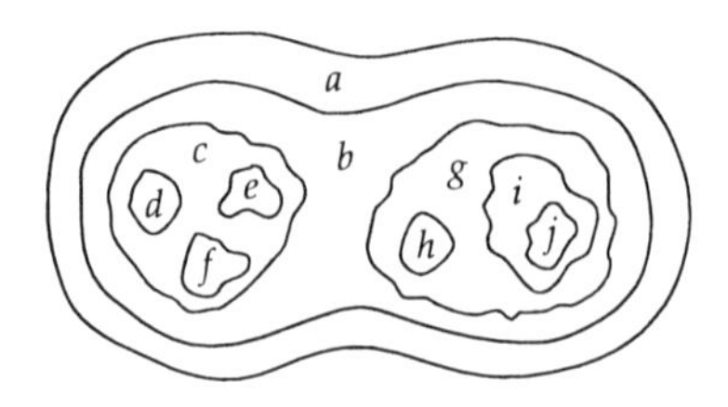

The nested parentheses corresponding to this tree are:

(((( )( )( ))(( )(( )))))

6.10

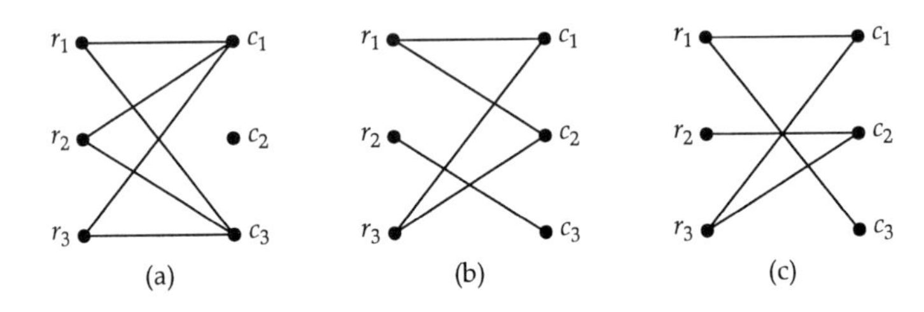

The bipartite graphs corresponding to frameworks (a) and (b) are disconnected; the bipartite graph corresponding to framework (c) is connected.

6.11

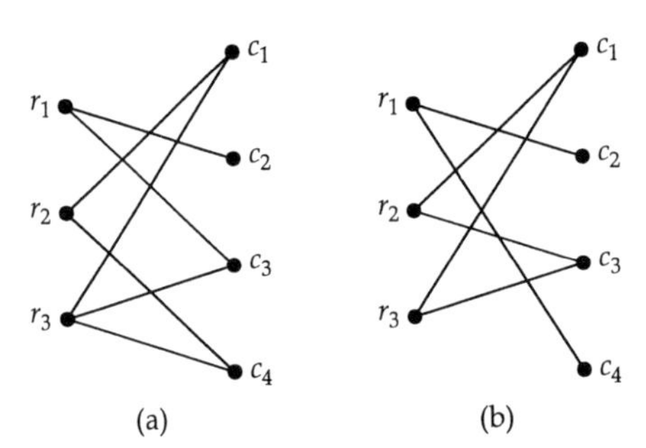

The bipartite graph corresponding to framework (a) is connected, so framework (a) is rigid.

The bipartite graph corresponding to framework (b) is disconnected – the path $c_2r_1c_4$ and the cycle $c_1r_3c_3r_2c_1$ are components of the graph – so framework (b) is not rigid.

6.12 (a)

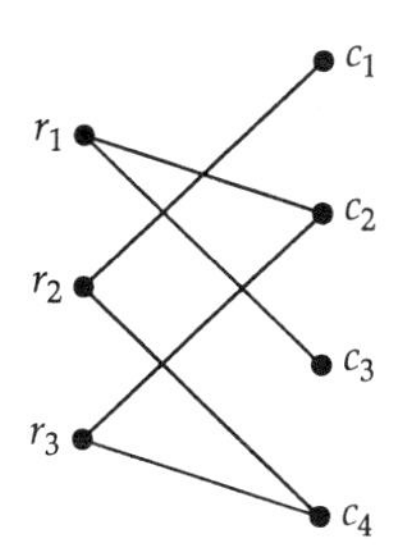

The bipartite graph is a spanning tree, so the braced framework is minimally braced.

(b) We find another spanning tree and construct the corresponding braced framework. For example:

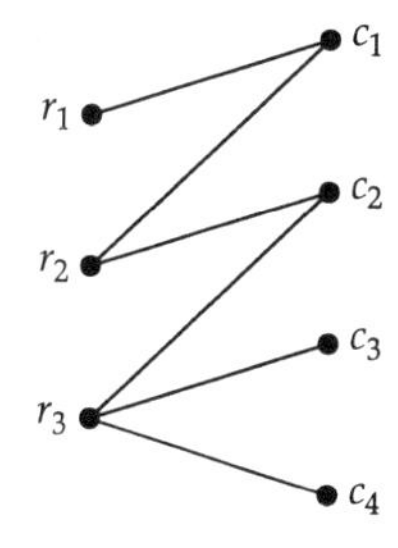

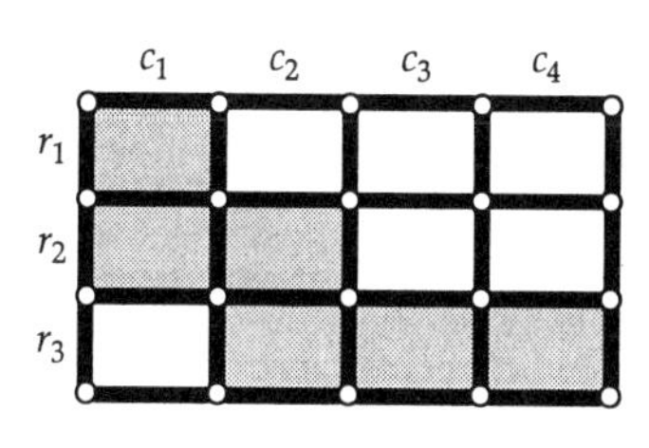

Chapter 7

7.1 The sixteen labelled trees with four vertices are as follows. The first four arise from labelling the complete bipartite graph $K_{1,3}$, and the others arise from labelling the path graph P_4.

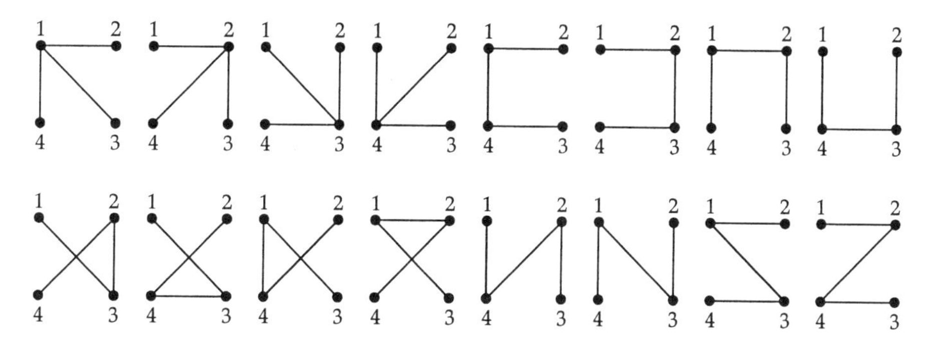

7.2 For each value of n, the number of labelled trees with n vertices is a power of n; it is in fact n^{n-2}.

7.3

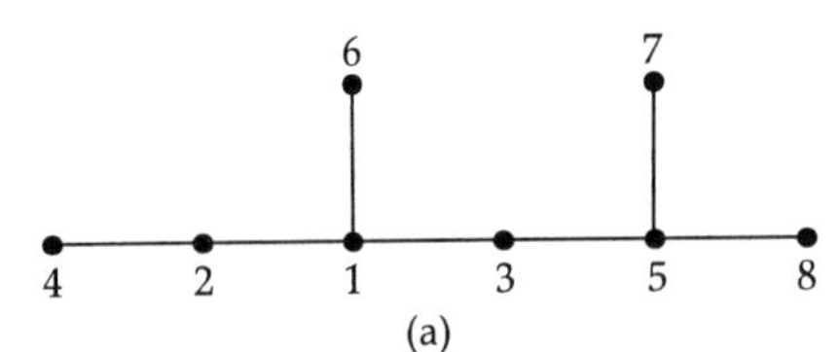

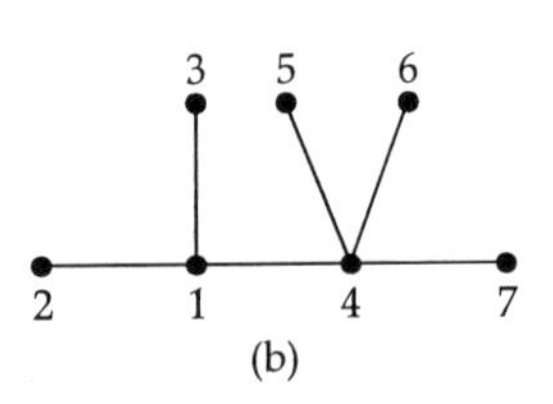

(a) Successively removing the edges 42, 21, 61, 13, 35, 75, we obtain the Prüfer sequence (**2**, **1**, **1**, **3**, **5**, **5**).

(b) Successively removing the edges 21, 31, 14, 54, 64, we obtain the Prüfer sequence (**1**, **1**, **4**, **4**, **4**).

7.4 (a) We start with the list (1, 2, 3, 4, 5, 6, 7, 8) and the sequence (**2**, **1**, **1**, **3**, **5**, **5**). Successively adding the edges 42, 21, 61, 13, 35, 75 leaves us with the list (5, 8). Joining the vertices with these labels, we obtain the labelled tree (a) in Problem 7.3.

(b) We start with the list (1, 2, 3, 4, 5, 6, 7) and the sequence (**1**, **1**, **4**, **4**, **4**). Successively adding the edges 21, 31, 14, 54, 64 leaves us with the list (4, 7). Joining the vertices with these labels, we obtain the labelled tree (b) in Problem 7.3.

7.5 The sixteen labelled trees with four vertices, and their associated Prüfer sequences are as follows.

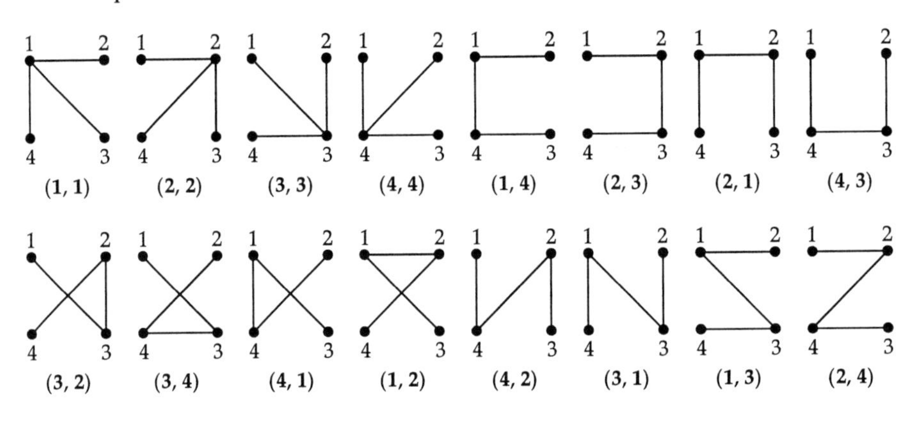

Note that each of the sixteen possible sequences occurs exactly once.

7.6 Each canal system corresponds to a labelled tree with eight vertices. By Cayley's theorem, there are $8^6 = 262144$ of these.

7.7 The fourteen binary trees with four vertices are:

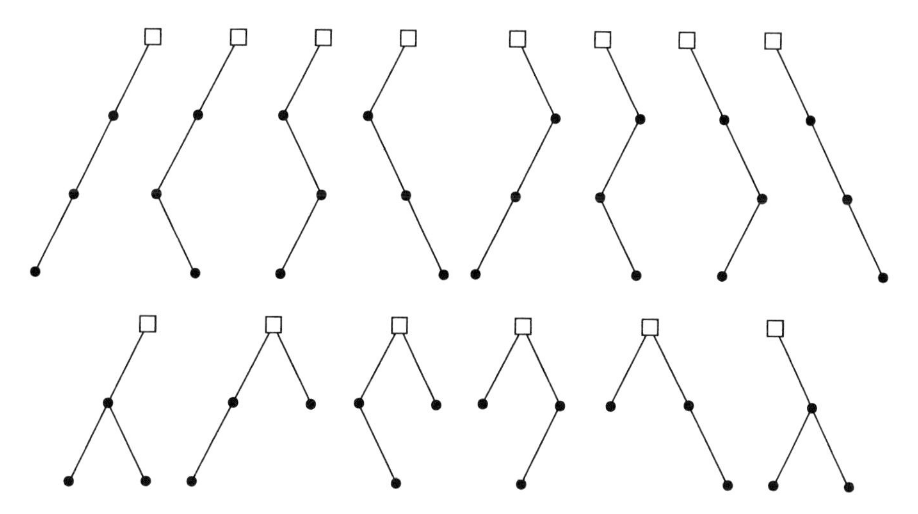

7.8 Substituting the value $n = 6$ in the recurrence relation, we obtain

$$u_6 = 2u_5 + (u_1u_4 + u_2u_3 + u_3u_2 + u_4u_1)$$
$$= (2 \times 42) + (1 \times 14) + (2 \times 5) + (5 \times 2) + (14 \times 1) = 132.$$

Thus there are 132 binary trees with six vertices.

7.9

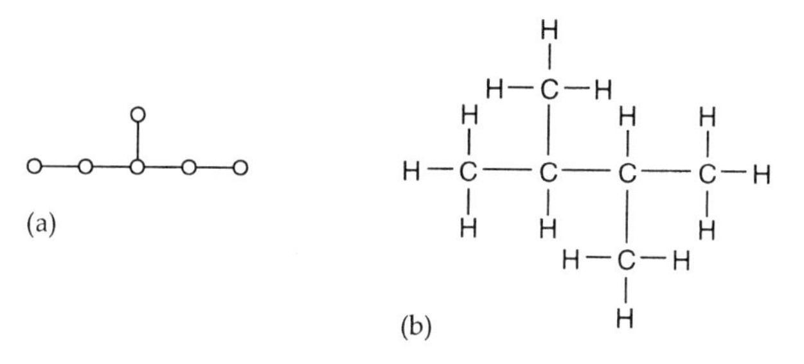

(a) (b)

7.10 The number of vertices in the graph of a molecule with formula C_6H_{14} is

$$6 + 14 = 20.$$

By the handshaking lemma, the number of edges is half the sum of the vertex degrees, that is,

$$\tfrac{1}{2}[(6 \times 4) + (14 \times 1)] = 19.$$

Since the graph is connected, and the number of vertices exceeds the number of edges by 1, the graph is a tree, by Theorem 7.1.

7.11 The number of vertices in the graph of any alkane with formula C_nH_{2n+2} is

$$n + (2n + 2) = 3n + 2.$$

By the handshaking lemma, the number of edges is

$$\tfrac{1}{2}[(n \times 4) + ((2n + 2) \times 1)] = \tfrac{1}{2}(4n + (2n + 2)) = 3n + 1.$$

Since the graph is connected, and the number of vertices exceeds the number of edges by 1, the graph is a tree, by Theorem 7.1.

7.12

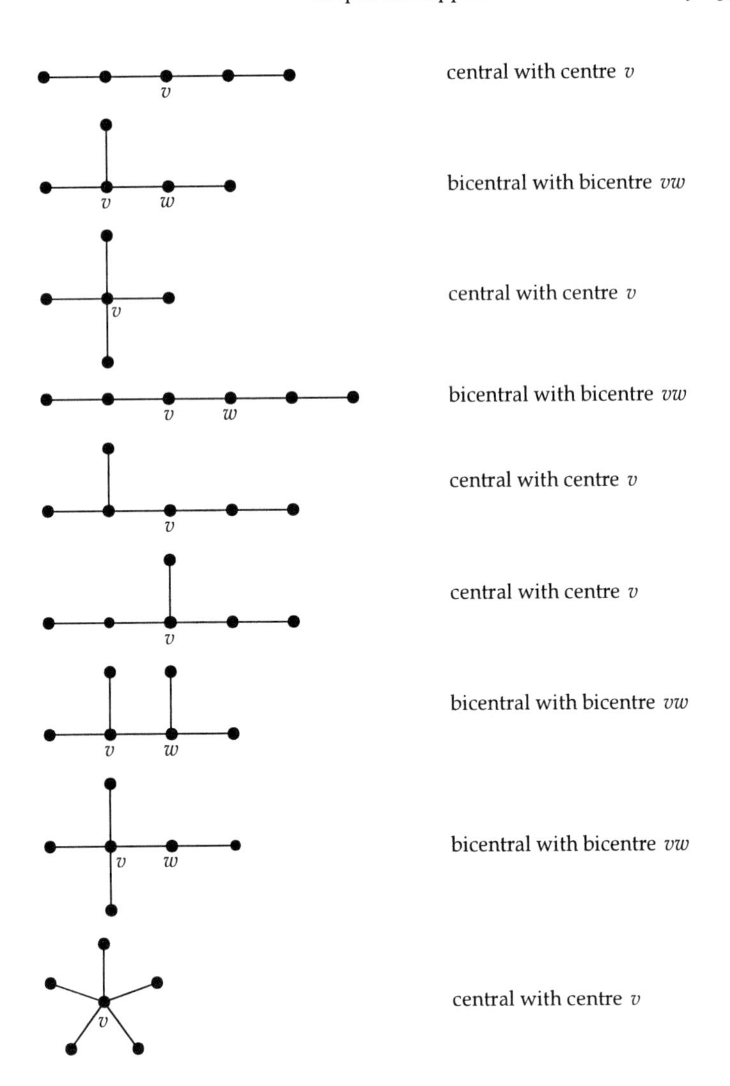

Chapter 8

8.1 We should have chosen edges in the order AE (length 2), AC (length 4), BC (length 5), DE (length 7), obtaining the following minimum spanning tree.

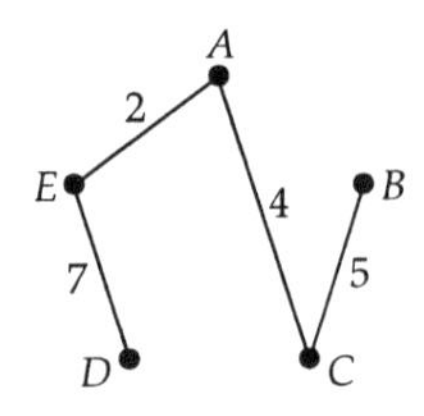

8.2 We apply Kruskal's algorithm as follows.

First edge We choose Athlone–Galway (weight 56).
Second edge We choose Galway–Limerick (weight 64).
Third edge We choose Athlone–Sligo (weight 71).
Fourth edge We cannot choose Athlone–Limerick (weight 73), as this creates a cycle, so we choose Athlone–Dublin (weight 78).
Fifth edge We cannot choose Galway–Sligo (weight 85), as this creates a cycle, so we choose Dublin–Wexford (weight 96).

This completes the required minimum spanning tree of total weight 365.

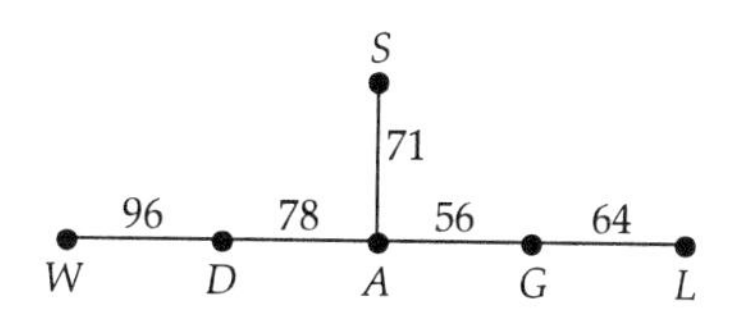

8.3 (a) We should have chosen edges in the order
Berlin–Paris (weight 7),
Paris–London (weight 3),
Paris–Seville (weight 8),
Paris–Rome (weight 9),
Berlin–Moscow (weight 11),
obtaining the following spanning tree of total weight 38.

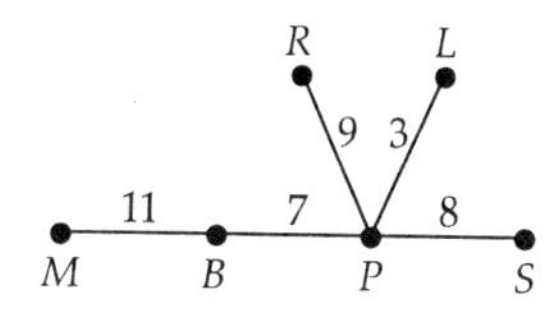

(b) We should have chosen edges in the order
Rome–Paris (weight 9),
Paris–London (weight 3),
Paris–Berlin or London–Berlin (weight 7),
Paris–Seville (weight 8),
Berlin–Moscow (weight 11),
obtaining a spanning tree of total weight 38 – either the same spanning tree as in part (a) or the spanning tree obtained in Example 8.3, depending on the choice of edge of weight 7.

8.4 We should have chosen vertices in the order

Rome,
Paris (distance 9),
London (distance 3 from Paris),
Berlin (distance 7 from London or Paris),
Seville (distance 8 from Paris),
Moscow (distance 11 from Berlin),

obtaining one of the following cycles, giving an upper bound of 66 or 73.

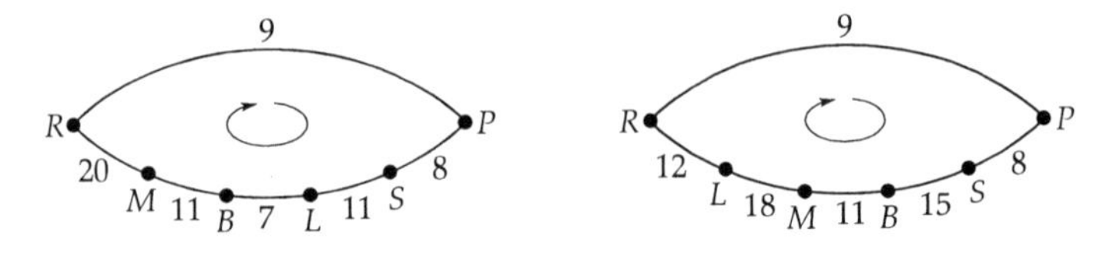

8.5 (a)

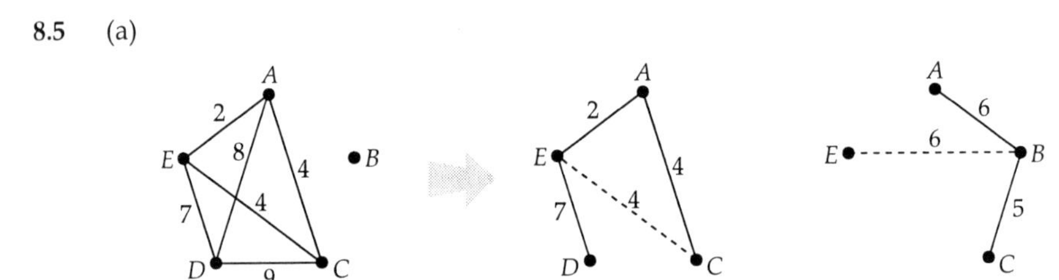

total weight $\geq$ $(7 + 2 + 4)$ + $(6 + 5) = 24$

The minimum spanning tree joining the vertices A, C, D, E is the tree with edges AE, DE and AC or CE, with total weight 13. The two edges of smallest weight incident with B are BC and BA, or BC and BE, with total weight 11. The lower bound is therefore $13 + 11 = 24$.

(b)

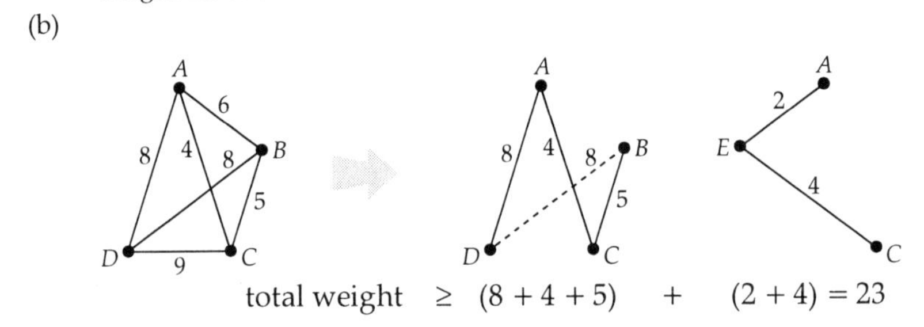

total weight $\geq$ $(8 + 4 + 5)$ + $(2 + 4) = 23$

The minimum spanning tree joining the vertices A, B, C, D is the tree with edges AC, BC and AD or BD, with total weight 17. The two edges of smallest weight incident with E are EA and EC, with total weight 6. The lower bound is therefore $17 + 6 = 23$.

The better lower bound is that given by part (a), that is, 24.

Chapter 9

9.1 Removing the edges uv and vz, we obtain the following graph:

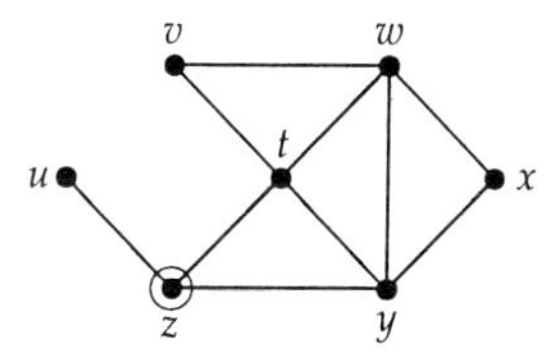

We cannot use the edge uz (which is a bridge), so we must use either zt or zy. There are now several possibilities. For example, we can traverse the edges zt, tv, vw and wy, giving the following graph:

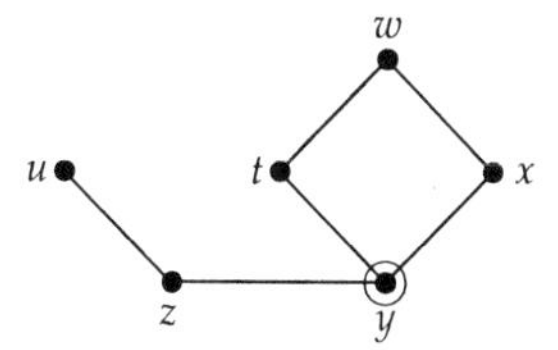

At this stage, we cannot use the edge yz (which is a bridge), so we traverse the cycle $ytwxy$, returning (since there is no alternative) by the bridges yz and zu. Thus we obtain the Eulerian trail $uvztvwytwxyzu$.

9.2 We obtain the following:

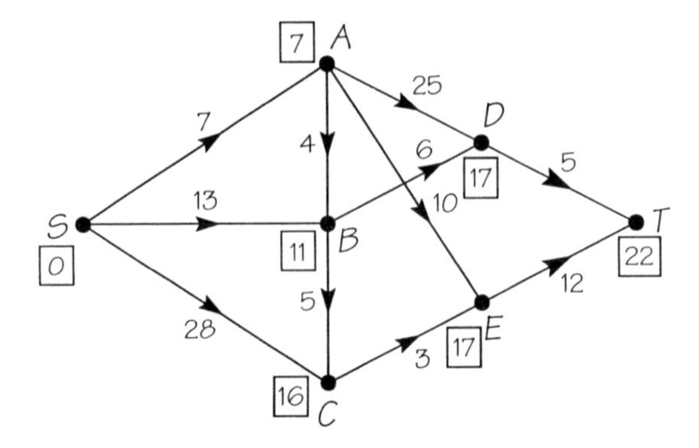

T has potential 22 assigned from *D*.

iteration	origin vertex	vertices assigned labels					
		A	B	C	D	E	T
1	S	7	~~13~~	~~28~~			
2	A		11		~~32~~	17	
3	B			16	17		
4	C				17	17	
5	D						22
	E						29

Tracing back from T, we find the shortest path $SABDT$, with length 22.

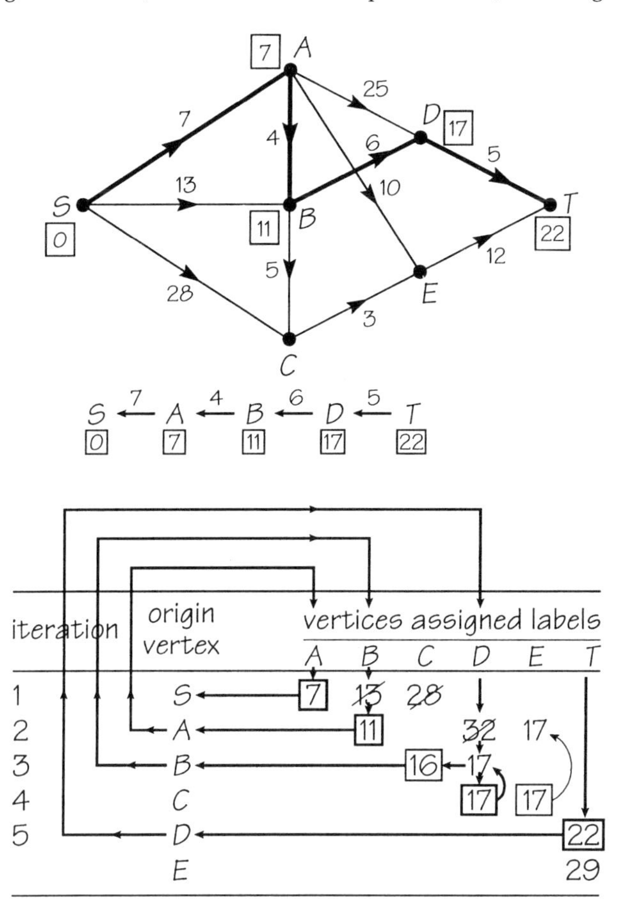

9.3

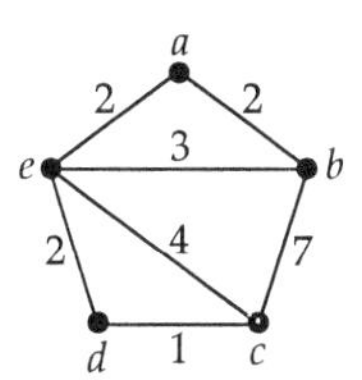

There are two vertices of odd degree – C and F.
A semi-Eulerian path from C to F is $CDECBEFBAF$.
The shortest path from F to C is $FEDC$.
Thus a suitable route is $CDECBEFBAFEDC$.

Chapter 10

10.1 (a), (c), (f) are cutsets;
(b) is not a cutset, since removal of the edges does not disconnect the graph;
(d) is not a cutset, since we can disconnect the graph by removing yt;
(e) is not a cutset, since we can disconnect the graph by removing xz and yz.

10.2 (a) $\lambda(G) = 2$ (for example, remove the edges vw and xy);
(b) $\lambda(G) = 1$ (remove any edge);
(c) $\lambda(G) = 3$ (for example, remove the edges uw, ux, vx).

10.3 (a) and (d) are vertex cutsets;
(b) is not a vertex cutset, since removal of its edges does not disconnect the graph;
(c) is not a vertex cutset, since we can disconnect the graph by removing u and x, or by removing y.

10.4 (a) $\kappa(G) = 2$ (for example, remove the vertices v and x), $\lambda(G) = 2$, $\delta(G) = 2$;
(b) $\kappa(G) = 1$ (for example, remove the vertex v), $\lambda(G) = 1$, $\delta(G) = 1$;
(c) $\kappa(G) = 2$ (for example, remove the vertices w and x), $\lambda(G) = 3$, $\delta(G) = 3$.

10.5

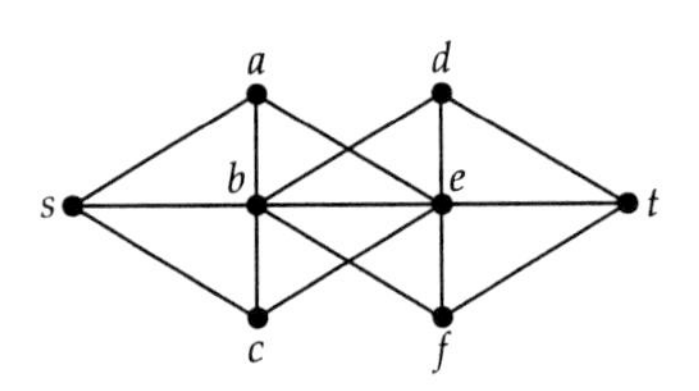

In each case there are several possibilities; for example,
(a) *saet, sbdt, sceft*; (b) *sbet, sabdt*; (c) *saet, sbft*.
This graph does not contain three vertex-disjoint st-paths, since every st-path must pass through at least one of the two vertices b and e.

10.6 (a) Suppose that two vertex-disjoint st-paths are not edge-disjoint; then they have an edge in common. This means that they have at least one vertex (other than s and t) in common, contradicting the fact that they are vertex-disjoint. This contradiction proves that two vertex-disjoint st-paths are edge-disjoint.
(b) There are many possibilities; for example,

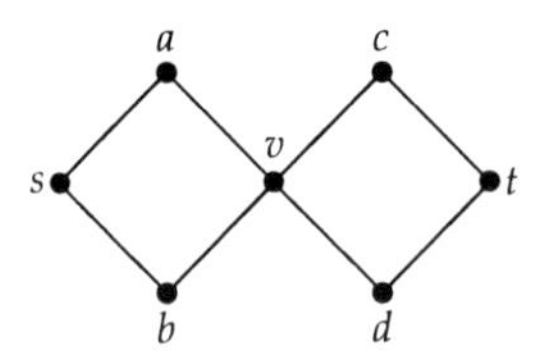

In the above graph, the only pairs of edge-disjoint st-paths are *savct* and *sbvdt*, and *savdt* and *sbvct*. In neither case are the paths vertex-disjoint, since they all pass through the vertex v.

10.7 (a)

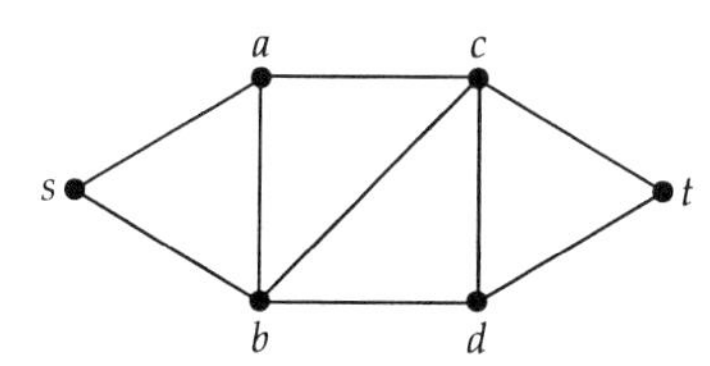

In this case, $k = 2$; two edge-disjoint st-paths are *sact* and *sbdt*, and two edges separating s from t are *sa* and *sb*. Thus the maximum number of edge-disjoint st-paths and the minimum number of edges separating s from t are both 2.

(b)

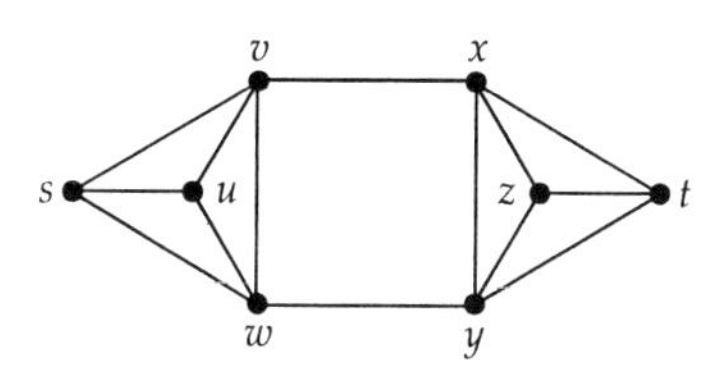

In this case, $k = 2$; two edge-disjoint st-paths are *svxt* and *swyt*, and two edges separating s from t are *vx* and *wy*. Thus the maximum number of edge-disjoint st-paths and the minimum number of edges separating s from t are both 2.

(c)

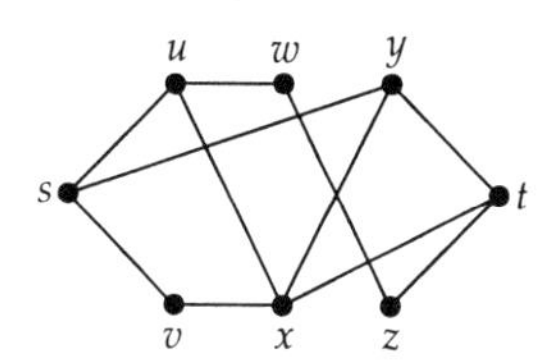

In this case, $k = 3$; three edge-disjoint st-paths are *suwzt*, *syt*, *svxt*, and three edges separating s from t are *su*, *sv*, *sy*. Thus the maximum number of edge-disjoint st-paths and the minimum number of edges separating s from t are both 3.

10.8 (a)

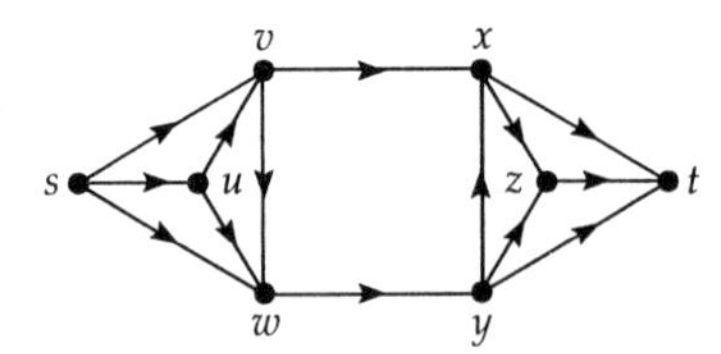

In this case, $k = 2$; two arc-disjoint st-paths are $svxt$ and $swyt$, and two arcs separating s from t are vx and wy. Thus the maximum number of arc-disjoint st-paths and the minimum number of arcs separating s from t are both 2.

(b)

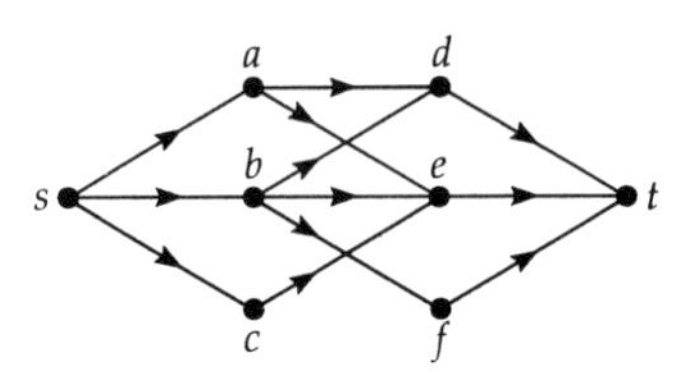

In this case, $k = 3$; three arc-disjoint st-paths are $sadt$, $sbft$, $scet$, and three arcs separating s from t are sa, sb, sc. Thus the maximum number of arc-disjoint st-paths and the minimum number of arcs separating s from t are both 3.

10.9

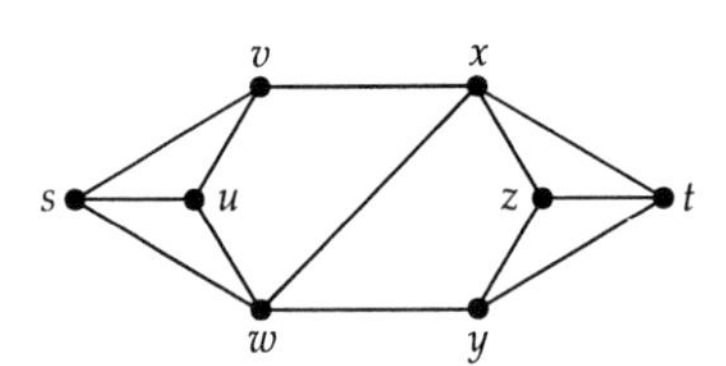

In this case, $k = 2$; two vertex-disjoint st-paths are $svxt$ and $swyt$, and two vertices separating s from t are v and w. Thus the maximum number of vertex-disjoint st-paths and the minimum number of vertices separating s from t are both 2.

10.10

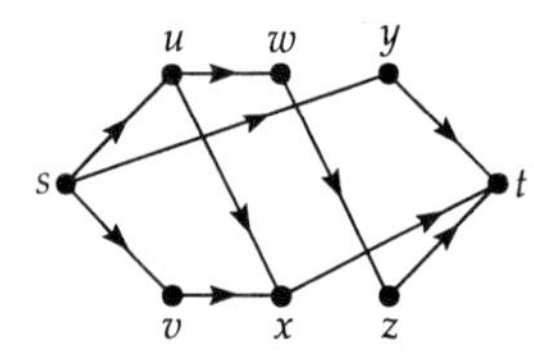

In this case, $k = 3$; three vertex-disjoint st-paths are $suwzt$, syt, $svxt$, and three vertices separating s from t are u, v, y. Thus the maximum number of vertex-disjoint st-paths and the minimum number of vertices separating s from t are both 3.

10.11 (a) For C_n,

$\kappa(C_n) = 2$ and $2m/n = 2n/n = 2$,

so C_n has optimal connectivity.

(b) For K_n, the number of edges is $\frac{1}{2}n(n-1)$, by a consequence of the handshaking lemma, so

$\kappa(K_n) = n - 1$ and $2m/n = 2 \times \frac{1}{2}n(n-1)/n = n - 1$,

so K_n has optimal connectivity.

(c) For $K_{r,r}$,

$\kappa(K_{r,r}) = r$ and $2m/n = 2r^2/(r + r) = r$,

so $K_{r,r}$ has optimal connectivity.

10.12 (a) There are only two regular graphs with 6 vertices and 9 edges:

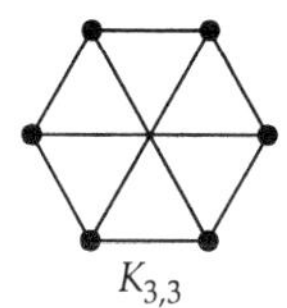

$K_{3,3}$

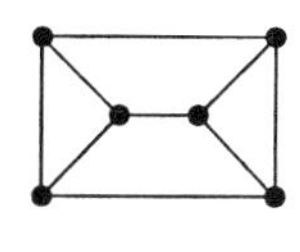

For each of these two graphs,

$\kappa(G) = 3$ and $2m/n = 3$,

so the graphs have optimal connectivity.

(b) There is only one possibility:

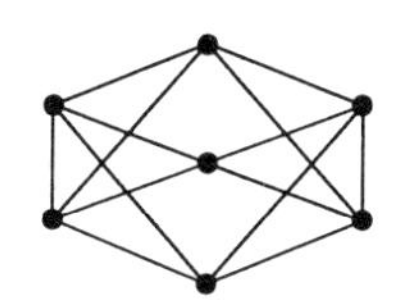

The removal of the middle three vertices disconnects the graph, so $\kappa(G) = 3$; since $\delta(G) = 4$, the graph does not have optimal connectivity.

Chapter 11

11.1

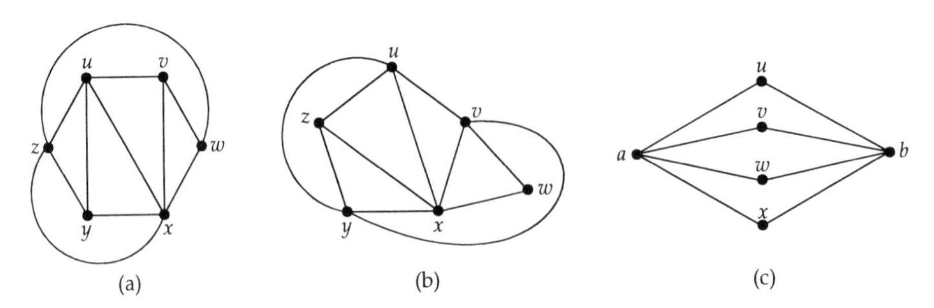

11.2 The nine connections are not possible, because the corresponding graph is $K_{3,3}$, which is non-planar.

11.3 

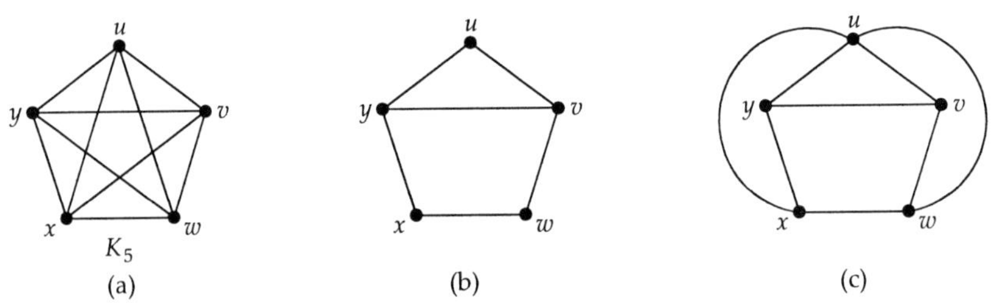

In any plane drawing of K_5, the cycle *uvwxyu* in diagram (a) must appear as a pentagon. The edge *vy* must lie either inside or outside the pentagon. Since the argument is similar in each case, we assume that *vy* lies *inside* the pentagon, as in diagram (b).

Since the edges *ux* and *uw* cannot cross *vy*, they must lie *outside* the pentagon, as in diagram (c). But the edge *vx* cannot cross *uw*, and the edge *wy* cannot cross *ux*, so both *vx* and *wy* must lie *inside* the pentagon, and must therefore cross. Since this is not allowed, we deduce that K_5 has no plane drawing – that is, K_5 is non-planar.

11.4 No. The corresponding graph is K_5, which is non-planar.

11.5 (a) This statement is TRUE, since if G is a planar graph, then we can draw G in the plane without crossings. If we now remove the vertices and edges not included in the subgraph, then we obtain a plane drawing of the subgraph.

(b) This statement is FALSE; for example, the graph $K_{3,3}$ is non-planar, whereas the cycle graph C_6, a subgraph of $K_{3,3}$, is planar.

(c) This statement is FALSE; for example, the graph C_6 is planar, whereas the graph $K_{3,3}$, which contains C_6, is non-planar.

(d) This statement is TRUE, since if G is a planar graph, then G cannot have a non-planar subgraph, by part (a).

11.6 (a) All trees are planar.

(b) All cycle graphs are planar.

(c) Using the results of Problem 11.5, parts (a) and (d), we deduce that:
since K_4 is planar and K_n is a subgraph of K_4 when $n \leq 4$,
K_n is planar when $n \leq 4$;
since K_5 is non-planar, and K_5 is a subgraph of K_n when $n \geq 5$,
K_n is non-planar when $n \geq 5$.
Thus K_n is planar for $n = 1, 2, 3, 4$.

(d) The bipartite graphs $K_{1,s}$ and $K_{2,s}$ are planar for all values of s, as illustrated in the following diagrams.

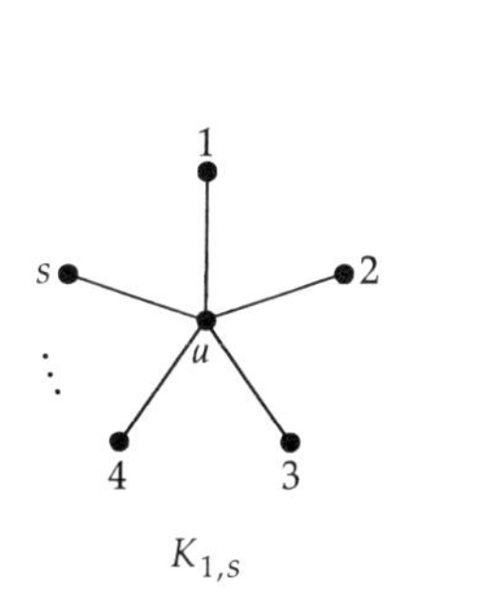

$K_{1,s}$

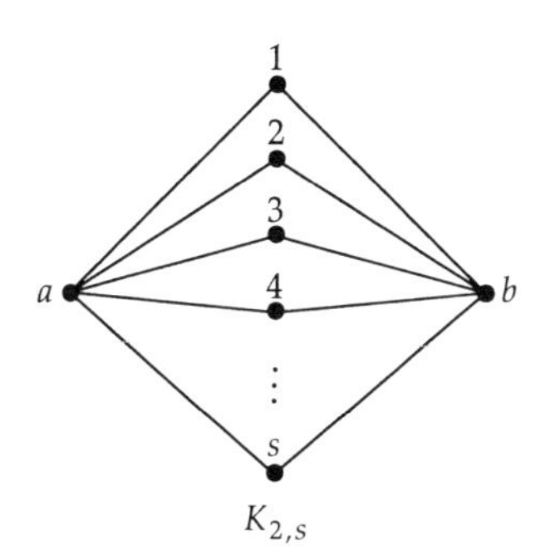

$K_{2,s}$

(e) The graphs $K_{1,s}$ and $K_{2,s}$ are planar for all values of s, by part (d). Since $K_{3,3}$ is non-planar and is a subgraph of $K_{r,s}$ when $r \geq 3$, it follows from Problem 11.5(d) that $K_{r,s}$ is non-planar when $r \geq 3$.
Thus $K_{r,s}$ is planar only when $r = 1$ or 2, for all values of s.

11.7

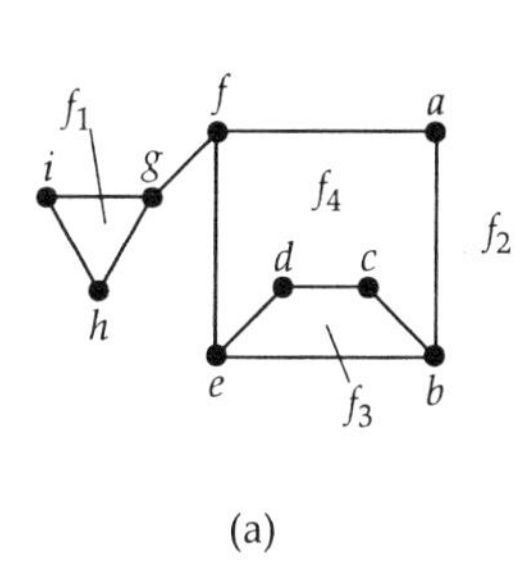

(a)

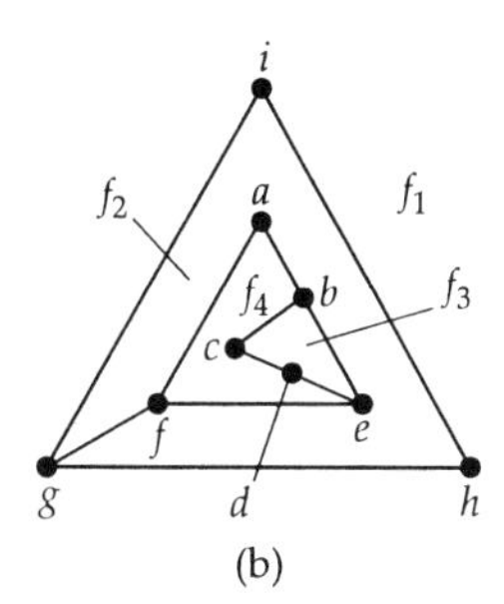

(b)

11.8 (a) There are 10 edges and six faces with face degrees 1, 2, 3, 3, 4, 7, and
$1 + 2 + 3 + 3 + 4 + 7 = 20 = 2 \times 10$.

(b) There are 11 edges and seven faces with face degrees 3, 3, 3, 3, 3, 3, 4, and
$3 + 3 + 3 + 3 + 3 + 3 + 4 = 22 = 2 \times 11$.

(c) There are 10 edges and six faces with face degrees 3, 3, 3, 3, 4, 4, and
$3 + 3 + 3 + 3 + 4 + 4 = 20 = 2 \times 10$.

11.9 (a) There are 6 vertices, 10 edges and 6 faces, so the required value is $6 - 10 + 6 = 2$.

(b) There are 6 vertices, 11 edges and 7 faces, so the required value is $6 - 11 + 7 = 2$.

(c) There are 6 vertices, 10 edges and 6 faces, so the required value is $6 - 10 + 6 = 2$.

11.10 (a) There are 6 vertices, 12 edges and 8 faces, and $n - m + f = 6 - 12 + 8 = 2$.

(b) There are $k + 1$ vertices, $2k$ edges and $k + 1$ faces, and $n - m + f = (k + 1) - 2k + (k + 1) = 2$.

(c) There are $k + 2$ vertices, $2k$ edges and k faces, and $n - m + f = (k + 2) - 2k + k = 2$.

(d) There are $(k + 1)^2$ vertices, $2k(k + 1)$ edges and $k^2 + 1$ faces, and $n - m + f = (k + 1)^2 - 2k(k + 1) + (k^2 + 1) = 2$.

11.11 In Corollary 11.1, equality occurs when $m = 3n - 6$ or $3n = m + 6$. Substituting for n in Euler's formula $n - m + f = 2$, we obtain $2m = 3f$. So equality occurs when G is face-regular of degree 3; for example, K_4 ($m = 6, f = 4$).

In Corollary 11.2, equality occurs when $m = 2n - 4$ or $2n = m + 4$. Substituting for n in Euler's formula $n - m + f = 2$, we obtain $2m = 4f$. So equality occurs when G is face-regular of degree 4; for example, the 3-cube ($m = 12, f = 6$).

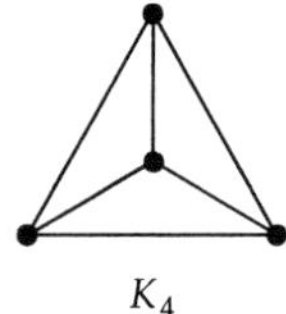
K_4

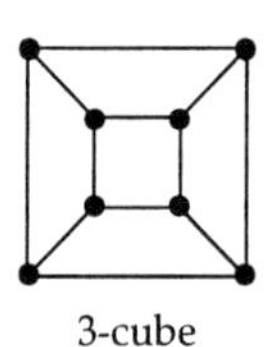
3-cube

11.12 (a) For a plane drawing of G with f faces, it follows from the handshaking lemma for planar graphs that

$$2m \geq 5f,$$

since the degree of each face is at least 5.

Substituting for f from Euler's formula $f = m - n + 2$, we obtain

$$2m \geq 5m - 5n + 10 \quad \text{or} \quad 3m \leq 5(n - 2),$$

and hence

$$m \leq \tfrac{5}{3}(n-2).$$

(b) Suppose that the Petersen graph is planar. Then the inequality in part (a) becomes (with $m = 15$ and $n = 10$):

$$15 \leq \tfrac{5}{3} \times (10-2) = 13\tfrac{1}{3},$$

which is FALSE. Thus the Petersen graph is non-planar.

11.13 Since G is a simple graph, we can apply Corollary 11.1 to deduce that, if G has n vertices and m edges, then

$m \geq 3n - 6.$

Suppose that each vertex of G has degree 6 or more. Then, by the handshaking lemma for graphs,

$2m \geq 6n$ or $3n \leq m.$

Combining these two inequalities, we obtain

$3n \leq 3n - 6,$

which is FALSE. Thus G must have at least one vertex of degree 5 or less.

11.14

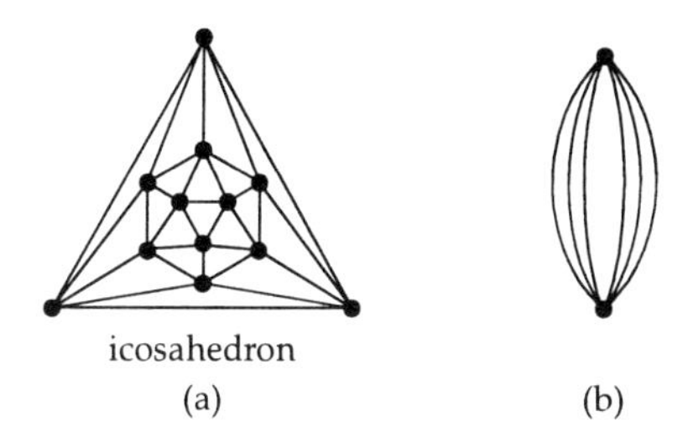

icosahedron
(a) (b)

11.15 (a) We choose C to be the cycle *abcdefga*.

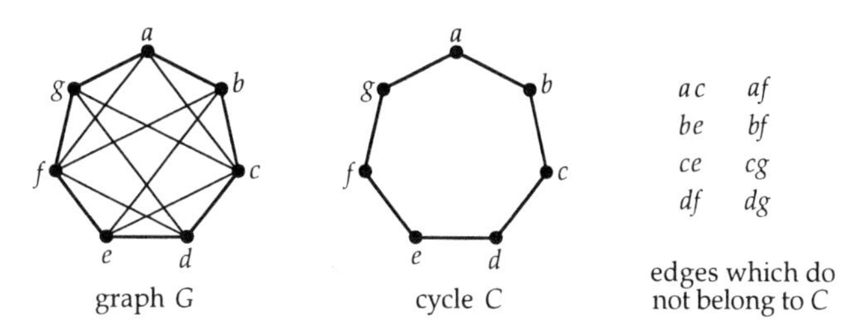

graph G cycle C edges which do not belong to C

We list the edges which do not belong to C:

list: *ac, af, be, bf, ce, cg, df, dg*

We put the first edge in the list, *ac*, in a set A and delete this edge from the list:

$A = \{ac, ...\}.$

list: *af, be, bf, ce, cg, df, dg*

The edge *ac* is incompatible with *be* and *bf*, so we put the edges *be* and *bf* in a set B:

$B = \{be, bf, ...\}.$

We check and find that the edges *be* and *bf* are compatible with each other.

We delete the edges *be* and *bf* from the list:

list: *af, ce, cg, dt, dg*

We consider the edge *be* in B.

The edge *be* is incompatible with *cg, df* and *dg*, so we put the edges *cg, df* and *dg* in A:

$A = \{ac, cg, df, dg\}.$

We check and find that the edges in A are compatible with each other.

We delete the edges *cg*, *df* and *dg* from the list:

list: *af*, *ce*

The edge *cg* in *A* is incompatible with *af* and *ce*, so we put the edges *af* and *ce* in *B*:

$B = \{be, bf, af, ce\}$

We check and find that all the edges in *B* are compatible with each other.

We delete the edges *af* and *ce* from the list.

The list is now empty, and we have:

$A = \{ac, cg, df, dg\}; \quad B = \{be, bf, af, ce\}.$

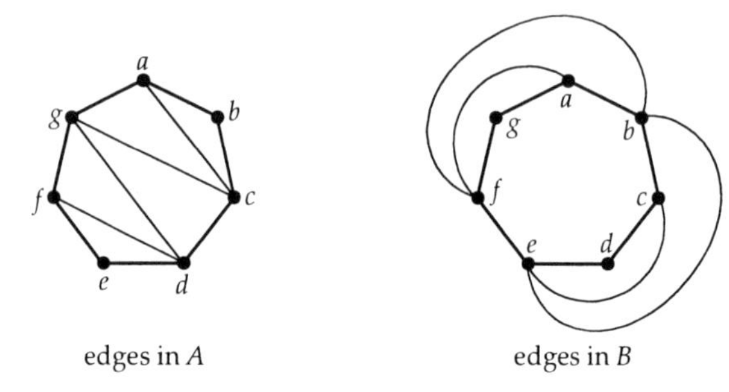

edges in *A* edges in *B*

All the edges in *A* are compatible and all the edges in *B* are compatible, so *G* is planar.

To obtain a plane drawing of *G*, we combine the above two figures as follows.

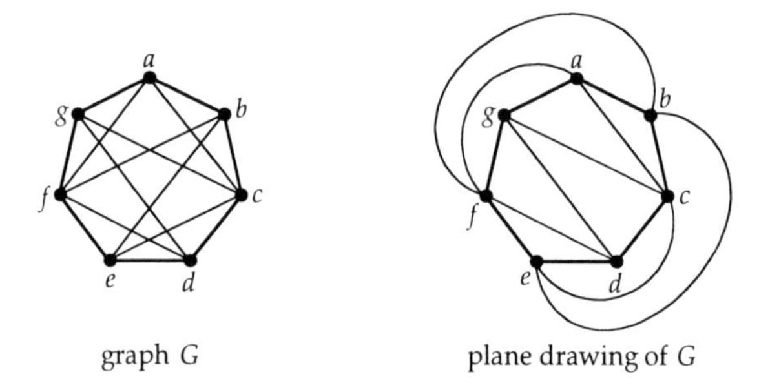

graph *G* plane drawing of *G*

(b) We choose C to be the cycle *abcdefghijklmna*.

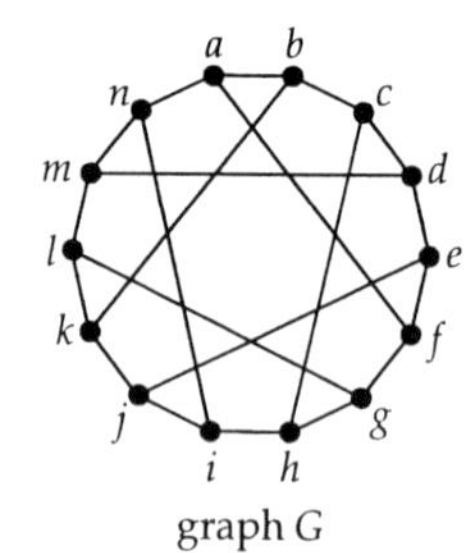

graph *G*

We list the edges which do not belong to C:

list: $af, bk, ch, dm, ej, gl, in$

We put the first edge in the list, af, into a set A and delete this edge from the list:

$A = \{af\}$

list: bk, ch, dm, ej, gl, in

The edge af is incompatible with bk, ch, dm and ej, so we put the edges bk, ch, dm and ej into a set B.

$B = \{bk, ch, dm, ej\}$

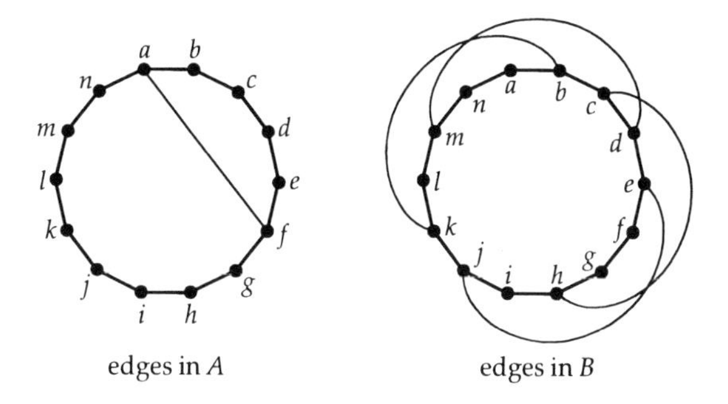

edges in A edges in B

We check the compatibility of the edges in B with each other, but find that bk and dm are incompatible, so G is non-planar.

11.16 (a) Deletion of the edge sw gives the following subgraph.

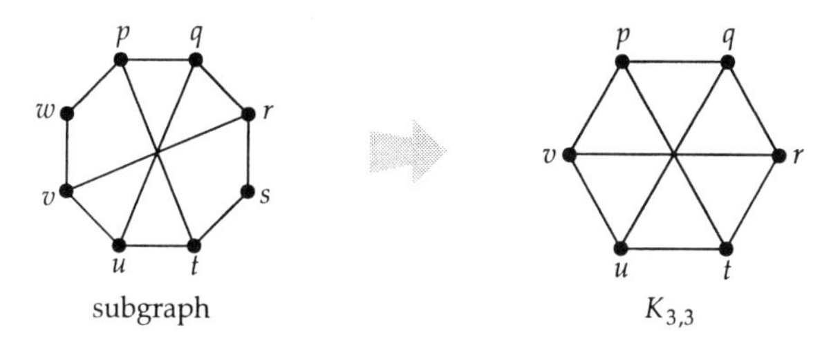

subgraph $K_{3,3}$

This is a subdivision of $K_{3,3}$. It follows from Kuratowski's theorem that the given graph is non-planar.

(b) Deletion of the two 'horizontal' edges gives the following subgraph.

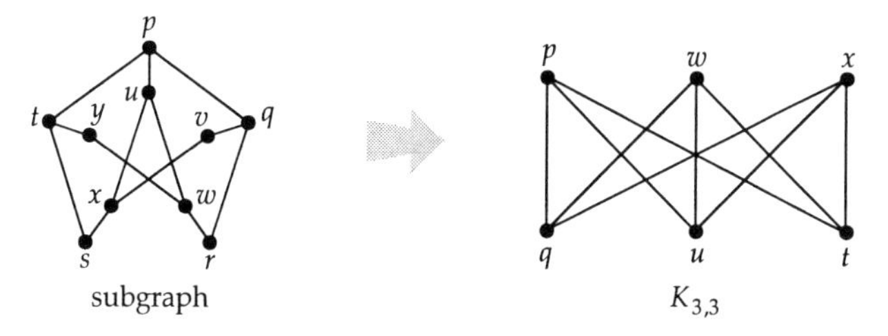

subgraph $K_{3,3}$

This is a subdivision of $K_{3,3}$. It follows from Kuratowski's theorem that the Petersen graph is non-planar.

Note that the Petersen graph does not contain a subdivision of K_5.

11.17

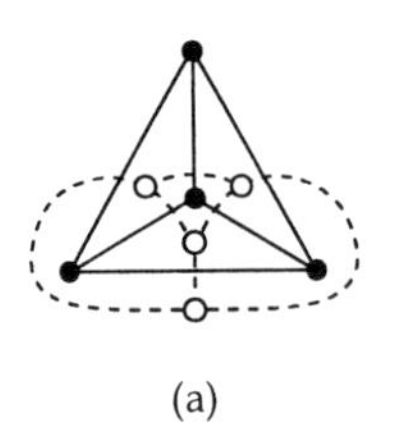

(a)

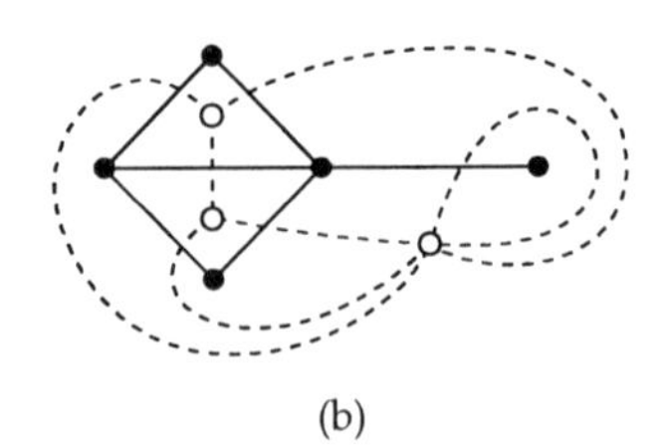

(b)

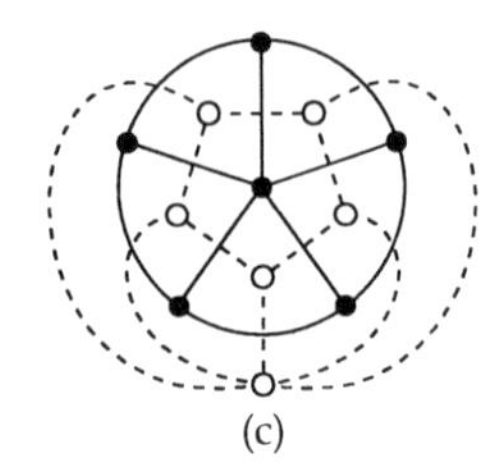

(c)

Notice that in parts (a) and (c) the dual graph is isomorphic to the original graph.

11.18 The dual graphs are as follows.

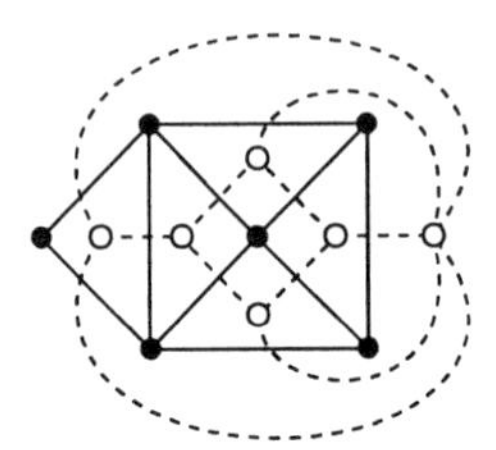

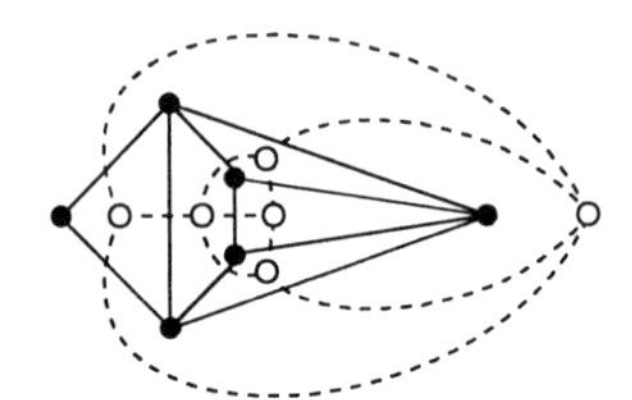

Their degree sequences are (3, 3, 3, 3, 3, 5) and (3, 3, 3, 3, 4, 4), so they are not isomorphic.

11.19 Since a triangle in G corresponds to a cutset with 3 edges in G^*, the dual statement is as follows.

Let G^* be a connected planar graph with f faces and m edges, and with no cutsets with 1, 2 or 3 edges. Then $m \leq 2f - 4$.

11.20 In each case, the dual is the solid with the same number of edges as the original, and with the numbers of vertices and faces interchanged:

the tetrahedron is its own dual;

the octahedron and the cube are duals of each other,

the dodecahedron and the icosahedron are duals of each other.

Chapter 12

12.1 Possible vertex colourings are given below.

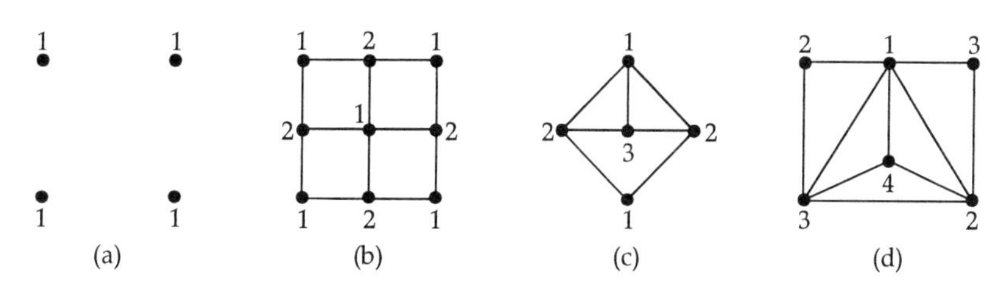

(a) The vertices can all be coloured with the same colour.
Thus $\chi(G) = 1$.

(b) Since the graph contains an edge, at least two colours are needed, so $\chi(G) \geq 2$.
A 2-colouring is shown above, so $\chi(G) \leq 2$.
Thus $\chi(G) = 2$.

(c) Since the graph contains a triangle (K_3), at least three colours are needed, so $\chi(G) \geq 3$.
A 3-colouring is shown above, so $\chi(G) \leq 3$.
Thus $\chi(G) = 3$.

(d) Since the graph contains K_4, at least four colours are needed, so $\chi(G) \geq 4$.
A 4-colouring is shown above, so $\chi(G) \leq 4$.
Thus $\chi(G) = 4$.

12.2 (a) The graphs with $\chi(G) = 1$ are the graph with no edges – the null graphs N_n.

(b) The graphs with $\chi(G) = 2$ are the bipartite graphs (other than N_n), since we can colour their vertices black and white in such a way that each edge joins a black vertex to a white vertex.

12.3 (a) n;

(b) 2;

(c) 2, if n is even; 3, if n is odd;

(d) 2, if the tree has at least two vertices;
1, if the tree has only one vertex.

12.4 (a) This statement is TRUE, because if G contains K_r as a subgraph, then G contains r mutually adjacent vertices, and these require r colours. So $\chi(G) \geq r$.

(b) This statement is FALSE; for example, the cycle graph C_5 has chromatic number 3, but does not contain a triangle (K_3).

12.5

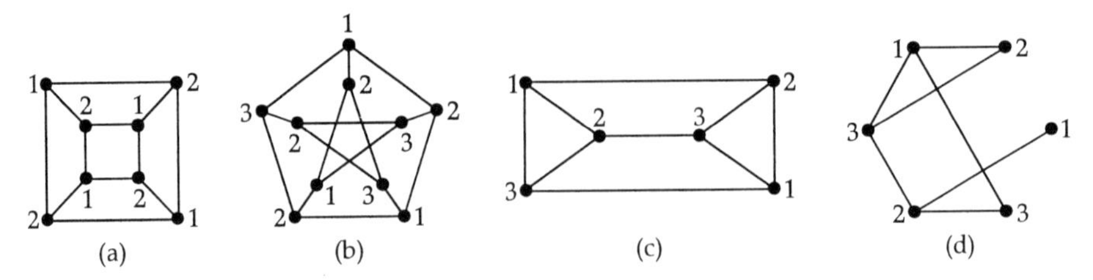

(a) lower bound: $\chi(G) \geq 2$; upper bound: $\chi(G) \leq 3$; actual value: $\chi(G) = 2$;
(b) lower bound: $\chi(G) \geq 2$; upper bound: $\chi(G) \leq 3$; actual value: $\chi(G) = 3$;
(c) lower bound: $\chi(G) \geq 3$; upper bound: $\chi(G) \leq 3$; actual value: $\chi(G) = 3$;
(d) lower bound: $\chi(G) \geq 3$; upper bound: $\chi(G) \leq 3$; actual value: $\chi(G) = 3$.

12.6 We obtain the following vertex colourings with 4, 3 and 2 colours.

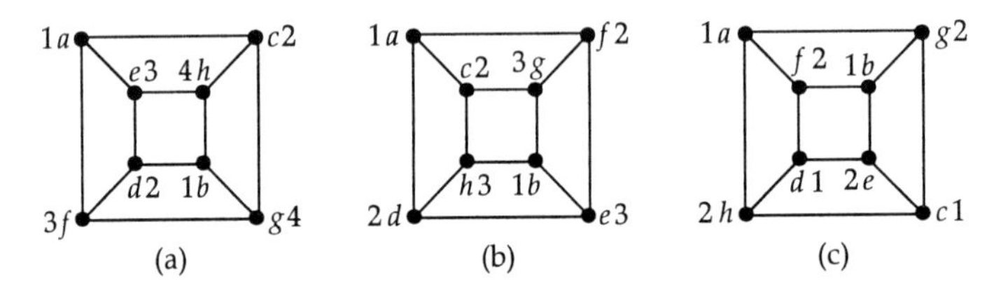

We showed that $\chi(G) = 2$ in Solution 12.5(a), so only colouring (c) uses $\chi(G)$ colours.

12.7 We showed that $\chi(G) = 3$ in Solution 12.5(c). A suitable labelling is shown below.

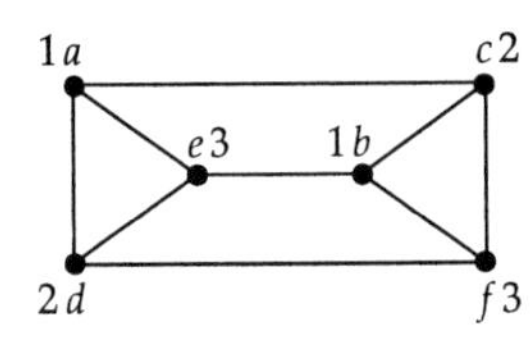

12.8

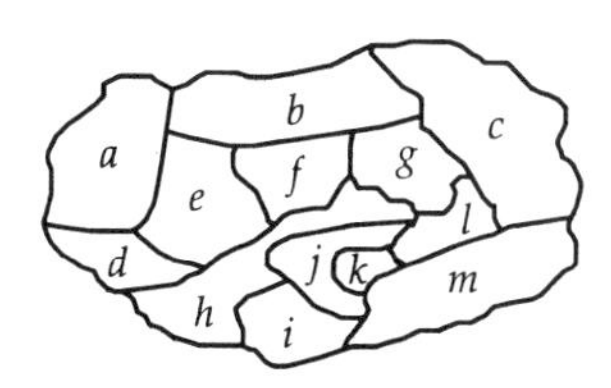

(a) There are many possibilities – for example, the colouring shown on the left below.

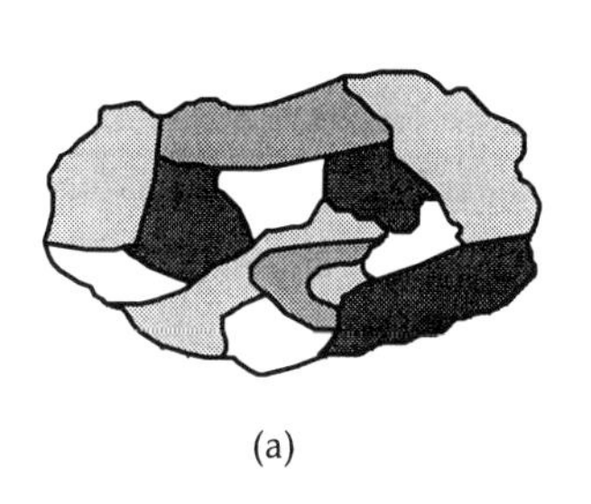
(a)

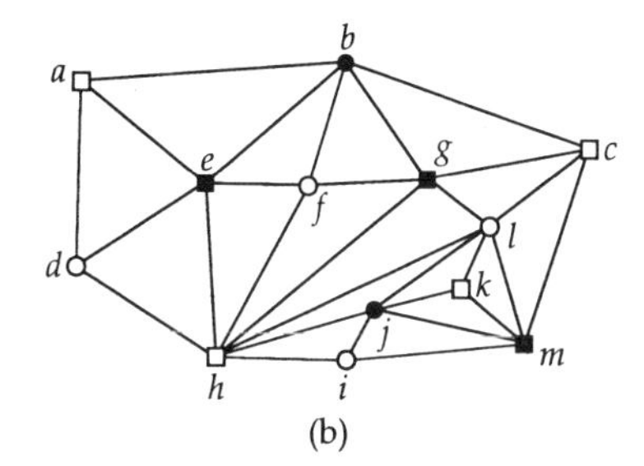

(b)

(b) The corresponding graph is shown above. The colouring in part (a) leads to a vertex decomposition of the required type:

$\{a, c, h, k\}$, $\{b, j\}$, $\{d, f, i, l\}$, $\{e, g, m\}$.

12.9 The tour graph is given below.

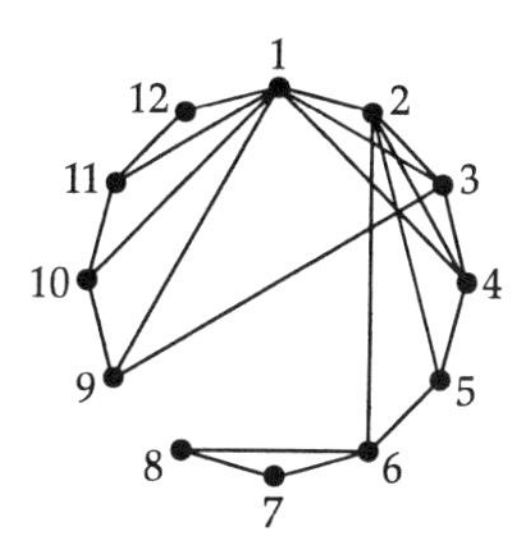

Since vertices 1, 2, 3, 4 are mutually adjacent, at least four colours are needed to colour the vertices of this graph so that neighbouring vertices are coloured differently. This means that at least four days are needed.

In fact, four days are sufficient, as the following vertex decomposition shows:

Monday	routes 1, 5, 7;
Tuesday	routes 2, 9, 12;
Wednesday	routes 3, 6, 11;
Thursday	routes 4, 8, 10.

Other vertex decompositions are possible.

12.10 There are several possibilities – for example:

	minimum dominating set	vertex decomposition
(a)	$\{a, c\}$	$\{a, b, e\}$, $\{c, d\}$;
	$\{b, d\}$	$\{b, a, e\}$, $\{d, c\}$;
(b)	$\{a, g\}$	$\{a, b, d, e\}$, $\{g, f, h, c\}$;
	$\{b, h\}$	$\{b, a, c, f\}$, $\{h, d, e, g\}$.

Chapter 13

13.1 Possible edge colourings are given below.

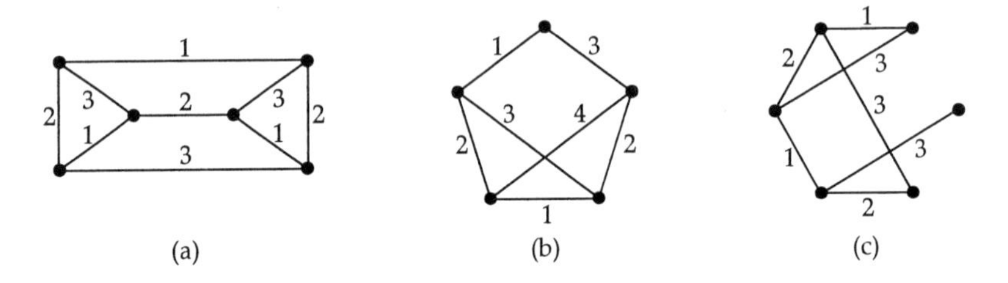

(a) Since the graph contains a vertex of degree 3, at least three colours are needed, so $\chi'(G) \geq 3$.
A 3-edge colouring is shown above, so $\chi'(G) \leq 3$.
Thus $\chi'(G) = 3$.

(b) Since the graph contains a vertex of degree 3, at least three colours are needed, so $\chi'(G) \geq 3$.
However, in this case there is no 3-edge colouring, because three colours are needed to colour the pentagon and a further colour is needed to colour one of the inside edges, so $\chi'(G) > 3$.
A 4-edge colouring is shown above, so $\chi'(G) \leq 4$.
Thus $\chi'(G) = 4$.

(c) Since the graph contains a vertex of degree 3, at least three colours are needed, so $\chi'(G) \geq 3$.
A 3-edge colouring is shown above, so $\chi'(G) \leq 3$.
Thus $\chi'(G) = 3$.

13.2 (a) The graphs with $\chi'(G) = 1$ are the graphs containing one component that is a single edge and in which each other component is either a single edge or an isolated vertex.

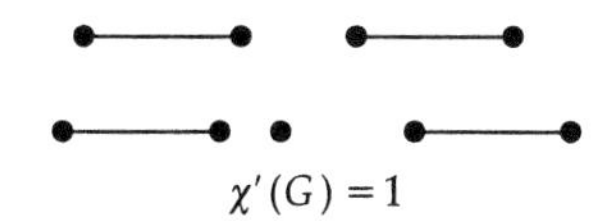

(b) The graphs with $\chi'(G) = 2$ are the graphs whose components are cycles of even length, path graphs or isolated vertices, and in which at least one component is a cycle or path graph.

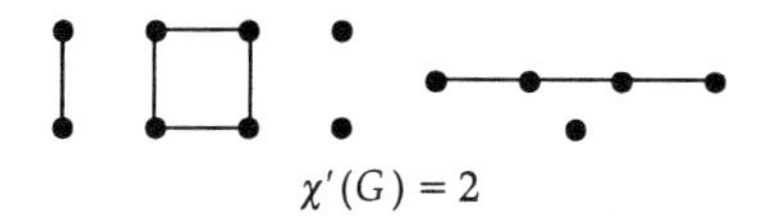

13.3 (a) 3; (b) 3; (c) 2.

13.4 (a) This statement is TRUE, because if G contains a vertex of degree r, then G contains r edges, and these require r colours. So $\chi'(G) \geq r$.

(b) This statement is FALSE; for example, the cycle graph C_5 has chromatic index 3, but does not contain a vertex of degree 3.

13.5

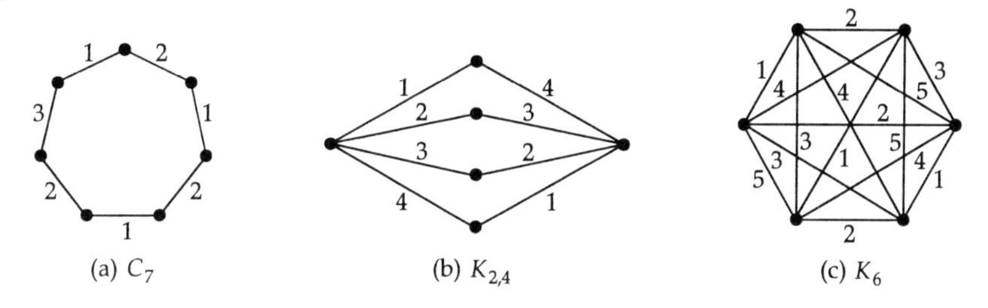

(a) lower bound: $\chi'(G) \geq 2$; upper bound: $\chi'(G) \leq 3$; actual value: $\chi'(G) = 3$;
(b) lower bound: $\chi'(G) \geq 4$; upper bound: $\chi'(G) \leq 5$; actual value: $\chi'(G) = 4$;
(c) lower bound: $\chi'(G) \geq 5$; upper bound: $\chi'(G) \leq 6$; actual value: $\chi'(G) = 5$.

13.6

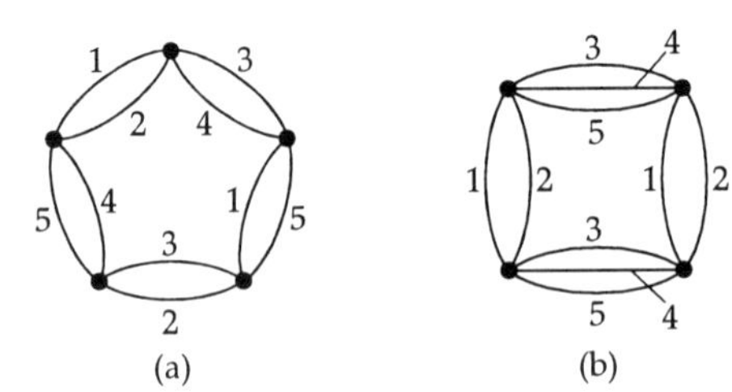

(a) (b)

(a) Vizing's theorem: lower bound: $\chi'(G) \geq 4$; upper bound: $\chi'(G) \leq 6$; Shannon's theorem: lower bound: $\chi'(G) \geq 4$; upper bound: $\chi'(G) \leq 6$; actual value: $\chi'(G) = 5$.

(b) Vizing's theorem: lower bound: $\chi'(G) \geq 5$; upper bound: $\chi'(G) \leq 8$; Shannon's theorem: lower bound: $\chi'(G) \geq 5$; upper bound: $\chi'(G) \leq 7$; actual value: $\chi'(G) = 5$.

13.7 In each part, we represent the competition by a complete graph K_n; the solution is then given by $\chi'(K_n)$.

(a) If $n = 31$, we have $\chi'(K_n) = 31$, by Theorem 13.4.

(b) If $n = 32$, we have $\chi'(K_n) = 31$, by Theorem 13.4.

Thus, in each case, 31 matches are necessary.

13.8 In each case, the chromatic index is the maximum vertex degree:
(a) s (b) 3; (c) k.

13.9 We obtain the following edge colourings with 4, 3 and 3 colours.

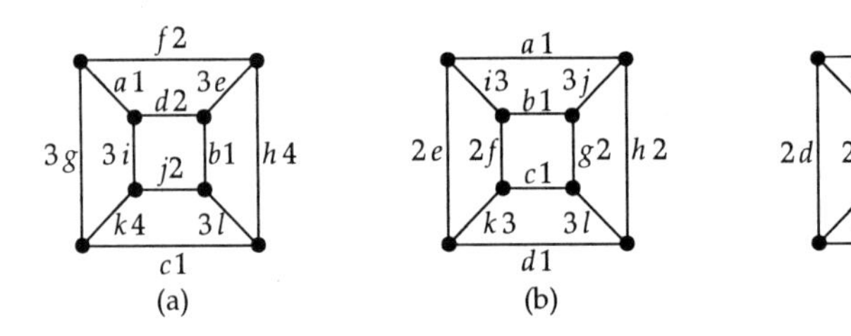

(a) (b) (c)

The actual value of $\chi'(G)$ is 3, so colourings (b) and (c) use $\chi'(G)$ colours.

13.10 We showed that $\chi'(G) = 3$ in Solution 13.1(a). A suitable labelling is shown below.

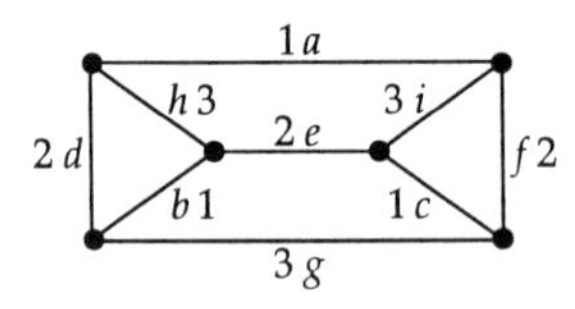

13.11 The bipartite graph representing this situation is shown below.

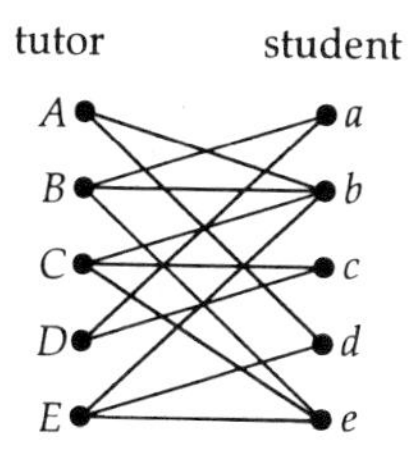

Since this graph has maximum degree 4, it follows from König's theorem that four examination periods are needed. One such schedule is as follows.

tutor	A	B	C	D	E
first examination period	b	a	c	–	e
second examination period	d	b	e	a	–
third examination period	–	e	b	c	d
fourth examination period	–	–	–	–	b

13.12 The graph K_6 can be 'printed' in two layers, as follows.

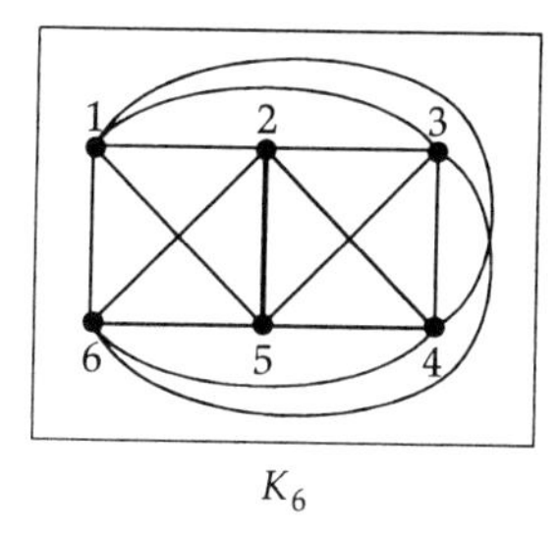

K_6

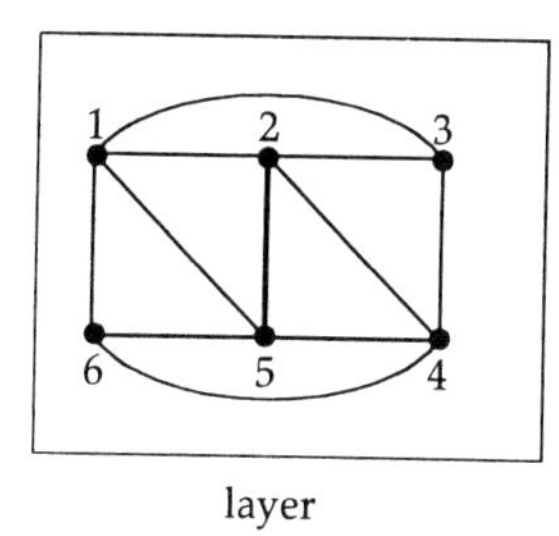

layer

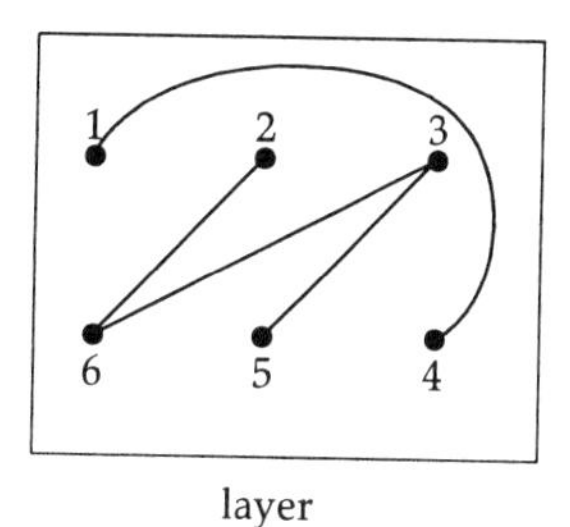

layer

The corresponding edge decomposition is:

$\{12, 13, 15, 16, 23, 24, 25, 34, 45, 46, 56\}$, $\{14, 26, 35, 36\}$.

Other solutions are possible.

13.13 Since each graph is non-planar, the thickness cannot equal 1.
But each graph can be 'printed' on two layers, as follows.

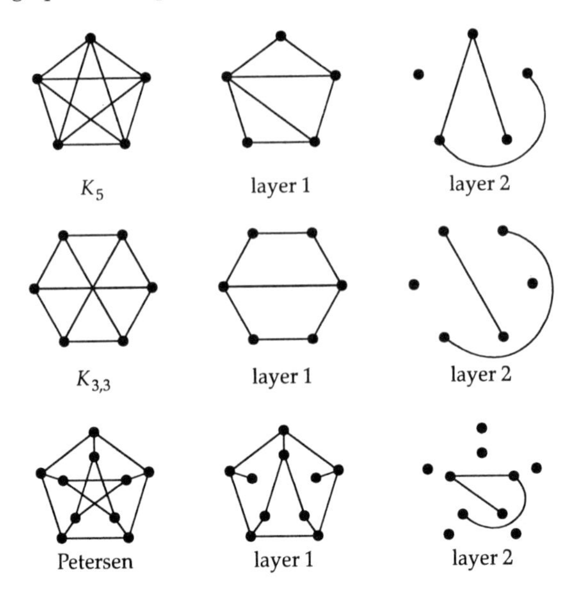

Other solutions are possible.
Thus, in each case, the thickness is 2.

13.14 The network has 13 towns and 28 roads, so $s(G) \leq 28/13$; it follows that $s(G) = 1$ or 2.
The following diagram shows that $s(G) = 2$.

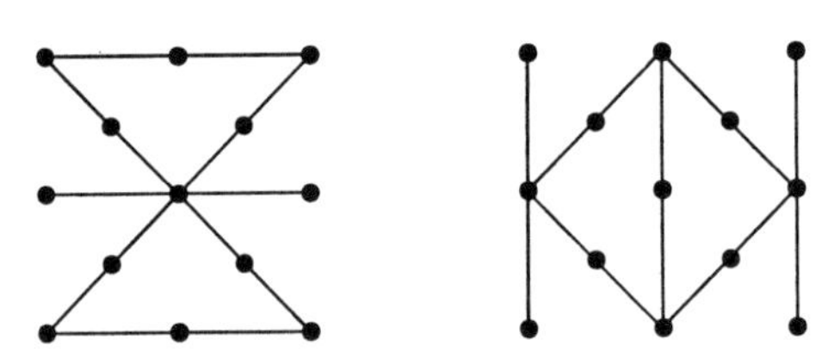

13.15 There are several possibilities – for example:

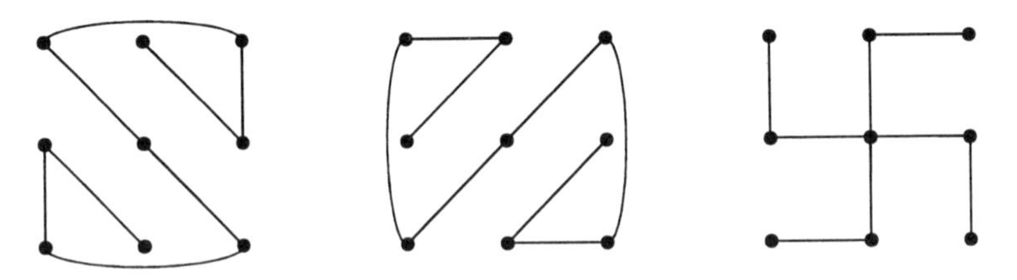

Note that $n = 9$ and $m = 24$, so m is a multiple of n 1.

Index

When several page references are given and they are not of equal status, the most important reference is shown in **bold**.